*P*HYSICS
in perspective

N. England C. Milward P. Barratt

Physical constants

gravitational constant, $G = 6.7 \times 10^{-11} \, \mathrm{N\,m^2\,kg^{-2}}$
mass of Earth, $m_E = 6.0 \times 10^{24} \, \mathrm{kg}$
radius of Earth $= 6.4 \times 10^6 \, \mathrm{m}$ (6400 km)
mass of proton $= 1.7 \times 10^{-27} \, \mathrm{kg}$
mass of electron $= 9.1 \times 10^{-31} \, \mathrm{kg}$
charge on electron, $e = 1.6 \times 10^{-19} \, \mathrm{C}$
constant in Coulomb's law, $k = 9 \times 10^9 \, \mathrm{N\,C^{-2}\,m^2}$
permittivity of free space, $\varepsilon_0 = 8.85 \times 10^{-12} \, \mathrm{F\,m^{-1}}$
permeability of free space, $\mu_0 = 4\pi \times 10^{-7}$ or $1.3 \times 10^{-6} \, \mathrm{T\,m\,a^{-1}}$
Earth–Moon distance $= 3.8 \times 10^8 \, \mathrm{m}$
Avogadro's constant, N_A or $L = 6.0 \times 10^{23} \, \mathrm{mol^{-1}}$
average wavelength of visible light $= 5 \times 10^{-7} \, \mathrm{m}$
speed of sound in air $= 340 \, \mathrm{m\,s^{-1}}$, in air $= 330 \, \mathrm{m\,s^{-1}}$
speed of light in vacuum $= 3.00 \times 10^8 \, \mathrm{m\,s^{-1}}$
Planck constant, $h = 6.6 \times 10^{-34} \, \mathrm{J\,s}$
molar gas constant, $R = 8.3 \, \mathrm{J\,mol^{-1}\,K^{-1}}$
Boltzmann constant, $k = 1.38 \times 10^{-23} \, \mathrm{J\,K^{-1}}$
specific heat capacity of water $= 4200 \, \mathrm{J\,kg^{-1}\,K^{-1}}$
1 MeV $= 1.6 \times 10^{-13} \, \mathrm{J}$

PHYSICS
in perspective

N. England   C. Milward   P. Barratt

Hodder & Stoughton
A MEMBER OF THE HODDER HEADLINE GROUP

British Library Cataloguing in Publication Data
England, Nick
    Physics in perspective.
    1. Physics
    I. Title   II. Milward, C. (Charlie)   III. Barratt, P. (Paul)
    530

    ISBN 0-340-40709-3

First published 1990
Impression number   10   9   8   7   6   5
Year   1999   1998   1997   1996   1995

Typeset by Wearset, Fulwell, Sunderland
Printed in Great Britain for Hodder & Stoughton Educational,
a division of Hodder Headline Plc, 338 Euston Road, London NW1 3BH
by The Bath Press, Avon.

Introduction

This book grew out of a need to fill gaps in the available literature. We wanted to write a fresh, accessible course that would provide a smooth transition from GCSE to A level, and cover *all* A level syllabuses.

We have adopted an approach which encourages students to read the text without feeling overwhelmed, and which replaces complex scientific jargon by simple, everyday English. The text is directed towards the students themselves – it concentrates on concepts and understanding rather than on factual detail, and each chapter has a sprinkling of worked examples to consolidate learning.

When writing the text, we have considered the impact of GCSE and Balanced Science on today's students and so for each chapter, you will find an explanation of the GCSE knowledge assumed within it. For some of the more advanced chapters, we have added topics from earlier parts of the book to the lists of GCSE knowledge. As a result, the order in which topics are covered can be very flexible, benefiting teachers *and* students.

The numerous questions at the ends of chapters encourage the students to use some of their knowledge and understanding in a variety of familiar contexts. We have tried to make these questions innovative and stimulating so that they form an essential part of the A level course. With this in mind, we devised three types of question:

(i) simple questions for practice purposes;
(ii) extended, structured questions to explore ideas in more depth;
(iii) teaching questions that cover important parts of the syllabus.

We are greatly indebted to David Gradwell of Clifton College, Bristol, who read the manuscript and made many useful suggestions. We are also very grateful to Wendy Rayner, Sue England and Yvonne Milward for converting our handwritten hieroglyphics into typescript.

N. England
C. Milward
P. Barratt
1990

Answers to all numerical questions are available to teachers free of charge from the publisher.
Please write to *Customer Services Department, Bookpoint Ltd, 39 Milton Park, Abingdon, Oxon OX14 4TD,* quoting ISBN 0 340 543760.

CONTENTS

Acknowledgements   vi
Introduction   vii

SECTION A   MECHANICS AND PROPERTIES OF MATTER

1 Statics   1
2 Dynamics   17
3 Collisions   37
4 Circular motion   50
5 Structure of solids   60
6 The mechanical properties of solids   64
7 Oscillations   77
8 Resonance and standing waves   95

SECTION B   CIRCUITS

9 Amp, volt and ohm   107
10 Resistivity   120
11 Circuits with resistors   129
12 Capacitors   146
13 Square pulses in circuits   159
14 a.c. circuits   165
15 Transistor electronics   177
16 Operational amplifiers   191

SECTION C   FIELDS

17 Gravitational fields   205
18 Gravitational potential   212
19 Orbits   222
20 Electric field   226
21 Electric potential   233
22 The parallel-plate capacitor   243
23 Electric flux   247
24 Magnetic fields near currents   251
25 Magnetic force on charged particles   264
26 Electromagnetic induction   271
27 The applications of electromagnetic induction   280

SECTION D   WAVES

28 Wave speeds   295
29 Electromagnetic waves   304
30 Diffraction   316
31 Interference   326
32 Reflection and refraction   345

SECTION E   GEOMETRICAL OPTICS

33 Rectilinear propagation of light   353
34 Lenses and prisms   361
35 Optical instruments   373

SECTION F   NUCLEAR PHYSICS

36 Radioactivity and decay   381
37 Nuclear physics   396

SECTION G   ATOMIC PHYSICS

38 Discovery of the nucleus   409
39 Light – wave or particle?   413
40 Electrons in atoms   420
41 Electrons – waves or particles?   428

SECTION H   HEAT

42 The kinetic theory of gases   437
43 Thermodynamics   449
44 Heat transfer   468

SECTION I   APPENDICES

45 Units and dimensions   483
46 Mathematics   489

INDEX   498

Acknowledgements

We are grateful to the following companies, institutions and individuals who have given permission to reproduce photographs in this book:

Action-Plus Photographic (293; 481); Professor E.H. Andrews, Department of Materials, Queen Mary College, London (70, bottom right); Heather Angel (352); Barnaby's Picture Library (326); Sir Lawrence Bragg and J.F. Nye (1947) *Proc. Roy. Soc. A 190* (62, left; 70, top left); The Building Research Establishment, Crown copyright (325, bottom); Camera Press Ltd (105; 294; 325; 380); CEGB (130; 435); C.B. Daish (317, top); Ealing Beck Ltd (70, bottom left); Dr Harvey Flower, Imperial College, London (62, bottom right); Philip Harris Ltd (133, middle and bottom two pictures); P. Lindley, Crystallography at Birkbeck College London (62, top right); JET Joint Undertaking (401, two pictures); Malaysian Rubber Producers' Research Association (70, top right); Marconi Radar Systems Ltd (322); Max–Planck–Gesellschaft–Pressebild (324); Metallurgy Department, Oxford University (3, two pictures); Mullard Radio Astronomy Observatory, University of Cambridge (343); NASA (cover; 225); National Grid Division of the CEGB (106); Oxford Scientific Films/Jonathon Watts (41); Palomar Observatory, California Institute of Technology (219; 220); Popperfoto (ix; 379); Popperfoto/Reuter (x); PSSC Physics, D.C. Heath and Co (40; 317, middle and bottom; 339; 346); Red Devils, The Parachute Regiment Free Fall Team (53); RS Components Ltd/Reg Arnold Photographers (126); Science Museum Library (49); Science Photo Library (29; 203; 204; 351; 407; 408; 482); Shell UK (436); Professor Charles A. Taylor (319); A.W. Trotter (17); Unilab (108; 109; 178; 280).

Every effort has been made to trace and acknowledge ownership of copyright. The publishers will be glad to make suitable arrangements with any copyright holders whom it has not been possible to contact.

SECTION A

MECHANICS AND PROPERTIES OF MATTER

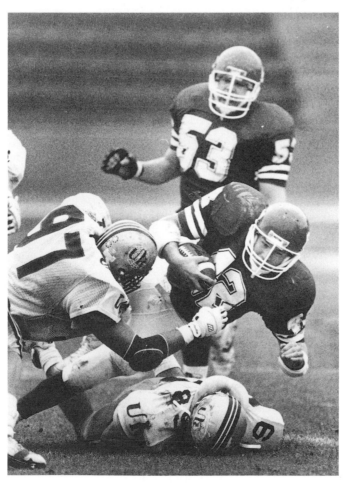

This section explains the action of forces in both dynamic situations...

. . . and static situations

1   Statics   1

2   Dynamics   17

3   Collisions   37

4   Circular motion   50

5   Structure of solids   60

6   The mechanical properties of
    solids   64

7   Oscillations   77

8   Resonance and standing
    waves   95

CHAPTER I

Statics

Introduction

In this chapter we begin by answering the question 'What do we mean by a force?' We examine in detail the various kinds of forces found in structures such as cranes and bridges, particularly weight, friction and contact forces, and we learn how the combined effect of several forces acting on one object can be calculated. We then reach the heart of the chapter: the principles which apply to objects which are stationary or 'in equilibrium'. We discuss in detail a large number of examples, to show you how the principles are applied in real situations.

GCSE knowledge

You should already be able to describe what a force is, and what effects it can have. You may also have met the idea of objects 'in equilibrium'; and in your particular GCSE course you may have solved problems which involved adding vectors such as forces together. All these ideas are thoroughly treated to A-level depth in this chapter.

Forces

For much of this chapter and the next one we will be looking at the effects of forces. Let us therefore consider how we recognise when forces are acting, and the types of force which occur.

A force has to act on some particular object; and it must be caused by some other object, or some kind of field. Thus any force can be described as 'The push (or pull) of object (or field) A on object B'. Object B experiences the force and its effect, while A causes the force. When considering forces and objects, it is best to describe every force in these terms, to clarify what effect each force is having.

For example, consider the question 'Why do you fall if you step off a cliff?' Many people would simply answer 'Gravity!' But a physicist might begin his answer by saying: 'The Earth exerts a gravitational pull on you (called your 'weight') . . .'

The unit of force is the newton (N). For example, the weight of an adult human is around 700 N, while a

loaf of bread has a weight of about 6 N. The definition of the newton is given on page 23.

A force is a vector quantity. This means that every force on an object acts in a particular direction, as well as having a particular strength.

The usual method of measuring force is to use the fact that the extension of a spring is proportional to the force applied to it. Thus if you calibrate the extension of a spring (as in a spring balance), you create a forcemeter or newtonmeter.

Forces on objects can have two possible effects:

I They can change the shape of the object. This occurs when a force on the object in one direction is balanced by another force in the opposite direction, putting the object into tension or compression. For example, a piece of thread, if pulled from both ends in your hands, may break when the tension in it reaches 10 N – see figure 1.

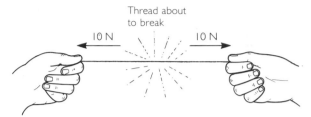

figure 1   Thread being pulled

2   They can alter the motion of the object. This occurs when the forces on the object add together to give a resultant which is not zero. There is then a net force on the object which makes it accelerate.

There are several types of forces (some of them are discussed in detail in later chapters):

☐   gravitational forces;

☐   contact forces, including frictional forces;

☐   viscous (fluid resistance) forces;

☐   electric forces;

☐   magnetic forces;

This chapter is mainly concerned with gravitational and contact forces.

Gravitational forces

Every object on or near the Earth's surface has a gravitational force acting on it, called its weight, W. The strength of the force is proportional to the object's mass m, and it is directed towards the Earth's centre, in other words vertically downwards.

The strength of the force can be calculated from the equation

$$W = mg$$

where g is the gravitational field strength at the position of the object. At the Earth's surface $g = 9.8\,\mathrm{N\,kg^{-1}}$; for many calculations, using the value $g = 10\,\mathrm{N\,kg^{-1}}$ is accurate enough.

The value of g is roughly constant over the Earth's surface. On the other hand, elsewhere in the universe objects may also experience gravitational forces, and g may have a quite different value: on the Moon's surface, for example, it is about $1.6\,\mathrm{N\,kg^{-1}}$, while in deep space, away from any other stars or planets, it is zero.

As an example, consider an astronaut of mass 50 kg. Wherever she goes her mass remains 50 kg. Her weight on the Earth is given by

$$W = 50\,\mathrm{kg} \times 10\,\mathrm{N\,kg^{-1}} = 500\,\mathrm{N}.$$

On the Moon it is

$$W = 50\,\mathrm{kg} \times 1.6\,\mathrm{N\,kg^{-1}} = 80\,\mathrm{N}.$$

In deep space it is zero.

Contact forces – normal reaction

Whenever two objects are touching, a contact force is exerted by each object on the other. This force has a component at right angles to the surfaces in contact, which is often called a 'normal reaction force' – 'normal ' meaning 'at right angles'.

For example, consider the forces acting on a coffee mug on a table: figure 2. Since the mug is stationary, we deduce that N and W are equal in strength.

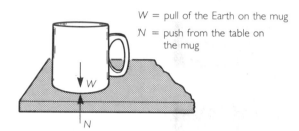

W = pull of the Earth on the mug
N = push from the table on the mug

figure 2   Coffee mug resting on table

The contact force is exerted by the table because of its elasticity. The table top is compressed very slightly by the contact force from the mug onto it; it then acts rather like a spring, pushing back on the mug with a force exactly equal to W.

Frictional forces

Frictional forces occur when two objects are in contact, and when their surfaces are sliding, or might slide, over one another. The frictional force is the component of the contact force acting *along* the surface of each object, in the direction which would help to stop the sliding.

For example, consider the contact between a brake drum and a brake shoe: figure 3. The frictional force on the drum causes it to slow down. (This in turn causes a frictional force from the road on the tyres which decelerates the vehicle.)

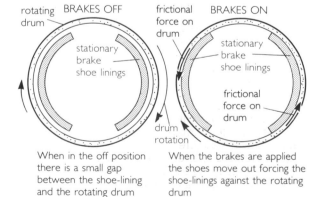

When in the off position there is a small gap between the shoe-lining and the rotating drum

When the brakes are applied the shoes move out forcing the shoe-linings against the rotating drum

figure 3   Brake drum and shoes

Frictional forces are caused by roughness of the two surfaces. When surfaces are viewed through a microscope, even the smoothest can be seen to be rough, like miniature sandpaper – see figure 4.

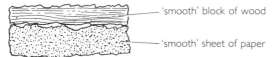

figure 4   Microscopic view of two 'smooth' surfaces in contact (see photographs of wood and paper below)

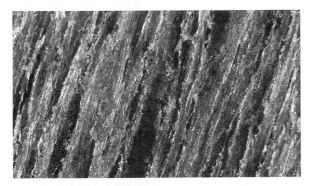

When two such surfaces are pressed together, contact forces between the points of roughness cause them to bond together; and if the surfaces are sliding, these links are continually being broken and reformed. Thus, whether sliding is actually occurring or not, there is a force along each surface. Inserting oil between the surfaces reduces the forces, because the surfaces are then kept slightly apart by the oil.

We distinguish between two situations where there are frictional forces.

1   *Static friction*: in this case the friction is preventing the two surfaces from sliding over one another. If there are external forces trying to slide the surfaces over each other, the static friction forces will exactly match the external forces and prevent sliding, up to a certain limiting value. If the external forces are bigger than this maximum limiting value, the surfaces will slip.

2   *Sliding friction* is the force which occurs when the surfaces are sliding over each other.

Experiments with a particular pair of surfaces show the following:

☐   the sliding frictional force F_{sl} and the limiting static frictional force F_{st} are both proportional to the normal contact force N;

☐ F_{sl} is approximately independent of how fast the surfaces are moving;

☐   neither F_{sl} nor F_{st} depends on the area of contact;

☐ F_{st} is bigger than F_{sl}.

The force being independent of the area of contact may seem surprising; but remember that if you keep the same contact force N, but double the area, then the force is spread over a doubled area, and the pressure between the surfaces is halved.

From the results $F_{sl} \propto N$ and $F_{st} \propto N$, we define *coefficients* of *sliding* and *limiting static friction*, μ_{sl} and μ_{st} respectively:

$$\mu_{sl} = \frac{F_{sl}}{N}$$

$$\mu_{st} = \frac{F_{st}}{N}$$

In applications, these versions of the equations are often more useful:

$$F_{sl} = \mu_{sl} \times N$$
$$F_{st} = \mu_{st} \times N$$

Example   A packing case of mass 100 kg rests on a horizontal roadway. Attached to it is a rope in which there is a forcemeter. A tractor exerts an increasing force on the rope. The reading on the forcemeter rises steadily to 300 N; at this moment the case begins to slide along the road, and the forcemeter reading settles to 200 N. What conclusions can we draw? See figure 5.

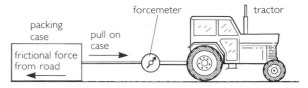

figure 5   Horizontal forces on a packing case as it is dragged along the road

From the description, the limiting frictional force is 300 N.

From $W = mg$, $W = 100 \, \text{kg} \times 10 \, \text{Nkg}^{-1} = 1000 \, \text{N}$.

Hence, from $\mu_{st} = \dfrac{F_{st}}{N}$,

$$\mu_{st} = \frac{300 \text{ N}}{1000 \text{ N}} = 0.3$$

Also, the sliding frictional force is 200 N.

Hence $\mu_{sl} = \dfrac{200 \text{ N}}{1000 \text{ N}} = 0.2$

Adding vectors

When we start to analyse the forces acting on objects, we will need to be able to add together several such forces. Forces are vectors, and must be added using methods which take account of their direction. Other vector quantities, such as velocity, acceleration, and momentum, are added using the same methods.

There are two ways of proceeding: by scale drawing and by resolving. To illustrate, let us solve the following problem using each method in turn.

Problem The horizontal forces on a sailing boat just as it sets off can be simplified to those in figure 6: 500 N at right angles to the sail, due to the wind; and 200 N at right angles to the boat, due to the centre-board. What is the resultant horizontal force R on the boat?

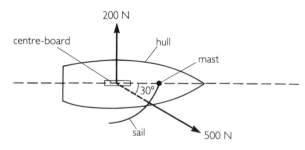

figure 6   Horizontal forces on a sailing boat

1 *Scale drawing.* We represent each force by a line, with the length of the line proportional to the strength of the force. We start by drawing a line representing the first force, in the same direction as the force (figure 7, stage 1); we then draw in the second force, in its particular direction, from the end of the first one (stage 2); (if there were more than two forces to consider, we would continue in this manner); finally we draw a line from the start of the first force line to the end of the last one, and its length and direction represent the resultant (stage 3).

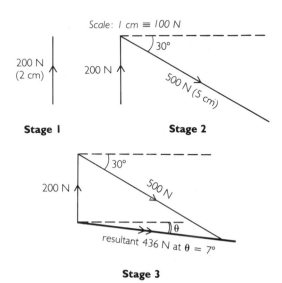

figure 7   Scale drawing

2 *Resolving.* Using this method, we split each force in turn into two components, which are at right angles to each other. The rules for resolving a force into components are illustrated in figure 8, using the 500 N force in our example.

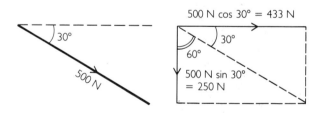

figure 8   Resolving a force into components

We choose the directions into which we want to resolve each force. In this case it is convenient to resolve along the line of the boat and at right angles to it, since the 200 N force then has very simple components: see figure 9.

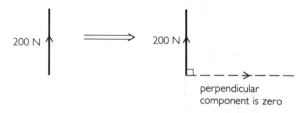

figure 9   Components of a force along and perpendicular to itself

In any particular problem, you can usually choose directions for resolving which will make the problem simpler, as in this case.

The next stage is to consider the two directions separately, and to add the components of the various forces in one direction and then in the other.

Along the boat:

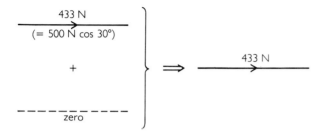

figure 10   Adding the horizontal components

At right angles to the boat:

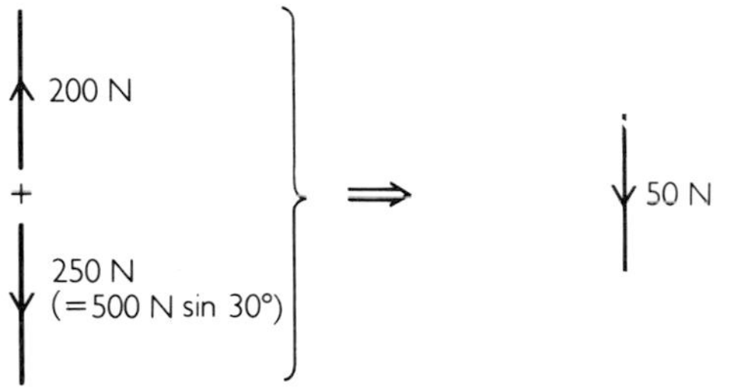

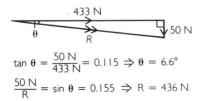

figure 11   Adding the vertical components

Finally, we recombine the two components. We sketch a triangle similar to the scale drawing method, and then use trigonometry. See figure 12.

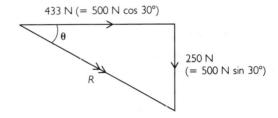

$$\tan \theta = \frac{50\text{ N}}{433\text{ N}} = 0.115 \Rightarrow \theta = 6.6°$$

$$\frac{50\text{ N}}{R} = \sin \theta = 0.155 \Rightarrow R = 436\text{ N}$$

figure 12   Recombining the components

(This procedure may seem complicated and laborious. But it is worthwhile taking the trouble to master it, as the ideas appear in other branches of physics throughout this book. There are often short-cuts you can take: these are illustrated in appropriate chapters.)

You may wonder why the resolving procedure is justified. Consider recombining the two components of 500 N by sketching a triangle – figure 13.

figure 13   Adding the components of 500 N

But this triangle can also be handled using trigonometry.

$$\tan \theta = \frac{250\text{ N}}{433\text{ N}}$$

$$= \frac{500\text{ N}\sin 30°}{500\text{ N}\cos 30°}$$

$$= \tan 30°$$

$$\Rightarrow \qquad \theta = 30°$$

Also

$$\frac{500\text{ N}\cos 30°}{R} = \cos 30°$$

$$\Rightarrow \qquad R = 500\text{ N}$$

In other words, the two components add together to give exactly the original force; thus their effect would be exactly that of the original force, and we are justified in imagining the original force replaced by them.

Free-body diagrams

Many situations require you to consider the effect of forces on an object: either a complex object such as a car, or a part of a complex object, such as one bone in the skeleton of an athlete. In any such problem, it is essential that you draw a picture of the object or part. You consider the object in isolation from its surroundings and you draw on the picture all the forces that you know are acting on the object. This picture is called a 'free-body diagram'. Having drawn it, you can go on to make deductions about the forces and their effects.

Here are some points about drawing free-body diagrams.

1 Mark the weight of the object as one force, acting at the centre of mass of the object. If the object is symmetrical, this point is at the centre of symmetry.

2 Be as specific as possible about the point of action of each force, and its direction.

3 Mark the direction of any motion beside the diagram.

4 Do not mark any forces which you are sure are negligible, for example, atmospheric pressure forces in most cases, or frictional forces when you are told to neglect friction.

5 Give each force a symbol.

Here are some examples.

☐ A car travelling along a road with the engine off.

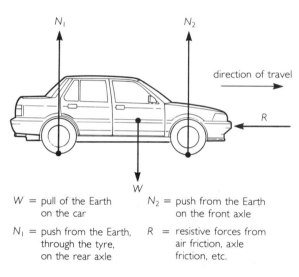

W = pull of the Earth on the car

N_1 = push from the Earth, through the tyre, on the rear axle

N_2 = push from the Earth on the front axle

R = resistive forces from air friction, axle friction, etc.

figure 14   Free-body diagram of a car

☐ The horizontal forearm of a human being holding a load in his hand.

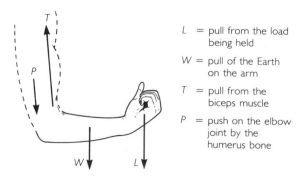

L = pull from the load being held

W = pull of the Earth on the arm

T = pull from the biceps muscle

P = push on the elbow joint by the humerus bone

figure 15   Free-body diagram of a forearm holding a load

☐ A ladder leaning against a smooth wall. ('Smooth' is a coded message from the question-setter meaning 'Ignore friction'.)

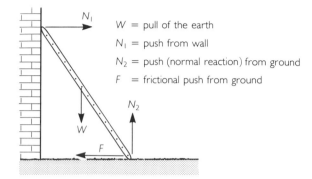

W = pull of the earth

N_1 = push from wall

N_2 = push (normal reaction) from ground

F = frictional push from ground

figure 16   Free-body diagram of ladder

Newton's third law

This law says that whenever one object exerts a force on a second object, the second also exerts a force on the first. These two forces are the same strength, but opposite in direction; and, most important to realise, they act on two different objects. Thus only one of this pair of forces can appear in any one free-body diagram. This law sounds quite simple but in fact we have to think very clearly when applying it, as the following two examples show.

(a)

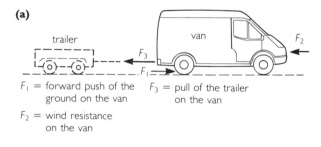

F_1 = forward push of the ground on the van

F_2 = wind resistance on the van

F_3 = pull of the trailer on the van

(b)

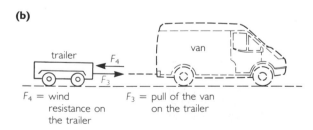

F_4 = wind resistance on the trailer

F_3 = pull of the van on the trailer

figure 17   (a) A van pulling a trailer (horizontal forces only); (b) a trailer being pulled by a van (horizontal forces only)

(c)

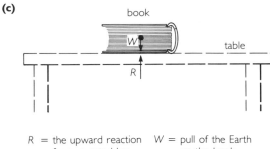

book

table

R = the upward reaction force exerted by the table on the book

W = pull of the Earth on the book

(d)

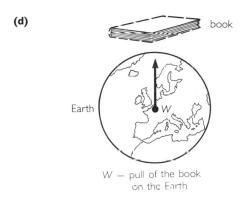

book

Earth

W — pull of the book on the Earth

figure 17 (c) A free-body diagram of a book resting on a table; (d) a free-body diagram of the Earth showing the force exerted by the book on the Earth

In the first example we consider a van towing a trailer along a road at constant speed. We will illustrate the pairs of forces acting by drawing free-body diagrams. (We do not include the weight of the van and trailer since we deal with such forces in the next example.) Figure 17(a) shows the forces on the van. F_1 is the force that the road exerts on the van. (The van exerts an equal force on the road to the left.) F_2 is the force due to wind resistance on the van and F_3 is the pull of the trailer on the van. The equal and opposite force to F_3 is the pull of the van on the trailer.

If the van moves at constant speed then $F_1 = F_2 + F_3$; if it is accelerating then $F_1 > F_2 + F_3$. Similarly, for the trailer $F_3 = F_4$ when the speed is constant, and if the trailer accelerates then $F_3 > F_4$: see figure 17(b).

The equal and opposite forces to the resistive forces F_2 and F_4 are the forces that the van and trailer exert on the air, in both cases to the right.

Next we think about a book on a table (figure 17(c)). We know that there is a downward pull on the book due to the Earth's gravitational pull, W; the equal and opposite force to W is the pull that the book exerts on the Earth: figure 17(d). The book

remains at rest, so the net force on it must be zero; this is so because the table exerts an upward reaction force, R, on the book, which is therefore the same size as W. The book exerts a force R on the table, in accordance with Newton's third law.

Moments

The moment of a force about a point measures how effective the force would be at rotating the object about that point. A force is more effective if it acts a long way from the pivot: for example, a spanner with a long handle is more effective than one with a short handle. It is therefore sensible to define the moment M of a force about a point by the equation

$$M = F \times d$$

where d is the perpendicular distance between the line of action of the force and the point in question – see figure 18.

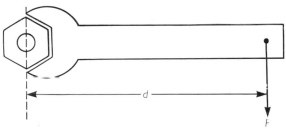

figure 18 Moment of a force about a point

Equilibrium

If there is a non-zero resultant force acting on an object, then the object accelerates. Conversely, if the resultant force is zero, the object has a constant velocity (which is often zero, that is, the object isn't moving). This idea is known as *Newton's first law*. (It is also true that if you know an object has constant velocity, then the resultant force on it must be zero.) To sum this up,

zero resultant force ⟺ constant velocity

(Note that velocity is a vector: constant velocity means an unchanging speed in a straight line – or no movement at all.)

For example, a car travelling along a straight road at steady speed must have zero resultant force acting on it. Its free-body diagram would be like figure 19. Since there can be no resultant force, $N = W$ and $T = R$.

But what about things which rotate, like fans and turbine blades? Such objects do not have linear velocity – that is, their centres of mass obey Newton's first law, so there is no resultant force on them. But the forces acting on such objects may have a resultant moment, which would cause them to rotate.

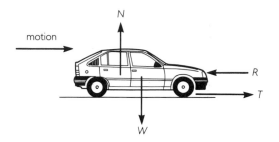

W = pull of the Earth

N = total normal push from road
onto wheels

T = forward push from
road onto wheels

R = resistive push
from air resistance

figure 19 Free-body diagram of a moving car in equilibrium

There is a rule for rotation which is analogous to Newton's first law: an object will rotate at a steady rate about a certain axis, provided the resultant moment of all forces about that axis is zero.

An object which has a net force of zero on it, and also net moment of zero about any axis, is said to be 'in equilibrium'. Such an object need not be stationary, as shown by the car example; but any object which is stationary must be in equilibrium.

Thus any system consisting of stationary parts may be analysed by using these two rules, expressed again below in algebraic form:

$\Sigma F = 0$. The net force on any part is zero.

$\Sigma M = 0$. The sum of the moments of all the forces on a part, about any axis, is zero. This statement is called the 'principle of moments'.

One further fact is sometimes useful. If an object is in equilibrium with just three forces acting on it, then either they must all be parallel (two in one direction and one in the other), or they must all act through the same point. See Example 3 below.

Example 1 The forces known to be acting on a crane jib are illustrated in figure 20. What additional force must be acting at the joint P, if the net force on the jib is zero? Let us answer this using a scale drawing. First we draw the three known forces, one after the other, (figure 21).

The net force on the jib must be zero; so the force F exerted at P must be just what is required to close up the diagram, as shown by the dotted line. Such a figure is sometimes called a *polygon of forces* – or *triangle of forces* if there are just three forces.

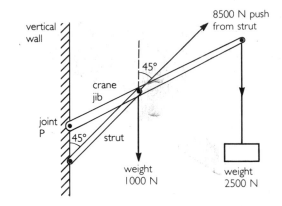

figure 20 Forces on a crane jib

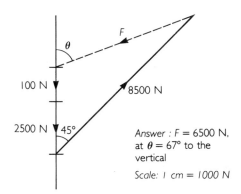

Answer : F = 6500 N, at $\theta = 67°$ to the vertical

Scale: 1 cm = 1000 N

figure 21 Scale drawing of crane-jib example

Example 2 A box of weight W rests on a sloping plank. The coefficient of static friction μ_{st} between the surfaces is 0.25. If the slope of the plank is gradually increased, at what angle of slope will the box begin to slide?

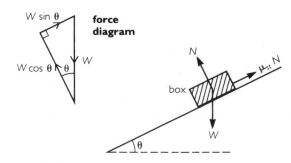

figure 22 Box just about to slide on slope

Resolving down the slope: $W \sin\theta = \mu_{st}N$ (i)

Resolving perpendicular to slope: $W \cos\theta = N$ (ii)

Dividing equation (i) by equation (ii) gives:

$$\frac{W \sin\theta}{W \cos\theta} = \mu_{st}$$

$\Rightarrow$ $\tan\theta = \mu_{st} = 0.25$

$\Rightarrow$ $\theta = 14°$

Example 3 A uniform drawbridge weighs 3000 N. It is hinged at one end; the other end is supported by a rope at an angle of 30° to the horizontal. What are the tension T in the rope and the force P exerted at the hinge?

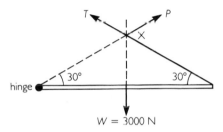

figure 23   Free-body diagram of drawbridge

Since there are only three forces acting on the bridge and W and T are not parallel, then W, T and P must all pass through the same point, X. Thus, by symmetry, P acts at 30° to the horizontal also, and T and P have the same strength.

Resolving vertically:

$$W = T \sin 30° + P \sin 30°$$

$$= 2T \sin 30°$$

$\Rightarrow$ $T = \dfrac{3000 \text{ N}}{2 \times \sin 30°}$

$$= 3000 \text{ N}$$

Thus $T = 3000$ N,

         $P = 3000$ N, at 30° to the horizontal.

Example 4 A massive iron ball (200 kg) is used to speed up demolition work. It hangs from a single steel cable. It is hauled back using a horizontal rope, then released, to swing into the doomed building. If the rope cannot safely tolerate a tension of more than 1000 N, what is the maximum angle from the vertical for the steel supporting cable?

Resolve horizontally:

$$P \sin\theta = T = 1000 \text{ N}$$

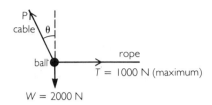

figure 24   Free-body diagram of demolition ball

Resolve vertically:

$$P \cos\theta = W = 2000 \text{ N}$$

Dividing:

$$\frac{P \sin\theta}{P \cos\theta} = \frac{1000 \text{ N}}{2000 \text{ N}}$$

$\Rightarrow$ $\tan\theta = \frac{1}{2}$

$\Rightarrow$ $\theta = 27°$

Example 5 A uniform diving board has mass 70 kg. It is attached at its inner end to a hinge H, and rests across another support S as shown in figure 25. A girl of mass 30 kg stands at the outer end. Calculate the force acting at each support.

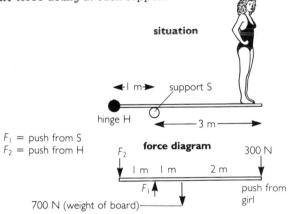

figure 25   Diving board

Taking moments about the hinge:

$$F_1 \times 1 \text{ m} = 700 \text{ N} \times 2 \text{ m} + 300 \text{ N} \times 4 \text{ m}$$

$\Rightarrow$ $F_1 = 2600$ N

Taking moments about the support S:

$$F_2 \times 1 \text{ m} = 700 \text{ N} \times 1 \text{ m} + 300 \text{ N} \times 3 \text{ m}$$

$\Rightarrow$ $F_2 = 1600$ N

Check by resolving vertically:

$$F_1 = F_2 + 700 \text{ N} + 300 \text{ N}$$

$\Rightarrow$ $F_1 = F_2 + 1000 \text{ N}$

as already found.

Complex structures

In many real engineering situations, a structure will be made up of a number of parts. The engineer must ensure that both the complete structure and each individual part are in equilibrium. It is also important to work out the compression and tension forces in each part; only then can he identify weak points and select the most suitable materials and dimensions for each part.

You can analyse such a structure by applying the rules for equilibrium to each part of it in turn; each joint between parts must also have zero net force on it, since the bolt or rivet has to be in equilibrium.

As an example, let us analyse the structure in figure 26. Six light rods are joined together at points A,B,C,D,E. ('Light' means that their weights should be neglected.) A and E are fixed to a vertical wall; a weight W of 100 N is suspended at C.

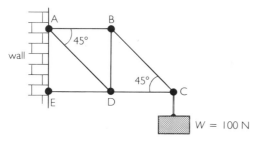

figure 26 Structure supporting a load

In a relatively simple structure like this, we should first examine whether each rod is in compression (exerting pushing forces on the joints at its ends) or in tension (exerting pulling forces). To do this, imagine each rod being removed in turn: ask yourself, when you remove it which way will the structure collapse? For example, if you were to remove BC, then DC and W would swing downwards: so BC is in tension.

Further examination shows that those rods in tension are AB, AD and BC. Those in compression are BD, DC and ED. We can now mark on the diagram the direction of the forces on each joint: so on the one picture we have now got a free-body diagram for each joint – figure 27.

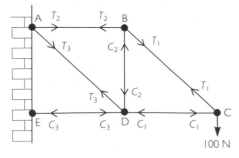

figure 27 Forces acting at each joint

Now starting at C, where there is a known force, we can calculate in order each of the unknown forces.

Resolving at C vertically:

$$T_1 \cos 45° = 100 \text{ N}$$
$$\Rightarrow \qquad T_1 = 141 \text{ N}$$

We can continue in a similar way, resolving vertically and horizontally at each joint, to find the following:

$$
\begin{aligned}
C_1 &= 100 \text{ N}\\
T_2 &= 100 \text{ N}\\
C_2 &= 100 \text{ N}\\
T_3 &= 141 \text{ N}\\
C_3 &= 200 \text{ N}
\end{aligned}
$$

Can you confirm these for yourself?

Centre of mass

In several examples we have marked the weight of an object on a free-body diagram. The objects have been described as uniform, and we have marked the weights as acting at the centre.

If an object is rigid, that is, it doesn't change shape, there is always a single point which we can call the *centre of mass*. The object behaves, for problems on statics and motion in a straight line, as if all its mass is concentrated there. If an object has symmetry about a point, that point must be the centre of mass. If it has symmetry about a line (two dimensions) or plane (three dimensions), the centre of mass C must be on that line or plane. See figures 28 and 29 for examples.

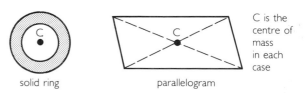

figure 28 Objects with point symmetry

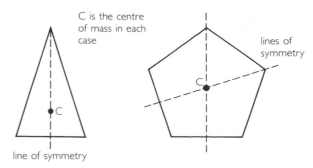

figure 29 Objects with line symmetry

As an example of the use of the idea of centre of mass, consider this simple experiment. A metre rule is pivoted at a point 20 cm from one end; a force of 1.2 N at that end then makes it balance (figure 30). What is the weight of the ruler?

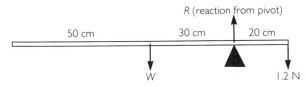

figure 30   Balancing a metre rule

Because the rule is uniform, we can mark its whole weight acting at its midpoint. Then taking moments about the pivot:

$$W \times 30 \text{ cm} = 1.2 \text{ N} \times 20 \text{ cm}$$

$$\Rightarrow \qquad W = 0.8 \text{ N}.$$

Another example is provided by 'executive' toys which balance on just one point. For normal situations this is impossible, because the centre of mass is higher than the point of contact (see figure 31). Then any movement of the centre of mass away from the vertical above the point of contact leads to toppling.

figure 31   'Executive' toy; the centre of mass, C, is above the point of contact, P, with the stand

However, if the toy's centre of mass is below the point of contact, then the toy is stable. This can be achieved if the toy includes massive parts which lie below the level of the point of contact, moving the centre of mass down (see figure 32).

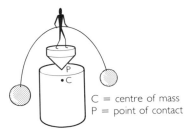

figure 32   'Executive' toy; the centre of mass, C, is below the point of contact, P, with the stand

Toppling

One special example of this was considered above. Let us now consider objects whose centres of mass are above the surfaces on which they rest.

An object resting on a surface will be in contact with the surface at a number of points, and when these points are joined together they define a base area: for example, a three-legged stool has a triangular base. The object will not topple over provided its centre of mass lies vertically over some point within the area of the base. If the centre of mass lies outside this area, then the weight of the object has a net moment about a point on the base, and equilibrium is not possible. See figure 33.

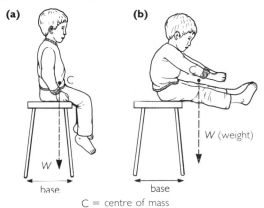

figure 33   (a) No toppling; (b) toppling occurs

A human being instinctively learns to balance at an early age. Balance is possible as long as your centre of mass is always over the base provided by your feet; since you are not a rigid body, but can change shape, you can achieve this by altering the position of your centre of mass (for example by moving your arms), as well as by moving your feet to form a different base. In many sports, for example karate, players train themselves to place their feet so as to have the optimum base for balance.

Stability

The idea of stable and unstable states occurs in many branches of physics and outside it, for example in discussing economic and political situations. One example, that of the balancing executive toy, has been discussed above. The toy of figure 31 can, in theory, balance and be in equilibrium; but if the balance is disturbed slightly then it will fall over completely. This is *unstable equilibrium*. The modified toy of figure 32, however, is in *stable equilibrium*. This is because, if disturbed slightly, it will return to its original balance position.

In physics, we more usually discuss stable and unstable states of systems in terms of energy. A stable state is one in which a small disturbance gives the system greater potential energy. For example, the ball in the saucer of figure 34(a) is in stable equilibrium, since disturbing it involves moving it uphill, giving it more gravitational energy. Like the ball, all systems tend to go to states of low potential energy if they can: so the ball rolls back to its former position.

Compare the behaviour of the ball in figure 34(b), which is in unstable equilibrium. In this case a disturbance reduces its gravitational potential energy, and it proceeds to fall even further.

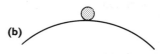

figure 34 (a) Stable equilibrium; (b) unstable equilibrium

Summary

1 A force is a push or pull on an object by another object or field.

2 A force can change the shape of an object, or affect its motion.

3 Force is a vector quantity.

4 Unit of force: the newton (N).

5 The weight W of an object means the gravitational pull on it, wherever it is.

6 $W = mg$.

7 Contact forces act between two objects in contact.

8 The normal reaction forces between two objects are the components of the contact forces at right angles to the surfaces in contact.
Frictional forces are the components of the contact forces which act parallel to the surfaces in contact, and help to prevent relative motion of the surfaces.

9 Limiting static friction forces are greater than sliding friction forces.

10 $F_{sl} = \mu_{sl} N$.

11 $F_{st} = \mu_{st} N$.

12 A vector may be resolved into two components at right angles.

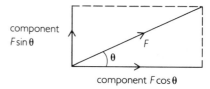

13 Vectors may be combined using scale drawing or by resolving them.

14 A free-body diagram shows all the forces acting on one particular object.

15 Newton's third law: 'If body A exerts a force F on body B, then body B exerts a force on body A which is equal to F in strength but opposite in direction.'

16 Moment M of a force = $F \times d$.

17 Newton's first law: Zero net force $\Longleftrightarrow$ constant velocity.

18 An object is in equilibrium if $\Sigma F = 0$ and $\Sigma M = 0$.

19 Complex structures may be analysed by considering the equilibrium of each part and joint in turn.

20 An object will topple if its centre of mass is not vertically over the area formed by its base.

21 An object is in stable equilibrium if disturbing it increases its gravitational potential energy.

Questions

1 On Jupiter, it is predicted that $g = 26\,\text{N kg}^{-1}$; on the Moon, $g = 1.6\,\text{N kg}^{-1}$. On the Earth's surface, $g = 9.8\,\text{N kg}^{-1}$.
   a A hammer weighs 8 N on the Earth. What is the hammer's mass (i) on the Earth, (ii) on Jupiter?

   b What will the hammer weigh (i) on the Moon, (ii) on Jupiter?

2 A paving slab rests on a concrete driveway. The slab has mass 20 kg. If you pull it with a rope, the force you need to move it is 160 N; to keep it

moving after that, you have to keep pulling with 90 N.

a Find the coefficient of limiting static friction.

b Find the coefficient of sliding friction.

c If you put five more paving slabs on top of the first one, what force would you need to exert on the bottom one to get the pile moving?

3 A towel lies over a rail – see figure 35 – with three-fifths of its length on one side and two-fifths on the other. What minimum coefficient of friction is needed between the rail and the towel to prevent the towel sliding off?

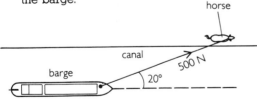

figure 35

4 A horse is pulling a barge along a canal. The barge moves at a steady $1\ \mathrm{ms^{-1}}$, straight ahead. See figure 36.
Calculate

a the resistive force of the water to the barge's forward motion;

b the sideways force exerted by the water on the barge.

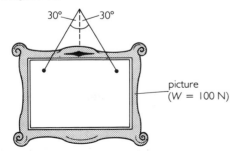

figure 36

5 A picture weighs 100 N. It hangs from two wires, each making an angle of 30° with the vertical. See figure 37.

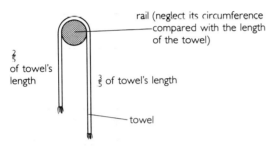

figure 37

a Calculate the tension in each wire.

b What qualitative effect would there be on the tension if (i) the wires were longer, (ii) their angle with the vertical was greater?

c The wires are likely to break under tension of 200 N. At what angle with the vertical will this occur?

6 Your car is stuck in a ditch. You have available only a rope, a single pulley and your own muscle power, which is capable of pulling the rope with a force of 400 N.

Your first thought is to pull directly, but this proves useless. You then decide to attach the rope to a tree, and pull it from inside the car. See figure 38.

a What force will this arrangement exert on the car?

This also proves useless, so you have another idea. You attach the pulley to the tree, and one end of the rope to the car. The other end you pass round the pulley, and then you pull from inside the car. See figure 39.

b What force will this arrangement exert on the car?

Your next idea works better. You tie one end of the rope to the car, and the other to the tree. Then you pull sideways on the middle of the rope. When the rope makes an angle of 10° at either end, your car comes out of the ditch. See figure 40.

c What pull is being exerted on the car by this arrangement?

d Explain why the pull was even bigger before the car moved and the angle became 10°.

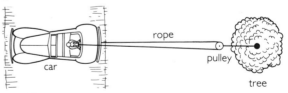

figure 38

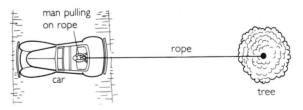

figure 39

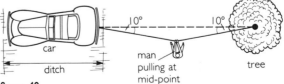

figure 40

7 You hold a book, which has the dimensions shown in figure 41, between your finger and thumb at the position shown, so that the book is upright.

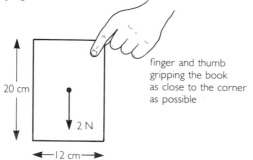

finger and thumb gripping the book as close to the corner as possible

20 cm

2 N

←12 cm→

figure 41

a What forces and couples are you exerting on the book?
b Explain why you cannot grip the book exactly *at* the top corner and hold it upright.

8 This question is about the transfer of pressure which occurs in the middle part of your ear. This can be analysed using a simple physical model of levers and pistons – figure 42.

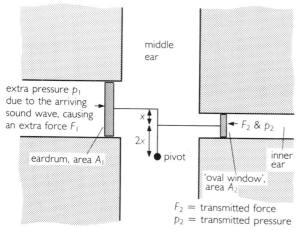

middle ear

extra pressure p_1 due to the arriving sound wave, causing an extra force F_1

F_2 & p_2

eardrum, area A_1

x
2x

pivot

inner ear

'oval window', area A_2

F_2 = transmitted force
p_2 = transmitted pressure

figure 42

$A_1 = 60 \, \text{mm}^2$

$A_2 = 3 \, \text{mm}^2$

Sound waves, which are fluctuations in atmospheric pressure, arrive at the eardrum. These exert force, which is passed on to the oval window of the inner ear by the lever (consisting of the three bones of the middle ear). The force results in change of pressure in the fluid there. The system can be treated as if it is static throughout, since none of the parts move any significant distance.

a Taking moments about the pivot, find a value for the ratio F_2/F_1.

b Write down an algebraic expression for the ratio p_2/p_1.
c Evaluate the ratio p_2/p_1. Comment on your value, and its possible contribution to your hearing process.

9 A waiter holds a tray at one edge with the thumb and fingers of one hand, so that the forces on it when empty are as in figure 43.

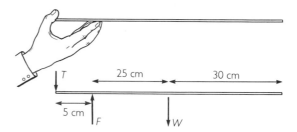

T

25 cm

30 cm

5 cm

F

W

figure 43 Illustration of a tray being held by a hand

a If the weight W of the tray is 5 N, calculate T and F while the tray is empty.
b The waiter now puts a drink, weight 5 N, on the tray. Calculate the new values for T and F if he puts the drink (i) in the middle of the tray, (ii) on the extreme right-hand end.
c Where should he put the drink so as not to alter T? Justify your answer.

10 Suppose you are standing on the point of one toe as a ballerina might. The free-body diagram of your foot would look like figure 44.

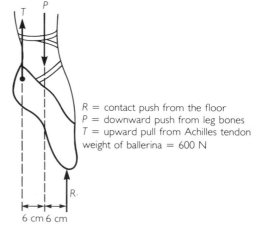

T
P

R = contact push from the floor
P = downward push from leg bones
T = upward pull from Achilles tendon
weight of ballerina = 600 N

R

6 cm 6 cm

figure 44 Free-body diagram of a ballerina's foot

a What must be the value of R? Explain.
b What are the values of T and P? (Assume your foot is static.)
c Comment on the value of P which you find.
d What can you say about the position of your centre of mass, assuming your whole body is static?

11 A uniform ladder (with its centre of mass at its centre) with weight W leans against a rough wall, at an angle θ to the vertical. Its base rests on rough ground. See figure 45.

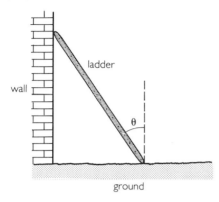

wall

ladder

θ

ground

figure 45

a Draw your own diagram of the ladder, marking on and labelling the *five* forces which act on it.
b Suppose that in fact there is no friction force from the ground. Is it possible for the ladder to remain stationary? Explain.
c Suppose there is no friction force from the wall. Can the ladder remain stationary now? Explain.

Consider further the situation where the wall exerts no frictional force.
d Redraw the diagram, marking all four remaining forces in terms of W and the coefficient of friction at the ground, μ.
e Show that μ is related to θ by the expression

$$\mu = \frac{1}{2}\tan\theta$$

(Hint: Take moments about the bottom end of the ladder.)

12 Harry is doing experiments on the frictional forces between a block of wood of mass 0.5 kg and a ramp. He can tilt the ramp to any angle θ between the ramp and the horizontal.
    He starts with the block stationary and the ramp horizontal. He finds that if he then increases θ from zero, there comes an angle at which the block suddenly starts to slide; the block then *accelerates* down the ramp.
a Harry thinks that after the block has started to slide it should move down the ramp at constant speed. Explain to him why it accelerates.
b Draw a free-body diagram of the stationary block when $\theta = 10°$.
c For the situation in (b), calculate
(i) the normal reaction force between the surfaces;
(ii) the frictional force acting on the block.

13 During a picnic at a roadside table, Charlie observes that the dimensions of the table are roughly as shown, with every distance marked x being the same (figure 46).

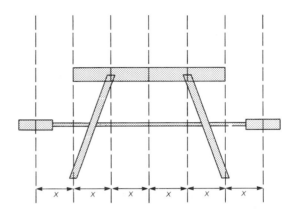

x x x x x x

figure 46

When Yvonne (50 kg) and Henry (70 kg) sit together on the right hand seat, the table topples over; but when Henry is replaced by Sue (60 kg), then the table doesn't topple.
What can you say about the mass of the table?

14 This question is about the forces which act on your lumbosacral disc, the lowest disc in your spinal column – first when you are standing upright, then when bending and lifting a load. See figure 47.

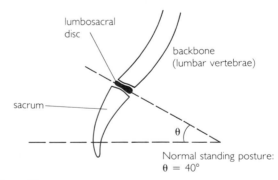

lumbosacral disc

backbone (lumbar vertebrae)

sacrum

θ

Normal standing posture: $\theta = 40°$

figure 47

The weight supported by your lumbosacral disc is normally about $0.6\,W$, where W is your total weight. Suppose $W = 600$ N.
a (i) Draw a free-body diagram of your lumbosacral disc when you are standing still.
(ii) Resolve the forces on your disc into compressive components (acting perpendicular to it) and shear components (acting along each side of it).

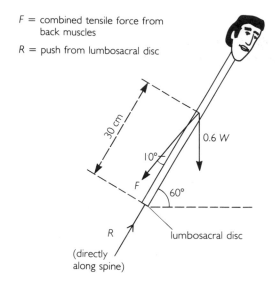

F = combined tensile force from back muscles

R = push from lumbosacral disc

0.6 W

30 cm

10°

F

60°

R

lumbosacral disc

(directly along spine)

figure 48

b Consider this free-body diagram, for your spinal column when you bend forward (figure 48).
The force F is the combined effect of all your back muscles, and acts at the centre of mass of your upper body, and angled at about 10° back from it, as shown.
(i) Explain why R must act straight along the line of your back.
(ii) Calculate the values of F and R.
(iii) Compare the value for R with the compressive forces on your lumbosacral disc when you are standing still and upright as in question a(ii).

c Now suppose that while in the position of figure 48 you pick up a load of 150 N with your arms. Effectively, we can take this as adding another force, of 150 N, a further 15 cm up your spinal column.
(i) Draw a free-body diagram to represent the new situation. (Note that R may no longer act straight along your spinal column.)
(ii) Find the new value of F. (Hint: take moments about the bottom end of your spine.)
(iii) Find the new value and direction of R. (Hint: Resolve along and perpendicular to your spine.)
(iv) Compare this value of R with the compressive forces in your lumbosacral disc when you are standing still and upright, as in a(ii).

15 This question is about the forces in a girder bridge. To make the analysis relatively simple, we make a number of assumptions: the girders have negligible weight compared to the load on the bridge; they are loosely jointed, so that the joints exert only reaction forces and not couples; the load is acting at a joint; the ends of the bridge are resting loosely on the end-supports, without friction.
This is the structure we will analyse. Each triangle is equilateral (figure 49).

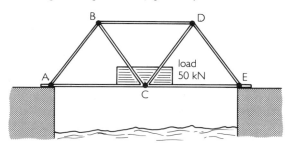

B   D

load
50 kN

A   E

C

figure 49

a What forces act on the end-supports at A and E?
b Guess which of the members (AB, BC, AC, BD, CE, DE) are in tension, and which are in compression.
c Draw a triangle of forces for point A. Hence find the forces in AB and AC.
d Similarly, draw a triangle of forces for point B, to find the forces in BD and BC.
e Now draw a polygon of forces for point C (there are five forces; but make use of the symmetry of the structure). Since you know the magnitudes of all the forces acting already, this should confirm that C is in equilibrium.

16 What force actually makes a car accelerate forward from being stationary? How is this force caused? (Hint: Newton's third law should come into your answer.)

17 Let us think about an aircraft at various stages during its flight. Draw a free-body diagram for the aircraft in each case.
a The aircraft is at the end of the runway, with engines at 50% power; but the brakes are holding, and the aircraft is stationary.
b The aircraft is in mid-flight, flying straight and level.
c The aircraft has touched down, the power is off, but the pilot is not yet applying the brakes.

In which of these cases is the aircraft in equilibrium?

CHAPTER 2

Dynamics

Introduction

In this chapter we look first at the ideas concerned with measuring movement: displacement, velocity and acceleration, and the relationships between them. We apply the ideas to find out how objects move in the Earth's gravitational field. We then examine quantitatively the effect of resultant forces on the motion of objects. Having studied the fundamental concepts of motion, we turn our attention to another key concept, which runs through all branches of physics: energy. The important ideas discussed include work, kinetic and gravitational potential energy, power and efficiency. As they are introduced, each new quantity is illustrated using simple worked examples.

GCSE knowledge

You will have met and discussed many of these ideas before in your GCSE course, at least in a qualitative way. This chapter develops the ideas further, and deals with them all in the more advanced quantitative way required for A-level. The concepts underlie many of the later ideas in the book, and will repay careful study.

Displacement

The displacement of an object means its distance and direction from some stated reference point. For example, 'The treasure is buried 400 paces north-north-west of the tallest tree on the island' tells us the displacement of the treasure relative to the tree.

Notice that displacement is thus a vector quantity: it has a direction as well as a size. This means that displacements must be added as vectors. For example, suppose you travel 3 km west, then turn north and go another 4 km; you have travelled a total distance of 7 km, but your displacement from your starting point is only 5 km. See figure 1.

Note also that if you run once round a 400 m track, you cover a distance of 400 m but you end up with zero displacement!

The unit of displacement is the same as that of distance: the metre (m). The metre is a fundamental unit of the SI unit system – see chapter 45.

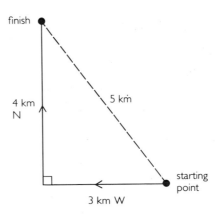

figure 1   Adding displacements as vectors

Velocity

Velocity means speed in a certain direction. Like displacement, it is thus also a vector.

For example, suppose you row a boat, and can propel it at $3\,\mathrm{ms^{-1}}$ in still water. If you aim straight across a river flowing at $4\,\mathrm{ms^{-1}}$, what is your velocity? You can see from figure 2 that vector addition of velocities gives your velocity as $5\,\mathrm{ms^{-1}}$.

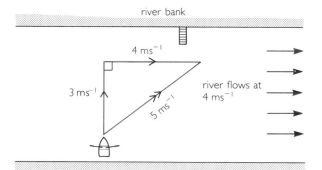

figure 2   Adding velocities

Speed can be described as 'rate of increase of distance'. Note that both speed and distance are scalars, without direction. Velocity can likewise be described as 'rate of increase of displacement' – both velocity and displacement are vectors.

Mathematically, velocity v at an instant in time is defined by the equation

$$v = \frac{\mathrm{d}s}{\mathrm{d}t}$$

where s is displacement and t is time. The unit of velocity is thus one metre/second ($\mathrm{ms^{-1}}$).

This means that if we have a graph of displacement against time, the velocity at any moment will be given by its gradient, or slope: see figure 3.

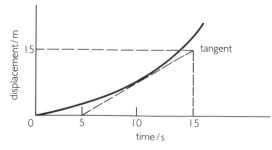

velocity at $10\,\mathrm{s} = \dfrac{\Delta s}{\Delta t} = \dfrac{(15-0)\mathrm{m}}{(15-5)\mathrm{s}} = \dfrac{15\,\mathrm{m}}{10\,\mathrm{s}}$

$= 1.5\,\mathrm{ms^{-1}}$, in the direction of the displacement

figure 3   Finding velocities from a displacement-against-time graph

In general we may define an average velocity for any journey, regardless of whether the velocity is constant, using the simpler equation:

$$\text{average velocity} = \frac{s}{t}$$

Most measurements of velocity involve measuring the speed of travel, and then observing the direction of movement separately. In a well-equipped school lab we have several methods available for measuring speeds or velocities.

1 *Using a ticker-timer, and tape attached to the moving object.* A stationary marker punches dots at regular time intervals onto the moving tape (usually at 50 dots per second, using the mains a.c. supply). You can then examine the marked tape and measure from it the distance travelled by the object in certain intervals of time.

Example   The tape in figure 4 shows that the object towing it moved at a constant speed, because the dots are evenly spaced. Calculate that speed.

figure 4   Tape punched by a ticker-timer

If the marker makes 50 dots/second, the 10 cm length was pulled through in $5 \times 0.02\,\mathrm{s} = 0.1\,\mathrm{s}$.

$\Rightarrow$ speed $= 10\,\mathrm{cm}/0.1\,\mathrm{s} = 100\,\mathrm{cms^{-1}}$.

This method is useful for speeds roughly in the range $0.1\,\mathrm{ms^{-1}}$ to $5\,\mathrm{ms^{-1}}$: that is, the speeds you commonly encounter with school laboratory equipment.

2 *Using multiflash or video photography.* You take successive pictures of the moving object, placing in the background a distance scale. If you know the time interval between successive pictures, you can examine the photographs and work out speeds. Using a video camera, the time intervals are usually 0.04 s (25 frames each second): you can step through the film frame by frame. You can use a still camera taking a single photograph, if you illuminate the object using a flashing stroboscope. Laboratories which specialise in photographing very rapid events such as the movement of insect wings use cine cameras which take a very large number of frames each second.

3 *By timing over a measured distance.* This sort of method covers a wide range of speeds. For example, a geographer might measure the speed of movement of a glacier by observing the few metres it moves each year. Athletes are timed over measured distances using stopwatches. At the fast extreme, the

speed of light was first measured accurately by
timing a light flash over a distance of about 70 km.
You may use this method in your lab, using a light
gate and electronic timer: you time how long a
measured length of card takes to traverse the gate,
and then calculate the speed of the card from $v = s/t$.

4 *Using a speedometer mounted on the moving
object.* This method is particularly useful when
you require an immediate value of speed which you
can read from a scale, for example when riding a
motorbike or driving a car. In the commonest design,
a steel cable runs inside a stationary sheath from the
gearbox up to the speedometer itself; at the gearbox
end, the cable is attached to the centre of one cog in
the linkage between gearbox and drive shaft, so that
the cable spins inside its sheath at a rate proportional
to the speed of the car. See figure 5.

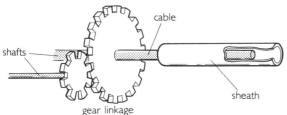

figure 5 Speedometer cable running off gearbox

At the speedometer end, the rate of spin of the
cable is converted to a reading on the meter by
electromagnetic induction – see page 271.

Although this instrument seems to give a speed
reading by magic, in fact it has to be calibrated
carefully in the design laboratories of the
manufacturer. At some stage during this process the
designers would, as in all other methods of measuring
speed, have to measure how far the car would travel
(s) in a certain time (t) with the cable spinning at a
certain rate, and then calculate the speed of the car
(v) from $v = s/t$.

5 *Using the Doppler shift in the frequency of
waves reflected by the moving object.* The
best-known example of this is in police measurement
of vehicle speeds using radar. The technique can
also be adapted to measure the speed of flow of
liquids in tubes, for example, the flow of blood in
blood vessels. See page 299 for discussion of the
Doppler effect.

Acceleration

Acceleration is a measure of how fast the velocity of
an object is changing: acceleration may be defined
as the rate of increase of velocity.

Mathematically, acceleration, a, at an instant in time
is given by the equation

$$a = \frac{dv}{dt}$$

Like velocity, therefore, acceleration is also a vector;
its direction is important.

The unit of acceleration is $1\,\mathrm{ms}^{-2}$.

If the acceleration of an object is constant over a
time interval Δt, then these equations are more useful:

$$a = \frac{\Delta v}{\Delta t} = \frac{v - u}{t}$$

Here Δv = change in velocity during the time interval
Δt,

  u = velocity at the beginning of the interval,
  v = velocity at the end of the interval.

Example A manufacturer states that a car
accelerates from $13.3\,\mathrm{ms}^{-1}$ (30 mph) to $22.2\,\mathrm{ms}^{-1}$
(50 mph) in 9.0 s. Calculate the acceleration, assuming
it is constant.

$$a = \frac{v - u}{t}$$
$$= \frac{22.2\,\mathrm{ms}^{-1} - 13.3\,\mathrm{ms}^{-1}}{9.0\,\mathrm{s}}$$
$$= 1.0\,\mathrm{ms}^{-2} \text{ in the direction of the}$$
velocity increase.

Note that assuming a constant acceleration is the
same as working out an average acceleration over
the time interval. We have no way of knowing how
the speed actually changed during the nine seconds;
all we can do is assume it changed steadily, or state
that the answer is an average value for the
acceleration.

There are two main ways of measuring the
acceleration of an object.

1 *Measuring velocities separated by time intervals
Δt.* We can then calculate acceleration using the
equation above. If we have a series of readings of the
velocity for the object, we can plot a graph of
velocity against time; we then find the acceleration at
a particular instant by measuring the gradient of the
graph – see example below.

2 *Using a direct-reading accelerometer.* This uses
the principle that the force required to
accelerate a particular mass is proportional to the
acceleration occurring. So if we attach a mass to a
spring, which in turn we fix to the accelerating
object, then the extension of the spring gives a
measure of the acceleration. We can calibrate a scale
to read directly in acceleration units. See figure 6.

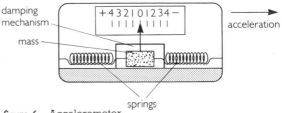

figure 6 Accelerometer

Velocity-against-time graphs

The velocity-against-time graph for an object can be particularly useful. As well as displaying velocity, you can use it to calculate acceleration, as mentioned above. Also, the area under the graph during any time interval gives the distance travelled during that interval.

Since
$$v = \frac{ds}{dt}$$

we also know that $s = \int v\,dt$.

This means that s can be obtained by calculating the appropriate area under the v-against-t graph.

Example The graph in figure 7 illustrates the speed of a jumbo jet as it accelerates during its take-off run. Let us use it to calculate the initial acceleration of the aircraft, and the length of runway used.

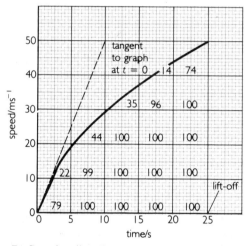

figure 7 Speed-against-time graph for jumbo jet

Acceleration. The dotted tangent at $t = 0$ shows the initial acceleration.

$$\text{acceleration} = \frac{50\,\text{ms}^{-1}}{10\,\text{s}} = 5.0\,\text{ms}^{-2}$$

Distance. The area under the graph is found by simply counting squares; the subtotals are shown on the graph.

$$\text{area} = 1563 \text{ little squares}$$

One little square represents $1.0\,\text{ms}^{-1} \times 0.50\,\text{s} = 0.50\,\text{m}$.

$$\text{distance travelled} = 1563 \times 0.50\,\text{m} \approx 780\,\text{m}.$$

Equations for constant acceleration

Constant acceleration of an object means that the gradient of its velocity-against-time graph is constant. Let us consider such a graph, shown in figure 8.

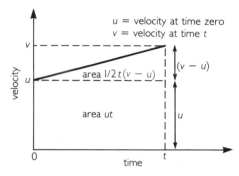

figure 8 Velocity-against-time graph for constant acceleration

The constant acceleration a is given by

$$a = \frac{v - u}{t}$$

$$\Rightarrow \qquad v = u + at \qquad \text{(i)}$$

The distance travelled, s, is given by the total area under the graph:

$$s = ut + \tfrac{1}{2}t(v - u)$$

Substituting for $(v - u)$ from equation (i), we obtain

$$s = ut + \tfrac{1}{2}at^2 \qquad \text{(ii)}$$

Taking equations (i) and (ii) and algebraically eliminating t between them gives a third equation:

$$v^2 - u^2 = 2as \qquad \text{(iii)}$$

Equations (i), (ii) and (iii) provide between them a valuable set of tools for handling problems in dynamics where the acceleration is constant.

Example For the example of the accelerating car on page 19, let us work out the distance it travels during this acceleration period. We have the following information:

$$u = 13.3\,\text{ms}^{-1}$$
$$v = 22.2\,\text{ms}^{-1}$$
$$t = 9.0\,\text{s}$$
$$a = 1.0\,\text{ms}^{-2}$$

The only unknown is s, and we could therefore use either of the equations (ii) or (iii) above. Let us use equation (ii).

$$s = ut + \tfrac{1}{2}at^2$$
$$= 13.3\,\text{ms}^{-1} \times 9.0\,\text{s}$$
$$+ \tfrac{1}{2} \times 1.0\,\text{ms}^{-2} \times 9.0\,\text{s} \times 9.0\,\text{s}$$
$$= 160.2\,\text{m}$$
$$\Rightarrow \quad s \approx 160\,\text{m}$$

Free fall

All objects near the Earth's surface, if allowed to fall freely, accelerate towards the Earth's centre with the same value of acceleration. 'Fall freely' means that the only force acting on them is their weight; other forces such as air resistance must be absent or negligible. You may think this statement is contrary to common experience, since we regularly see leaves fluttering down whereas stones fall much more quickly; but the difference is because air resistance is acting significantly on the leaf. If we pump the air out of a glass tube in which we have placed both a leaf and a stone, then turn the tube over, we can see both objects falling together. See figure 9.

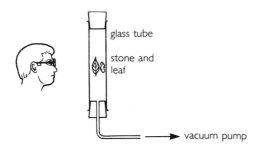

figure 9 Stone and leaf (or 'guinea and feather') experiment

We give this acceleration of free fall the symbol g. We can measure g using the experiment illustrated in figure 10.

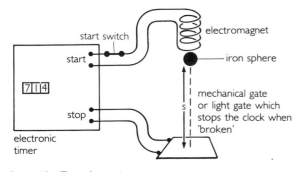

figure 10 Experiment to measure g

To calculate g we can use the constant acceleration equation

$$s = ut + \tfrac{1}{2}at^2$$

We measure s and t (doing several repeat runs, of course, and averaging the most consistent results); $u = 0$, so you can calculate the acceleration of the sphere, that is, g.

A better technique is to vary s and measure the best value of t for each s. We then plot a graph of $\sqrt{s}$ against t, which may look like figure 11.

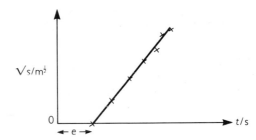

figure 11 Graph of $\sqrt{s}$ against t

The intercept e on the time axis could be caused by delay in the sphere leaving the electromagnet, due to the magnet not losing its magnetism instantaneously.

The true motion of the sphere would then be described by the equation

$$s = \tfrac{1}{2}a(t-e)^2$$
$$\Rightarrow \quad \sqrt{s} = \sqrt{\frac{a}{2}}\,(t-e)$$

Thus the gradient of the graph gives you the value of $\sqrt{a/2}$; hence we calculate a, or g.

The value of g is about $9.81\,\text{ms}^{-2}$. The approximate value $10\,\text{ms}^{-2}$ is sufficiently accurate for many calculations. The value in any case varies slightly over the Earth's surface; this is due to local variations of rock densities and altitude, and also to the fact that the Earth is not truly spherical, being slightly flatter at the Poles. For example, at the top of Mount Everest g is about 0.3% lower than at sea level: that is, it has a value of about $9.78\,\text{ms}^{-2}$. The value of g also varies with latitude around the Earth due to the effect of the Earth's rotation on a falling body.

Projectiles

You may have seen the 'monkey and hunter' demonstration during your GCSE course. See figure 12.

Provided the bullet or projectile is fired straight at the 'monkey', and the monkey is released at the same moment as the projectile, then they inevitably collide later.

figure 12   'Monkey and hunter' demonstration

This shows that an object travelling with a horizontal component of velocity (the projectile) has exactly the same vertical acceleration (g) as the monkey which falls straight down. This in turn means that for any projectile we can analyse the two components of velocity (the horizontal and the vertical) separately, provided we can neglect air resistance. Vertically, the projectile has acceleration g; horizontally, it has a constant velocity component. (This results in flight trajectories which are parabolic. See figure 13.)

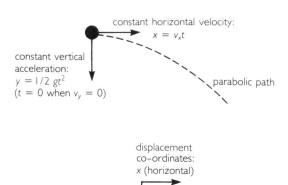

figure 13   Parabolic path of projectile

Example   Screwball, the great human cannonball, is fired from his cannon with a muzzle velocity of $16\,\mathrm{m\,s^{-1}}$ at an angle of 30° to the horizontal. How far away from the cannon should he place his arresting net (assuming he wants to land at the same level as he left)?

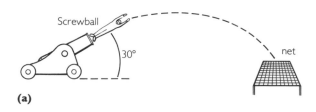

(a)

(Hint: In projectile questions, the key quantity which links the horizontal and vertical motions is the time of flight, t).

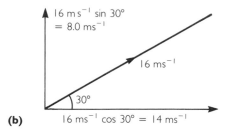

(b)

figure 14   (a) Screwball's situation; (b) initial velocity components

Vertical motion: $u = -8.0\,\mathrm{m\,s^{-1}}$ (taking downwards as positive)
$$v = +8.0\,\mathrm{m\,s^{-1}}$$
$$a = +9.8\,\mathrm{m\,s^{-2}}$$

$$v - u = at$$
$$\Rightarrow \quad 8.0\,\mathrm{m\,s^{-1}} - -8.0\,\mathrm{m\,s^{-1}} = 9.8\,\mathrm{m\,s^{-2}} \times t$$
$$\Rightarrow \quad t = 1.6\,\mathrm{s}$$

Horizontal motion: $\quad v = 14\,\mathrm{m\,s^{-1}}$ (constant)
$$t = 1.6\,\mathrm{s}$$
$$\Rightarrow \quad s = vt$$
$$= 14\,\mathrm{m\,s^{-1}} \times 1.6\,\mathrm{s}$$
$$= 22.4\,\mathrm{m}$$
$$\Rightarrow \quad s \approx 22\,\mathrm{m}$$

Force and acceleration

We have already discussed the idea that if an object has zero net force acting on it, then it has no acceleration (Newton's first law). Such objects include anything which is stationary, and also objects moving with constant velocity. We discussed a number of examples in chapter 1. Another example is a spacecraft travelling through deep space. With no rocket motors in operation, and therefore no forces acting on it at all, the craft continues to move with constant velocity through the universe. The converse is also true: a finite (that is, non-zero) resultant force on an object causes it to accelerate.

You probably did experiments in your GCSE course to find out how the acceleration of an object depends on the net force acting on it, and also on the mass of the object. A typical experiment is illustrated in figure 15.

In this experiment, you start by arranging the runway so that the trolley will run down it at a steady velocity, once you have given it a push (that is, 'friction-compensating' the runway); in this situation the trolley must have zero resultant force on it. Thus any resultant force acting on the trolley is due only to any extra force you exert using the elastic strings.

The results of these experiments, and many others which have been performed, are as follows:

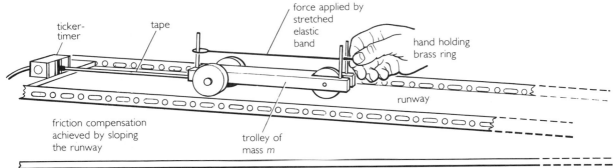

figure 15   Accelerating a trolley by applying a constant force

1   For a fixed mass of trolley, the acceleration a is proportional to the resultant force F: if we double F, by adding another elastic string, then a doubles.

Mathematically, $a \propto F$

2   For a fixed F, the acceleration a is inversely proportional to the mass m being accelerated: if we double m by stacking another trolley on top, then a halves.

Mathematically, $a \propto \dfrac{1}{m}$

Combining these two results mathematically gives

$$a \propto \frac{F}{m}$$

and hence $F \propto ma$

or $F = kma$

where k is a constant of proportionality.

The quantity k depends on the units we use for F, m and a. The equation is simplest if we choose units which make $k = 1$. To do this, we must have the same units for F as for the product ma.

The unit of m is $1\,\text{kg}$, and that of a is $1\,\text{ms}^{-2}$. Thus if the unit we use for F is $1\,\text{kgms}^{-2}$, then $k = 1$.

The unit $1\,\text{kgms}^{-2}$ is one of those units to which we give a special name. It is called one newton, $1\,\text{N}$.

The equation linking F, m and a is then

$$F = ma$$

Example   Let us consider the accelerometer illustrated on page 19. If the mass is 10 g, and the restoring force per metre of displacement, k, is $2\,\text{Nm}^{-1}$, then what displacement on the scale corresponds to an acceleration of $6\,\text{ms}^{-2}$?

The force required to accelerate the mass is given by

$$F = ma$$
$$= 0.01\,\text{kg} \times 6\,\text{ms}^{-2}$$
$$= 0.06\,\text{N}.$$

This force will be exerted on the mass when the springs are stretched a distance x given by

$$x = \frac{F}{k}$$

$$= \frac{0.06\,\text{N}}{2\,\text{Nm}^{-1}}$$

$$= 0.03\,\text{m}$$

$$= 3\,\text{cm}$$

Thus 3 cm displacement on the scale corresponds to $a = 6\,\text{ms}^{-2}$.

Inertial mass

You may have heard the mass of an object being explained with a phrase like 'the resistance of the object to being accelerated'. This arises from the finding mentioned above, that $a \propto 1/m$ for fixed F. In fact, this equation is used as the fundamental way of comparing the masses of different objects, and of defining mass.

The procedure we should follow to measure an unknown mass from fundamental principles is as follows.

1   Obtain a standard mass m_0, which has been calibrated using the 'prototype' kilogram (see chapter 45).

2   Apply a certain force F to it, and measure the resulting acceleration a_0.

3   Apply the same force F to the unknown mass m_x, and measure the resulting acceleration a_x.

4 Calculate m_x, using this theory:

$$F = m_0 a_0 \text{ and } F = m_x a_x$$

$$\Rightarrow \quad m_0 a_0 = m_x a_x$$

$$\Rightarrow \quad m_x = m_0 \times \frac{a_0}{a_x}$$

In reality, the existence of the Earth's gravitational field provides us with a much easier method of comparing masses, using a lever balance. See figure 16.

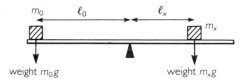

figure 16   Lever balance

Using the principle of moments:

$$m_0 g \ell_0 = m_x g \ell_x$$

$$\Rightarrow \quad m_x = \frac{m_0 \ell}{\ell_x}$$

Weight

We should now bring together two ideas which have been presented to you as though they were unrelated: g the acceleration of free fall, and g the gravitational field strength.

Consider an object of mass m falling freely (figure 17).

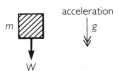

figure 17   Free-body diagram of falling object, without air resistance

We know it has acceleration g. So the resultant force F is given by

$$F = ma = mg$$

But the only force acting is the object's weight W. So $W = mg$.

Now this equation, $W = mg$, was stated as a fact without justification on page 2, where g was called 'gravitational field strength'. Now that we have found the equation $F = ma$, and defined one newton, the statements on that page are justified.

Terminal velocity

Let us consider the motion of an object falling near the Earth's surface, in conditions where air resistance is important: for example, a sky-diver in free fall. Air resistance now matters because he falls for quite a long time, and reaches such a speed that the air resistance force becomes the same order of magnitude as his weight.

The air resistance force F_r builds up with speed v. Roughly, $F_r \propto v^2$. So for a particular sky-diver, in a particular orientation of fall, we can write

$$F_r = kv^2$$

Now consider his free-body diagram: figure 18.

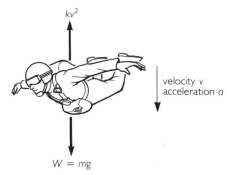

figure 18   Free-body diagram of falling sky-diver, with air resistance

The resultant force on him F is given by $F = mg - kv^2$.
Thus from $F = ma$,

$$mg - kv^2 = ma$$

He will thus gather speed, initially at acceleration g (when $v = 0$), then gradually more slowly, until he eventually reaches a steady, terminal velocity. See figure 19.

When this velocity is reached, the resultant force is zero, and the air resistance force exactly counteracts his weight. In practice, this may happen for a sky-diver at a speed of about $80 \, \text{m s}^{-1}$.

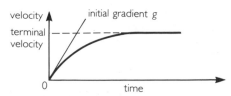

figure 19   Velocity-against-time graph to illustrate terminal velocity

Work and energy

We now consider one particular consequence of one object exerting a force on another and changing its motion: the exchange of energy between them.

Consider a diesel-driven crane with a heavy box suspended from its hook. If the box is simply held suspended while the driver has a cup of tea, the ratchet mechanism exerts the necessary force without the crane motor being needed – it can be switched off. See figure 20(a).

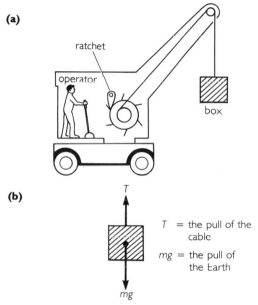

(a)

ratchet

operator

box

(b)

T

T = the pull of the cable

mg = the pull of the Earth

mg

For the load stationary or moving up with a constant speed: $T = mg$

figure 20 (a) Crane with ratchet mechanism; (b) free-body diagram of the box

When the driver wishes to raise the box, however, he has to start the motor and engage it – and diesel fuel starts being used up.

The pull T required from the cable is the same, whether the box is stationary or moving up (once it has been set in motion). But in the latter case, the crane motor is winding in the cable and we say that the crane is doing work. In formal language, we say that a force (T) does work when its point of application (the hook) is moved along the direction of the force. The greater the load, the more fuel is used to move the object a given distance; also, more fuel is used to move the object through a greater distance. The work done, W, by a force F is defined by the equation:

$$W = F \times s$$

where s is the distance moved in the direction of the force F.

If F varies with the movement, we would calculate W by finding the area under a graph of F against displacement.

The unit of work has a special name, the joule (J). $1\,J = 1\,Nm$.

We say that an object possesses energy if it is capable of causing a force on another object and doing work on it. In the above case the crane initially has energy. When the crane does some work on the box, the box gains energy (by being lifted up) and the crane loses energy (by using up fuel).

The amount of energy exchanged is equal to the work done. The joule is thus the unit of energy as well.

An object can possess energy in a number of different ways. We classify types of energy as follows:

☐ kinetic energy – possessed by an object in motion;

☐ gravitational potential energy – possessed by the raised box;

☐ thermal (internal or heat) energy;

☐ wave energy;

☐ other types of potential energy:

   chemical;

   elastic;

   electric field;

   magnetic field;

   nuclear;

There are many examples of energy being transferred from one object to another. You have no doubt discussed some during your GCSE course. For example, in the crane example above, chemical potential energy (in the fuel) becomes gravitational potential energy (in the box) plus thermal energy (in the exhaust gases, pulley wheels, etc.).

In many simple situations the law of conservation of energy can be used:

'The total energy of the universe remains constant.'

A mathematical statement of this is ΣE = constant.

A more complete law takes account of the fact that mass and energy are interconvertible: that is, mass can disappear, and energy be created instead, according to the equation

$$\Delta E = c^2 \Delta m$$

where c is the velocity of electromagnetic waves in a vacuum.

This happens notably in nuclear reactions.

A more complete equation, covering all systems we know about, would be

$$\Sigma E + c^2 \Sigma m = \text{constant}$$

See page 398 for examples of mass-to-energy conversions in nuclear reactions; and page 454 for discussion of energy conservation as applied in the first law of thermodynamics.

Gravitational potential energy

Referring again to the crane example on page 25, the work done on the box was equal to the gravitational potential energy E_p gained by the box. If the box rose through a height h, then

$$E_p = \text{work done}$$

$$= \text{force} \times \text{distance}$$

$$= mg \times h$$

Thus, provided we are restricting our attention to regions where g does not change appreciably, gravitational potential energy is given by

$$E_p = mgh$$

In situations when g may change, that is, if h becomes large, then the calculation is more complicated, although the principles are the same.

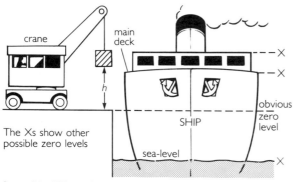

figure 21   Where is E_p measured from?

The Xs show other possible zero levels

Note that whenever you calculate a value for E_p, you are assuming some zero level from which you are measuring h. Usually the zero level will be obvious – see figure 21. On the other hand, driving the crane forward so that the box is over the water will alter the zero level: will the box then have greater E_p? If so, where has the extra energy come from? This point is discussed further in chapter 18.

Elastic potential energy

Elastic potential energy is stored when a spring or elastic band is stretched – for example, in a clockwork mechanism or rubber-band aeroplane.

If we stretch a spring from its natural length, the force F we have to apply gradually increases, according to Hooke's law:

$F \propto x$ where x is the distance the spring is stretched,

$\Rightarrow$ $F = kx$ where k is called the spring constant of the spring.

At the beginning of the stretching process, therefore, the work you must do to extend the spring by a

certain small distance is less than later on in the process: because the force is less.

The work ΔW to increase the extension by a small amount Δx is approximately $F \times \Delta x$, where F, the force being applied at that extension, hardly changes. $F \times \Delta x$ is thus roughly equal to the area of a small column under the force-against-extension graph: see figure 22.

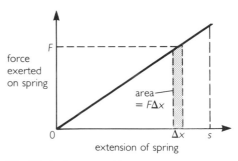

figure 22   Work done stretching a spring

You can probably now see that the work you do in stretching the spring to any particular extension x is the area under the graph as far as that value of extension. Algebraically, for the spring which has a straight-line graph,

$$W = \text{area of triangle} = \tfrac{1}{2} x \times F$$

$$\Rightarrow \quad W = \tfrac{1}{2} Fx$$

or, using $F = kx$,

$$W = \tfrac{1}{2} kx^2$$

If the graph is not a straight line, we must evaluate the area numerically.

If you have an equation relating F and x, then you can evaluate the work done using the relationship

$$\Sigma \Delta W = \Sigma F \Delta x$$

$$\Rightarrow \quad W = \int F dx$$

Kinetic energy

We will first use a piece of theory to decide a sensible formula for working out kinetic energy E_k; then we will examine an experiment which verifies the formula.

If an object is to gain kinetic energy, then work has to be done on it to give it that energy. Let us assume that a force F gives a body of mass m a uniform acceleration a over a distance s. The body starts at rest and reaches a speed v at the end of its accelerating period.

$$\text{gain in kinetic energy} = \text{work done}$$

$$= F \times s$$

$$= mas$$

$$= ma(\tfrac{1}{2} at^2)$$

$$= \tfrac{1}{2} m a^2 t^2$$
$$= \tfrac{1}{2} m v^2$$

Thus we may write kinetic energy $E_k = \tfrac{1}{2} m v^2$.

[In general the formula above may be derived for non-uniform accelerations as follows:

Gain in kinetic energy = work done

$$= \int F \, \mathrm{d}s$$
$$= \int ma \, \mathrm{d}s$$
$$= \int m \frac{\mathrm{d}v}{\mathrm{d}t} \, \mathrm{d}s$$
$$= \int m v \, \mathrm{d}v$$
$$= \tfrac{1}{2} m v^2]$$

One experiment to verify this formula is as follows:

1 Arrange a rubber-band catapult over a linear air track – see figure 23(a).

(a)

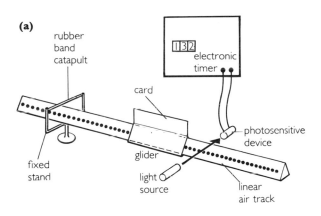

(b)

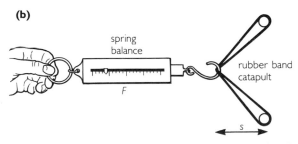

figure 23 (a) Experimental arrangement; (b) measuring F against s

2 Using a spring balance and ruler, obtain a series of readings of force F applied against pull-back distance s – see figure 23(b). Plot these readings on a graph – figure 24.

3 The area under the graph from $s = 0$ up to $s = $ (a certain distance) gives the work done by whoever pulls the band back to that distance. If we assume that the band will impart all this energy to the air-track glider when the time comes, then by evaluating the area we obtain the amount of kinetic energy the glider will receive. Calculate an energy from the graph for a suitable value of s.

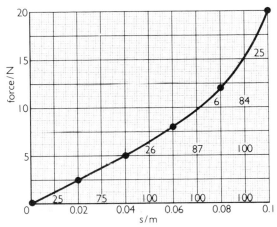

figure 24 Force against pull-back distance for rubber-band catapult

4 Fire the glider of mass m, using the same value of s as in (3). Measure its speed v using the light-gate assembly.

5 Calculate $\tfrac{1}{2} m v^2$. Compare this with the energy expected from the graph. It will probably be somewhat less, because of losses and residual E_k in the rubber band and oscillation of the glider.

6 Repeat the procedure (3)–(5) for other values of s.

Example Using the catapult of figure 23, with a pull-back of 0.10 m the glider (mass 0.10 kg) leaves the catapult at $3.3 \, \mathrm{m s}^{-1}$. Check that this is consistent with the formula for E_k.

The area under the graph up to $s = 0.10$ m is 728 little squares. Each little square represents:
$0.50 \, \mathrm{N} \times 0.0020 \, \mathrm{m} = 0.0010 \, \mathrm{J}$
$\Rightarrow$ energy stored $= 728 \times 0.0010 \, \mathrm{J} = 0.73 \, \mathrm{J}$

Formula: $\tfrac{1}{2} m v^2 = \tfrac{1}{2} \times 0.1 \, \mathrm{kg} \times (3.3 \, \mathrm{m s}^{-1})^2$
$= 0.54 \, \mathrm{J}$

Since energy losses are bound to occur in the catapulting process, we can say that these values are in reasonable agreement.

Power

We describe energy-transferring devices by the general term 'machine'. Examples include engines, like the one in the crane which transfers chemical potential energy in the fuel into gravitational potential energy; the human body, which transfers chemical potential energy from food in a variety of ways; turbine/dynamo systems, which transfers kinetic energy into electrical potential energy; batteries; pulley systems, levers and gear boxes; transformers; and many others.

One important quality of a machine is the rate at which it can transfer energy. This is called its power, P, and is defined by the equation

$$P = \frac{\mathrm{d}W}{\mathrm{d}t}$$

where W is energy being transferred.

If the rate remains constant, a more useful equation is

$$P = \frac{W}{t}$$

The unit of power is the watt (W). $1\,\mathrm{W} = 1\,\mathrm{J\,s^{-1}}$.

Example Suppose our crane lifts a box of mass $100\,\mathrm{kg}$ through $8.0\,\mathrm{m}$ in $5.0\,\mathrm{s}$. We calculate the power output of the crane like this:

$$P = \frac{W}{t}$$

$$= \frac{mgh}{t}$$

$$= \frac{100\,\mathrm{kg} \times 10\,\mathrm{N\,kg^{-1}} \times 8.0\,\mathrm{m}}{5.0\,\mathrm{s}}$$

$$\Rightarrow \qquad P = 1600\,\mathrm{W}$$

An old but still-used unit of power is the horsepower. 1 horsepower $\approx 750\,\mathrm{W}$.

Efficiency

A machine takes energy in from one object and transfers it to another. An ideal machine would give out exactly as much as it took in, without any waste. (It couldn't give out more, by the law of conservation of energy.)

Real machines can never quite achieve this perfection. But, as a measure of how close they get to it, we can calculate their efficiency, as defined by either

$$\text{eficiency} = \frac{\text{useful energy out}}{\text{energy in}}$$

or $\qquad \text{efficiency} = \dfrac{\text{useful power out}}{\text{power in}}$

(These two equations are equivalent.)
This fraction is often converted to a percentage.

Example A typical coal-burning power station is 30% efficient. If the station can output $1.5\,\mathrm{GW}$ at full load, at what rate is it extracting heat energy from the coal?

$$\text{efficiency} = \frac{\text{power out}}{\text{power in}} \times 100\%$$

$$\Rightarrow \qquad 30 = \frac{1.5\,\mathrm{GW}}{\text{power in}} \times 100$$

$$\Rightarrow \quad \text{power in} = 1.5\,\mathrm{GW} \times \frac{100}{30} = 5.0\,\mathrm{GW}$$

This coal input rate, incidentally, is about one tonne of coal every six seconds!

Summary

1 Displacement s is a vector quantity.

2 Velocity $v = ds/dt = s/t$; also a vector.

3 Speed may be measured using tickertape, multiflash photography, a speedometer, the Doppler effect, or by direct measuring of distance and time.

4 Acceleration $a = dv/dt$ or $(v - u)/t$; also a vector.

5 Acceleration may be found from measurements of velocity, or using an accelerometer.

6 Velocity-against-time graph: gradient represents acceleration, area under graph represents distance.

7 Equations for constant acceleration:
$$v - u = at$$
$$s = ut + \tfrac{1}{2}at^2$$
$$v^2 - u^2 = 2as$$

8 Free fall; measurement of g, the acceleration of an object near the Earth.

9 Projectiles have constant vertical acceleration (g), constant horizontal velocity, hence a parabolic path.

10 Experiments lead to $F = ma$.

11 $1\,N = 1\,kgms^{-2}$.

12 Measuring mass using its inertial property, or with a lever balance.

13 Weight $= mg$, where g is the gravitational acceleration.

14 Terminal velocity of an object falling through air.

15 Work $W = F \times s$, or $W = \int F\,ds$.

16 $1\,J = 1\,Nm$.

17 Work is one way in which energy is converted from one form to another.

18 An object has energy if it can do work.

19 Energy conservation throughout the universe:
$$\Sigma E = \text{constant}$$

20 Improved energy conservation equation:
$$\Sigma E + c^2 \Sigma m = \text{constant}$$

21 Gravitational potential energy $E_p = mgh$.

22 Elastic energy stored in a spring $= \tfrac{1}{2}kx^2$.

23 Kinetic energy $E_k = \tfrac{1}{2}mv^2$.

24 Power $P = dW/dt$ or W/t.

25 $1\,W = 1\,Js^{-1}$.

26 Efficiency = useful energy out/energy in or power out/power in.

Questions

figure 25 Photo (multiflash) of tennis serve

1 For the multiflash photograph of a tennis serve shown in figure 25, the interval between flashes is 0.04 s. Assume that the distance between the player's feet and his racquet-head when he is at full stretch is 3 m. Estimate, showing your reasoning,
   a the speed of the racquet-head when the ball is struck;
   b the initial speed of the ball.

2 The following data give speed against time for an accelerating car.

t/s	0	2	4	6	8	10	12	14	16	18
v/ms^{-1}	0	9.5	18	25.5	32	37.5	42	45.5	48	49.5

   a Plot the data as a graph.
   b Find how far the car has travelled after (i) 10 s, (ii) 18 s.
   c Find the acceleration of the car (i) initially, (ii) after 10 s.

3 An Intercity 125 train accelerates from $30\,\mathrm{ms}^{-1}$ to $50\,\mathrm{ms}^{-1}$ while covering a distance of 800 m. Assuming the acceleration is constant, find
  a its acceleration;
  b the time over which this occurs.

4 A racehorse of mass 500 kg sprints at about $20\,\mathrm{ms}^{-1}$, achieving this speed from a standing start in about 2 s. Assuming constant acceleration, calculate
  a its acceleration,
  b the distance it covers in this time,
  c the average power it develops in this time.

5 At the time of writing, the men's world sprint records stand as follows:
  100 m   9.92 s
  200 m   19.72 s
  Assume, for the sake of a simple model, that in each case the sprinter concerned accelerated steadily to the same top speed, then maintained that speed to the end of the race.
  a Sketch the appearance of the 200 m runner's speed-against-time graph.
  b Calculate his (and the 100 m runner's) top speed.
  c Why is the 200 m runner's average speed greater than that of the 100 m runner?
  d Calculate, on the basis of the model we are using, the time taken for both runners to reach top speed. (Hint: Consider the velocity-against-time graph you sketched above.)
  e Calculate the acceleration as each runner accelerates. Comment on its magnitude.
  f Criticise the model constructively, commenting on (i) the ways in which the reality probably differs from the model, (ii) the ways in which these differences would affect your answers to parts a, b, d and e.

6 The following is adapted from a competition run by British Airways in 1986 to celebrate 10 years of Concorde services.
  'The Concorde Challenge. We challenge you to plot as exactly as you can the position of Concorde at 13.00 hrs.
  The Concorde fact sheet:
  1. Concorde departs from Heathrow at 10.30 a.m. sharp.

  2. Concorde's flight from London Heathrow to Kennedy Airport takes 3 hours 40 minutes exactly.

  3. The distance between the airports is 5800 km.

  4. Concorde flies for the first 15 minutes to a point 235 km from Heathrow, over the Bristol Channel. At this point it is flying at $990\,\mathrm{km\,h}^{-1}$. It then accelerates for 11 minutes over a distance of 250 km to a cruising speed of $2150\,\mathrm{km\,hr}^{-1}$.'

How far from Heathrow will Concorde be at the time of 13.00 hrs?
(Hints: Assume that it flies in a straight line, and that the deceleration into Kennedy airport is uniform. Try plotting a graph of speed against time.)

7 One of the most spectacular tricks performed by stuntmen is to jump on a motorbike over a line of cars (figure 26).

stuntman on motorbike

10 cars, each 2 m wide

figure 26

  a At what speed must this cyclist leave the ramp in order just to clear the 10 cars?
  b If there were *n* cars, what minimum speed would he need?

8 Boris the cat is sitting on a tree-branch 5 metres above ground level when a small rabbit darts out from a burrow at the base of the tree, heading at $10\,\mathrm{ms}^{-1}$ for another burrow 20 m away. See figure 27.

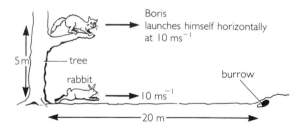

Boris launches himself horizontally at $10\,\mathrm{ms}^{-1}$

5 m

tree

rabbit

$10\,\mathrm{ms}^{-1}$

burrow

20 m

figure 27

  Boris immediately launches himself horizontally from his branch, with the same direction and speed as the rabbit, intending to descend on him somewhere along his path. Will Boris intercept the rabbit? If so, where? If not, how far away would the burrow have to be to give Boris a chance?

9 In a well-known television commercial for beer, a full pint glass is slid from one end of the bar to the other, a distance of about 5 m. If the glass leaves the sender at $4\,\mathrm{ms}^{-1}$, and just comes to rest at the other end of the bar, find the coefficient of sliding friction between the glass and the surface.

10 In a large car, a balloon filled with helium is attached by string to the floor, and a balloon filled with carbon dioxide is attached to the ceiling. How does each one behave when the car accelerates? (Hint: Being accelerated by a force is rather like experiencing weight, as Einstein emphasised in his general theory of relativity.)

11 A box of mass 2000 kg is being towed over a rough surface using a winch which can maintain a constant tension in the towing cable. Initially the box is stationary. As the tension builds up, the box starts to move when the tension reaches 5000 N, and it then continues to accelerate at 1.5 ms⁻². Find

a the coefficient of limiting static friction, μ_{st};
b the coefficient of sliding friction, μ_{sl}.

12 Comment on the correctness, or otherwise, of each of the three sentences in this paragraph, which are about a book of mass m on the floor of a lift.

'When the lift is stationary, by Newton's third law the force R exerted by the floor on the book is equal to mg. When the lift moves downwards with acceleration $g/4$, R is equal to $3mg/4$. Since R no longer equals the weight of the book, Newton's third law is not obeyed when the lift is accelerating.'

13 A man stands on some scales while ascending in a sky-scraper lift. He watches the scales, and makes the following sketch graph of the reading on them against time (figure 28).

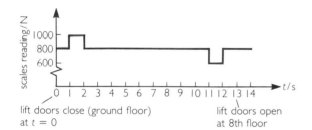

figure 28

Make suitable calculations so that you can derive a graph showing the lift's speed upwards against time. Deduce the height of one floor above another. (Hint: It may be useful to calculate the following: his mass; accelerations; speeds of lift; distance it travels.)

14 Pilots of large jet aircraft usually put their engines into 'reverse thrust' after landing, to slow them down. Figure 29 shows how the exhaust, instead of leaving the engines backwards as in normal flight, is deflected towards the front of the aircraft, at an angle of about 45°, both above and below the engines.

Use this data for the calculations which follow.

Thrust from four engines at full power if in normal flight = 8×10^5 N

Touch-down speed of aircraft = 65 ms⁻¹

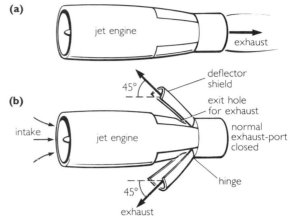

figure 29 (a) Normal flight; (b) reverse thrust arrangement

Mass of aircraft at touch-down = 2×10^5 kg.

a Explain in terms of basic physical principles why putting the engines into reverse thrust slows the aircraft down.
b Calculate the retarding force on the aircraft with reverse thrust operating, assuming the engines are run at full power.
c What is the deceleration of the aircraft on the runway?
d Assuming that the pilot operates reverse thrust immediately on touch-down, and keeps it on till the aircraft comes to a halt, what distance does the aircraft cover before stopping? What time does it take for the aircraft to stop?

15 Paul is weighing himself in his front hall, prior to going out jogging, when Boris the cat drops off the banister above into his arms. Boris has mass 2 kg, and drops a distance of 2 m. Paul has mass 90 kg. During the catching process, Paul's arm moves down a distance of 20 cm.

a Calculate the reading of the scales (in newtons) (i) before Boris arrives, (ii) after Boris has come to rest in Paul's arms.
b What kinetic energy does Boris have when he arrives at Paul's arms?
c Assuming Boris is brought to rest smoothly, what is the force on the scales during this process?
d How long does this 'coming to rest' take?

16 This question is about the terminal velocities reached by raindrops of different size.

The drag force F_L on a sphere moving through air is given approximately by

$$F_L = \tfrac{1}{2} C_w \rho_a A v^2$$

where C_w is the drag coefficient, with value 0.2 for a sphere,
ρ_a is the density of air, 1.3 kgm⁻³,
A is the frontal area of the sphere,
v is its speed through the air.

a Write down the mass of the drop in terms of its radius r, and density ρ_w.

b What is the condition for the drop to be travelling downwards at a constant speed?

c Use (a) and (b) to write an expression for the terminal velocity of a drop in terms of its radius. (Density of water = $1000\,\mathrm{kg\,m^{-3}}$.)

d Find a value for the terminal velocity of a drop with radius (i) 1 mm, (ii) 0.1 mm, (iii) 0.01 mm.

e Comment on the relevance of these calculations to the meteorological study of fog, mist and rain.

17 On a fairground ride, the vehicle is hauled up to point A, then released. It runs down to B, then nearly up to C; it is helped up by a winch to C, from where it runs down again to D. The lengths of track AB, BC, and CD are all equal; A and C are on the same horizontal level; so are B and D.

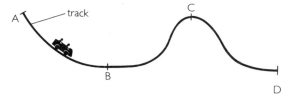

figure 30

a An observer says that the vehicle reaches the same speed at D as at B. Comment on this.

b The same observer says that the vehicle takes longer to go from C to D than it did from A to B. Comment.

c Why does the vehicle need help up the last part of BC?

d The observer says finally that whether the vehicle is carrying people or not, it still covers the distance AB in the same time. Comment.

18 Some design students are making an egg-protecting case, to prevent an egg from breaking if it falls onto a hard floor from a height of 1.5 m. The average egg has mass 60 g, and is known to break if subject to any force in excess of 30 N (steady or instantaneous).

a Calculate the energy which the case must absorb as the egg is brought to rest.

b David has a design with 2 cm of foam rubber in front of the egg. Show that his egg will break when the case hits the floor.

c Mary has a design in which the egg is supported inside a rather weak paper cylinder, which crumples on impact. Tests with a 60 g potato show that the cylinder will shorten by 5 cm on impact. Show that this design will probably work.

19 This question is about the distance covered by a car as it stops, and the separation which cars should keep when following each other. We can use a simple model for calculating how far a car travels before stopping, from the moment the driver senses the need to stop. Suppose the relevant parameters are as follows:

mass of car + occupants $\qquad m$
initial speed $\qquad v$
available braking force $\qquad F$
driver's reaction time $\qquad t_r$

a Write down an expression for the 'thinking distance' travelled by the car during the driver's reaction time.

b Write another expression for the distance it takes for the car to stop, once he applies the brakes: the 'braking distance'.

The total stopping distance is the sum of the answers to (a) and (b). The table below shows some possible values for thinking, braking and stopping distances.

initial speed $v/\mathrm{ms^{-1}}$	thinking distance/m	braking distance/m	stopping distance/m
10	7.5	5.0	12.5
20	15	20	35
30	—	—	—
40	—	—	—

c What value of t_r is being used to construct the table?

d Work out the values to go in the blank lines in the table.

e If the car has mass 800 kg, what is the value of F?

f The *Highway Code* suggests a rule of thumb for drivers: 'Keep one car's length gap between you and the car in front for every 10 mph ($5\,\mathrm{ms^{-1}}$) of speed.' How is this rule related to the model we have been looking at? (Make your comments quantitative, using suitable values where necessary for quantities involved.)

20 This question is about Bernoulli's theorem. This relates to pressures in moving fluids, and explains such effects as the lift of an aircraft wing and the effects of spin on the flight of balls.

Consider first the situation illustrated in figure 31. Liquid with density ρ is moving through a pipe which narrows. An initial volume $A_1\ell_1$ becomes a longer narrower volume $A_2\ell_2$.

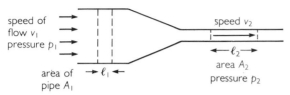

figure 31

a Explain why the liquid must speed up in the narrower length of pipe.

The extra kinetic energy of the faster-flowing liquid must be provided by the force exerted by the pressure p_1 on the area A_1.

b What force is acting on the left hand end of the liquid in figure 31?

c How much work is done by the force in moving a distance ℓ_1?

Some of this work is required to move the volume in the narrower pipe against pressure p_2 through distance ℓ_2.

d How much work is required for this?

e What is the excess energy, or work, given to the liquid as it moves into the narrower region of pipe?

f Write down an expression in terms of A_1, ℓ_1, v_1, A_2, ℓ_2, v_2 and ρ for the increase in kinetic energy of the liquid.

If all the excess work goes into kinetic energy, then (e) and (f) are equal. Putting them equal you should obtain

$$p_1 A_1 \ell_1 - p_2 A_2 \ell_2 = \tfrac{1}{2} A_2 \ell_2 \rho v_2{}^2 - \tfrac{1}{2} A_1 \ell_1 \rho v_1{}^2$$

But $A_1 \ell_1 = A_2 \ell_2$

So we then obtain

$$p_1 - p_2 = \tfrac{1}{2}\rho v_2{}^2 - \tfrac{1}{2}\rho v_1{}^2$$

or $p_1 + \tfrac{1}{2}\rho v_1{}^2 = p_2 + \tfrac{1}{2}\rho v_2{}^2$

or in general $p + \tfrac{1}{2}\rho v^2 = $ constant

This is Bernoulli's theorem for the streamline flow of an incompressible fluid.

g Show that p and $\tfrac{1}{2}\rho v^2$ have the same unit, J m^{-3}.

h What do you think is the meaning of (i) 'streamline', (ii) 'incompressible'?

An aircraft wing is shaped so that the air has to travel slightly further over the top of it than underneath it. See figure 32.

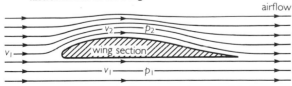

figure 32   Air velocity and pressure above and below a wing

i Explain why this gives rise to a lifting force.

Let's try some numbers. Assume, roughly, that air is incompressible. Here is some data for a Jumbo jet.

    total wing area $440\,\text{m}^2$
    air speed (v_1) $250\,\text{ms}^{-1}$
    weight $3.0 \times 10^6\,\text{N}$
    density of air (at cruising height) $0.50\,\text{kg m}^{-3}$

j What average difference in pressure is needed between the top and bottom of the wings to generate the necessary lift?

k what must be the speed of the air over the wing, v_2, to give rise to this pressure difference? Does this seem a reasonable increase in speed to expect?

Finally, let us consider how a ball in flight is affected by spin. Suppose we are looking from above at a ball travelling to the right, and spinning clockwise (figure 33).

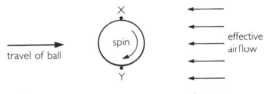

figure 33

The surface of the ball takes a layer of air round with it. The velocity of this layer adds, as a vector, to the velocity of the air through which the ball moves.

l At which point is the pressure lower, X or Y? Explain.

m Hence deduce and explain which way the ball swerves in its flight.

21 Sky-divers who jump from aeroplanes experience an air-resistance force upwards on them as they hurtle earthwards, given very roughly by

$$F = 0.06 v^2 A \qquad \text{(SI units)}$$

where v is their speed through the air and A is the surface area they present. This equation is roughly true whether their parachute is deployed or not. If it is deployed, then of course A is the effective area of the parachute.

a The number 0.06 is not a pure number, it must have a unit. Find the simplest SI unit for 0.06 in terms of fundamental units. The number 0.06 is related to a property of air – guess what this property might be.

b Make a sensible estimate of the area presented by a sky-diver who is falling face downwards with her arms by her sides. Hence calculate the terminal velocity she reaches, if she weighs 800 N. Comment on the value you find.

c With parachute deployed, the same woman falls at about $8\,\text{ms}^{-1}$. Estimate the area presented by the parachute. Comment on the value you find.

d Suppose that while she is in free fall, her parachute *instantaneously* becomes deployed to its full area. Calculate, using your previous answers, the instantaneous deceleration she would experience. Comment on your value.

e In free fall, the woman is losing gravitational potential energy at a considerable rate. Estimate the rate of energy conversion. Where does this energy go?

22 An extract from a flying manual for jet aircraft states the following: 'The typical drag of a jet transport aircraft of mass $110\,000\,\text{kg}$, flying at $240\,\text{ms}^{-1}$ at height $12\,000\,\text{m}$, is $92\,\text{kN}$. On a four-engined aircraft this means that each engine must be delivering $23\,\text{kN}$ thrust; and under these conditions $23\,\text{kN}$ of thrust is equivalent to just over $5.6\,\text{MW}$.'

a Show how the author has assumed Newton's first law, that the forces on the aircraft are balanced.

b Show how the figure of $5.6\,\text{MW}$ follows from the earlier data.

23 Data on one particular modern car are given in the handbook as follows:

rolling friction $F_R = mg[(2.02 \times 10^{-2})$
$+ (7.04 \times 10^{-4}v) + (2.04 \times 10^{-5}v^2)]$
$\qquad$ (F_R in newtons if v in ms^{-1})

aerodynamic drag $F_L = 0.47v^2$
mass m $\qquad$ $1000\,\text{kg}$
top speed $\qquad$ $118\,\text{mph}$ ($52\,\text{ms}^{-1}$)

What is the car engine's power output when at top speed?

24 Consider the following data for a top-class sprinter:

frontal area $\qquad$ $1\,\text{m}^2$
speed $\qquad$ $10\,\text{ms}^{-1}$
density of air $\qquad$ $1.3\,\text{kgm}^{-3}$

Let us make the crude simplifying assumption that as he moves through the air, all the air through which he passes is set in motion with the same speed as himself. Use this idea to estimate the following:

a the kinetic energy transferred to the air each second;

b the power that he must generate just to overcome this air resistance effect;

c the force exerted on him by this air resistance.

25 This series of questions is about the dynamics of a car built by the Austin-Rover Company, the 1986 Montego 2.0 HL. Examine the manufacturer's data below, and then answer the questions which follow.

I. Acceleration through the gears (this data is converted from speeds in miles per hour)

Speed/ms^{-1}	13.3	17.8	22.2	26.7	31.1	35.6	40.0
Time to reach that speed from rest/s	3.4	5.1	7.1	9.8	13.5	18.2	27.0

II. Graph of available force at the wheels against speed

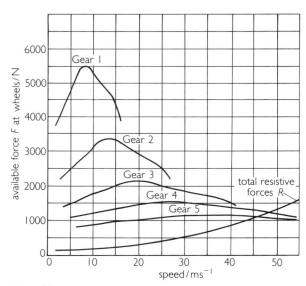

figure 34

III. Mass of vehicle under test = $1160\,\text{kg}$
Air resistance drag $F_L = \frac{1}{2}\rho C_w A v^2$
where ρ = density of air = $1.2\,\text{kgm}^{-3}$
C_w = drag coefficient = 0.37
A = frontal area = $2.0\,\text{m}^2$
v = speed of vehicle (in ms^{-1})

IV. Gear ratios (engine rotation speed/propellor shaft rotation speed)

1st	2.92
2nd	1.75
3rd	1.14
4th	0.85
5th	0.65

a (i) Plot a graph of speed against time for the car as it accelerates through the gears. (This means that at each speed the gear giving the maximum force and power at that speed is in use.)

(ii) Show from the speed-against-time graph that the manufacturer's claim that the car covers $400\,\text{m}$ from rest in $17.8\,\text{s}$ is correct.

(iii) From the speed-against-time graph, obtain a value for the acceleration at a speed of $20\,\text{ms}^{-1}$.

b Consider the graphs showing available force F and resistive forces R against speed v for the car.
(i) What is the unbalanced (accelerating) force on the car in terms of F and R?
(ii) At $20\,\mathrm{ms^{-1}}$, which is the optimum gear for maximum acceleration?
(iii) Calculate the theoretical acceleration at that speed, and compare this with your answer to a(iii).
(iv) Make a similar comparison for $30\,\mathrm{ms^{-1}}$ between the acceleration measured from the speed-against-time graph and that calculated from the forces-against-speed graph.
(v) What is the maximum speed which the car should reach? In what gear will it be reached?
c We will now consider the resistive forces.
(i) Show that the expression given for air resistance drag has the units of force.
(ii) Calculate the resistive force due to air drag alone, at (1) $10\,\mathrm{ms^{-1}}$, (2) $30\,\mathrm{ms^{-1}}$, (3) $50\,\mathrm{ms^{-1}}$.
(iii) How does the fraction of the total drag which is due to air resistance change as the speed increases?
(iv) The other component of the total resistive forces is called 'rolling friction'. Explain how this force arises.
(v) From the data available to you, does the rolling friction appear to be speed-dependent, or is it essentially a constant value? Justify your answer quantitatively.
d (i) How is the force at the wheels at a given speed related to the power at the wheels?
(ii) Use the graphs of force against speed for each gear to construct graphs, on one set of axes, for power against speed.
(iii) Add to the axes in (ii) a line showing the power expended against resistive forces.
(iv) Comment on the values of peak power available in each gear.
(v) In any particular gear, does the peak power occur at the same speed as the maximum force available? Comment on your answer.
(vi) Show that the peak power occurs at roughly the same engine speed for each gear.

26 The following is an extract from *The Bosch Automative Handbook* on the subject of aerodynamic drag (air resistance). Study it, then answer the questions which follow.
Aerodynamic drag (air resistance) F_L is given by
$$F_L = 0.5\rho C_w A v^2$$
where ρ, the density of air, is $1.2\,\mathrm{kgm^{-3}}$,
A is the frontal area of the car,
C_w is the drag coefficient of the particular car,
v is its speed (in $\mathrm{ms^{-1}}$).

Drag power P_L is given by
$$P_L = F_L \times v$$

Drag coefficient and drag power of various vehicle designs

| Vehicle | C_w | Drag power in kW for $A = 2\mathrm{m}^2$ at different speeds | | | |
		40 $\mathrm{kmh^{-1}}$	80 $\mathrm{kmh^{-1}}$	120 $\mathrm{kmh^{-1}}$	160 $\mathrm{kmh^{-1}}$
Open convertible 	0.5–0.7	1	7.9	27	63
Modern saloon 	0.3–0.4	0.58	4.6	16	37
Optimum streamlined design 	0.15–0.2	0.29	2.3	7.8	18

figure 35

a (i) From the expression for F_L, can you attach a physical meaning to the product $\rho A v$?
(ii) Show how the expression for F_L can be interpreted using the idea 'force = rate of change of momentum'.
(iii) What is the significance of the coefficient C_w?
(iv) What factors affect the value of C_w for a particular car?
b (i) From the information at the beginning of the extract, show that $P_L \propto v^3$.
(ii) For each of the three vehicles in the table, show that the drag power values are in accordance with $P_L \propto v^3$.
(iii) Consider the modern saloon travelling at $120\,\mathrm{kmh^{-1}}$. Work out the value of C_w (between 0.3 and 0.4) which has been used in the calculation of the values of P_L for the modern saloon.
(iv) Work out the values of C_w which have been used for the other two cars.
(v) Write a few lines of advice to a friend who is designing a vehicle to enter a fuel-economy competition. The vehicle will have to travel a pre-set course, at any speed the competitor wishes, provided he completes it within a time-limit.

27 This question is about the hill-climbing performance of one particular Austin-Rover car, the Maestro 1.3 HL. The mass of the Maestro is approximately $1000\,\mathrm{kg}$.
Study the table and graphs on the next page and answer the following questions.

a Draw a free-body diagram of the car on a gradient of angle α.
b What forward force must the engine exert to counteract the effect of the component of gravity which acts down the slope?
c In the table this force is called the 'climbing resistance', F_{st}. Verify for 10°, 20° and 30° that the table gives the right values for climbing resistance.

The total resistive force F_w to the car's motion is given by

$$F_w = F_{Ro} + F_L + F_{st}$$

where F_{Ro} is rolling resistance (of tyres, etc.) and F_L is aerodynamic resistance. The graph shows how F_L varies with v; and also how F_w varies with v both on level ground and at three different climbing gradients.

d Say what you can about the value of F_{Ro} at different speeds.
e Show how the four graphs for F_w are related to each other and to the values in the table.
f What would be the car's top speed on slopes of (i) 10%, (ii) 20%, (iii) 30%?

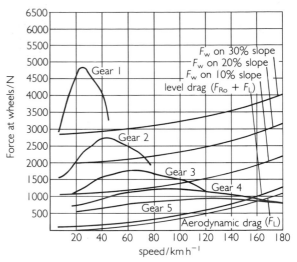

figure 37

g Calculate the maximum possible acceleration of the car (i) on a 10% slope at 60 km h^{-1}, (ii) on a 30% slope at 40 km h^{-1}.
h While travelling at full speed on the level, the car suddenly moves onto a 10% uphill slope. What will be its initial deceleration?

Angle of gradient α	Gradient p %	Climbing resistance (with m = 1000 kg) N

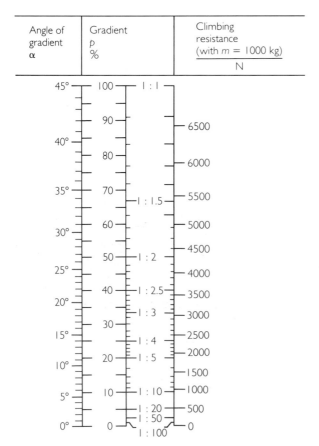

figure 36

CHAPTER 3

Collisions

Introduction

In this chapter we consider how Newton's laws can be applied to predict the effect one object will have on another. The range of situations governed by these laws is vast: from objects on an astronomical scale, through man-sized events, down to sub-atomic interactions. We develop the concepts of momentum, impulse, momentum conservation and elastic collisions, starting from Newton's third law and the way in which animals and vehicles propel themselves. We then apply these ideas to predict the results and various special cases of objects interacting or colliding; these cases provide models from which any particular real problem may be approached. For example, the impacts between two snooker balls or two sub-atomic particles can be analysed in very similar ways.

GCSE knowledge

You may have used the idea of momentum in your GCSE course, but it is not essential that you have done so: this chapter develops it from scratch. Kinetic energy, as discussed in detail in chapter 2, is the other quantity developed in more detail in this chapter.

Vehicle propulsion

What do the following all have in common: human beings, birds, motor-boats, hovercraft, aircraft and rockets? Answer: each one can propel itself forwards, and in order to move forward, each one has to push backwards on something. An aircraft accelerates, for example, by thrusting backwards quantities of air from its engines; this air in return exerts a force of equal size on the aircraft: pushing the aircraft forwards. This is an example of Newton's third law.

Let us consider the case of the aircraft more carefully. Its free-body diagram, and that of a mass of air m which it is in the process of thrusting backwards, are shown in figure 1.

If the mass m of air is accelerated by the force F from speed u to speed v in time t with constant acceleration a, then we can use an equation of constant acceleration:

$$v = u + at$$

$$\Rightarrow mv - mu = mat \quad \text{(multiplying by } m\text{)}$$

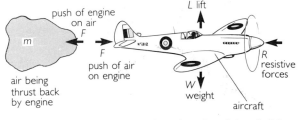

figure 1 Free-body diagram of an aircraft and the air it is pushing backwards

But
$$F = ma$$

$$\Rightarrow mv - mu = Ft \quad \text{(i)}$$

$$\Rightarrow F = \frac{m(v-u)}{t}$$

$$= \frac{m}{t}(v-u) \quad \text{(ii)}$$

In this equation (ii), we can reinterpret m/t: it means the mass per unit time of air or exhaust gas being thrust backwards (because mass m is thrust backwards in time t), and is measured in $kg\,s^{-1}$.

Thus the equation can be expressed in words like this:

thrust on craft (F) = (rate of air expulsion) × (increase in speed of air)

Also, it no longer matters whether the air has constant acceleration or not, because we are no longer really interested in its detailed motion: the force depends only on how much we push out, and by how much we change its speed.

Impulse and linear momentum

Let us consider equation (i) above again:

$$mv - mu = Ft$$

The product $F \times t$ on the right hand side is called the 'impulse' I of the force F. It gives us a measure of how much effect the force F has: if the force acts on the aircraft for a longer time, it will have a greater effect on the aircraft's velocity.

On the left hand side, ($mv - mu$) means in words 'the change in the quantity mass × velocity'. We give this quantity mass × velocity the special name *momentum*, with symbol p. In symbols again,

$$p = mv$$

Momentum is a vector, because velocity is a vector. Therefore impulse also is a vector.

Impulse and momentum must have the same unit, because they appear as terms in the same equation. Unit: 1 Ns (which may also be stated as $1\,kgms^{-1}$).

The equation then reads in words:

change in momentum of object = impulse of force acting on it.

Again, it no longer matters whether the acceleration is constant or not. If the force (thus acceleration) varies, we have to evaluate a sum to find the impulse.

$$I = \Sigma F \Delta t$$

This works because over a very short time interval Δt the force must be nearly constant. I is thus the area under a graph of F against t.

Example   The force acting on a golf ball from a club varies with time according to the graph in figure 2. If the ball has mass 46 g, how fast does it leave the club?

Area under graph = 709 little squares
One little square represents:
$0.50 \times 10^3\,N \times 0.10 \times 10^{-4}\,s$

$$= 0.0050\,Ns$$

$\Rightarrow$ Impulse = 709×0.0050 Ns

$$= 3.545\,Ns \text{ or } 3.545\,kgms^{-1}.$$

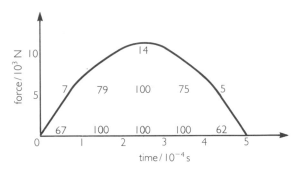

figure 2   Graph of force against time for a golf ball

Impulse $I = mv - mu$

$$m = 0.046\,kg$$

$$u = 0$$

$\Rightarrow$ $3.545\,kgms^{-1} = 0.046\,kg \times v$

$\Rightarrow$ $v = 77\,ms^{-1}$

An alternative treatment of the momentum/impulse relationship uses calculus, as follows. We define impulse by

$$I = \int_{t_0}^{t_1} F\,dt$$

But $F = ma$ at any instant,

$\Rightarrow$ $I = \int_{t_0}^{t_1} ma\,dt$

$$= m\int_{t_0}^{t_1} a\,dt$$

$$= m(v - u)$$

$$= mv - mu$$

Hence it doesn't matter whether a is constant or not.

Momentum and collisions

Momentum is a particularly useful quantity to know about when you want to analyse what happens in collisions, because the changes in velocity of the colliding objects occur very quickly. It may be difficult to find out much about the forces between the objects, because they also occur very quickly; but it turns out that momentum can tell us most of what we want to know, without needing to know about the forces at all!

Consider the situation of a collision in one dimension between two objects: for example two gliders on an air track, with magnets on them which create repulsive forces between them. See figure 3.

When the collision occurs, an impulse acts on each glider: I_1, to the left on glider 1, I_2 to the right on glider 2.

At any moment during the collision, the sizes of F_1 and F_2 must be equal, by Newton's third law;

therefore I_1 and I_2 must be equal in size also. This means that objects 1 and 2 have equal changes in momentum: 1 to the left, 2 to the right. So in effect 1 has given 2 some momentum; but they still have the same momentum between them.

This is a very simple argument to suggest a law which physicists consider most important: the law of *conservation of momentum*. This states:

'For any system of interacting objects, the total linear momentum remains constant in any direction, so long as there is no resultant external force acting on the system'.

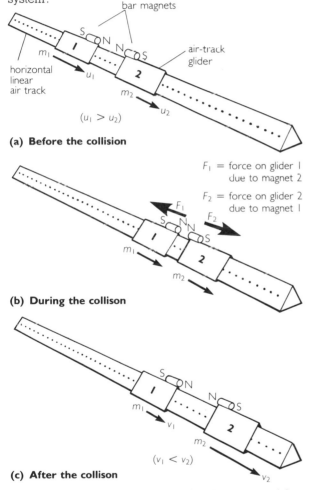

$(u_1 > u_2)$

(a) Before the collision

F_1 = force on glider 1 due to magnet 2

F_2 = force on glider 2 due to magnet 1

(b) During the collison

$(v_1 < v_2)$

(c) After the collison

figure 3 The general case of a collision along one straight line

☐ Note 1. 'Linear' momentum is to distinguish it from rotational or angular momentum.

☐ Note 2. In algebraic form, this could be written: Σp = constant.

☐ Note 3. As argued above, this law is only a theoretical hypothesis, and must be verified by experiment.

We can use an algebraic treatment for the situation of figure 3.

For object 1: $I_1 = m_1 v_1 - m_1 u_1$ (to the left)

For object 2: $I_2 = m_2 v_2 - m_2 u_2$ (to the right)

As I_1 and I_2 are equal in size,

$$I_2 \text{ (to the right)} = I_1 \text{ (to the left)}$$

$$\Rightarrow m_2 v_2 - m_2 u_2 \text{ (to the right)} = m_1 v_1 - m_1 u_1 \text{ (to the left)}$$

$$\Rightarrow m_2 v_2 - m_2 u_2 \text{ (to the right)} = -(m_1 v_1 - m_1 u_1) \text{ (to the right)}$$

$$\Rightarrow m_1 v_1 + m_2 v_2 = m_1 u_1 + m_2 u_2 \quad \text{(iii)}$$

The left hand side of this equation is the sum of the two momenta after the collision; while the right hand side is the sum of the two momenta beforehand. Hence, the equation states algebraically that the total momentum is unchanged.

Experiments on momentum

The outline of an experiment in one dimension is suggested in figure 4. The general idea is to measure each of the quantities in equation (iii), over a range of different collisions, and to verify that the equation is correct, within experimental error, in each case. Collision conditions should range from ones in which the gliders stick together on impact through to ones in which magnets are used to create repelling forces and there is no impact at all.

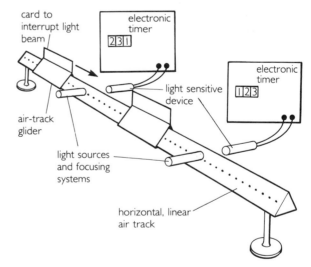

figure 4 Collision experiment on a linear air track

A different experiment involves using dry-ice or other frictionless pucks on a smooth surface. We can then verify the law for two-dimensional collisions. The best way of measuring the velocities in this case is to use multiflash or video photography. See figure 5.

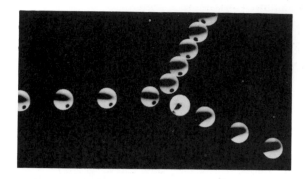

figure 5

Example A moving puck of mass 100 g collides with one of 200 g which is initially stationary. The resulting multiflash motion is shown, to scale, in figure 6. Use it to verify that momentum is conserved in the collision.

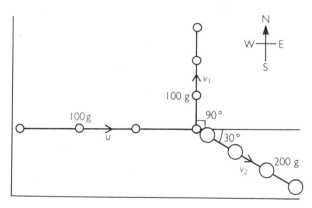

figure 6 Two-dimensional collision between frictionless pucks

Measuring from the diagram, and using the distance travelled in three flash-intervals as an arbitrary speed unit:

$$u = 10.4 \text{ units}$$
$$v_1 = 6.00 \text{ units}$$
$$v_2 = 6.00 \text{ units}$$

Since one mass is just twice the other, it is convenient if we use arbitrary momentum units as well:

initial momentum of 100 g = 1×10.4 units = 10.4 units to the east
final momentum of 100 g = 1×6.00 units = 6.00 units to the north

final momentum of 200 g = 2×6.00 = 12.0 units 30° south of east

We now add the two amounts of final momentum as vectors. Let us use scale drawing: figure 7.

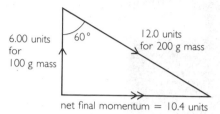

figure 7 Adding momenta after a two-dimensional collision

Thus the net final momentum is 10.4 units, to the east: in other words, it is equal to the initial momentum of the system, as the law predicts. We started this discussion on momentum by stating that when two bodies collided they would exert equal but opposite forces on each other. This application of Newton's third law led us to the principle of the conservation of momentum. It is worth pointing out that we could have tackled the problem from the other direction. We could have started with the experimental observation that 'momentum is conserved'; this would then lead us to Newton's third law.

Energy in collisions

The total linear momentum is always conserved in collisions (if there is no resultant force from outside). The total kinetic energy of the colliding objects is never completely conserved, except in collisions where no actual contact occurs. In all other situations, some of the kinetic energy possessed initially by the objects is changed during the collision into non-kinetic forms: most of the lost kinetic energy becomes internal energy in the objects; some becomes sound energy; sometimes some is emitted as light, as for example when sparks appear.

In a collision between a tennis racquet and a ball, for example, both the racquet strings and the ball become distorted (see figure 8). Their initial kinetic energies become converted into elastic potential energy in the compressed ball and stretched strings (figure 9a). The force between them alters the velocity of both of them, and they soon speed apart; however, the graph of the force moving them apart again is invariably lower than that required to push them together: see figure 9(b). Thus less energy is given back than was put in, and the total kinetic energy is less afterwards. The energy required to push them together is represented by the area under graph 9(a). The energy given back as kinetic energy as they move apart is represented by the area under graph 9(b). Thus the loss is represented by the area

figure 8

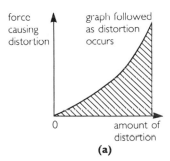

(a)

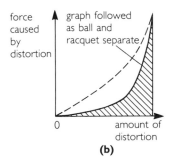

(b)

figure 9 Graphs of force against displacement for a racquet and ball impact

between the two graphs. This amount of energy has been converted into internal energy in the racquet and ball, and the noise of the impact.

The paragraph above describes the situation in which the tennis player holds the racquet firmly and allows the ball to rebound; work is done in distorting the strings and kinetic energy of the ball is lost. However the player can return the ball with extra kinetic energy if he wishes, by doing work on it

during the impact. (This process is normally called hitting the ball back.)

Here is a numerical example of another everyday though less pleasant event. A car, a Renault 18, of mass 900 kg, was travelling along the M25 towards junction 12, which is the intersection between M25 and M3. As he approached the junction the driver was rather slow to notice a traffic jam, applied his brakes too late, and slid into the back of a stationary Volkswagen of mass 600 kg. Moving into the impact the Renault was travelling at $5.0\,\text{ms}^{-1}$ and immediately after the impact it was travelling at $2.0\,\text{ms}^{-1}$. Let us calculate (i) the velocity of the Volkswagen immediately after impact and (ii) the proportion of the total energy E lost during collision.
(i) Momentum is conserved:

initial momentum of Renault = final momentum of Renault + momentum of Volkswagen

$$\Rightarrow\quad 900\,\text{kg}\times5.0\,\text{ms}^{-1} = 900\,\text{kg}\times2.0\,\text{ms}^{-1}+600\times v$$

$$\Rightarrow\quad v = \frac{900\,\text{kg}\times3.0\,\text{ms}^{-1}}{600\,\text{kg}}$$

$$= 4.5\,\text{ms}^{-1}$$

(ii) The kinetic energy, E_k, of the Renault before impact $=\frac{1}{2}\times900\,\text{kg}\times(5.0\,\text{ms}^{-1})^2$

$$\Rightarrow\quad E_k = 11250\,\text{J}$$

After the collision the remaining kinetic energy, E_k' is:

E_k of Renault + E_k of Volkswagen.

$$E_k' = \frac{1}{2}\times900\,\text{kg}\times(2.0\,\text{ms}^{-1})^2+\frac{1}{2}\times600\,\text{kg}\times(4.5\,\text{ms}^{-1})^2$$
$$= 7875\,\text{J}$$

So the fraction of the energy lost during the collision is:

$$\frac{11250\,\text{J}-7875\,\text{J}}{11250\,\text{J}} = 0.30$$

The lost kinetic energy has mainly been used to bend the two cars. As a result of the crash both cars will be a little warmer. Also some sound energy has been emitted.

Perfectly elastic collisions

A perfectly elastic collision (often referred to simply as an elastic collision) is one in which the total kinetic energy E_k of the colliding objects is the same after the collision as it was before.

The conservation of a quantity means that its value is *always* constant. Therefore, technically kinetic energy is *not* conserved in this case, but it is common usage to refer to this situation as conservation of kinetic energy.

Even though E_k is only conserved in collisions between everyday-sized objects when they do not make contact, such as when they interact magnetically, it is still important to develop a theory for such collisions. There are two reasons.

☐ Elastic collisions occur very often between atom-sized particles – for example, between the molecules of the air.

☐ Collisions between everyday objects can come near to being elastic: for example, the impact between two snooker balls.

The general situation of a one-dimensional collision, as occurs for example on an air track, it is shown again here in figure 10.

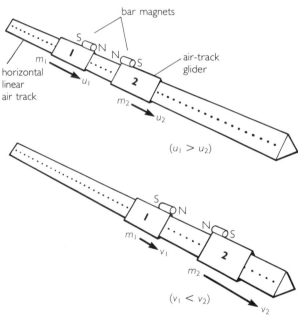

figure 10 General situation of a one-dimensional collision

If the collision is elastic, an equation expressing this is:

$$\tfrac{1}{2} m_1 v_1^2 + \tfrac{1}{2} m_2 v_2^2 = \tfrac{1}{2} m_1 u_1^2 + \tfrac{1}{2} m_2 u_2^2$$

For example, in the dry-ice puck collision of figure 6, was kinetic energy E_k conserved or not?

$m_1 = 100\,\text{g}$	$m_2 = 200\,\text{g}$
$u_1 = 10.4$ units	$u_2 = 0$
$v_1 = 6.00$ units	$v_2 = 6.00$ units

In units of $\text{kg} \times (\text{velocity units})^2$,

E_k before $= \tfrac{1}{2} \times 0.1 \times (10.4)^2$ units $+\, 0$ units

$\qquad\qquad = 5.4$ units

E_k after $= \tfrac{1}{2} \times 0.1 \times 6.00^2$ units $+ \tfrac{1}{2} \times 0.2 \times 6.00^2$ units

$\qquad\qquad = 5.4$ units

Thus within the accuracy of the measurements we can conclude that the collision was elastic.

Coefficient of restitution

There is a simpler way of analysing both elastic and inelastic collisions along one line. This uses the relative speeds of the objects as they approach and then separate.

Experiments show that for particular colliding conditions (that is, the same colliding objects and surfaces each time), the ratio of these relative speeds stays constant. This ratio is called the coefficient of restitution, e. Thus

$$e = \frac{\text{relative speed of separation}}{\text{relative speed of approach}},$$

or

$$e = \frac{v_2 - v_1}{u_1 - u_2}$$

A special and useful result is that $e = 1$ for an elastic collision. For a simple inelastic collision $e < 1$. For example, in our motorway crash the speed of approach of the vehicles was $5\,\text{ms}^{-1}$ and the speed of separation was $2.5\,\text{ms}^{-1}$.

So

$$e = \frac{v_2 - v_1}{u_1 - u_2}$$

$$= \frac{4.5 - 2.0}{5.0 - 0}$$

$$= 0.50$$

If $e > 1$, this means that the system gained net E_k during the collision: this could occur if, for example, the collision triggered an explosion, or released a compressed spring.

Special cases of elastic collisions

The following four situations all consider a moving object colliding elastically with a stationary object. In no case is a real-life problem as simple as these analyses suggest; but they provide useful models for considering more complex real situations.

1 *Two equal masses, in one direction*

figure 11 Elastic collision between two equal masses, in one dimension

Momentum conserved:

$$mv_1 + mv_2 = mu$$

$$\Rightarrow \qquad v_1 + v_2 = u \qquad \text{(i)}$$

Restitution:

$$\frac{v_2 - v_1}{u} = 1$$

$$\Rightarrow \qquad v_2 - v_1 = u \qquad \text{(ii)}$$

Adding equations (i) and (ii):

$$2v_2 = 2u$$

$$\Rightarrow \qquad v_2 = u$$

$$\Rightarrow \qquad v_1 = 0$$

The second object moves off with the velocity, momentum and kinetic energy which the first had before the collision; the first object becomes stationary. This situation occurs (nearly) during a two-ball Newton's cradle demonstration; or in a straight pot on a snooker table. (Snooker-ball collisions are, however, complicated by the rolling motion of the balls, which gives them rotational energy as well as ordinary translational kinetic energy.)

2   *Two equal masses, in two dimensions*

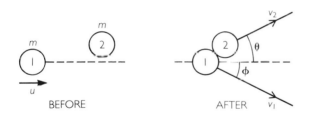

figure 12   Elastic collision between two equal masses, in two dimensions

The kinetic energies before and after the collision are equal (elastic collision):

$$\tfrac{1}{2} mu^2 = \tfrac{1}{2} mv_1{}^2 + \tfrac{1}{2} mv_2{}^2$$

$$\Rightarrow \qquad u^2 = v_1{}^2 + v_2{}^2$$

We can add the two momenta after the collision using a vector addition triangle (figure 13).

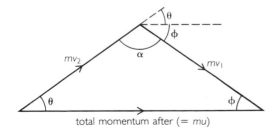

figure 13   Vector addition of momenta

But the sides of the triangle in figure 13 are in proportion to the velocities v_1, v_2 and u; and also $u^2 = v_1{}^2 + v_2{}^2$, from above. Thus the triangle must be a Pythagoras triangle, with a right-angle at α, and $\theta + \phi = 90°$. In other words, the objects separate at $90°$.

This situation is close to what occurs with snooker balls: whatever the direction of the pot, if there is little spin then the balls separate at nearly $90°$. It also occurs when an α-particle collides with a nearly stationary helium nucleus.

Much information in nuclear physics is gained by observing collisions between particles. Often they are not of equal mass. The general rule for the angle at which the objects then separate is this:

$$\text{if } m_1 > m_2 \text{ then } (\theta + \phi) < 90°$$
$$\text{if } m_1 < m_2 \text{ then } (\theta + \phi) > 90°$$

3   *A small object hitting a large object (one-dimensional elastic collision)*

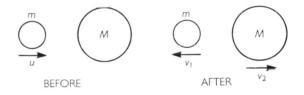

figure 14   Elastic one-dimensional collision between small object and large object

Using momentum and coefficient of restitution equations again, we arrive at these results:

$$v_2 = u \cdot \frac{2m}{M + m}$$

$$v_1 = u \cdot \frac{M - m}{M + m}$$

Thus if $M \gg m$ then v_2, the speed of M after impact, is very small; also v_1, the rebound speed of m, is nearly equal to u. This means that m keeps almost all the kinetic energy it possesses when it strikes M.

Example   Let us consider an α-particle of mass 6.8×10^{-27} kg and velocity $1.0 \times 10^7 \,\text{ms}^{-1}$ colliding head on with a stationary gold nucleus of mass 3.3×10^{-25} kg. What fraction of the α-particle's original energy is lost?

Figure 15 summarises the velocities before and after the collision using the formulae of the last paragraph.

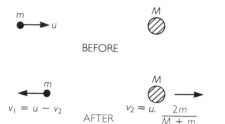

figure 15 Velocities before and after collision

Energy of α-particle before collision, E_1, is:
$$E_1 = \tfrac{1}{2} \times 6.8 \times 10^{-27}\,\mathrm{kg} \times (1.0 \times 10^7\,\mathrm{ms}^{-1})^2$$
$$= 3.4 \times 10^{-13}\,\mathrm{J}$$
Speed of gold nucleus after collision is:
$$v_2 = \frac{u \times 2m}{M+m}$$
$$= \frac{1.0 \times 10^7\,\mathrm{ms}^{-1} \times 2 \times 6.8 \times 10^{-27}\,\mathrm{kg}}{(330 + 6.8) \times 10^{-27}\,\mathrm{kg}}$$
$$= 4.0 \times 10^5\,\mathrm{ms}^{-1}$$
⇒ speed of recoil of α-particle is:
$$v_1 = u - v_2$$
$$= 1.0 \times 10^7\,\mathrm{ms}^{-1} - 4.0 \times 10^5\,\mathrm{ms}^{-1}$$
$$= 0.96 \times 10^7\,\mathrm{ms}^{-1}.$$
The kinetic energy of the α-particle after the collision, E_2, is:
$$E_2 = \tfrac{1}{2} \times 6.8 \times 10^{-27} \times (0.96 \times 10^7\,\mathrm{ms}^{-1})^2$$
$$= 3.1 \times 10^{-13}\,\mathrm{J}.$$
The fraction of the energy lost from the α-particle is:
$$\frac{3.4 \times 10^{-13} - 3.1 \times 10^{-13}}{3.4 \times 10^{-13}} = 0.09$$

In two dimensions, this result still holds true: the small object retains most of its kinetic energy when colliding elastically with a large object. In a glancing blow with a nucleus, the α-particle would lose even less energy. Another good example in which very little energy is transferred to the larger object is that of an electron colliding with a gas atom.

4 *A large object hitting a small object (one-dimensional elastic collision)*

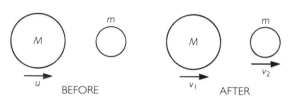

figure 16 Elastic one-dimensional collision between large object and small object

Using momentum and coefficient of restitution equations once again, we obtain:
$$v_2 = \frac{2Mu}{M+m}$$
$$v_1 = u.\frac{M-u}{M+u}$$
If $M \gg m$, then $v_2 \approx 2u$,
and $v_1 \approx u$.

Thus the motion of M is virtually unaffected, while m sets off at twice the speed of M. This is approximately what happens when a golf driver hits a golf ball.

Summary

1 Animals and vehicles move by exerting force backwards on some object, which by Newton's third law exerts force forwards on them.

2 Thrust = mass of material moved per unit time × change in velocity of material.

3 Momentum $p = mv$.

4 Impulse $I = Ft\ (= \Sigma F \Delta t)$.

5 Impulse = change of momentum
$$Ft = mv - mu$$

6 Law of conservation of momentum:
'For any system of interacting objects on which no resultant force acts, the total linear momentum in any direction remains constant.'

7 Kinetic energy may or may not be conserved in a collision.

8 An elastic (or perfectly elastic) collision is by definition one in which kinetic energy is conserved.

9 In a one-dimensional collision, the coefficient of restitution e is given by

$$e = \frac{v_2 - v_1}{u_1 - u_2}$$

$$= \frac{\text{relative speed of objects after collision}}{\text{relative speed of objects before collision}}$$

10 For an elastic collision, $e = 1$.

11 For an inelastic collision, $e < 1$.

12 Special cases of elastic collisions between moving object m_1 and stationary object m_2:
1 If $m_2 = m_1$, in one dimension m_1 stops, m_2 proceeds with initial velocity of m_1.
2 If $m_1 = m_2$, in two dimensions m_1 and m_2 separate at an angle of 90°.
3 If $m_1 \ll m_2$, m_1 bounces off at almost the same speed, and m_2 hardly moves.
4 If $m_1 \gg m_2$, m_1 is hardly affected, and m_2 sets off at twice the arrival speed of m_1 (if the collision is in one dimension).

Questions

1 The jet engines of a large aircraft take in air through openings with area about $2\,m^2$. They cruise at about $250\,ms^{-1}$, at a height where the air has density about $0.5\,kgm^{-3}$. Each engine gives a thrust of about $2 \times 10^5\,N$. Use this information to calculate approximate values for
   a the mass of air accelerated backwards each second;
   b the increase in speed of the accelerated air.

2 The Saturn V rockets which launched the Apollo space missions had the following specifications:
   initial mass on launch pad $3 \times 10^6\,kg$
   initial rate of fuel consumption $3 \times 10^3\,kgs^{-1}$
   velocity of exhaust gases $1.1 \times 10^4\,ms^{-1}$
   a Calculate the thrust provided by these first-stage engines.
   b Calculate the initial upward acceleration of the rocket.
   c Explain why this acceleration increased as the flight progressed. (There are at least three reasons you might think of.)
   d Considering only the change in mass of the rocket, write down an equation giving an expression for the acceleration a in terms of the time t after launch.

   This equation is rather hard to solve mathematically. ('Solve' means to find an equation relating distance travelled, s, and time, t.) However, you may be able to write a computer program to find s at any time t, using a method which makes calculations at repeated short time intervals Δt.

3 Suppose a physicist is doing experiments to investigate the dynamics of a high-jumper's art. The jumper stands on a pair of scales, and does a standing jump off one foot. From this position the 50 kg jumper can clear a height of 1.5 m. (Assume her centre of mass rises to this height, having started at 1.0 m above the ground.)

   a With what upward speed is her centre of mass moving as she leaves the scales?
   b What do the scales read while she is standing still on them?
   c If the upward thrust takes 0.2 s, what is the average reading on the scales during this time?
   d Estimate what the maximum reading on the scales would be likely to be.

4 A girl is standing on a set of newton scales, and carrying a 2 kg dumb-bell in one hand. Her own weight is 400 N. She moves the dumb-bell vertically upwards, so that it takes 0.2 s to move a distance of 1 m. The scales show readings which vary as in figure 17.

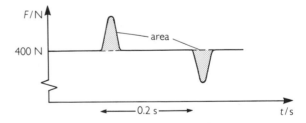

figure 17

   a Account for the shape of the graph as fully as you can.
   b Estimate the value that the two shaded areas should have, from the data in the first paragraph.

5 If a fast serve is returned during a game of tennis, the following are approximate values for the relevant parameters:
   arrival and departure speeds of
   ball $30\,ms^{-1}$
   duration of impact $5 \times 10^{-3}\,s$
   mass of ball $0.06\,kg$
   effective mass of racquet head $0.42\,kg$

a Calculate
  (i) the change in momentum of the ball,
  (ii) the average force between racquet and ball,
  (iii) the change in velocity of the racquet head during the stroke.
b Estimate the maximum force acting between the racquet and the ball. What can you say about the motion of the racquet and the ball at the instant of maximum force?
c Look at the photograph of figure 8 (page 41), showing a tennis racquet and ball at the instant of maximum force between them. Do you think a steady force with the value you calculated in part (b) would produce this distortion in the ball? Explain your ideas.

6 In the book by George Gamow, Mr Tomkins finds himself on an asteroid which is quite small – let us say 1000 times his own mass of 60 kg (figure 18). The gravitational field strength, however, is the same as on our Earth – and we will assume it remains the same even if he jumps upwards, distancing himself from the asteroid.

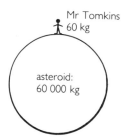

Mr Tomkins
60 kg

asteroid:
60 000 kg

figure 18

Mr Tomkins bends his knees and springs upwards. He leaves the asteroid surface with a speed of $2\,\mathrm{ms}^{-1}$.
a With what speed does the asteroid recoil downwards due to the push of his feet?
b For what length of time are Mr Tomkins and the asteroid moving apart?
c How far will the asteroid move in that time?
d How far will Mr Tomkins move in that time?
e With how much total kinetic energy do they collide again?
f Will the asteroid (+ Mr Tomkins) be moving up, down, or not at all after he has landed again? Justify your answer.

7 We tend to assume that nothing we do on the Earth's surface will ever affect its motion through space. Let us do a little calculation to see if this is right. Consider a golf ball, mass 0.046 kg, dropped from a height of 5 m onto the Earth, and bouncing elastically. The mass of the Earth is 6×10^{24} kg.

Calculate
a the speed of the ball just before impact;
b the speed of the Earth just before impact;
c the distance moved by the Earth during the fall of the ball.
d If the ball is caught and held stationary at its release point as it returns, what can you say about the motion of the Earth from then on?
e Comment quantitatively on the idea sometimes mooted, that if all the human beings alive on Earth were to walk round the Earth in the same direction, its rate of rotation could be affected noticeably. (There are about 5×10^9 people on the Earth.)

8 Boris, the 2 kg cat, is capable of jumping onto a table 1 m high.
a What vertical take-off speed is he capable of attaining?

He can also attain this speed horizontally if he does a leap from a fixed platform. Knowing this, he attempts to jump to land from a floating canoe of mass 4 kg, failing to realise that the canoe is not a 'fixed platform'.

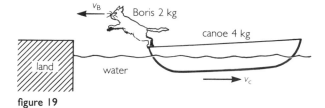

v_B Boris 2 kg

canoe 4 kg

land   water v_c

figure 19

b This time v_B and v_C are the speeds of Boris and the canoe respectively away from their common centre of mass: in other words, relative to the land. Write down an equation relating v_B and v_C.
c Assuming that Boris can still impart 20 J to himself and the canoe together, find a value for v_B. Hence explain why Boris could easily misjudge how far he can jump and land in the water.

9 The nuclide $^{144}_{60}\mathrm{Nd}$ emits an α-particle of energy 1.83 MeV. Take the masses of the nuclides to be whole-number multiples of the mass of a proton, which is 1.67×10^{-27} kg. Calculate
a the emission speed of the α-particle;
b the recoil speed of the daughter nuclide;
c the energy carried by the daughter nuclide;
d the energy carried by the daughter nuclide as a fraction of the total energy released.

Mention any assumptions which are implied in your calculations (a)–(d).
(1 MeV = 1.6×10^{-13} J)

10 You have probably come across the Rutherford/ Geiger/Marsden experiment, in which α-particles

collided elastically with gold nuclei ($^{197}_{79}$Au). The occasional α-particle ($^4_2$He) came straight back very close to its path of arrival. Take the initial energy of the arriving α-particle as 5 MeV, and assume it comes straight back along its path.

a Calculate the speed of the arriving α-particle.

b It is usual to assume, on a simple analysis, that the α-particle comes back with no loss of kinetic energy. On this assumption, calculate the recoil speed of the gold nucleus after the collision.

c Calculate the kinetic energy of the gold nucleus after the collision.

d Compare the answer to (c) with 5 MeV, and comment on the validity of the assumption that the α-particle loses no kinetic energy in the collision.

11 Consider a head-on elastic collision between a moving neutron and a stationary deuterium ($^2_1$H) nucleus. Such collisions occur when heavy water is used as a moderator for slowing neutrons down in a nuclear reactor.

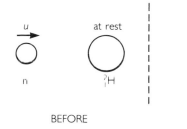

BEFORE   AFTER

figure 20

a Draw a suitable 'after' picture.

b Write down a momentum conservation equation.

c Write down a restitution equation.

d Solve these equations to find what happens after the collision.

e Calculate the fraction of the neutron's energy given up to the $^2_1$H.

f The neutrons emitted from a fission event emerge with energy about 2 MeV; to cause another such event they have to be slowed to an energy of about 0.05 eV. Estimate the number of collisions with $^2_1$H nuclei which an individual neutron would have to make.

12 One specification for a tennis ball is that it should rebound to a height of between 53 and 58 inches when dropped onto a concrete floor from a height of 100 inches.

a The ball retains a certain fraction of its kinetic energy after the impact. Calculate the range within which this fraction must lie. State any assumptions you make.

b Calculate the range of values of e for the impact.

13 In this question, we will apply the equations of momentum conservation and restitution to the collision between a tennis racquet and a stationary ball during a serve, and use the result to draw some conclusions about the optimum mass of the racquet.
Data:
  mass of ball   0.060 kg
  effective mass of normal racquet   0.42 kg
  coefficient of restitution   e
Assume that, for the collision between racquet and ball, we can treat the racquet as a point mass, with the effective mass given above.

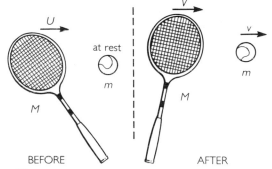

BEFORE   AFTER

figure 21

a Write down an equation, in terms of the letters in figure 21, expressing the conservation of momentum during the collision.

b Write down another equation using the coefficient of restitution.

c Eliminate V between these two equations; you should arrive at the expression

$$v = U(1+e) \cdot \left(\frac{M}{m+M}\right)$$

d Assuming for the moment that U and e will not change, the speed of the serve depends on the fraction $M/(m+M)$.
Evaluate this using the data given.

e It is sometimes suggested that strong players benefit from using heavier racquets. Suppose a player chose a racquet with double the usual mass. Calculate the percentage increase he could expect for his service speed, assuming a constant U and e.

f Comment on why you think players in fact do not use racquets with masses as large as this.

14 This question is about a Newton's cradle, a toy (or important physics demonstration) which you may have seen. It consists of several very elastic steel spheres, hanging so that they all just touch when stationary. For this question, assume that the spheres all have the same mass, m, and all collisions are perfectly elastic.

For the first part of the question, we consider just two spheres. We analysed this situation before; here we look at it again, to show that for an elastic collision in a straight line $e = 1$. See figure 22.

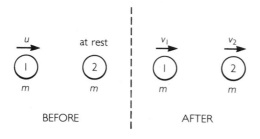

figure 22

a Write down the equation for momentum conservation.
b Write down an equation for total kinetic energy being conserved.
c Show from these equations that there are two possible outcomes:
$$v_2 = u \text{ and } v_1 = 0$$
or
$$v_2 = 0 \text{ and } v_1 = u.$$
(Hint: Take the momentum equation, cancel the m's, square both sides, then compare the result with the energy equation.)
d Does the second solution have a physical meaning? If so, what?

For the second part of the question, we will consider a three-sphere cradle. The general picture is this:

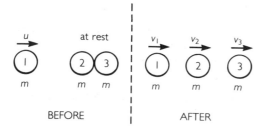

figure 23

e Write down equations for (i) momentum conservation, (ii) energy conservation.

f Solve them together (the same hint applies). You should be able to show the familiar result: that sphere 3 continues on, at speed u, 2 stays stationary, 1 stops.

15 The following collision experiments are carried out in a laboratory, using steel ball bearings with $e = 0.9$. In each case the balls are hung on V-suspensions, so they can only move in one dimension.
Predict what will happen to each ball in each case.

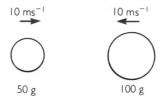

figure 24

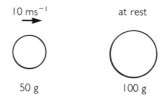

figure 25

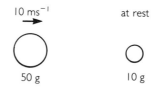

figure 26

16 In a cloud chamber photograph, an event is observed in which two particles move apart (presumably after a collision) at 90°. See figure 27(a). It is reasonable to assume that before the collision one of the particles was stationary, and the other was moving at high velocity. Suppose that on one such occasion the two emerging particles have velocities as shown in figure 27(b).
a What can you say about the masses of the two particles?
b Deduce from the information given the direction of motion of the high-velocity particle before the collision.

figure 27 (a)

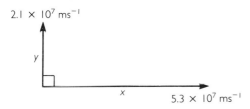

figure 27 (b)

17 A cannon is mounted on a firing turret as shown in figure 28.

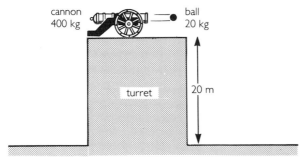

figure 28

The gunners are rather inexperienced, and do not realise that the gun will recoil. The cannon ball leaves the barrel horizontally with a speed of $60 \, \text{ms}^{-1}$. At the same moment the cannon leaves the turret, travelling backwards.

a With what speed does the cannon recoil?
b Which hits the ground (20 m below) first, the ball or the cannon? Justify your answer.
c How far back from the turret does the cannon land?

18 Imagine wind blowing at $20 \, \text{ms}^{-1}$ at a vertical wall, 10 m long and 2.0 m high. The density of the air is $1.3 \, \text{kgm}^{-3}$. Assume that all the forward momentum of the moving air is reduced to zero when the air collides with the wall.
a What is the force exerted on the wall by the wind? (Hint: Try calculating, for one second, the volume, mass and momentum of the arriving air.)
b What couple is necessary at the base of the wall to prevent it falling over?

19 A rolling-mill is a machine for rolling hot metal out from a thick sheet into a thinner one. Assume for this question that the width of the sheet is unaltered, but its length and thickness change as it passes through the rollers.

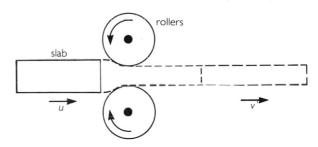

figure 29 A rolling-mill

Data:
    initial thickness of steel slab   6.0 mm
    final thickness of steel slab   5.0 mm
    approach speed, u $10 \, \text{ms}^{-1}$
    initial length of slab   2.5 m
    width of slab   0.80 m
    density of steel $9000 \, \text{kgm}^{-3}$
Calculate from these data:
a the time for the slab to traverse the rollers;
b the speed of the emerging slab, v;
c the mass of the slab;
d the change of momentum of the slab (including its direction);
e the net force exerted on the rollers (including its direction).

CHAPTER 4

Circular motion

Introduction

This chapter deals with the rather special case of objects moving in circles. We consider why centripetal force must be required to keep an object on a circular path; work out how big the force must be; and then consider actual examples, and what causes the centripetal force in each case. The scientists' unit of angle, the radian, is introduced; and, using angle as the rotational equivalent to displacement, a range of rotational quantities such as angular velocity and torque is developed by analogue with linear motion.

GCSE knowledge

No knowledge from your GCSE course is specifically required for this chapter, though you may have discussed qualitatively the forces on objects moving in circles. However, a good understanding of chapter 2, dynamics, will certainly help.

Centripetal force and acceleration

Let us begin by discussing a very familiar situation: an object, such as a conker, being whirled in a circle on the end of a string (figure 1), at a fixed speed. At any particular moment, the direction of its velocity (a vector, remember) is along the tangent to the circle. Since this direction changes constantly, the conker's velocity is changing constantly. To have a changing velocity, the conker must be accelerating; to be accelerating, it must have a net resultant force acting on it.

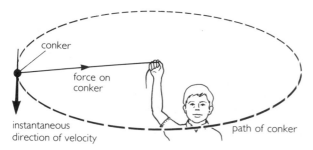

instantaneous direction of velocity

conker

force on conker

path of conker

figure I   Path of a conker being whirled on a string

Clearly, without the string the conker would not go round in a circle; it would fly off along the tangent if the string broke or was released. Furthermore, the string is in tension. We conclude that the string is supplying the required net force on the conker; and that the force must be along the string, that is, towards the centre of the circle, since the pull of the string on the conker is in that direction. Any object travelling in a circle must have a net force on it, acting towards the centre of the circle. We call this force a *centripetal force*.

The equation F = *ma* applies to this situation. The acceleration *a* caused by the net centripetal force *F* is called centripetal acceleration. Note that this acceleration does not change the speed of the object, but only its direction of motion.

A force which always acts at right angles to the velocity of an object can never change the speed of the object. The force has no component along the velocity direction, and therefore does no work; the force cannot increase the kinetic energy of the object, and its speed remains constant if no other forces act. See figure 2.

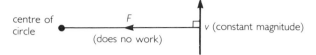

figure 2 The centripetal force F cannot change the speed v of the object

How big is the centripetal acceleration?

First let us consider what factors are likely to appear in an expression for the centripetal force needed to keep an object moving round a circular path (rather than going straight on, as it would if no force made it go round). Think of driving in a car round a corner.

☐ The greater the speed v with which you go round the bend, the more force you feel from the seat: hence with high v, a high F is needed.

☐ If the bend is tight, that is, the radius of curvature r is small, then you also feel more force from the seat: we need a high F if r is small.

☐ A more massive object will need a bigger force, since $F = ma$. In fact, if we can find an expression for a, then the equation will tell us F.

To find an expression for the value of the centripetal acceleration a, we must consider how the velocity of the object changes in a short time interval Δt. Figure 3(a) shows how the direction of the velocity changes as the radius line from the object to the centre moves through an angle θ, in going from A to C.

(a)

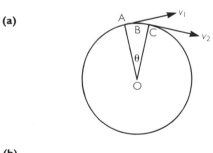

(b)

v_1 and v_2 have the same magnitude v

figure 3 Change of velocity in time Δt: (a) overall view; (b) vector addition of velocities

We can work out the change in velocity in going from A to C using the vector triangle shown in figure 3(b). Here the change in velocity is given by the vector subtraction $\Delta v = v_2 - v_1$. The diagram shows us that Δv is directed towards the centre of the circle, and therefore the acceleration is in that direction too.

The size of the acceleration is given by $a = \Delta v/\Delta t$, where Δt is the time taken to go from A to C. It is important to realise that this acceleration is only an average acceleration between the points A and C. To work out the exact acceleration at B we must take points A and C very close indeed on either side of B.

When AC is a very small distance we can make a useful approximation which is shown in figure 4.

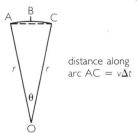

figure 4 Part of the circular motion from figure 3

When θ is small the arc length AC is very nearly equal to the straight line AC (shown as a dotted line). Now we have two similar isosceles triangles, XYZ in figure 3(b) and OAC in figure 4. The distance AC is equal to $v\Delta t$, since an object travelling with a speed v will travel that distance in a time Δt.

Using the similar triangles we may now write:

$$\frac{\Delta v}{v} \text{ (from figure 3(b))} = \frac{v\Delta t}{r} \text{ (from figure 4)}$$

$$\Rightarrow \qquad \frac{\Delta v}{\Delta t} = \frac{v^2}{r}$$

$$\Rightarrow \qquad a = \frac{v^2}{r}$$

This is the centripetal acceleration of the object towards the centre of the circle.

Then, using $F = ma$, we obtain an expression for the size of the centripetal force:

$$F = \frac{mv^2}{r}$$

If you look back to the beginning of this section, you can see that this expression is compatible with what we reasoned qualitatively: the force F needed is big when m and v are big and r is small.

Angles in radians

We are familiar with measuring angles in degrees: a right angle is 90°, a complete rotation is 360°, and so on. When we look more quantitatively at objects moving in circles, it will be convenient to use another unit for angle: the *radian*. The definition of this unit uses the fact illustrated in figure 5.

If you draw an arc of a circle with its centre at the apex of the angle θ, then the length of the arc (s, or $2s$, or $3s$) is proportional to the radius (r, or $2r$, or $3r$), so long as your arc always subtends the same angle θ.

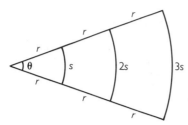

figure 5 Arc length is proportional to radius

Thus arc length α radius

or $\dfrac{\text{arc length}}{\text{radius}}$ = constant, for given θ.

But if you double θ you must double the constant, since the arc length would double. It is therefore sensible to say that the constant is equal to the angle θ, provided θ is measured in suitable units. This unit is called the radian (rad). Thus

$$\theta \ (\text{in radians}) = \frac{\text{arc length}}{\text{radius}} = \frac{s}{r}$$

Also $s = r\theta.$

How big is one radian? Consider a complete circle. The arc length is $2\pi r$, for radius r. Thus the angle θ at the centre of a complete circle is given by

$$\theta = \frac{s}{r} = \frac{2\pi r}{r} = 2\pi \text{ radians}.$$

We also measure this as 360°. Thus,

$$2\pi \text{ radians} = 360°$$

so $1 \text{ radian} = 57.3°$

Angular velocity

It is often useful to consider the movement of a rotating object in terms of the angle through which it turns. Consider again a point object moving at constant speed in a circle – figure 6.

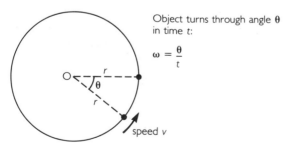

Object turns through angle θ in time t:

$$\omega = \frac{\theta}{t}$$

speed v

figure 6 The angular velocity of a circular motion

The angular velocity (symbol ω) of the object about O is defined by the equation

$$\omega = \frac{\theta}{t}$$

Thus also, $\theta = \omega t.$

Provided θ is measured in radians, we also have $s = r\theta.$

Thus $$\frac{s}{t} = v = \frac{r\theta}{t} = r\omega$$

so $$v = r\omega \text{ or } \omega = \frac{v}{r}$$

It can be useful to know the expressions for centripetal acceleration and force in terms of ω:

$$a = \frac{v^2}{r} = \frac{(r\omega)^2}{r} = r\omega^2$$

and $$F = mr\omega^2.$$

Solving problems about circular motion

There are four particular areas of physics which give rise to circular motion problems, as outlined below.

1 *The orbital motion of bodies within the universe: satellites or moons round planets, planets round stars such as the Sun, stars round the centre of galaxies.* The centripetal force is provided in all cases by a gravitational pull.

2 *The circular motion of everyday-sized objects, such as the spinning of machine parts or wheels, or the motion of cars or aircraft round bends.* In most of these cases the centripetal force is provided by some parts of a structure being in tension or compression; by some kind of contact force for a car; or an aerodynamic lift force for an aircraft.

3 *The motion of charged particles in a magnetic field.* A moving charged particle in a magnetic field receives a force perpendicular to its velocity. Its speed is therefore unaffected, and its path is circular or helical. The centripetal force is caused by the magnetic field.

4 *In early models of the atom – including Bohr's famous model – it was thought that electrons orbited round the atomic nucleus.* Though this model has now been modified, it is important as a forerunner of modern theories. The electrostatic attraction between the positive nucleus and the electron was thought to provide a centripetal force so that the electrons would move in circular orbits.

Notice that in all these four cases, some object moves in a circle and we can identify a force acting upon it. In handling quantititive situations, and applying the equation

$$F = \frac{mv^2}{r}$$

it is usually advisable to consider first the motion of the object, mv^2/r, and then to consider separately what information you may have available about the forces on the object – their resultant must be the centripetal force F.

Example

An aircraft of mass 1000 kg is flying at $100\,\mathrm{ms}^{-1}$. The pilot executes a correctly-banked horizontal turn at an angle of 30°. Calculate (a) the lift force on the wings, (b) the radius of the curve in which he flies. See figure 7(a).

We need a free-body diagram – figure 7(b). There are also thrust and drag forces, in and out of the paper. But these balance each other and can be left out.

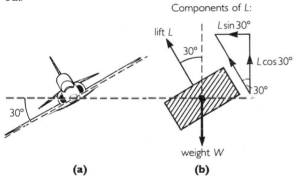

figure 7   (a) Aircraft in banked turn; (b) free-body diagram

Vertical forces are balanced, since there is no vertical motion. Thus

$$L \cos 30° = W = 1000\,\mathrm{kg} \times 10\,\mathrm{N\,kg}^{-1}$$

$$\Rightarrow \qquad L = \frac{10^4\,\mathrm{N}}{\cos 30°}$$

$$= 1.2 \times 10^4\,\mathrm{N}$$

The horizontal component of L is unbalanced. This is therefore the resultant force on the aircraft, and is providing the centripetal force causing the aircraft to follow a circular path.

Thus
$$F = L \sin 30° = \frac{mv^2}{r}$$

$$\Rightarrow \qquad r = \frac{mv^2}{L \sin 30°}$$

$$= \frac{1000\,\mathrm{kg} \times (100\,\mathrm{ms}^{-1})^2}{1.2 \times 10^4\,\mathrm{N} \times \sin 30°}$$

$$\Rightarrow \qquad r = 1.7 \times 10^3\,\mathrm{m}$$

Weightlessness

'Weightlessness' is the experience felt by astronauts who appear to float about inside their orbiting spacecraft. The word is scientifically misleading: the astronauts must still be acted upon by the gravity of the Earth, since otherwise they would have no centripetal force on them making them orbit. The experience of floating is felt because the astronauts and their craft are both moving under gravity, along exactly the same path (which happens to be circular).

We experience the feeling of weight in normal life on Earth because of the contact forces between us and other objects – the floor when we are standing, the chair-seat when sitting. There is no contact force between the astronaut and the craft, and therefore he feels no weight.

For short periods of time it is possible for us on Earth to experience weightlessness. One such occasion is during descent from a high diving board towards a swimming pool – or a free-fall before opening a parachute. Another way to obtain temporary weightlessness is to fly an aircraft so that it follows the same path as a freely-falling projectile.

figure 8

As a measure of the weight a person experiences, 'g-forces' are sometimes quoted. This is defined as

$$\frac{\text{normal contact force acting on person}}{\text{person's weight on Earth's surface}}$$

In the aircraft example, the pilot and the aircraft have the same motion. Thus the g-force experienced by the pilot is equal to the ratio L/W.

$$g\text{-force} = \frac{L}{W}$$

$$= \frac{1.2 \times 10^4\,\mathrm{N}}{10\,000\,\mathrm{N}}$$

$$= 1.2g$$

Rotational dynamics

In chapter 2, we dealt with many aspects of linear dynamics – objects moving in straight lines. The significant quantities in linear motion are displacement (and hence velocity and acceleration), force, mass, work and energy. When objects are rotating, it is more appropriate to think in angular displacement, and hence angular velocity (see above) and angular acceleration. We can build up an analogous system of equations and ideas for rotating objects based on angular displacement.

To begin this task, consider again a point mass moving in a circle – figure 9. Its kinetic energy is given by $E_k = \frac{1}{2}mv^2$.

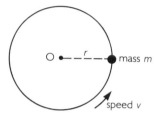

figure 9   Point mass moving in a circle

But energy is a universal quantity: the object must have the same kinetic energy whether we consider its angular or its linear motion. Let us therefore convert the expression for E_k into angular motion quantities:

$$E_k = \frac{1}{2}mv^2 = \frac{1}{2}m(r\omega)^2$$
$$\Rightarrow \quad E_k = \frac{1}{2}(mr^2)\omega^2$$

Now if ω is analogous to v, then (mr^2) must be analogous to mass, so that $\frac{1}{2}(mr^2)\omega^2$ is still an

expression for kinetic energy in the same form as $\frac{1}{2}mv^2$.

This point mass we have considered could be part of a more extensive structure. In that case the equivalent quantity of mass would have to be obtained by summing mr^2 for each particle in the structure. This sum is called the *moment of inertia* (symbol I) for the structure, for rotation about the specified axis O.

$$I = \Sigma mr^2$$

Reverting once more to the point mass of figure 9, consider the effect of a force F acting along the motion of the object. Using linear dynamics,

$$F = ma$$
$$\Rightarrow \quad a = \frac{dv}{dt} = \frac{F}{m}$$

But $v = r\omega$,

$$\Rightarrow \quad a = r\frac{d\omega}{dt} = \frac{F}{m}$$
$$\Rightarrow \quad \frac{d\omega}{dt} = \frac{F}{mr} = \frac{Fr}{mr^2} = \frac{Fr}{I}$$

In words, we can say

$$\text{angular acceleration} = \frac{\text{turning moment of } F \text{ about O}}{I}$$

You can see that this is analogous to

$$\text{linear acceleration } a = \frac{\text{resultant force } F}{m}$$

The table below shows the analogy between rotational and linear motions.

Linear motion		Rotational motion (about specified axis)	
Quantity	Symbol	Quantity	Symbol
displacement	s	angular displacement	θ
velocity	$v\left(=\frac{ds}{dt}\right)$	angular velocity	$\omega\left(=\frac{d\theta}{dt}\right)$
acceleration	$a\left(=\frac{dv}{dt}\right)$	angular acceleration	$\frac{d\omega}{dt}$
mass	m	moment of inertia	$I(=\Sigma mr^2)$
force	F	turning moment (torque)	$T(=\Sigma Fr)$
momentum	mv	angular momentum	$I\omega$
impulse	Ft	angular impulse	$L(=Tt)$
translation E_k	$\frac{1}{2}mv^2$	rotational E_k	$\frac{1}{2}I\omega^2$
work	Fs	work	$T\theta$
power	Fv	power	$T\omega$

Example As an example of the use of rotational dynamics ideas, consider this problem. A roundabout has moment of inertia $500\,\mathrm{kg\,m^2}$. (This would be equivalent to a one-tonne disc with a radius of 1 metre. The quantity I can in general be calculated from the details of a structure.) It rotates at one revolution every 2 seconds. How long will it take to stop if the brakes can exert a maximum frictional force of 3000 N on a rim 10 cm out from the axis of rotation?

We will use the equation

angular impulse = change of angular momentum.

$$Tt = I\omega$$
$$\Rightarrow \quad Frt = I\omega$$
$$\Rightarrow \quad t = \frac{I\omega}{Fr}$$
$$= \frac{500 \times \pi}{3000 \times 0.1}\,\mathrm{s}$$
$$\Rightarrow \quad t = 5\,\mathrm{s}$$

$I = 500\,\mathrm{kg\,m^2}$
$\omega = \frac{1}{2}\,\mathrm{revs^{-1}}$
$\quad = \frac{1}{2} \times 2\pi\,\mathrm{rads^{-1}}$
$F = 3000\,\mathrm{N}$
$r = 0.1\,\mathrm{m}$

Conservation of angular momentum

Angular momentum received brief mention above, in both the table and the example. The law of conservation of angular momentum is analogous to that for linear momentum. It is one of the most fundamental laws in physics – in other words, a law that is found to hold for a very wide range of conditions and systems. It can be stated as:

'The total angular momentum of a system about any axis remains constant, provided no resultant external torque is exerted about that axis.'

$I\omega$ is the angular momentum for an extended object. It is often useful to consider L for a particle in, for example, orbit situations.

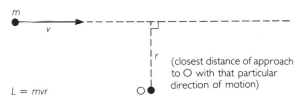

$L = mvr$

(closest distance of approach to O with that particular direction of motion)

figure 10 To illustrate angular momentum of a point mass about a point

In figure 10, the angular momentum L of the particle of mass m about O is given by $L = I\omega$.
However, $I = mr^2$ for a point mass
$$\Rightarrow \quad L = mr^2\omega = mvr.$$

r is the closest distance of approach of the object to O. This expression for L is true, whether m is actually orbiting around O or not. If the only force on m is directed towards O, then the product mvr is constant.

Summary

1 To move in a circle, an object needs a resultant force to act on it. The force is called a centripetal force.

2 A centripetal force acts towards the centre of rotation of an object.

3 An object moving in a circle is accelerating towards the centre, although its speed is not changed by this acceleration.

4 Centripetal acceleration $= \dfrac{v^2}{r}$

5 Centripetal force $= \dfrac{mv^2}{r}$

6 Angles in radians are defined by $\theta = \dfrac{s}{r}$

7 Angular velocity is given by $\omega = \dfrac{\theta}{t}\left(\text{or } \dfrac{\mathrm{d}\theta}{\mathrm{d}t}\right)$

8 $\omega = \dfrac{v}{r}$

9 There is a close relationship between the equations describing linear and rotational dynamics.

10 In an isolated system, angular momentum about any axis is conserved.

Questions

1 Read the passage below. Comment on the errors it contains, and then write a correct version.

> 'A satellite has a tendency to fly outward, away from the Earth. This is caused by centrifugal force. This centrifugal force is balanced by the pull from the Earth's gravity. The satellite thus has balanced forces acting on it, and therefore retains a constant velocity.'

2 The minute hand of Big Ben is about 2 m long. Calculate
  a the time it takes to turn through 1 rad,
  b the average speed at which its tip moves.

3 A rotating bicycle wheel has 24 equally-spaced spokes. A flashing stroboscope makes the wheel appear stationary when flashing at 300 Hz, but not at any higher flashing rate. Calculate the angular velocity of the wheel
  a in degree s^{-1},
  b in rad s^{-1}.

4 A proton can travel round a synchrotron of diameter 1 km at $3 \times 10^8 \, ms^{-1}$. Calculate
  a its time of orbit,
  b its angular velocity.

5 A record-player turntable turns at $33\frac{1}{3}$ rpm (rev$\,min^{-1}$). The diameter of the turntable is 20 cm. Find
  a the time it takes to turn through 1 degree,
  b the speed of a point on its rim.

6 The Moon completes an orbit round the Earth in 27.3 days, in a circle of radius $3.8 \times 10^8 \, m$. Calculate
  a its angular velocity,
  b its speed.

7 A racing car going at $100 \, ms^{-1}$ travels along a straight road 50 m from a TV camera. At what angular velocity must the cameraman swing his camera on its pivot as the car passes nearest to him?

8 The blades of a rotary lawnmower each have mass 200 g, and rotate with their centres of mass 20 cm from the central pivot. If the screw holding a blade will shear off under a force 100 N, what is the greatest possible angular velocity?

9 The radius of the Earth is $6.4 \times 10^6 \, m$, and it rotates once in 24 hours.
  a What is the centripetal acceleration of a person at the equator?
  b By what factor would the Earth's rate of rotation have to increase to make people standing at the equator 'fly off'?
  c Assuming the Earth is exactly spherical, by what percentage is the value of g, as you would measure it, different at the equator from at the north pole? Is your measured g larger or smaller at the equator?

10 One blade in one compressor-fan of a jet engine has mass 300 g. It rotates at 300 rev s^{-1} in a circle of radius 0.50 m. Calculate
  a the centripetal acceleration of the blade;
  b the force on the root of the blade.

11 The limiting frictional force between a mass and a particular turntable surface is equal to half the weight of the mass. If the mass is 5 cm from the centre of the turntable when it starts to rotate, at what angular velocity will the mass start to slide?

12 A gymnast of mass 70 kg is swinging round a high bar: at the fastest point of the swing (the bottom), she is rotating at 7 rad s^{-1}; her centre of mass is 1.5 m from her hands on the bar. What is the force between her hands and the bar?

13 When a golf-club strikes a ball, it may be travelling at a speed of about $45 \, ms^{-1}$. It has a mass of 200 g, and is effectively travelling in a circle with radius about 1.5 m. What force must the player be holding it with to prevent it flying out of his hands? Comment on the size of this force.

14 One ride in a fairground consists of 'boats' hanging on lengths of wire 6 m long, on a rim of radius 6 m. The rim revolves round its centre, and the 'boats' fly outwards. See figures 11(a) and (b).
    At its fastest rate of rotation, the wires are at angles of 50° to the vertical . Calculate
  a the rate of rotation of the rim in rev s^{-1};
  b the acceleration of the boats;
  c the factor by which the tension in the cables has been increased, compared to when the ride was stationary.

15 A spectacular fairground ride consists of a boat on the end of stiff supporting rods, which swings higher and higher until it eventually goes right over the top. See figure 12. The boat plus passengers have weight W.

(a)

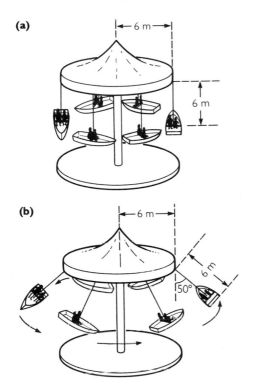

(b)

figure 11   (a) 'Boat' ride stationary; (b) 'boat' ride in motion

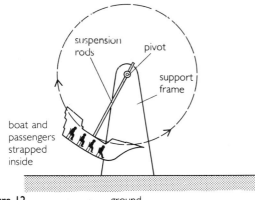

figure 12   ground

Suppose the boat reaches the top, momentarily coming to a halt there. At that moment there is a net compressive force of W in the rods. Suppose the boat then swings, under gravity alone, through a circle.
a What net force will there be in the rods at the bottom of the swing?
b At a particular point on this swing, the compressive force in the rods becomes a tensile force. Instantaneously there is zero force in the rods. Find the angle the rods make with the vertical at this moment.

16 A pilot of mass 70 kg takes his aircraft through a loop at a constant speed of $100 \, \mathrm{ms}^{-1}$. The radius of the loop is 400 m. See figure 13.

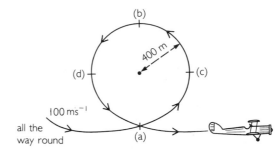

figure 13   Aeroplane looping the loop

Calculate the force exerted on the pilot by the aircraft
a at the bottom of the loop;
b at the top of the loop;
c half-way up the loop;
d half-way down the loop.
e In (d), how is the force actually transmitted to the pilot's body?

17 Consider the pilot (mass 60 kg) of a glider pulling up into a loop. Assume that during the manoeuvre the speed changes only as a result of changes in gravitational potential energy. She enters the loop at $60 \, \mathrm{ms}^{-1}$, and maintains a fixed radius of 50 m all the way round it. See figure 14.

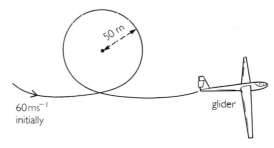

figure 14   Glider looping the loop

Calculate
a the force exerted on the pilot by the glider at the bottom of the loop;
b the speed at the top of the loop;
c the force exerted on her by the glider at the top of the loop.

18 Read the passage below. Comment on the errors and ideas it contains, and then write a correct version.
'An astronaut circling the Earth in an orbiting space station is weightless because the centrifugal and centripetal forces on him are balanced. In the same way a diver is weightless under water because the gravity force is balanced by the upthrust from the water. On the other hand, an astronaut in deep space has weight when his craft accelerates, because otherwise the force from the engines would produce infinite acceleration.'

19 This question is about the Earth's rotation. Use the following information:

moment of inertia of a sphere (mass m, radius r) rotating about a diameter $\frac{2}{5}mr^2$;
mass of Earth 6.0×10^{24} kg;
radius of Earth 6.4×10^6 m.

Find values for
a the moment of inertia of the Earth;
b the angular velocity of the Earth;
c the rotational kinetic energy of the Earth.

Suppose the entire energy requirements of the world's population (currently about 10^{21} J year^{-1}) were to be met by harnessing tidal power. The energy then effectively comes from the Earth's rotation. Estimate the effect this would have in one year on
d the angular velocity of the rotating Earth;
e the length of one day.

20 Here is an interesting fact: it is easier to balance a broomstick on your hand than a matchstick. This question explores the physical reason for this fact.

Let us assume that the broomstick is a scaled-up version of a matchstick, made of the same wood but 100 times bigger in its linear dimensions.

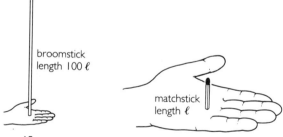

figure 15

a By what factor is the mass of the broomstick greater than the mass of the match?
b By what factor is the moment of inertia of the broomstick greater? ($I = m\ell^2/3$)
c Suppose each one tilts over slightly, at the same angle. See figure 16.
There is now a torque tending to rotate each of them clockwise. What is the general expression for the torque for a stick of mass m, length ℓ, at angle θ?

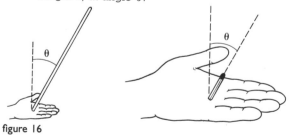

figure 16

d By what factor is the torque greater for the broomstick?

Angular acceleration of each stick is given by

$$\frac{d\omega}{dt} = \frac{\text{couple}}{I}$$

e Which of our two sticks has the greater angular acceleration?
f By what factor is it greater?
g Now explain why it is much easier to balance the broomstick.

21 A flywheel with $I = 0.050$ kg m^2 is arranged with a string wrapped round its axle, which has radius 1.0 cm. A 200 g mass is hung on the string. Assume the bearing of the flywheel is frictionless.

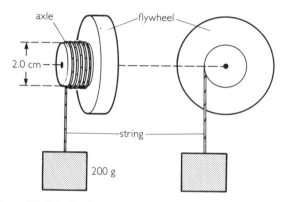

figure 17   Flywheel

Calculate
a the torque exerted by the mass-plus-string;
b the angular acceleration of the flywheel;
c the linear acceleration of the mass;
d the time taken for the mass to fall through 1.0 m;
e the angular velocity of the flywheel at this time.

The fraction of the energy lost by the 200 g mass which is transferred into its own translational kinetic energy has been ignored.
f Show that this fraction is very small, and therefore negligible.

22 Two circular gearwheels are cut from the same thickness of steel sheet; they have radii 10 cm and 20 cm, and masses 5 kg and 20 kg respectively. The large one spins at 10 Hz, while the small one is stationary; then they mesh together. (Moment of inertia of disc $= \frac{1}{2}mr^2$.)
a What is the angular momentum of the large disc initially?
b What is the initial rotational kinetic energy of the large disc?
c After they have meshed together, what is the ratio of their angular velocities (small disc : large disc)?

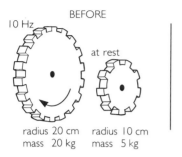

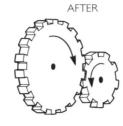

BEFORE   AFTER

10 Hz

at rest

radius 20 cm   radius 10 cm
mass 20 kg   mass 5 kg

figure 18

    d What quantity will remain conserved during this operation?
    e Calculate the new angular velocity of each disc.
    f Calculate the total rotational kinetic energy in the system now.
    g Why has energy been lost by the system?

23 The hammer is perhaps the most spectacular of athletic throwing events. The 7.3 kg steel ball is whirled on the end of a wire, in a radius of about 2 m. When he reaches a turning rate of about 2 rev s^{-1} the athlete lets go. The optimum angle for release is at 45° to the horizontal.

    a Calculate a value for the speed of the ball in its circular path.
    b Treat the ball as a projectile, and find out how far it will go.

24 Read the extract below, then answer the questions which follow it.
    a From the data given (16 000 rpm, 900 mph), calculate values in SI units for (i) angular velocity ω, (ii) rim speed v, (iii) the radius of the rotating flywheel. Is this compatible with the dimensions suggested in the article?
    b From the data given in the second paragraph, calculate in SI units (i) the mass of the bus, (ii) its speed (equivalent to 20 mph), (iii) the kinetic energy which Kess can store.
    c Assume that the flywheel is a disc. The moment of inertia I of a disc is given by $I = \frac{1}{2} mr^2$, where m is the mass of the disc and r its radius. Using values already calculated, and a suitable expression for the kinetic energy of a rotating disc, calculate a value for the mass of the flywheel. Does your answer seem reasonable?
    d Explain, using appropriate vocabulary, why a flywheel might fail and break up. What features of Kess make it capable of operating successfully when previous designs have not?

Apply brakes to boost speed

EVERY time the brakes are applied in a moving vehicle, energy of motion is converted to heat and lost for ever. If this wasted energy could be stored in some way and then used, big fuel savings would result, particularly with buses and other vehicles that are continually stopping and starting.

A Leyland bus is now on trial near Preston which puts this principle into practice, with the help of a flywheel system called Kess (Kinetic Energy Storage System) developed by BP. When the driver brakes, energy is transferred via the transmission system to the flywheel; when he accelerates energy is transferred in the opposite direction. The Kess unit is only about the size of a small dustbin yet it can store enough energy to accelerate a 16 ton bus from rest to 20 mph.

The engine and flywheel cut in and out under the control of a microprocessor, so the driver has nothing extra to do, except get used to putting his foot down and moving off in almost complete silence, an experience which first-time passengers also find a bit disconcerting.

The idea of flywheel energy storage for vehicles is not new, but BP believes Kess is a breakthrough, 'the first practicable lightweight system'. Materials are the key to its success. The use of advanced glass fibre composites, and careful design to exploit their properties, has led to a flywheel that can withstand the stresses of continuous operation at speeds of up to 16,000 rpm (when the rim is travelling at 900 mph) without breaking up as other designs have tended to do.

Even if a flywheel does fail, there is no danger of external damage. A casing which acts like a break drum in the event of a failure can bring the flywheel safely to rest, and a 'bandage' of ultra strong woven Kevlar fibre provides additional protection. 'The casing successfully contained a deliberately induced failure at 25,000 rpm though the design speed is 16,000,' said the project leader.

Bryan Silcock
The *Observer*

CHAPTER 5

Structures of solids

Introduction

The purpose of this short chapter is to introduce you to the way in which atoms pack together in solids. First we review the evidence for the existence of atoms and molecules which has been collected over the last two or three hundred years. Then we look at two simple models for the structure of a crystalline solid; and finally we see how the crystalline nature of some solids may be confirmed by the techniques of X-ray diffraction and electron microscopy.

GCSE knowledge

Matter can exist in three different states, either as a gas or a liquid or a solid. As a gas, atoms or molecules are widely separated and they are moving at very high speeds such as $300 \, \text{ms}^{-1}$. As a gas is warmed the molecules move faster. We explain the pressure that a gas exerts on the walls of its container by suggesting that these fast-moving molecules are continually colliding with the container walls, thereby exerting a force. A gas will occupy any volume that we allow; if we expand the size of a container the molecules are free to move into the additional volume, although the spacing between the molecules will increase.

If a gas or vapour is cooled sufficiently it will turn into a liquid; even helium turns into a liquid at 4 K. In a typical gas at room temperature and atmospheric pressure the spacing of the molecules is about ten times the molecular diameter. Intermolecular attractions in the gaseous state are small because the molecules move quickly and are not often close together. However, as the gas cools the molecules slow down and the intermolecular attractions become more important. When a substance becomes a liquid the volume of it is well defined. The molecules are now not free to move into any available space because they are held back by the attractive forces of neighbours. However, the molecules have enough energy to move around at random within the volume of the liquid.

When a substance is cooled even further it becomes a solid. The difference in volume between a solid and liquid is very small – in both cases the molecules are in contact. However, in a solid the atoms or molecules are not free to move around within the entire volume of the solid; they are confined to particular positions, although they do vibrate around those fixed positions. The atoms pack together in well-defined structures.

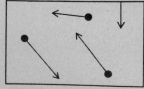

(a) Gas
molecules are widely spaced and move rapidly and randomly

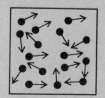

(b) Liquid
molecules are confined to a small volume due to attractive forces, but move at random

(c) Solid
molecules are fixed in position and arranged in well-defined structures. They vibrate about these fixed positions

figure 1

Evidence for the existence of atoms

The idea of individual atoms was a matter of debate in classical times; could matter be divided indefinitely or not? The Roman poet Titus Lucretius Carus, 98–55 BC, wrote that atoms were never annihilated but were hard, solid and individually devoid of colour, taste or smell. Considering the complete lack of experimental evidence available to him at the time, his remarks either showed considerable scientific insight or were a lucky guess.

More recently some firm evidence for the existence of atoms has been produced. First, Louis Joseph Guy-Lussac (1778–1850) showed that, when gases react chemically, the volumes of each constituent gas required are in a simple ratio, for example, two parts of hydrogen to one part of oxygen. This led to Avogadro's hypothesis that a given volume of any gas (at a fixed temperature and pressure) contains the same number of molecules or atoms. Similarly, electrolysis of aqueous solutions produces two volumes of hydrogen to one volume of oxygen. John Dalton (1766–1844) performed a series of experiments in which he showed that, in all chemical reactions, the masses of each substance required are also in a simple ratio, for example, 4 g of oxygen to 3 g of carbon. We now understand the significance of this last result: oxygen has a relative atomic mass of 16 and carbon a relative atomic mass of 12. The result above suggests that one particle (atom) of oxygen reacts with one particle of carbon to form a new substance. The new compound is of course carbon monoxide, CO.

Further evidence is provided by the behaviour of gases. For example, the diffusion of one gas through another is only satisfactorily explained in terms of molecules. We account for the *slow* diffusion of bromine through air as follows. Bromine and air molecules move very quickly but the bromine molecules are continually in collision with air molecules so that they cannot travel in a direct path. The most famous evidence for the existence of molecules is provided by Brownian motion. You will probably have observed the random motion of smoke particles through a microscope. Our explanation for the haphazard motion of the particles is that they are being bombarded by rapidly moving air molecules. We cannot see the air molecules because they are too small.

How big is an atom?

We can obtain at least an idea of the size of one atom of a solid element by using Avogadro's constant, together with data about the element. For example, consider this data for copper:

> density of copper $8900 \, \text{kg m}^{-3}$
> molar mass of copper $63.5 \, \text{g}$
> Avogadro's constant $6.0 \times 10^{23} \, \text{mol}^{-1}$

From this, we can calculate as follows.

One atom of Cu has mass $63.5 \, \text{g} \div 6.0 \times 10^{23}$

$$= 1.06 \times 10^{-22} \, \text{g}$$

$$= 1.06 \times 10^{-25} \, \text{kg}$$

The volume occupied by this atom is thus

$$\frac{1.06 \times 10^{-25} \, \text{kg}}{8900 \, \text{kg m}^{-3}}$$

$$= 1.19 \times 10^{-29} \, \text{m}^3$$

If we make the very crude assumption that this volume is cubical, then the side of the cube is given by

$$\sqrt[3]{1.19 \times 10^{-29}} \, \text{m}$$

which is $2.3 \times 10^{-10} \, \text{m}$.

Interestingly, similar calculations for other elements suggest that most atoms are of very comparable size: between $2 \times 10^{-10} \, \text{m}$ and $3 \times 10^{-10} \, \text{m}$ across.

Packing of spheres – a model of a solid

A theme that occurs time and again in physics is that of modelling. This is particularly useful when trying to picture matter on an atomic scale. We will start with Titus Lucretius Carus' model of the atom – hard, solid and spherical – and see where this takes us. Since most solid substances are difficult to pull apart, it is reasonable to assume that attractive forces exist between atoms, which might cause them to pack together. It is interesting to investigate the packing of many identical spheres. If we start by packing spheres in a square array as shown in figure 2, we can then add another layer above, putting the next layer of spheres into the gaps marked with a cross.

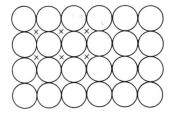

figure 2

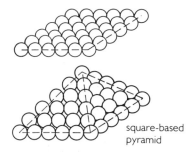

square-based pyramid

figure 3

In this way a pyramid of spheres can be constructed as shown in figure 3. The remarkable thing about this shape is that many crystals grow with an identical shape, suggesting that they might be constructed in a similar way by the addition of millions of microscopic spheres, or atoms.

Bubble raft

Blowing bubbles can be an amusing way of demonstrating the way in which atoms might pack together in one layer of a solid. Soap bubbles stick together due to strong cohesive forces and so provide a useful model for a solid which has only one sort of atom, such as a metal like copper. Bubbles can be blown in a soap solution by attaching a hypodermic syringe to the gas mains. Figure 4 shows a bubble raft of hexagonally-close-packed bubbles.

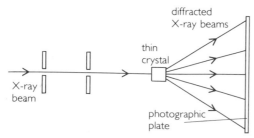

figure 4 Hexagonally close-packed soap bubbles

The forces of attraction between bubbles act equally in all directions and so there is a high degree of symmetry in the layer. In discussing inter-bubble and intermolecular forces, we must not forget repulsive ones. The fact that solids are very hard to compress suggests that if atoms get too close together, large repulsive forces act.

X-ray diffraction photography

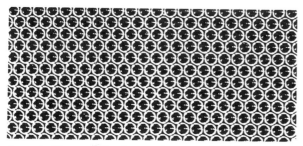

figure 5 Taking an X-ray diffraction photograph of a single crystal

Figure 5 shows the arrangement for X-ray photography of a single crystal. Figure 6 shows a diffraction photograph of a single crystal of sodium chloride. The striking features are the sharpness and symmetry of the dots in the photograph, which results from the symmetry of the crystal itself.

In general, any X-ray photograph of a solid which shows a regular pattern is evidence that there is regularity of some sort in the solid. A single crystal gives a pattern of dots, as in figure 6. A metal sample, which consists of a large number of small crystals in random orientations, gives a pattern of rings.

By careful analysis of the X-ray patterns obtained, physicists can build a complete model for the structure of any solid they are studying. The best measurements of Avogadro's constant and of the sizes of atoms are also made using this technique.

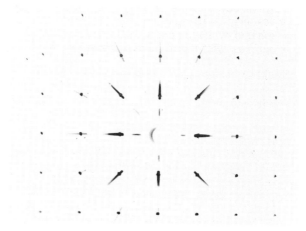

figure 6 X-ray diffraction pattern of a single crystal of sodium chloride

Electron microscopy

Electron microscopes make use of the short wavelength of travelling electrons (see chapter 41) to obtain very high magnifications. Figure 7 shows an electron micrograph of the surface of a sample of titanium dioxide. Individual rows of atoms are clearly visible. This technique gives further strong evidence that crystalline solids have atoms which are arranged in regular layers, and allows physicists to confirm structures previously worked out by analysing X-ray diffraction patterns.

figure 7 Electron micrograph of the surface of titanium dioxide

Summary

1 Evidence for the existence of individual atoms and molecules comes from
   a the laws of chemical combination,
   b Brownian motion,
   c diffusion,
   d the shapes of crystals,
   e X-ray diffraction,
   f electron microscopy.

2 Simple models for the packing of atoms in a crystal may be made using
   a spheres, which form a pyramid structure;
   b bubbles, which form a hexagonal layer.

3 X-ray diffraction analysis confirms the regular nature of the structure of many solids, and can be used to make detailed models of the structures of solids.

4 Electron microscopes can show up individual rows of atoms on the surface of a sample of solid material.

Questions

1 Bubbles floating on water are sometimes used as a model of one layer of atoms in certain solids.
   a What similarities are there between atoms of, say, copper and the bubbles of the raft?
   b List some of the limitations of the model: features where there is no analogue between bubbles and atoms.

2 What two features in particular make X-rays suitable for investigating crystal structures?

3 Estimate the number of atoms in a lump of sugar.

4 Aluminium has density $2700\,\mathrm{kg\,m^{-3}}$ and atomic mass number 27. One mole contains 6.0×10^{23} atoms. Calculate
   a the number of atoms of aluminium in one cubic metre of the substance,
   b the approximate 'diameter' of an aluminium atom.

5 Estimate the number of layers of atoms lost from the surface of a car tyre each time it makes contact with the road surface.

CHAPTER 6

The mechanical properties of solids

Introduction

In this chapter we investigate the mechanical properties of solids. We learn a lot about materials by stretching and breaking them. In particular we examine the properties of five materials: steel, copper, glass, polythene and rubber. We then link the observed properties of the material with the atomic structure.

GCSE knowledge

In this chapter we deal with the elasticity of solids so you need to understand the stretching of springs. The following points are useful knowledge.

☐ When a spring is stretched the extension is proportional to the force that causes the stretching. This is known as Hooke's law. When the force is removed from the spring it returns to its original length. This is elastic behaviour. If the spring is overloaded then it can be deformed permanently; then we say that it has exceeded its elastic limit. Hooke's law can be written mathematically:

$$F = kx$$

where k, the constant of proportionality, is the spring constant. If k is large the spring is stiff, if k is small the spring is floppy.

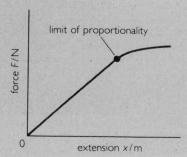

figure 1 Extension of a spring

☐ Figure 2(a) shows a spring being stretched by a load of 1 N; its extension is 1 cm.

Figure 2(b) shows two springs, each identical to the first, in series, being stretched by another load of 1 N. The same force acts through both springs, so each is stretched by 1 cm. The *total* extension of the two springs is 2 cm.

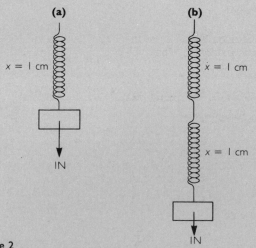

figure 2

☐ Figure 3 now shows two of the springs placed in parallel. They support between them a load of 1 N. This means that each supports a load of 0.5 N and the extension of each spring is now only 0.5 cm.

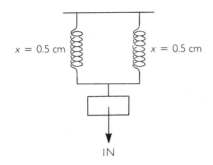

figure 3

Example Figure 4 shows three springs supporting a load of 3 N. X has a spring constant of $3\,\mathrm{N\,cm^{-1}}$; this means 3 N will extend it by 1 cm. The spring constants of Y and Z are $2\,\mathrm{N\,cm^{-1}}$ and $1\,\mathrm{N\,cm^{-1}}$ respectively. How far down does point A move when the 3 N load is attached?

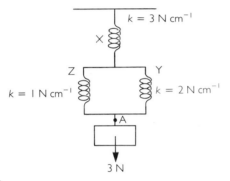

figure 4

Springs Y and Z are in parallel and so have the same extension. They will each extend by 1 cm. Thus A moves down 2 cm.

Stress, strain and the Young modulus

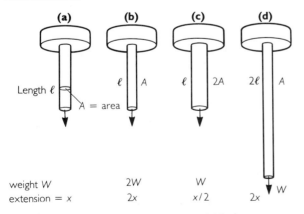

figure 5 Four samples of the same material being stretched

Figure 5 illustrates four samples of a material being stretched. Sample (a) has a length ℓ, cross-sectional area A, and is extended a distance x by a load W. Figure 5(b) shows that if the load is doubled the extension is doubled. In figure 5(c) the same load is applied to twice the cross-section and the extension is half the original; this is like applying a load to two springs in parallel. In figure 5(d) the specimen is twice as long as the original and the extension is also doubled. (What happened when a load was applied to two identical springs in series?)

We may summarise our findings:

$$x \propto W$$

$$x \propto \frac{1}{A}$$

$$x \propto \ell$$

$$\Rightarrow \qquad x \propto \frac{W\ell}{A}$$

or $\qquad \dfrac{x}{\ell} \propto \dfrac{W}{A}$

This last statement is very important. It means that the fractional change in length is proportional to the force per unit area.

We now define the following useful quantities:

$$\frac{\text{force}}{\text{area}} \left(\frac{F}{A}\right) = \text{tensile stress (unit: } \mathrm{Nm^{-2}})$$

$$\frac{\text{extension}}{\text{original length}} \left(\frac{x}{\ell}\right) = \text{tensile strain (strain is a ratio of lengths so has no unit)}$$

Tensile stress is a useful quantity because it measures the force applied to a particular number of inter-atomic bonds. Tensile strain is useful because it measures proportional increase in length. (Sometimes strain is quoted as a percentage increase in length.) In other words, both quantities take account of the physical dimensions of the sample being tested, and are therefore more useful for making comparisons than just using 'force' and 'extension'.

Note that other stresses can also be important in a material: in particular, compressional stress and shear (twisting) stress.

Example A wire has a length of 2 m and a cross-sectional area of $1\,\mathrm{mm^2}$. When a force of 2 N is applied it stretches by 1 mm. What stress is the wire subjected to and what strain does it cause?

$$\begin{aligned} \text{stress} &= \frac{\text{force}}{\text{area}} \\ &= \frac{2\,\mathrm{N}}{10^{-6}\,\mathrm{m^2}} \\ &= 2 \times 10^6\,\mathrm{Nm^{-2}} \end{aligned}$$

$$\text{strain} = \frac{\text{extension}}{\text{length}}$$

$$= \frac{10^{-3}\,\text{m}}{2\,\text{m}}$$

$$= 5 \times 10^{-4}$$

The conclusion that we drew from our stretching experiment was that

$$\frac{x}{\ell} \propto \frac{F}{A}$$

or that strain ∝ stress. This is shown graphically in figure 6 for two materials A and B.

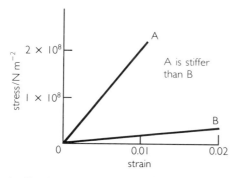

figure 6 Graph of stress against strain for two materials

We can now define a quantity that describes the 'stretchiness' of a material.

The Young modulus $(E) = \dfrac{\text{tensile stress}}{\text{tensile strain}}$

(units: Nm^{-2})

Example Calculate the Young modulus of material A in figure 6.

$$E = \frac{\text{stress}}{\text{strain}}$$

$$= \frac{2 \times 10^8\,\text{Nm}^{-2}}{0.01}$$

$$= 2 \times 10^{10}\,\text{Nm}^{-2}$$

Note that the Young modulus can be calculated from a stress-against-strain graph by measuring the gradient. However, it must be emphasised that this can only be done while the material is within its elastic limit. Once a material has exceeded its elastic limit the gradient of a stress-against-strain graph is *not* equal to the Young modulus.

The Young modulus can be thought of as 'the stress required to produce unit strain for a material'. A stiff material requires a very big stress to stretch it and so

has a high Young modulus; a floppy material is easily stretched and so has a low value for its Young modulus. Steel has a Young modulus of about $2 \times 10^{11}\,\text{Nm}^{-2}$ whereas the value for rubber is only about $5 \times 10^7\,\text{Nm}^{-2}$.

Example A lift cable which is 50 m long, is made from high-tensile steel of Young modulus $2.0 \times 10^{11}\,\text{Nm}^{-2}$. The cable is made from 100 wires each of radius 2.0 mm. Calculate the extra extension in the cable when 10 passengers pile into the lift; the total mass of the passengers is 500 kg.

The area of one wire $= \pi r^2$

$$= \pi (2.0 \times 10^{-3}\,\text{m})^2$$

$$= 1.3 \times 10^{-5}\,\text{m}^2$$

Area of cable $A = 100 \times 1.3 \times 10^{-5}\,\text{m}^2$

$$= 1.3 \times 10^{-3}\,\text{m}^2.$$

Now $\qquad E = \dfrac{\text{stress}}{\text{strain}}$

$\Rightarrow \qquad \text{strain} = \dfrac{\text{stress}}{E}$

$\Rightarrow \qquad \dfrac{x}{\ell} = \dfrac{F}{AE}$

or $\qquad x = \dfrac{F\ell}{AE}$

$\Rightarrow \qquad x = \dfrac{(5000\,\text{N}) \times (50\,\text{m})}{(1.3 \times 10^{-3}\,\text{m}^2) \times (2.0 \times 10^{11}\,\text{Nm}^{-2})}$

$$= 1.0 \times 10^{-3}\,\text{m}$$

Energy stored

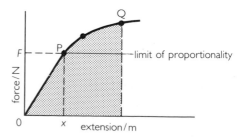

figure 7 Force-against-extension graph for a metal

Figure 7 shows a force-against-extension graph for a certain metal. To calculate the energy stored up to the limit of proportionality (P) we use the expression:

work done   = average force × extension
(or energy stored) $= \frac{1}{2}F . x$ (= area under graph)

The energy stored per unit volume $= \frac{1}{2}\dfrac{F.x}{V}$

But $V = A \times \ell$.

So energy stored per unit volume $= \frac{1}{2}\dfrac{F}{A}\cdot\dfrac{x}{\ell}$

Energy stored per unit volume $= \frac{1}{2}$ stress $\times$ strain

If we wished to estimate the energy needed to stretch the wire to Q, we would need to work out the shaded area under the curve but it should be noted that not all of this energy is recoverable because the elastic limit has been exceeded.

Example   Calculate the energy stored in a steel cable of cross-sectional area $1\,\text{cm}^2$, length $100\,\text{m}$, under an elastic strain of 0.1%. The Young modulus is $2 \times 10^{11}\,\text{Nm}^{-2}$.

$$\text{Energy stored} = \frac{1}{2} \times \text{stress} \times \text{strain} \times \text{volume}$$
$$= \frac{1}{2}\,\text{Young modulus} \times (\text{strain})^2 \times \text{volume}$$
$$= \frac{1}{2} \times (2 \times 10^{11}\,\text{Nm}^{-2}) \times (0.001)^2$$
$$\times (100\,\text{m}) \times (10^{-4}\,\text{m}^2)$$
$$= 1000\,\text{J}$$

Properties of materials

Elastic describes substances with this property: when stress is removed from it, the substance returns to its original length. *Strength* or *breaking stress* is the stress that can be applied to a substance before it breaks. *Stiffness* refers to how easily a substance can be deformed; a stiff substance is hard to distort and has a high Young modulus. A *ductile* material is one that can be rolled out or drawn out into a wire; most metals are ductile. A *malleable* material can be hammered (malleted) into shape; soft metals such as lead, gold and hot steel are malleable. A *brittle* substance cracks and breaks without any plastic deformation. A *tough* material is not brittle, it has high strength and will deform plastically before breaking.

Figure 8 shows five sketches of stress-against-strain graphs for five different types of material. (Note that the scales are different.)

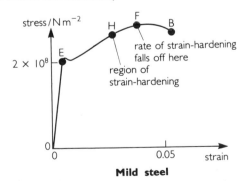

figure 8(a)   *Mild steel* is stiff (high Young modulus), and has a large region of plastic flow. It is also tough and ductile. (The properties of a given sample of steel depend very much on its impurity content and its past treatment)

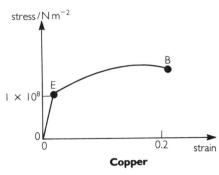
Copper

figure 8(b) *Copper* has a small region over which it exhibits elastic behaviour, but a large plastic region. It is readily ductile or malleable

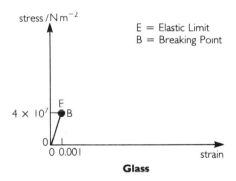
Glass

figure 8(c)   *Glass* is brittle and shows elastic behaviour all the way to its breaking point; it has a high Young modulus. (Very thin whiskers can have extremely high strength)

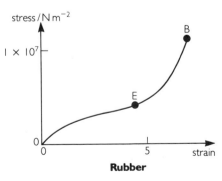
Rubber

figure 8(d)   *Rubber* is relatively hard to extend for very low strains, becomes very easy to extend for moderate strains, then becomes harder or stiffer at large strains

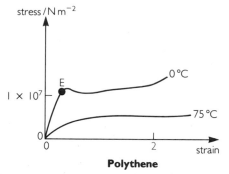

figure 8(e) *Polythene* is harder to extend than rubber for low stresses, but then yields and shows large strains before breaking

In the following section the behaviour of all these materials is examined on the atomic scale.

Hooke's law behaviour

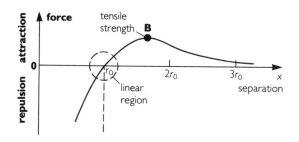

figure 9 Variation of force between two atoms with the distance between them

Figure 9 shows how the force between two atoms might vary with separation; r_0 is the equilibrium separation. The linear region of this force-against-separation graph explains Hooke's law at small stresses. A steep gradient of this graph in the region of r_0 would mean a large Young modulus.

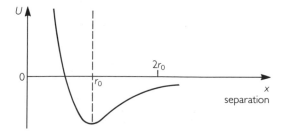

figure 10 Variation of potential energy of two atoms with the distance between them

Figure 10 shows how the potential energy, U, of two atoms might vary with separation. Note that U has its minimum value at the equilibrium separation. The potential energy is negative and this means that work has to be done to pull the atoms apart or indeed to push them closer together than their equilibrium separation.

The Young modulus – a microscopic version

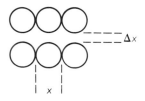

figure 11 The atomic separation x is increased by an amount Δx when force F is applied

The force-against-separation graph of figure 9 suggests that we can use the relationship $F = k.\Delta x$ to describe the extension of an atomic bond when $x \approx r_0$.

So

$$\text{strain} = \frac{\Delta x}{x}$$

$$\text{stress} = \frac{F}{A}$$

$$\approx \frac{k\Delta x}{x^2}$$

Thus

$$E = \frac{\text{stress}}{\text{strain}}$$

$$= \frac{k.\Delta x.x}{x^2.\Delta x}$$

$$= \frac{k}{x}$$

This result means that the Young modulus of a material is directly related to the strength of a particular molecular bond. We will calculate the strength of k for steel: $E = 2.0 \times 10^{11}\,\text{Nm}^{-2}$, $x = 2.5 \times 10^{-10}\,\text{m}$.

Thus $k = Ex$

$$= (2.0 \times 10^{11}\,\text{Nm}^{-2}) \times (2.5 \times 10^{-10}\,\text{m})$$

$$= 50\,\text{Nm}^{-1}$$

Plastic behaviour

Plastic behaviour suggests some sort of irreversible flow of atoms past each other. Figure 12 shows a possible mechanism, but currently the idea of *dislocations* is preferred.

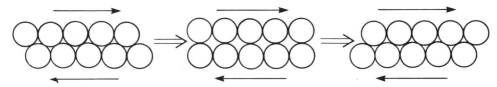

figure 12   Layers of atoms moving past each other might explain plastic flow

In dislocation theory, imperfections in the crystal structure can move through the crystal as shown in figure 13. Plastic deformation may now take place, but only one atomic bond is broken at a time – this explains why plastic deformation takes place at such low stresses.

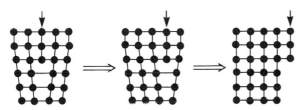

figure 13   Imperfections or 'dislocations' in the crystal structure can move through the crystal

The stresses required to produce slip as shown in figure 12 can be calculated (see question 21). These stresses are far greater than those that actually cause plastic flow. The theory of dislocations is more consistent with experimental results.

The strength of a metal depends on how easily dislocations can move through its lattice structure (see figure 13). A single perfect crystal of copper would be free from dislocations and would be extremely strong, until scratched. Every bond across a plane through the crystal would have to break simultaneously for the specimen to give way. Whereas, if there were dislocations in a single crystal of pure copper, they would be able to move across the crystal, so small forces would cause the specimen to stretch permanently. A metal is strengthened by restricting the movement of its dislocations.

There are three kinds of barriers against the movement of dislocations:

(i) *Extra dislocations* cause distortion of the lattice and block the movement. The repeated stressing and relaxing of steel generates further dislocations. The dislocations become entangled with one another preventing any further movement. (*Work-hardened* steel is stronger than freshly-made steel, see figure 8(a), region H).

(ii) *Impurity atoms* in the crystal will distort the crystal lattice and impede the movement of dislocations (brass is stronger than pure copper due to the presence of zinc atoms);

(iii) *Crystal boundaries* prevent the movement of dislocations. Most metals are made out of many small crystals. When a dislocation reaches the edge of one crystal it cannot move on to another one.

Ductile fracture

In the early stages of plastic deformation a wire thins uniformly. Later it develops a waist or neck. Cracks in ductile materials are not as serious as they are in brittle ones, because dislocations tend to flow into cracks, thus blunting them (figure 15).

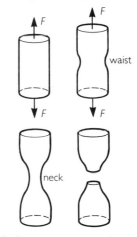

figure 14   Ductile fracture

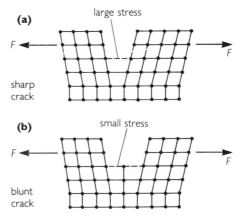

figure 15   Cracks in (a) a brittle material; (b) a ductile material

Glass – a brittle substance

Glass has no long-range order but only short-range order. It is often described as a 'frozen liquid'. As a result it does not display plastic properties because dislocations cannot flow. Thus any surface cracks will seriously weaken the structure. Cracks will quickly propagate right through the structure until fracture occurs, rather than being blunted by dislocation flow as in a ductile material.

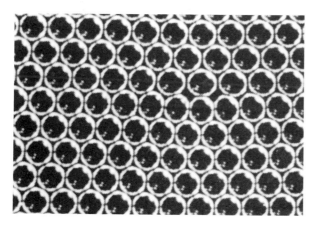

figure 16   This photograph shows a dislocation in a raft of soap bubbles

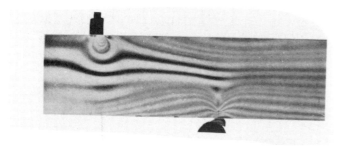

figure 17   Stress lines in Perspex, produced when viewed in polarised light

Rubber

Electron diffraction photographs figures (18, 19) of rubber reveal a disordered structure when it is unstressed, and an ordered structure when it is stretched. Rubber is polymerised isoprene. It is thought that under no stress the molecules tend to coil up randomly, but under stress they will be stretched out in a more ordered way. A certain amount of cross-linking between the chains gives the substance strength under tension. While the

molecules are uncoiling they are easy to stretch, but once they are fully stretched it is very difficult to stretch them further. This behaviour accounts for the form of the stress-against-strain graph of figure 20.

figure 18   Electron diffraction pattern of unstretched rubber

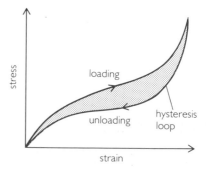

figure 19   Electron diffraction pattern of stretched rubber

figure 20   Stress/strain graph for rubber

Rubber displays a certain amount of hysteresis, as shown in figure 20. Imagine the force applied to a specimen of rubber being systematically increased then gradually reduced. The length of the specimen does not return to its former value for each value of the applied force. This lagging behind is known as hysteresis. The area enclosed by the hysteresis loop gives the amount of energy lost as heat during the cycle.

Polythene

Polythene is polymerised ethene (C_2H_4). The polythene molecules are long chains of carbon and hydrogen atoms. The making of a strong and rigid polymer is really a problem in chemical engineering.

Figure 22 shows the type of arrangement that the polythene molecules adopt. Groups of chains run parallel to one another and so the structure has a certain amount of order. The chains are weakly linked by intermolecular forces, but the combined effect of these is to make the structure quite stiff. However, if a sufficiently large stress is applied these weak molecular bonds are broken and there is little to stop the long molecules sliding past each other until fracture occurs (figure 23).

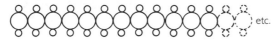

figure 21   Part of a polythene molecule

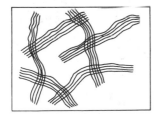

figure 22   The arrangement of molecules in polythene

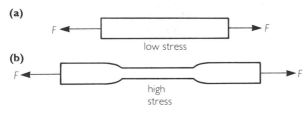

figure 23   Polythene under stress

Composite materials

We have been using composite materials for a long time; primitive people used a mixture of twigs and mud to build their huts. The modern construction engineer is interested in materials which give him the greatest possible strength and stiffness for the least weight. If a bridge is built from very dense materials, such as concrete or metals, then the weight of the bridge itself puts the girders under great stress. However, materials such as carbon and glass can provide great strength while being a lot less dense. These materials have their atoms joined by covalent bonds. The covalent bonds are very strong and so produce materials with a high Young modulus. In addition, these bonds do not allow hexagonal close packing, because the atoms bond in well-defined directions only. As a result the spacing of the atoms is large and the density is low.

The first modern, strong composite was fibreglass or glass reinforced plastic (GRP). The strength of the glass is exploited by placing it in the form of fibres in the more flexible plastic. The plastic performs these duties:

☐ it prevents the glass fibres from getting scratched. Scratches make the glass weak.

☐ it holds the glass fibres together, so that they carry the load.

☐ the plastic is ductile and so does not allow cracks to spread through it. Glass by itself is brittle.

The table below allows a comparison between steel and two composites, glass reinforced plastic (GRP) and carbon fibre reinforced plastic (CFRP).

	Young modulus/ $10^{10}\,N\,m^{-1}$	Breaking stress/ $10^8\,N\,m^{-2}$	Density/ $10^3\,kg\,m^{-3}$
Mild steel	20	4	7.9
GRP	2	5	2.3
CFRP	50	10	2.5

The table reveals the shortcoming of GRP: too low a stiffness (Young modulus). Thus it is not a suitable material for making very large structures. CFRP is used widely in the aircraft industries but is very expensive to produce.

The use of concrete

Concrete is brittle and is weak under tension but strong under compression. Figure 24 suggests why this is so. The upper surface is under tension and any surface cracks open up; the lower surface is compressed and surface cracks close.

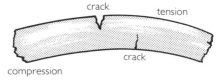

figure 24   The effect of tension and compression on concrete

Concrete is thus often used for pillars. Concrete is also used in composite form, either by prestressing or by reinforcing. Prestressed concrete contains steel rods that are held in tension by external forces while the concrete is poured into its mould and allowed to set. When the external forces are removed, the rods try to contract, and the concrete is held in a permanent state of compression – even when subjected to tensile forces, as when used for building.

Reinforced concrete works on the same principle as GRP. The concrete acts like the plastic and the steel bars take the stress. Figure 25 shows how the steel bars are bent to ensure there is no slip between the steel and concrete. This composite allows the steel to take the tensile forces and the concrete the compressive ones.

Glass may also be prestressed to reduce the likelihood of crack propagation. In thermal toughening, jets of air cause the outside of the glass

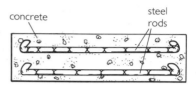

figure 25   Reinforced concrete

to solidify whilst the inside is still molten. Later the inside contracts and pulls the outside inwards, thus causing it to be in a permanently compressed state. Car windscreens are prestressed and malleable. The pattern left by the air jets becomes visible when viewed through polarising sunglasses.

Summary

1   $\text{stress} = \dfrac{\text{force}}{\text{area}}$

2   $\text{strain} = \dfrac{\text{extension}}{\text{length}}$

3   The Young modulus $= \dfrac{\text{stress}}{\text{strain}}$

4   Energy stored per unit volume $= \frac{1}{2} \times \text{stress} \times \text{strain}$.

    (This formula only holds true if the extension is proportional to the applied force.)

5   You should be familiar with terms such as elastic, plastic, ductile, brittle, tough, strong. These are explained on page 67.

6   You should be familiar with the stress-against-strain graphs for steel, copper, glass, rubber and polythene. You should be able to account for the behaviour of these materials in terms of their molecular structures.

7   A composite material blends the desirable properties of two materials. For example, glass is strong but brittle; it can be embedded in a plastic resin to make tough fibre glass.

Questions

1   Calculate the strain in the following cases.
    a   A steel bar of length 1.0 m is extended by 3 mm.
    b   A rubber band of length 10 cm is stretched to a length of 40 cm.
    c   An iron hoop of radius 30.0 cm is warmed until its radius increases to 30.1 cm.

2   Calculate the tensile strength in $N\,m^{-2}$ in these cases.
    a   A force of 200 N is applied to a steel bar of cross-sectional area $10^{-4}\,m^2$.
    b   A force of 10 N is applied to a copper wire of radius 1 mm.

3   An experiment is carried out to measure the Young modulus of a human hair. The radius of the hair used is 0.1 mm and when a weight of 1.0 N is applied to it the hair (of initial length 10 cm) extends by 0.1 mm. What is the Young

modulus for this hair? What assumptions have you made in your calculation?

4   Copper reaches its elastic limit at a strain of about 1%; the Young modulus is $1.3 \times 10^{11}\,Nm^{-2}$. A sample of a certain gauge of copper wire is strained by 0.5% when a load of 100 N is placed on it. What can you say about the strain that would result in a similar piece of wire with loads of (i) 30 N, (ii) 300 N?

5   The following questions require you to use the data on page 71, which compares mild steel, GRP and CFRP.
    a   Compare the strength/weight ratios of the three materials.
    b   Compare the stiffness/weight ratios of the three materials.
    c   If samples of each material were tested, each sample having the same length and cross-

section, which material would stretch furthest before breaking?

d If samples of each material were tested, each sample having the same length and mass, which material would stretch least before breaking?

6 Consider the situations shown in figure 26, in which different samples of the same material were subjected to different forces. Hooke's law applies in all the situations.

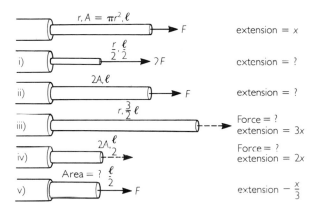

r = radius of cross-section
A = area of cross-section
ℓ = length of rod
F = applied force

figure 26

a Write down values for the quantities asked, in terms of the initial situation.

b Which of the samples in figure 26 would yield at the smallest applied force?

c Which of the samples in figure 26 would have the greatest extension when it yields as the applied force is gradually increased?

7 Using the graphical information for mild steel, glass, copper, rubber, and polythene given in figure 8 (page 67), answer the following questions. (In most cases, common experience should provide a good check for your answers.)

a What is the Young modulus for glass?

b What tensile force would be required to break a piece of steel with cross-section $1\ \text{cm}^2$? Give an example of some object with roughly this weight.

c Which material has the highest tensile strength?

d If identically shaped pieces of each material were tested, which would break at the smallest extension?

e Which material remains elastic over the greatest range of strain?

f Which material remains elastic over the greatest range of stress?

g What original length of copper would stretch by about 2 m before finally breaking?

h What cross-section of copper wire would be needed in a copper cable if it is to be able to take tensions up to 40 N without passing its elastic limit?

i What load should a $1\ \text{mm}^2$ glass thread be able to carry without breaking?

8

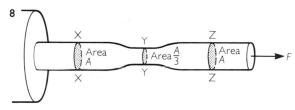

figure 27

Figure 27 shows a magnified section of ductile wire under tension; a 'neck' has appeared at section YY, as the tension F was gradually increased from zero.

a Sketch a graph showing how the displacement of ZZ probably increased as F increased from zero.

b Write down expressions for the stress at (i) XX, (ii) YY.

c Where along the wire is the strain now increasing fastest with increasing stress? Explain.

9

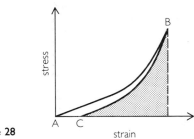

figure 28

Figure 28 shows a graph of stress against strain as a sample of a certain material is increasingly stressed, then the stress is relaxed.

a Describe in a few words the behaviour of the material, and speculate about what sort of material it is.

b Which section of graph shows the effect of *increasing* the stress?

c Explain the significance of (i) the shaded area, (ii) the area enclosed by ABC, (iii) the length AC.

d On a copy of figure 28, sketch how you think the material would behave if, after the initial stress is relaxed, the material is gradually stressed again to the same value of stress as at B.

10

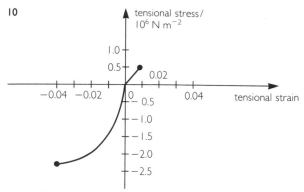

figure 29

Figure 29 shows a stress-against-strain graph for a type of concrete.

a Estimate the maximum weight which a column 1 m by 10 cm could carry without crumbling.

b What cross-sectional area of column would be needed to support material of mass 1 tonne if the maximum strain is to be 1%?

11

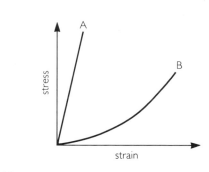

figure 30

Figure 30 shows the stress-against-strain graphs for high-grade steel (A) and spider's silk (B). Both specimens were stretched until they broke.

a Which material has the higher breaking stress?

b Which material requires more energy to break it?

c Explain why it is important for spider's silk to have a lot of give in it.

d What other useful properties does spider's silk have?

12 Suppose a steel wire of cross-sectional area 1 mm² is subjected to a force of 1000 N. The diameter of a steel (iron) atom is 2.3×10^{-10} m, and the Young modulus is 2×10^{10} Nm⁻².

a What stress is there in the wire?

b What strain occurs (assuming the elastic limit is not passed)?

c If adjacent layers of atoms perpendicular to the length of the wire are packed in a square array (see figure 31), what force is placed on each bond between atoms in two adjacent layers?

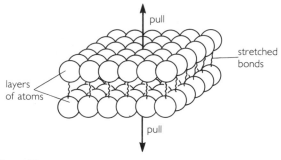

figure 31

d How far do the adjacent layers of atoms move apart from each other?

e What is the effective spring constant k for each bond?

13 The idea of bonds behaving like springs is a model. Does the model continue to be useful in predicting what will happen if the material is subjected to

a compression stress;

b shear stress;

c increased temperature?

Explain your answers.

14 a Estimate the energy stored in a rubber band stretched as in figure 32. Assume the band is 5 cm long when slack.

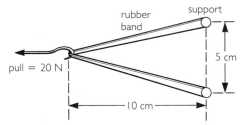

figure 32

b If this energy is all converted to the kinetic energy of a paper pellet of mass 10 g, estimate the initial velocity of the pellet.

c Estimate the pellet's range if fired horizontally at this speed from 1.5 m above floor level. Comment on the reality of this estimate.

d If Hooke's law was being obeyed in figure 32, estimate the Young modulus for the band, if the cross-section of the band was 1 mm². (Assume, incorrectly, that the cross-section remains constant.)

e Now assume that the *volume* of the material stays constant, so that the area goes down as the length goes up. Repeat the estimate for the Young modulus.

f Which of (d) and (e) is 'right'? Explain.

g Does reduction of cross-sectional area occur in metal wires in tension? Does it matter? Explain.

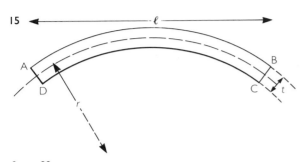

figure 33

figure 34   Boris   Vladimir

15 A certain structure is designed so that one beam within it, of thickness t, may flex elastically into a circular shape with radius r (figure 33).

  a If only three forces act on the beam, sketch possible positions and directions of action for them.

  b Part AB of the beam is in tension, and part CD is in compression. Assuming $\ell \ll r$, find an expression for the tensile strain in the part AB.

16 Sometimes when fibre composites are tested to failure, the fibres just pull out of the plastic without breaking. What limits the strength of the material in this case?

17 Fibres in fibre-glass are usually laid down in a cross-ply formation, rather than all parallel. Why?

18 During the design of a hydroelectric power plant, a scale model is made of the supply lake, dam, and downhill pipe with all *linear* dimensions reduced by a factor of 100. By what factor would the following change:

  a the pressure at the bottom of the dam when the lake is full,

  b the gravitational potential energy stored when the lake is full of water,

  c the mass of the downhill pipe,

  d the speed at which the water leaves the open end of the downhill pipe?

19 Sketch stress-against-strain graphs for chewing gum (i) before chewing, (ii) after chewing; use the same axes for each. Which of the two samples do you think is more likely to show (i) creep, (ii) fatigue?

20 If it were true that the height to which an animal could jump depended on its size, then since fleas can jump 0.5 m or so, it would appear that indeed the cow might jump over the moon. However, to a good approximation all animals can lift their centres of gravity between 0.5 m and 1.0 m (obviously there are some exceptions). The following questions explain this observation. Boris and Vladimir are two cats, identical in every respect apart from the (obvious) fact that Boris is twice as tall, twice as long, and twice as wide as Vladimir.

  a What is the ratio of their masses?

  b Given that all cat sinews have the same Young modulus, justify the assumption that, at the instant of springing from the ground, the muscles in each of Boris' and Vladimir's legs are subjected to the same stress and strain.

  c Calculate the ratio of the elastic energy stored in the two cats' muscles.

  d Now explain why each cat should be able to lift its centre of gravity the same height above the ground.

21 **a** Show that if a solid of Young modulus Y is subjected to a tensile stress of σ, then the elastic energy stored per unit volume is $E = \sigma^2/2Y$.

  b It has been suggested that when a wire begins to flow plastically the stored elastic energy $E = L$, where L is the latent heat of fusion. Explain why this idea is reasonable.

  c Use this argument to estimate the elastic strain at which copper begins to yield. ($E = 1.3 \times 10^{11}\,\mathrm{Nm^{-2}}$, $L = 2 \times 10^9\,\mathrm{Jm^{-3}}$)

  d In practice it is discovered that plastic flow in copper occurs at strains of only 1%. Comment in the light of your answer to (c).

22 Two oxygen atoms bond together to form a molecule. In a simple model of the molecule, the bond may be considered as a spring, with $k = 1020\,\mathrm{Nm^{-1}}$ (measured using the infrared spectrum of the molecule); the graph of interatomic force against distance near the equilibrium separation r_0 thus looks like figure 35(a).

We can write $F = +k(r_0 - r)$ for varying values of the separation, r. Also, U, the energy, $= -\int F\mathrm{d}r = A + \frac{1}{2}k(r_0 - r)^2$.

This shape is parabolic, and is illustrated by the solid line in figure 35(b). For $r < r_0$, the parabolic shape is roughly followed by real molecules; as r increases above r_0, however, the model breaks down, following instead the dotted line in figure 35(b). By definition, $U = 0$ when $r = \infty$.

Given that the dissociation energy E of the molecule is 5.1 eV, and $r_0 = 1.2 \times 10^{-10}$ m, find a value for x (the separation less than r_0 at which $U = 0$) in terms of r_0.

(a)

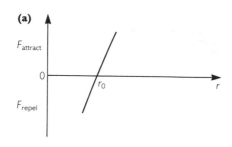

(b) U

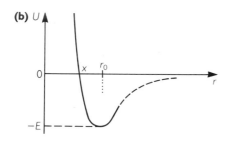

figure 35

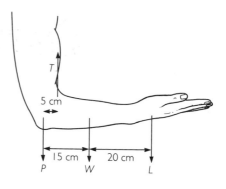

L = pull from load being held
W = weight of forearm + hand
P = push down from pivot
T = pull from bicep muscle

figure 36

23 In chapter 1, a free-body diagram was drawn for a forearm. This is the drawing again, with lengths marked on:
Suppose $W = 20$ N and $L = 100$ N.
a Calculate (i) T, (ii) P. (Assume the forearm is stationary.)

b Guess at the largest load L you think you could hold stationary with your hand for a short period, with forearm horizontal and elbow tucked into your side.
c Hence estimate the largest T your bicep can stand.
d Estimate, by feeling it, the cross-sectional area of your bicep when it is in tension.
e Estimate, from (c) and (d), the greatest tensile stress your bicep can take. Compare your value with the ultimate tensile stress of rubber, which is about 4×10^6 Pa. Is it reasonable to expect the values to be about the same? Justify your answer.

CHAPTER 7

Oscillations

Introduction

We start the chapter by looking at the widespread occurrence of oscillatory motion. We examine how knowledge of the way displacement changes with time can yield other interesting information. After this we look in considerable detail at how a whole family of oscillators moves with sinusoidal, or simple harmonic motion; and we develop equations describing this motion, and graphs to represent it, referring particularly to the standard example of a tethered trolley. Finally we look briefly at the effects of friction (damping) on oscillations.

GCSE knowledge

We assume you know that:

☐ an unbalanced force causes a system to accelerate in the direction of the force;

☐ force = mass × acceleration ($F = ma$);

☐ Hooke's law is obeyed when springs are stretched slightly (Hooke's law states that the increase in length of a spring is proportional to the force applied to it);

☐ frictional forces do work, and therefore remove kinetic energy from a moving system.

Oscillations everywhere

An oscillating object is one which moves backwards and forwards in some sort of repetitive way. Among the many everyday examples we could consider are:

☐ the movements of an insect's wing,

☐ vibrations of a guitar string,

☐ the movement of the piston in an engine,

☐ the movement of windscreen-wiper blades,

☐ the rise and fall of tides.

If we extend the discussion to include branches of science which are less easy to visualise, we could also consider:

☐ the movement of electrons in a wire carrying alternating current,

☐ the movement of atoms in a solid,

☐ the ebb and flow of ice ages across northern Europe,

☐ the variations of electric and magnetic fields in an electromagnetic wave.

The remarkable feature of many of these examples, which cover a very wide range of events, is that they can be described with reasonable accuracy by the same mathematical model: that is, the same set of equations. The model is called *simple harmonic motion* (s.h.m.), and much of the rest of the chapter is devoted to working out or discovering the equations.

Displacement-against-time graphs for oscillators

You can deduce more or less everything you ever need to know about a particular oscillator if you have a detailed displacement-against-time graph for it. This is a graph showing its position at each instant of its motion: preferably during several repeats, or *cycles*.

As an example of this, let us consider a windscreen-wiper blade. For displacement we will take the angle through which the blade is turned from its rest position. Essentially the blade moves at a constant rate, turns round quickly, and moves at the same rate back again. The displacement-against-time graph might therefore look like figure 1.

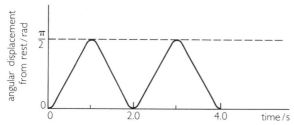

figure 1 Displacement-against-time graph for a windscreen-wiper blade

Using the information on the graph, we can work out several important parameters, or quantities, which help to describe the motion.

1 *Time-period (T).* This is the time taken for one complete cycle. In this example, $T = 2.0$ s.

2 *Frequency (f).* This is the number of complete cycles per unit time. The unit of frequency is one hertz (Hz): 1 Hz = 1 cycle/second. A useful relationship for oscillations, and indeed any repetitive event, is:

$$f = \frac{1}{T}$$

In this example, $f = 0.50$ Hz.

3 *Amplitude (A).* This is the maximum displacement in either direction of the oscillating object from its central position. (In many cases, though not in the wiper-blade example, the central position is the same as the rest position.) In this example, $A = \pi/4$ rad.

From the displacement-against-time graph, we can deduce another important graph: the velocity-against-time graph, or for our example, the graph of angular velocity against time. This is given by the gradient at each moment of the displacement-against-time graph, using the definition of velocity:

$$v = \frac{ds}{dt}$$

(In fact we have already used a qualitative knowledge of the motion of the blade to sketch its displacement-against-time graph.) See figure 2.

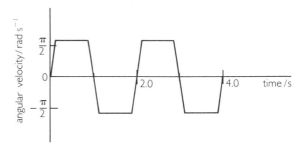

figure 2 Angular velocity-against-time graph for wiper blade

From this graph we can deduce one further important graph: the acceleration, or angular acceleration, against time. The gradient of a velocity-against-time graph at any instant gives the acceleration, according to the definition of acceleration:

$$a = \frac{dv}{dt}$$

Thus, for the wiper blade, angular acceleration against time is shown in figure 3.

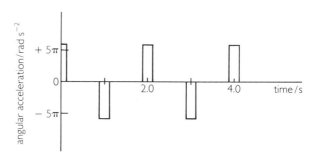

figure 3 Angular acceleration-against-time graph for wiper blade

Two points are worth stressing before we turn our attention to other examples.

1 At every instant, the acceleration of an oscillator must be caused by a resultant force acting on it, according to the law $F = ma$. Likewise, an angular acceleration, as in the example above, must be caused by a resultant torque: in this case exerted by the wiper motor.

2 The wiper-blade example was *not* simple harmonic motion; but the definitions of time-period, frequency and amplitude are common to all types of oscillation – as are the relationships between the three graphs.

Some experiments

In this section we will suggest three methods of determining displacement-against-time graphs by experiment.

1 *Using multiflash or video photography.* This technique would be very suitable for plotting out the wiper-blade graph. A video camera takes 25 frames each second; so one cycle would occupy 50 frames. This is suitable for any oscillator with a period between, say, 0.5 s and 10 s, e.g. a pendulum, a large balance wheel, water slopping in a bath, liquid in a U-tube, a mass on a spring, a marble rolling in a bowl, a tethered trolley, etc.

2 *Using electronic data-logging.* An example of this technique is illustrated in figure 4.

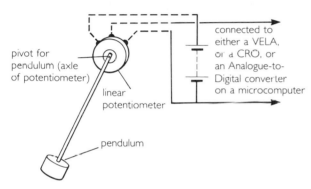

figure 4   An electrical method for obtaining a displacement-against-time graph

The signal going to the data-logger is an analogue voltage which is directly related to the angular displacement of the pendulum.

3 *Drawing a direct graph.* A way for doing this for a tethered trolley is illustrated in figure 5.

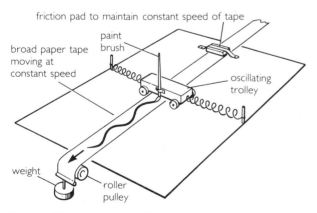

figure 5   Drawing a direct displacement-against-time graph

Sinusoidal oscillations

As mentioned earlier, a large number of oscillators follow very similar patterns of oscillation. Among them are the popular laboratory experiments: pendulums, masses on vertical springs, balance wheels, liquid in U-tubes.

The displacement-against-time graphs for all these oscillators look something like figure 6.

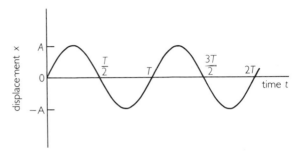

figure 6   Displacement-against-time graph for many oscillators

The shape of this graph is very similar to the mathematical graph of $\sin\theta$ against θ (figure 7). For this reason, oscillators giving a graph like that of figure 6 are sometimes called 'sinusoidal' oscillators.

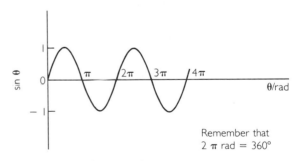

Remember that
2π rad = 360°

figure 7   Graph of $\sin\theta$ against θ

For a particular set of readings of displacement x and time t from one oscillator, we can test to see if its oscillations really are sinusoidal using the following procedure.

1 Plot x against t using our data.

2 Find the period T and the amplitude A from the graph.

3 To obtain a sine function which varies with time, we need to convert t values into angle values. The sine graph repeats at $\theta = 2\pi$ rad; the oscillator repeats at $t = T$ (seconds). Therefore we multiply values of t by $2\pi/T$. This converts t values into angle values so that when t reaches T the angle reaches 2π.

4 Find values for the sines of the angles calculated in step 3.

5 The values of $\sin(2\pi/T)t$ found in step 4 range between 1 and -1. We need to scale them to range between A and $-A$, in order to compare them directly with our experimental x values; so multiply them all by A, to obtain values of the function $A\sin(2\pi/T)t$.

6 Plot these values on the same graph, on top of the x-against-t line. If the two lines are identical, then we have confirmed that the original oscillations were indeed sinusoidal.

In carrying out this confirmation process, we have also derived a mathematical expression for a sinusoidal oscillation:

$$x = A\sin\left(\frac{2\pi}{T}\right)t$$

Points to note:
1. The factor $(2\pi/T)$ is called *pulsatance* – symbol ω. Unit: $1\,\mathrm{rad\,s^{-1}}$.

$$\omega = \frac{2\pi}{T} = 2\pi f$$

Here ω does <u>not</u> represent an angular speed, it is <u>only</u> a constant of the motion.

2. Our 'sinusoidal' oscillations can equally well be described by the mathematical cosine function:

$$x = A\cos\left(\frac{2\pi}{T}\right)t$$

The only difference between this expression for the oscillator and the sine expression is in the position of the oscillator at the moment $t = 0$. This is illustrated in figure 8.

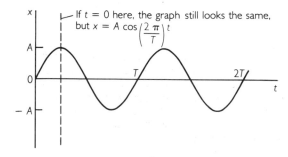

figure 8 Sine oscillations and cosine oscillations have the same graph

From now on, we will use the equation

$$x = A\cos\omega t$$

as the description of a 'sinusoidal' oscillator.

Sinusoidal oscillations and circular motion

There is an interesting link between an object moving in a circle and one oscillating sinusoidally. We can observe this link using the apparatus of figure 9.

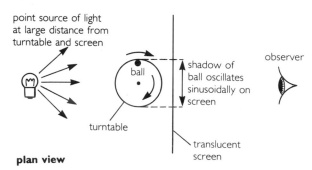

plan view

figure 9 A ball moving in a circle appears from the side to be oscillating sinusoidally

If we hang a pendulum above the turntable, then it is possible to adjust the apparatus so that one complete oscillation on the pendulum takes exactly the same time as one rotation of the turntable. If we then start them off at the same point, we observe that the two shadows, of the ball and the pendulum, move in precisely the same manner. Now a pendulum does move sinusoidally (see page 84); therefore the shadow of the ball does so as well.

Phase differences

The term *phase* refers to the position of an oscillator within its cycle of oscillation. A *phase difference* occurs when two oscillators which have identical or very similar time-periods are, at a given instant, at different positions within their cycles. We consider what this means in the examples below.

The two oscillators illustrated in figure 10 are in phase: the maximum displacements in each direction occur at exactly the same time.

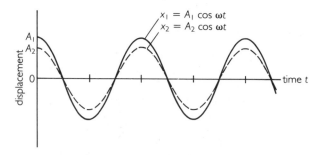

figure 10 Two oscillators in phase

Two oscillators which are exactly out of phase (in 'antiphase') are illustrated in figure 11.

Two oscillators in antiphase both move through zero displacement at the same moment; but they are travelling in opposite directions, so that when their displacements are maximum they are on opposite sides of zero.

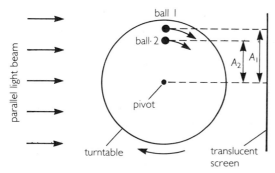

$$x_1 = A_1 \cos \omega t$$

$$x_2 = -A_2 \cos \omega t$$

figure 11   Two oscillators in antiphase

Mathematically, a phase difference between two oscillators is usually expressed as an angle, ϕ. Then the expressions for the two oscillators are written

$$x_1 = A_1 \cos \omega t$$

$$x_2 = A_2 \cos(\omega t + \phi)$$

We can visualise the meaning of this angle by using the experiment of figure 9 again.

1   Two oscillators in phase (figure 10) would appear on the screen if two balls were fixed to the turntable as shown in figure 12.

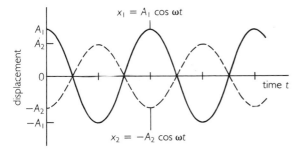

figure 12   Shadows in phase

2   We will see two shadows oscillating in antiphase on the screen (figure 11) if the balls are fixed to the turntable as in figure 13.

3   For a general phase difference of ϕ, we should fix the balls to the turntable so that there is an angle of ϕ between them at the centre of the turntable. Figure 14 illustrates this.

Note that in the antiphase case, figures 11 and 13, this phase difference ϕ is 180°, or π radians.

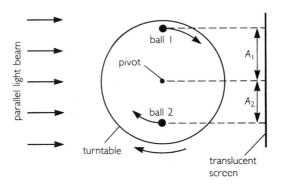

figure 13   Shadows in antiphase

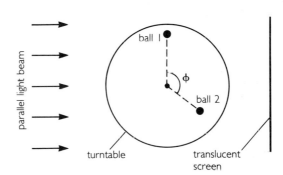

figure 14   Phase difference of ϕ

Velocity and acceleration of a sinusoidal oscillator

If we have a displacement-against-time (x/t) graph for an oscillator, then we can deduce its velocity and acceleration graphs. The velocity at any moment is the gradient of the x/t graph ($v = \mathrm{d}x/\mathrm{d}t$); the acceleration is the gradient of the v/t graph ($a = \mathrm{d}v/\mathrm{d}t$).

Let us now consider carefully one cycle of the x/t graph. Figure 15 considers the gradient at different moments, so that we can build up the v/t graph.

The graph we end up with (figure 15(d)) for v against t looks like an upside-down sine graph, with the same time period T as the x/t graph. The mathematics on page 84 confirm this. We can thus write an expression for v:

$$v = -v_0 \sin \omega t$$

v_0 is the maximum value of v in either direction, in other words the maximum gradient of the x/t graph. Figure 15(a) shows that this maximum gradient is given by

$$v_0 = \frac{A}{T/2\pi} = A\omega$$

This result is also confirmed by the maths on page 84.

The complete expression for v now becomes

$$v = -A\omega\sin\omega t$$

The acceleration-against-time (a/t) graph can be deduced from the v/t graph in exactly the same way. See figure 16.

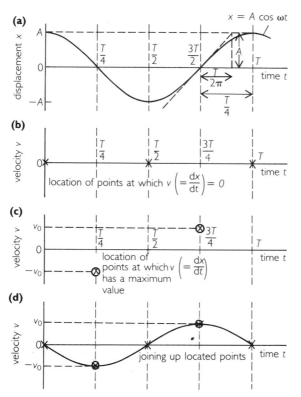

(a)

(b)

(c)

(d)

figure 15 Deduction of v/t graph from x/t graph

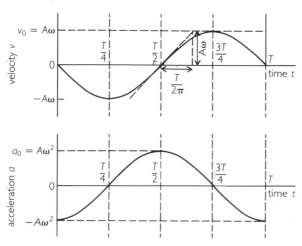

figure 16 Deduction of a/t graph from v/t graph

The graph for a is an upside-down cosine graph. So the mathematical expression for a is

$$a = -a_0\cos\omega t$$

A similar measurement of the maximum gradient of the v/t graph gives

$$a_0 = A\omega^2$$

(confirmed mathematically on page 84), so the final expression for a is

$$a = -A\omega^2\cos\omega t$$

Simple harmonic oscillation

After all this rather abstract graph work, it is time we returned to something concrete. We will take as our standard example the tethered trolley illustrated in figure 17.

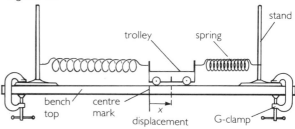

figure 17 A tethered-trolley system

Every kind of oscillator has to possess two specific features in order for oscillations to occur. These are:

☐ some *restoring force*, which acts on it when it is displaced from its central rest position, and is directed towards the central position;

☐ some *inertia*, which makes it pass through the central position when it arrives there, so that it acquires displacement in the other direction.

In the tethered-trolley example, the restoring force is supplied by the springs. Provided neither of them is ever slack, and both operate within their individual Hooke's law regions, then their combined effect on the trolley also obeys Hooke's law: the restoring force F on the trolley is proportional to its displacement x, with constant of proportionality k:

$$F = kx$$

The inertia is simply the mass of the trolley, m.

Now let us develop an equation of motion for the trolley, that is, an equation based on the application of the law $F = ma$. If the trolley is displaced a distance x to the right, then the restoring force has magnitude kx, and acts to the left. So in terms of the convention positive-to-the-right, the force F in $F = ma$ has to be put in as $-kx$. The equation of motion is thus

$$-kx = ma$$

or

$$a = -\frac{k}{m}x$$

Now let us return for a moment to the equations we developed for sinusoidal oscillators in general:

$$x = A\cos\omega t \qquad\text{(i)}$$
$$v = -A\omega\sin\omega t \qquad\text{(ii)}$$
$$a = -A\omega^2\cos\omega t \qquad\text{(iii)}$$

Equations (i) and (iii) combine together to give

$$a = -\omega^2 x$$

Now an acceleration requires a resultant force ($F = ma$ again). So what this equation is saying is: a sinusoidal oscillator has an acceleration, and therefore resultant force, proportional to its displacement, and in the opposite direction, i.e. restoring it to its central position. The converse is also true: an oscillator which has

$$\text{restoring force} \propto \text{displacement}$$

or $\qquad F \text{ (hence } a) \propto -x$

has sinusoidal motion.

This last equation is the definition of *simple harmonic motion* (s.h.m.). As well as being sinusoidal, this type of oscillator has one other very interesting feature: its time-period T is not dependent on its amplitude A.

It is important to realise that these equations represent a simplified model of reality. Real situations almost invariably involve more complex equations: terms have to be added to the equation of motion for, for example, resistive forces (constant axle friction, air resistance proportional to v or v^2, etc.), effective mass of springs, failure of restoring forces to obey Hooke's law for large displacements and so on.

Finally, let us compare these two equations:

$$a = -\omega^2 x \qquad \text{for s.h.m. in general;}$$
$$a = -\frac{k}{m}x \qquad \text{for our tethered trolley.}$$

Clearly, the tethered trolley obeys the general requirement for s.h.m., with

$$\omega^2 = \frac{k}{m}$$

If you recall from page 80 that $\omega = 2\pi/T$, then the time-period for the tethered trolley is given by

$$T = \frac{2\pi}{\omega} = 2\pi\sqrt{\frac{m}{k}}$$

As always in physics, we ask ourselves: does this seem to tally with our experience? A more massive trolley (larger m) would have more inertia, take longer to accelerate, and hence have larger T; stiffer springs (larger k) would accelerate the trolley faster, and lead to a smaller T. Thus the equation does seem sensible. The next section explores this relationship a little further.

Changing the time-period, T, of oscillation

1. *Mass increase.* Suppose that for some particular mass increase the time-period is doubled. Assume the amplitude remains the same. Then, since the trolley travels the *same distance* as before, it must be travelling at *half the average speed.* That is, the trolley reaches half of its original maximum speed in *double the original amount of time*, therefore it must be accelerating at a *quarter of its original acceleration.* This can be achieved by quadrupling the mass.

Thus, *quadrupling the mass doubles the time period.*
This fits in with $T \propto \sqrt{m}$.

2. *Stiffer springs.* The argument used above, that the doubling of the time-period is due to the acceleration being quartered, still applies. Springs which are four times weaker would have this effect. Equally well, the time-period will be halved if springs which are four times stiffer are used, or four parallel springs on either side of the trolley, as illustrated in figure 18, are used.

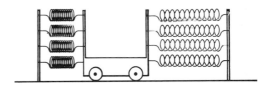

figure 18

Therefore, *quadrupling the stiffness of the springs halves the time period.*

This fits in with $T \propto \sqrt{\frac{1}{k}}$.

Time-period independent of amplitude

If the trolley is pulled to one side a distance e, and then released, it experiences a net force F which causes it to accelerate. It will take a certain time to go from the extreme displacement to the centre of the motion, i.e. to perform one quarter of an oscillation. Now suppose that the initial displacement is doubled to 2e. The initial net force is now $2F$, which produces a new acceleration of *double* the original value. If the average acceleration is doubled then in an equal time interval the trolley will acquire twice as much speed. Therefore, in order to return to the equilibrium position the trolley has *twice* the original distance to travel, but it does so at *twice* the average speed. Clearly, the trolley will complete one quarter of an oscillation in exactly the same amount of time, whether the original displacement is *e* or 2*e*.

Thus, *the time-period is independent of the amplitude of oscillation.*

This feature of s.h.m. is particularly useful for the creating of clocks and watches. The time-keeping part of the system is almost always a simple harmonic oscillator: pendulum, balance wheel or quartz crystal.

Mathematical treatment of the tethered trolley

Let us start from the equation of motion of the trolley, as derived before:

$$a = -\frac{k}{m}x$$

Since
$$a = \frac{dv}{dt} = \frac{d}{dt}\left(\frac{dx}{dt}\right) = \frac{d^2x}{dt^2},$$

this equation is a differential equation with variables *x* and *t*:

$$\frac{d^2x}{dt^2} = -\frac{k}{m}x \qquad \text{(i)}$$

A general solution to this equation is

$$x = P\cos(\omega t + Q)$$

where *P*, ω and *Q* are constants.

If *t* = 0 is taken to be when *x* has its maximum value to the right, then *Q* = 0 and *P* = *A*, the maximum value of displacement, since cos θ has its maximum value when θ = 0.

Thus the solution of equation (i) which fits our needs is our old friend

$$x = A\cos\omega t \qquad \text{(ii)}$$

Differentiating this gives velocity *v*:

$$v = \frac{dx}{dt} = -A\omega\sin\omega t \qquad \text{(iii)}$$

Differentiating this gives acceleration *a*:

$$a = \frac{dv}{dt} = -A\omega^2\cos\omega t \qquad \text{(iv)}$$

This is also d^2x/dt^2, so substituting equations (ii) and (iv) into equation (i) gives

$$-A\omega^2\cos\omega t = -\frac{k}{m}.A\cos\omega t$$

$$\Rightarrow \qquad \omega^2 = \frac{k}{m} \quad \text{as before}$$

The simple pendulum (small oscillations)

The simple pendulum is a common example of s.h.m. Figure 19 shows a pendulum displaced a small distance to the left of its equilibrium position. The distance that the heavy bob on the end of the light, rigid rod will move through in returning to its equilibrium position is marked *s* on the diagram. It is the distance along the arc. The resultant force acting on the bob, which causes it to move along this arc is the tangential component of the gravitational pull on the bob, namely $F = W\sin\theta$. The pull of the rod on the bob has no component in the direction in which the bob moves. We shall assume that the pendulum is friction-free and that it will go on oscillating with the same amplitude for ever.

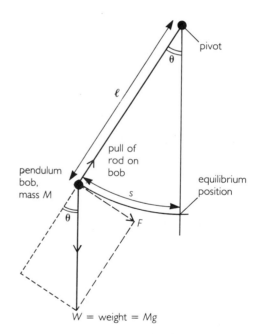

figure 19 A simple pendulum

Therefore, we can write

resultant force = mass × acceleration

But   resultant force = $W\sin\theta$

$$= Mg\sin\theta$$

$$\Rightarrow \qquad Mg\sin\theta = M\times\left(-\frac{d^2s}{dt^2}\right) \qquad \text{(i)}$$

The acceleration is written as $(-(d^2s/dt^2))$ because it is making the displacement smaller. We have three different variables in equation (i) above. θ and *s* are related, therefore equation (i) can be simplified. Working in radians, the angle θ is defined as $\theta = s/\ell$. Also, if we limit the oscillations of the pendulum to small angles, such that $\theta < 10°$, then to a very close approximation $\sin \theta \approx \theta$. This can be seen from figure 20, where the straight line plot of $y = \theta$ and the graph of $y = \sin \theta$ are almost identical for angles of less than 10°. There is approximately 0.5% disagreement at 10° (0.17 rad).

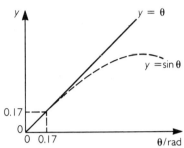

figure 20

Therefore, equation (i) can be simplified to the following, for small oscillations:

$$g\left(\frac{s}{\ell}\right) = -\frac{d^2s}{dt^2}$$

$$\Rightarrow \qquad \frac{d^2s}{dt^2} = -\left(\frac{g}{\ell}\right)s \qquad \text{(ii)}$$

This should be compared with the definition of s.h.m., namely

$$a = \frac{d^2x}{dt^2} = -\omega^2 x$$

Since the acceleration of the bob is proportional to the displacement s along the arc from the equilibrium position, the motion of the bob must be simple harmonic motion. From page 80, the pulsatance, ω, is given by

$$\omega = \frac{2\pi}{T}$$

From equation (ii) we see that for a simple pendulum $\omega^2 = (g/\ell)$. Hence, for a simple pendulum we can conclude that for small oscillations the motion is s.h.m. with a time-period of

$$T = 2\pi\sqrt{\frac{\ell}{g}} \qquad \text{(iii)}$$

Example (a) Prove that a mass, M, suspended from a fixed point by a helical spring, which obeys Hooke's law, undergoes simple harmonic motion when it is displaced vertically from its equilibrium position.
(b) If $M = 0.10$ kg and the extension of the spring is 4.0 cm when the mass is in its equilibrium position, determine the time period, T, for the oscillations.

(a) Figure 21 shows the unloaded spring (A), the loaded spring in equilibrium (B), and the displaced spring during oscillation (C).
    Consider the mass in figure 21(C).

$$\text{restoring force} = \text{mass} \times \text{acceleration}$$

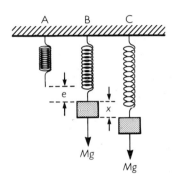

figure 21

i.e. pull of spring − weight = mass × acceleration

i.e. $\Rightarrow \qquad k(x+e) - Mg = -Ma$

But in the equilibrium position, figure 21(B), the pull of the spring, ke, is equal to the weight, Mg. Therefore,

$$kx + Mg - Mg = -Ma$$

$$\Rightarrow \qquad kx = -Ma$$

The negative sign indicates that the direction of the acceleration is opposite to that of the displacement.

Therefore, acceleration, $a = -\dfrac{k}{M} \cdot x$.

Compare this with $a = -\omega^2 x$ which is the definition of s.h.m. Since k and M are both constant it means that the mass performs s.h.m.
Also it shows that

$$\omega^2 = \frac{k}{M}$$

$$\Rightarrow \qquad T = 2\pi\sqrt{\frac{M}{k}}$$

(b) We know that the spring constant,

$$k = \frac{1.0\,\text{N}}{0.040\,\text{m}}$$

$$= 25\,\text{Nm}^{-1}$$

Substituting the values gives

$$T = 2\pi\sqrt{\frac{(0.10\,\text{kg})}{(25\,\text{Nm}^{-1})}}$$

$$= 0.40\,\text{s}$$

Energy in s.h.m. for a friction-free system

Let us now consider how the energy of an oscillating system varies with the displacement of the object vibrating. We will use the notation of our standard example, the tethered-trolley oscillator.

Potential energy is stored in a simple harmonic oscillating system as a result of the work done against the restoring force. This energy could be stored, for example, in a stretched spring, or it could be gravitational potential energy; or it could be a combination of both.

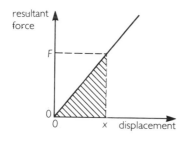

figure 22 Potential energy stored is the area under the F/x graph

work done = average force × distance

$\Rightarrow$ potential energy stored = work done

$$= \tfrac{1}{2} Fx \quad \text{(area under } F/x \text{ graph).}$$

However, the resultant restoring force is $F = kx$, where k is the force per unit displacement.

i.e. potential energy: $E_p = \tfrac{1}{2} kx^2$ (i)

Conservation of energy

$$E_k + E_p = \text{constant}$$
$$= \text{maximum } E_p$$

$\Rightarrow$ $\tfrac{1}{2} mv^2 + \tfrac{1}{2} kx^2 = \tfrac{1}{2} kA^2$

$\Rightarrow$ $v^2 = \dfrac{k}{m}\{A^2 - x^2\}$

$\Rightarrow$ velocity $v = \pm\sqrt{\dfrac{k}{m}(A^2 - x^2)}$

$$= \pm \omega \sqrt{(A^2 - x^2)} \quad \text{(ii)}$$

where $\omega = 2\pi f = \dfrac{2\pi}{T} = \sqrt{\dfrac{k}{m}}.$

Kinetic energy ($\tfrac{1}{2} mv^2$) at a displacement x, by using equation (ii) can be written as:

kinetic energy $E_k = \tfrac{1}{2} m\omega^2 (A^2 - x^2)$ (iii)

Total energy is the sum of the kinetic energy and the potential energy, or can be written as the maximum potential energy.

total energy of oscillation $= \tfrac{1}{2} kA^2$ (iv)

Figure 23 shows how E_p and E_k of the oscillator vary with the oscillator's displacement.

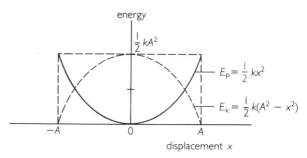

figure 23 Graph of potential energy and kinetic energy against displacement

It is also interesting to consider how potential energy and kinetic energy vary with time t.

$$E_p = \tfrac{1}{2} kx^2 \quad = \tfrac{1}{2} kA^2 \cos^2 \omega t$$
$$E_k = \tfrac{1}{2} mv^2 \quad = \tfrac{1}{2} m . A^2 \omega^2 \sin^2 \omega t$$
$$= \tfrac{1}{2} kA^2 \sin^2 \omega t$$

Graphs of these relationships are shown in figure 24.

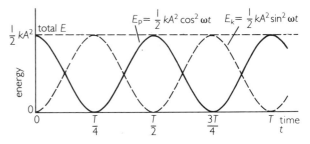

figure 24 Graphs of energy against time

The effect of damping (friction) in s.h.m.

A real oscillator is invariably subject to resistive forces of some sort, in addition to the restoring force which makes it oscillate. Consider again, for example, the experiment for tracing out directly the displacement-against-time graph for a tethered trolley – figure 5.

The resulting paper tape will have a wavy line drawn along it by the paint brush attached to the trolley. Figure 25 depicts a typical result. The points to note about this tape are that (i) the amplitude is decreasing, and (ii) the time-period is approximately constant.

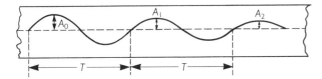

figure 25

The frictional forces acting in the mechanical, oscillating system produce a resistance to motion, which causes energy dissipation and a decay of the amplitude of oscillation. An oscillation whose amplitude decays is known as a *damped* oscillation.

Decreasing amplitude in damped s.h.m.

There are many different types of frictional force, such as axle friction, wind resistance, etc. Each produces a different form of decrease in amplitude. The oscillating trolley and springs lose energy to

☐ the surrounding air, which gains molecular kinetic energy as it is stirred up by the passing trolley;

☐ the axles of the trolley, which become hot due to the frictional forces;

☐ the twisting metal of the springs, which warms up.

All these energy losses will be greatest when the trolley is moving most rapidly, that is, during the first few cycles of its motion. Therefore, the reduction in amplitude of our trolley will become progressively smaller as the oscillations decrease in amplitude.

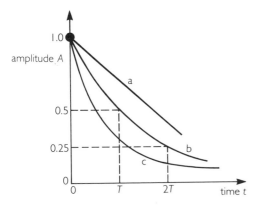

figure 26

Figure 26 illustrates some possible forms of amplitude reduction for different oscillating systems. It should be noted that all decrease rapidly to begin with. Mathematically, systems in which the amplitude varies with time according to the curve labelled (b) in figure 26 are easiest to analyse, i.e. systems in which the amplitude decays exponentially.

Exponential decay of amplitude

In some cases the amplitude decays exponentially as illustrated in figure 27. This happens when the frictional force is proportional to the velocity.

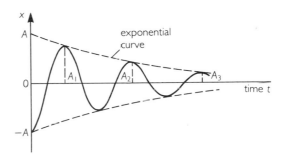

figure 27   Exponential decay of amplitude

The amplitude in figure 27 can be expressed as

$$\text{amplitude} = A e^{-\lambda t}$$

Also, since the time-period T is constant, the oscillatory aspect of this motion can still be expressed as $\cos(\omega t)$ where $\omega = 2\pi/T$. Hence, the full curve drawn in figure 27 must have the equation

$$x = -A e^{-\lambda t} \cos \omega t$$

We can make a simple test for this exponential property, by examining the successive maximum displacements – A_1, A_2, A_3 in figure 27.

If
$$\frac{A_1}{A_2} = \frac{A_2}{A_3}$$

then the decay of amplitude is indeed exponential.

Summary

1 There are many examples of oscillations, from all branches of physics.

2 A displacement-against-time graph contains much of the information about any particular oscillation.

3
$$f = \frac{1}{T}$$

f = frequency T = time period

4 Amplitude A is the maximum displacement of the oscillator from its central (or rest) position.

5 Many oscillations are approximately sinusoidal:

$$x = A \cos \omega t$$
$$v = -A\omega \sin \omega t$$
$$a = -A\omega^2 \cos \omega t = -\omega^2 x$$

where pulsatance $\omega = \dfrac{2\pi}{T} = 2\pi f$

6 Graphs of x, v, and a appear as follows, for a trolley released from its maximum displacement:

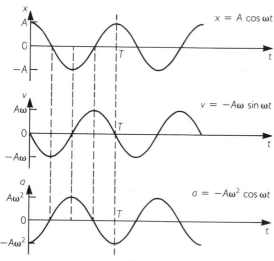

figure 28

7 The shadow of an object moving in a circle oscillates sinusoidally.

8 If two oscillators are described by these displacements:

$$x_1 = A_1 \cos \omega t$$
$$x_2 = A_2 \cos (\omega t + \phi)$$

then oscillator 2 leads oscillator 1 with a phase difference ϕ.

9 The equation $a = -\omega^2 x$, or the statement 'restoring force ∝ displacement' define simple harmonic motion.

10 A tethered trolley has the equation of motion

$$a = -\frac{k}{m}x$$

It thus oscillates with simple harmonic (sinusoidal) motion, with

$$\omega = \sqrt{\frac{k}{m}}$$

and
$$T = 2\pi\sqrt{\frac{m}{k}}$$

11 This table summarises the changes in x, F, v, and a for a tethered trolley.

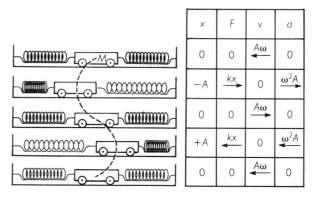

		x	F	v	a
		0	0	$A\omega$ ←	0
		$-A$	kx →	0	$\omega^2 A$
		0	0	$A\omega$ →	0
		$+A$	kx ←	0	$\omega^2 A$
		0	0	$A\omega$ ←	0

figure 29

12 S.h.m. has time-period T independent of amplitude A.

13 For a simple pendulum,

$$T = 2\pi\sqrt{\frac{\ell}{g}}$$

14 Potential energy of a tethered trolley system (stored in the springs) is given by
$$E_p = \tfrac{1}{2} kA^2 \cos^2 \omega t$$

15 Kinetic energy for the trolley is: $E_k = \tfrac{1}{2} kA^2 \sin^2 \omega t$

16 Total energy is $\tfrac{1}{2} kA^2$.

17 Damping causes energy dissipation; the amplitude of oscillations decreases, sometimes exponentially.

Questions

1 Consider a trolley tethered between two springs. When the trolley is in the undisturbed position each spring is stretched by 4.0 cm. The following questions apply to the case when the trolley is released from a displacement of 3.0 cm to the left of the equilibrium position. In the calculations take the mass of the trolley to be 0.50 kg and the force constant of each spring as $2.5 \, \text{Nm}^{-1}$.

Displacement-against-time graph (approximate graphical solution)

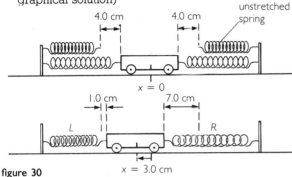

figure 30

The procedure followed here is to *assume that the velocity of the trolley is constant for short intervals of time*, and that at the end of each time interval it changes suddenly and discontinuously to its new value.

When the trolley is initially released its speed will start from zero, therefore the speed will be assumed to stay constant at zero during the first time interval. This is illustrated by the line AB drawn on the graph paper in figure 31.

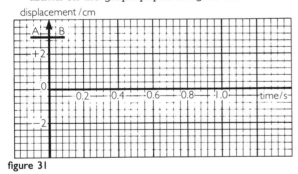

figure 31

The length of the time interval is 0.20 s and it is centred on the time $t = 0$. The next 0.20 s interval centres around the time 0.20 s from the start, and each successive interval will centre around 0.40 s, 0.60 s, 0.80 s etc.

a *step 1*: Calculate the *resultant* force acting on the trolley at the beginning of the next time interval (i.e. at $t = 0.10$ s, which is marked B on the graph paper).

step 2: Calculate the corresponding acceleration of the trolley.

step 3: Assuming that the acceleration calculated in step 2 is constant during the next time interval, calculate the speed of the trolley at the end of this time interval.

step 4: Calculate how far the trolley will travel during this next time interval, i.e. during the next 0.20 s.

step 5: Mark, on the graph, the new displacement of the trolley at the end of this time interval (i.e. at $t = 0.30$ s).

b Repeat steps 1 to 5 for the next time interval, i.e. 0.30 to 0.50 s.

c Repeat steps 1 to 5 for the next three time intervals, i.e. up to 1.10 s.

d Join the points on the graph with straight lines.

2 Repeat question 1, but use time intervals of 0.10 s.

a In the above two cases, compare the times taken by the trolley to reach a displacement of zero for the first time.

b Compare these two values with the time taken by an actual friction-free trolley of the same mass and with the same springs.

3 Four long, similar springs are attached to either end of a freely-running, friction-free trolley, the other ends being firmly fixed as shown in figure 32.

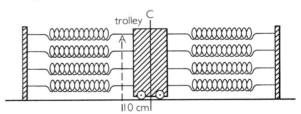

figure 32

Initially the trolley is displaced 10 cm to the left of its rest position, C, and then released.

a Describe, with some detail, how the trolley moves after it is released.

b If the trolley takes 0.5 s to travel from A to C, what is the time taken by the trolley in travelling from A back to A?

c Where is the trolley when it is travelling fastest to the right?

d What is the speed of the trolley 5 seconds after it was released?

e Discuss the energy changes which occur during the first 2 seconds of motion.

f In what way would the motion of the trolley be different if the upper six springs were removed and the trolley again released from A? In answering this question discuss what happens to (i) the initial accelerating force, (ii) the initial potential energy stored in the springs, (iii) the maximum speed of the trolley, (iv) the time period for one oscillation.

g In what way would the motion of the trolley be different if the mass of the trolley were to be halved, whilst still using eight springs? [Hint: Refer to items (i), (ii), (iii), and (iv) as in part (f).]

h In what way would the motion of the trolley be different from the original motion if the release point were to be half-way between A and C when using eight springs as in the diagram above. [Hint: Use (i), (ii), (iii) and (iv) as in (f).]

i Where is the trolley when it is accelerating most rapidly to the left?

j The motion of the trolley is called simple harmonic motion. Describe the characteristics of this motion.

k In what way will the real motion of a trolley differ from that you have described in part (j)?

4 All the springs in figure 33 are identical.
   a What is the value of L? Give your answer in centimetres.
   b What is the value of M? Give your answer in kilograms.

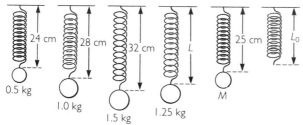

figure 33(a)

c What is the value of L_0?
d Whose law have you assumed in answering parts (a), (b) and (c) above?

Assuming that the springs in the following diagrams are identical to the ones in figure 33, what is the value of the force in each of the following diagrams?

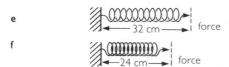

e

f

figure 33(b) and (c)

g What is the value of the resultant force acting on the trolley in the following diagram? (Give both magnitude and direction.)

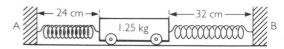

figure 33(d)

h What is the acceleration of the trolley shown in part (g)? (Magnitude and direction)

i In part (g), what was the displacement of the trolley from its equilibrium position?

j Use the information given in figure 33(a) to work out the spring constant, k.

k What is the value of E_p stored in the springs in the case shown in figure 33(d)?

l If figure 33(d) represents the starting point of the motion, i.e. the point of release, what will be the value of E_k when the trolley passes through its equilibrium position?

m What is the velocity of the trolley when it passes through its equilibrium position?

n Calculate the time period T of oscillation for the above trolley.

o In fact the motion of such a trolley would not be pure s.h.m., but would be damped s.h.m. Discuss in some detail where the lost energy would go to.

5 Figure 34 illustrates a block of mass M tethered between two fixed supports by springs. It is resting on a friction-free surface.

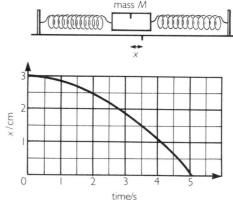

figure 34

The block is pulled to one side and then released. The graph shows how the displacement, from the central equilibrium position, varies with time.

a Copy this curve onto a sheet of graph paper. Then, using different coloured pencils, draw on the same sheet of graph paper the curves which would be obtained if the same springs were used but the mass were increased first to $2M$, then to $3M$.

b The following nine graphs show various relationships between the time period T and the mass M. Which could be correct for the above experimental situation, that is, for masses tethered between springs?

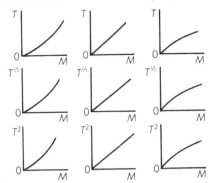

figure 35

6 Figure 36 illustrates a block of mass M tethered between two fixed supports. It is resting on a friction-free surface.

The block is pulled to one side and then released. The graph shows how the displacement of the block, from the central equilibrium position, varies with time.

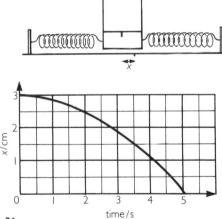

figure 36

a Copy this curve onto a sheet of graph paper. Then, using different coloured pencils, draw on the same axes the curves which would be obtained if the same mass were used, but the number of springs (identical) were first doubled and then trebled, as illustrated in figure 37.

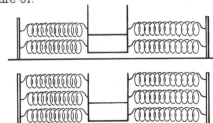

figure 37

b The following twelve graphs show various relationships between the time period T and the number of springs n in parallel on each side of the block. Which could be correct for the above experimental situation, that is, for masses tethered between springs?

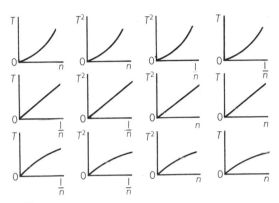

figure 38

7 A mass on the end of a spring vibrates through an amplitude of 5.0 cm with a time-period of 0.20 s. Calculate:
   a the frequency f of the vibrations;
   b the pulsatance ω of the motion;
   c the maximum velocity of the mass;
   d the maximum acceleration of the mass;
   e the velocity of the mass when it is 2.0 cm from the centre;
   f the acceleration of the mass when it is 2.0 cm from the centre.

8 The end of a hacksaw-blade oscillates with a frequency of 10 Hz and an amplitude of 3.0 cm. Calculate:
   a the time period of the oscillations;
   b the pulsatance ω of the motion;
   c the velocity of the blade-end 0.010 s after it passes through the mid-point;
   d the acceleration of the blade-end 0.020 s after it leaves its greatest displacement.

9 Each prong of a tuning fork, frequency 340 Hz, vibrates with an amplitude of 3.0 mm. Calculate:
   a the maximum velocity of each prong;
   b the maximum acceleration of each prong.

10 A mass of 100 g executes s.h.m. as it oscillates on the end of a spiral spring. It moves with an amplitude of 4.0 cm and takes 0.70 s to complete one oscillation. Calculate:
   a the pulsatance ω of the mass's motion;
   b the maximum kinetic energy of the mass;
   c the force constant k of the spring;
   d the kinetic energy of the mass 0.10 s after it passes through the centre of motion;

e the elastic potential energy of the spring when the mass is 1.0 cm above its lowest position.

f Sketch two graphs on the same axes to illustrate how the kinetic energy and elastic potential energy of the system vary with time. Put numerical values on your axes.

11 A spiral spring of force constant k equal to $40\,\mathrm{N\,m^{-1}}$ supports a mass of 200 g. The mass is pulled down 3.0 cm below its equilibrium position and released. Calculate:

a the pulsatance ω of the motion;

b the time-period of the oscillations;

c the equilibrium extension of the spring when the mass is not oscillating;

d the maximum velocity of the mass during its oscillations.

e sketch a graph to show how the velocity of the mass would vary with time if it were to be released from 7.0 cm below its equilibrium position. Put numerical values on your axes. Would the motion still be s.h.m.?

12 Figure 39 shows a U-tube containing a liquid. The U-tube is of uniform cross-sectional area. The second diagram shows the liquid displaced slightly.

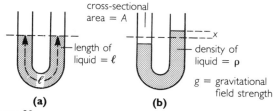

figure 39

a What is the force trying to return the liquid in figure 39(b) to its equilibrium position shown in figure 39(a)? That is, write an expression for the force in terms of A, x, ρ and g.

b Write down an expression for the total mass of liquid in the U-tube in terms of A, ℓ and ρ.

c Use your answers to parts (a) and (b), and the expression

resultant restoring force = mass × acceleration

to show that the liquid oscillates with simple harmonic motion. (In order to do this you will have to assume that the liquid has negligible viscosity, i.e. friction forces within the flowing liquid are zero.)

d Use your expression in part (c) to show that a liquid with negligible viscosity will oscillate with a time period T given by the expression:

$$T = 2\pi\sqrt{\frac{\ell}{2g}}$$

13 Figure 40 shows a weighted fishing float floating vertically in water. The second diagram shows the float displaced vertically downwards by a

small amount x. The float has a uniform cross-sectional area A and a mass M. Take the density of the water to be ρ.

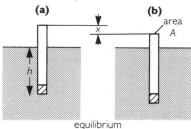

figure 40

a Write down an expression, in terms of A, x, ρ and g, for the resultant upward force on the float when it is displaced downwards a small distance x as shown in figure 40(b). [Hint: Archimedes' Principle states that 'upthrust equals the weight of fluid displaced'.]

b Use the expression 'resultant restoring force = mass × acceleration' and your answer to part (a) to show that the float oscillates with simple harmonic motion. (In order to do this you will have to assume that the viscosity of water is zero.)

c Use the expression that you obtained in part (b) to show that such a fishing float will oscillate with a time period T given by the expression

$$T = 2\pi\sqrt{\frac{h}{g}}$$

d Discuss, and explain, the effects produced by the non-negligible viscosity of the water.

e Put some reasonable values into your expression in part (c) and obtain a numerical value for the time period T.

14 Figure 41 represents a small mass and a large mass attached to either end of a spring. Figure (a) shows the situation when the spring is compressed by applying equal and opposite forces to the two masses. When the forces are removed the spring will cause the masses to oscillate backwards and forwards about their mean positions. Figure (b) illustrates the situation when the spring has maximum extension.

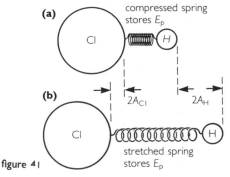

figure 41

This system of masses and spring is used as a model for understanding the behaviour of diatomic molecules such as hydrogen chloride (HCl).

a Work out the relative amplitudes of the oscillations of the hydrogen and chlorine atoms in the HCl molecule, given that the atomic mass of the hydrogen atom is 1 and the atomic mass of the chlorine atom is 37.

b HCl molecules absorb strongly infrared radiation of wavelength 33×10^{-7} m. Given that the mass of a hydrogen atom is 1.7×10^{-27} kg, work out the 'spring constant k' for the electronic bond between the two atoms.

c Given that the energy of infrared radiation is quantised, as is all electromagnetic radiation, calculate the amplitudes of the oscillating hydrogen and chlorine atoms. (You will find the equation $E = hf$ useful, where E = energy of photon, h = the Planck constant and f = frequency of oscillations. $h = 6.6 \times 10^{-34}$ Js.)

15 Figure 42 represents the front wheel and suspension spring of a motor-car. Figure (a) shows the situation when the spring is compressed, and figure (b) depicts the case when the spring is stretched.

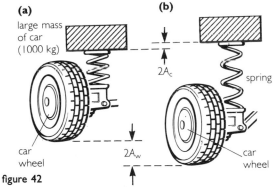

figure 42

a an 80 kg man sitting over one of the front wheel arches causes that arch to depress by 2 cm. Use this information to determine the spring constant k for the suspension spring.

b Imagine that for this particular car the tyre, wheel and axle have an effective mass of 120 kg. Use this information and your answer to part (a) to calculate the natural frequency with which the wheel will oscillate, when disturbed.

c Consider what happens when a front wheel has not been balanced properly. That is, when there is effectively a small mass off-centre on the rotating wheel. If the above front wheel has a diameter of 60 cm, at what speed will the car be travelling when the driver feels violent vibrations in the steering wheel? What is this process known as? *Explain* what the driver would observe if he were to increase his speed up to 100 mph.

16 The extension of a spiral spring which obeys Hooke's law is directly proportional to the extending load. A mass M attached to the end of the spring produces an extension e as illustrated in figure 43. Figure (b) shows the equilibrium position where the tension in the spring is balanced by the weight of the mass M. If the spring constant is k then

$$Mg = ke$$

Suppose that the mass M is now pulled down a further distance x, and then released, as shown in figure (c).

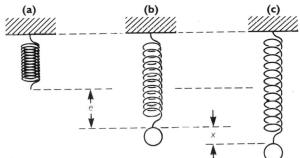

figure 43

a Show that the mass oscillates with simple harmonic motion, with a time period T given by the equation

$$T = 2\pi \sqrt{\frac{M}{k}}$$

b In science it is most useful to plot experimental data in such a way as to obtain a straight-line graph. Consider an experiment in which the mass M is varied and the corresponding time periods T are taken. If the experimenter decides to plot the mass M as his x-axis, what must he plot for the y-axis if he is to obtain a straight line (assuming that the expression given in part (a) is correct)?

c The following data was obtained by a student who did the above experiment.

Mass M/g	100	200	300	400	500
Time for 10 0scillations/s	7.8	10	11.8	13.4	14.8

Use this data to plot the graph that you decided upon in part (b). From your graph, determine the value of the spring constant k. You will have discovered that your graph does not go through the origin. In other words, the student's data does not agree with the theoretical equation given in part (a). This should not surprise you as the mass of the spring was completely ignored in part (a). Use your graph to determine the effective mass m of the spring. The actual mass of the spring was 150 g; how does this value compare with the effective mass?

17 These questions are about the natural walking or running action of a human being.

a Leg and arm movements both involve, among other things, effective rotation about a vertical axis. Explain why in the natural running action a person moves forward his left arm at the same time as his right leg, and vice versa. (Try to discuss this in terms of appropriate physical quantities.)

b Whether walking or running, the *time* to execute one step is similar (the time is less for running, but the change is by a much smaller factor than the change in speed). Why should there be a natural time for a person's leg to swing forward? By treating it as half a swing of a pendulum, and making suitable estimates for quantities, estimate the likely natural time for one pace. Is this time likely to be nearer to the natural time for walking or running? Explain.

CHAPTER 8

Resonance and standing waves

Introduction

We start with driven oscillations. The response of a large, slow-moving system is analysed. We consider the amplitude and phase of the system, and how the response varies with damping. Then we study the resonances (standing waves) on wires. This leads on to musical instruments. First of all we consider string instruments, and then we extend the discussion to wind instruments.

GCSE knowledge

We assume you know

☐ that oscillations are repeated motion;

☐ that the *frequency* of an oscillation or wave is the number of oscillations performed per unit time;

☐ that the distance between successive crests in a wave-train is called the *wavelength*, λ;

☐ that for waves

$$\text{speed} = \text{wavelength} \times \text{frequency},$$

i.e. $v = \lambda f$;

☐ that the maximum displacement in an oscillation is called the *amplitude*.

Driven oscillations

The systems we shall be considering in this section are systems that perform damped simple harmonic motion when disturbed. We shall start with a system which oscillates so slowly that it is easy to see all the points we shall be making. The apparatus is illustrated in figure 1.

First of all let us consider what happens when the drive wheel is turned *very slowly* by hand. The string attached to the pin will pull the spring and plate up and down by an amount equal to the radius at which the pin is rotating. A piece of paper stuck to the string at A will help us to see what is happening. When the point A moves up and down very slowly the aluminium plate moves up and down in phase with A, and with the same amplitude as A.

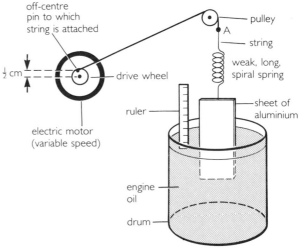

figure 1   Resonance apparatus

Amplitude of driven oscillations

If the electric motor is switched on and its speed is gradually increased, we will see that the amplitude of the oscillations of the plate increases, even though the amplitude of the oscillations at A remains the same. But when the speed of the electric motor exceeds a certain value we will find that further increases in the motor's speed cause the amplitude of the plate's oscillations to decrease. When the electric motor rotates at very high speeds, the point A on the string will be oscillating up and down rapidly but the plate will remain stationary. These results are summarised by the curve in figure 2.

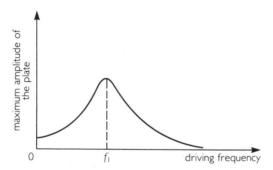

figure 2   Sharpness of resonance

At all driving frequencies there is initially a transient period of behaviour: during this time the amplitude of the plate is not constant, but varies in a manner illustrated typically by figure 3. The amplitude settles down in due course to a steady value, with the plate oscillating at the driving frequency.

We say that the system *resonates* at frequency f_1. That is, the amplitude of the plate oscillations takes its maximum value when the driving frequency equals f_1. f_1 is equal to the frequency at which the plate/spring system oscillates naturally on its own.

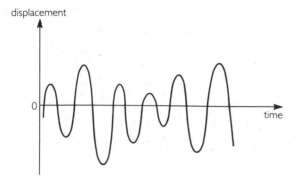

figure 3   Transient period of oscillations

Phase of driven oscillations

If we adjust the speed of the motor to produce resonance, we will observe a phase difference between the motion of point A on the string and the motion of the plate. We will observe that the plate lags behind A by one quarter of a period ($\pi/2$). If the driving frequency (the speed of the electric motor) is increased or decreased we will observe the *phase lag* changing.

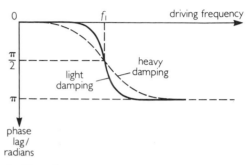

figure 4   Phase lag

The variation of phase lag with applied driving frequency is illustrated in figure 4. The curve labelled *heavy damping* is obtained with engine oil in the drum of figure 1, whereas *light damping* represents what would happen if a less viscous liquid, such as water, were to be used. With light damping the plate responds with a greater amplitude. This is illustrated in figure 5.

Barton's pendulums, figure 6, provide a good demonstration of the variation in phase and amplitude depicted by figures 4 and 5 respectively. The paper cones should be loaded with Plasticine to reduce the effect of damping. Photographs of Barton's pendulums help to illustrate these points. A long-exposure photograph illustrates the maximum amplitudes of each cone, and an instantaneous photograph shows the phase lag; see figure 7.

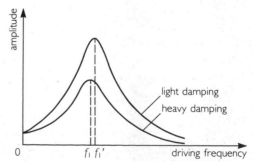

figure 5   Sharpness of resonance; the change in resonant frequency with damping is very small

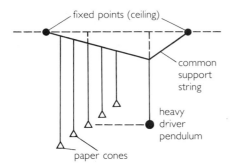

figure 6 Barton's pendulums

Waves and wave motion

A wave carries energy from one point to another without any particles of the intervening medium travelling from the first point to the final point. For example, consider a long straight rope lying across the floor of a classroom. If one end of the rope passes close to a tennis ball, then a wave sent along the rope from the other end will cause the ball to move.

We shall be dealing with transverse and longitudinal waves. A *transverse* wave is one in which the oscillations are *perpendicular* to the direction of travel of the wave (direction of energy flow). A *longitudinal* wave is one in which the oscillations are parallel to the direction of travel of the wave. Both types are illustrated in the diagrams of 'slinky' springs in figure 8.

Examples of resonance occur in almost all branches of physics.

□ *Mechanics*
   (i) The oscillations of a child's swing pushed by a parent.
   (ii) The wind-induced destruction of the Tacoma Bridge, Washington, 1940.
   (iii) The loose part of a car rattling when the car travels at a constant, particular speed.

□ *Electricity*
   (i) Transmitting oscillations at radio and television stations.
   (ii) Tuning circuits at the input stages of radio and television sets are forced to resonate by the incoming electromagnetic waves.

□ *Sound*
   (i) An opera singer shattering a wine glass by forcing the glass to vibrate at its natural frequency.

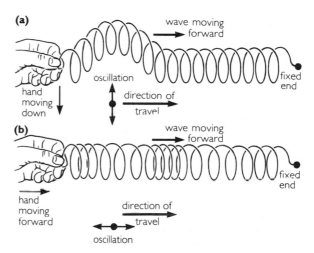

figure 8 (a) Transverse wave; (b) longitudinal wave

figure 7 Barton's pendulums. The first two photos use prolonged exposure to show the sharpness of the resonance peak with heavy damping (left) and light damping (middle). The third photo shows the phase lag of the driven pendulums

Wavelength, frequency and speed

Consider transverse waves travelling across the surface of water in a large tank, figure 9. When the vibrating rod, which is the source of the waves, has completed one extra oscillation then the whole wave-pattern will have moved forward by one *wavelength*, λ, as shown in figure 9(b).

(a)

(b)

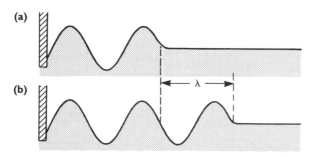

figure 9   Water waves

If the source vibrates with a *frequency* f then f new waves will be generated in unit time. Therefore, the wavepattern will move forward by $f\lambda$ in unit time.

$$\Rightarrow \qquad \text{speed } v = f\lambda \qquad (i)$$

This relationship holds for all types of waves.

Standing waves

Consider the length of thin rubber tubing illustrated in figure 10. One end is fixed; the other end passes over a friction-free pulley and carries a weight. A vibrator supports the rubber tubing a few centimetres from one end.

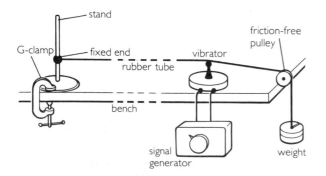

figure 10   Apparatus to demonstrate standing waves

The vibrator will cause the rubber tube to move up and down. This will cause a transverse wave to travel along the tubing. As the frequency driving the vibrator is slowly raised, the rubber tubing will resonate in increasing numbers of sections as depicted in figure 11.

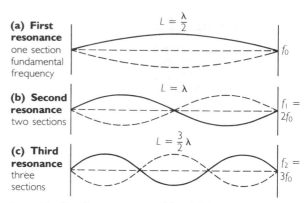

(a) First resonance
one section
fundamental
frequency
$L = \frac{\lambda}{2}$
f_0

(b) Second resonance
two sections
$L = \lambda$
$f_1 = 2f_0$

(c) Third resonance
three sections
$L = \frac{3}{2}\lambda$
$f_2 = 3f_0$

figure 11   Standing waves on rubber tubing

If the scale reading on the signal generator feeding the vibrator is noted each time a resonance occurs, it will turn out that each successive resonance is related to the initial, fundamental resonance as shown on the right hand side of figure 11. That is, the second resonance will occur at a frequency equal to twice the fundamental frequency, and so on. The wave-patterns depicted in figure 11 are known as *stationary* or *standing* waves. They are called stationary or standing waves because the waveform does not seem to be travelling along the rubber tubing in either direction. In order to see the oscillations in detail the tubing should be illuminated by a stroboscopic lamp in a darkened room. When the flashing rate of the stroboscope has been adjusted to be almost equal to that of the vibrator, the tubing will be seen moving up and down. Successive illuminations are labelled, 1, 2, 3, etc., in figure 12.

figure 12   Stroboscopic view of standing waves

The following points should be noted about the standing waves shown in figure 12.

☐   There are points on the tubing where the displacement is always zero. These points are called *nodes*.

☐   Points on the tubing where the amplitude is a maximum are referred to as *anti-nodes*.

☐   Between two adjacent nodes each part of the tubing oscillates in phase with every other part between those nodes.

☐   The oscillations in one section (between adjacent nodes) are 180° out of phase with the oscillations in the neighbouring section. That is, the oscillations in adjacent sections are in antiphase with one another.

☐   The distance between two adjacent nodes is equal to half the wavelength of the wave.

Explanation of standing waves

Let us think about what will happen to transverse waves travelling along the tubing from the point driven by the vibrator. They will strike the fixed end and be reflected back. Therefore, stationary waves result from the superposition of two trains of progressive waves, of equal frequencies, travelling with equal speeds in opposite directions. In figure 13 are depicted five successive positions of these progressive wave-trains, one travelling to the left and the other to the right. The continuous curves in figure 13(c) are the results of superposition at these instants. The successive times t_1, t_2, t_3, etc., represent increments of $T/8$, where T is the time-period of the driving oscillations.

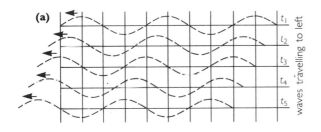

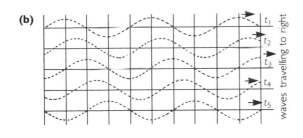

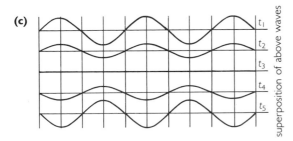

figure 13   Superposition of waves

If the mathematical result of superposing these wave-trains, which is illustrated in figure 13(c), is compared with the experimental observations obtained with the stroboscope, which are shown in figure 12, we see that the curves labelled t_1, t_2, t_3, t_4 and t_5 are identical to the curves labelled 1, 2, 3, 4 and 5 respectively. Thus, standing waves are an example of the superposition of waves.

Overtones

The frequency of the lowest resonance of the tubing is known as the *fundamental* frequency, f_0. For the fundamental, we saw in figure 11 that the wavelength $\lambda = 2L$, where L was the length of the tubing between the vibrator and the fixed end. We also obtained resonances at the higher frequencies $2f_0$, $3f_0$, $4f_0$, $5f_0$, etc. These frequencies are known as the *overtones* of this system.

Musical instruments

The musical notes produced by

☐   string instruments, such as the violin, guitar, etc.,

☐   wind instruments, such as the flute, organ, etc., and

☐   rod instruments, such as the xylophone, tuning fork etc.,

are all due to vibrations set up in the air by the standing waves in these instruments.

String instruments

When the string of one of these instruments is plucked it will vibrate as a standing wave. If a high-quality microphone and cathode ray oscilloscope (CRO) are used to 'see' the note, a curve similar to the one drawn in figure 14 might be seen. We explain the shape of this curve by saying that the string vibrates simultaneously in a mixture of standing waves of different frequencies. That is, that the note produced is a mixture of the fundamental frequency and the overtones. The overtones of string instruments are identical to those of the rubber tubing (figure 11).

figure 14   Waveform produced on a CRO by a violin

This interpretation of figure 14 can be justified by using modern electronics: it is basically what happens in electronic organs. A signal generator is set to oscillate, as a sine wave, at the fundamental frequency. Part of the signal is fed directly to a mixer unit at the output, but other parts are fed through frequency doubling, trebling, quadrupling, etc., circuits, and then into the mixer unit. This is illustrated in figure 15. By adjusting the amplitudes A_0, A_1, etc. of the overtones before they are added

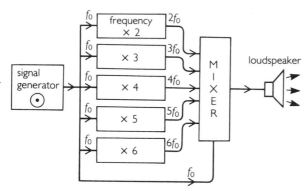

figure 15 Electronic imitation of a musical instrument

together in the mixer, it is possible to produce a note that is identical to the one shown in figure 14. This superposition, or addition, of sine waves is illustrated in figure 16. If the waves in (a), (b), (c), (d), and several other higher overtones are added then figure 16(e) is the result.

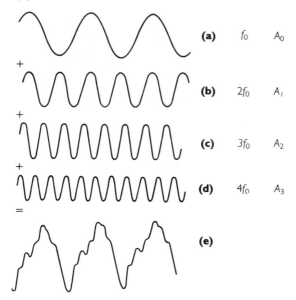

figure 16 Superposition of overtones

Wind instruments

Open pipes
An 'open' pipe is one that is open at both ends as shown in figure 17. Air blown into the instrument sets up vibrations in the column of air between B and A within the tube. In some instruments, for example the clarinet, B is a vibrating reed; in others, for example the flute, it is simply an open hole; but in either case vibrations pass down the instrument and are reflected from the open end at A.

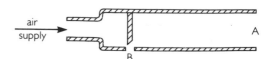

figure 17 Open organ pipe

Resonance occurs at particular frequencies, hence a stationary wave is set up within the air in the pipe. Since the ends of the pipe are open, the air molecules will be free to vibrate with large amplitudes. So the ends of the pipe must be antinodes. The fundamental mode and possible overtones of an open pipe are given in figure 18. The sine curves drawn within each tube of figure 18 represent the variation in amplitudes and phase of the air molecules as they oscillate longitudinally along the length of the pipe.

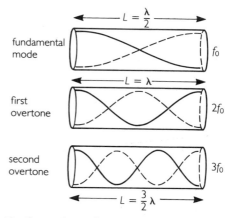

figure 18 Open pipe – displacement amplitudes

Closed pipes
A *closed* or *stopped* organ pipe is illustrated in figure 19. Vibrations travel from B to A and are reflected back again. Resonance occurs at particular frequencies, setting up stationary waves within the air in the pipe. Since the end of the pipe at A is closed, and the waves are longitudinal, the air molecules will not vibrate at A.

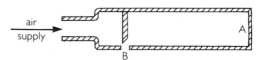

figure 19 Closed organ pipe

The left-hand end, B, is open, therefore the fundamental mode and possible overtones of a closed pipe must be as shown in figure 20. Hence only odd multiples of the fundamental frequency, f_0, are possible. We say that the pipe is only capable of giving the odd *harmonics*. The fundamental is known as the first harmonic. The sine curves drawn within each tube again represent the variation in the amplitude and phase of the longitudinal oscillations of the air molecules in the tube.

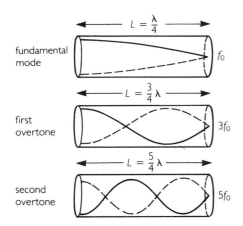

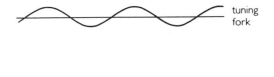

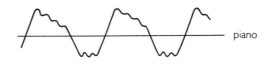

figure 20   Closed pipe – displacement amplitudes

Subjective aspects of sound

Musicians refer to the *pitch*, *quality* or *loudness* of a note; physicists try to quantify these qualitative descriptions. Figure 21 shows the waveforms obtained on the screen of a CRO when different instruments all play the same note into the microphone attached to it. The physicist says that they are all of the same *frequency*, whereas the musician says that they are all at the same *pitch*. Quantitatively we say that the *amplitude* of the oscillations is small for the tuning fork, but to the

figure 21   The same note on different instruments

musician the *loudness* is quiet. Anyone listening to these three notes would be able to tell that they came from different instruments. We say that the *quality* or *timbre* of the notes is different. Clearly you would have to add together different overtones in order to produce the piano and trumpet notes. Thus, we say that the *overtones* present in a note determine its *quality*.

Summary

1   Driven oscillations:
 a Amplitude variation of response;
 b frequency of resonance increases slightly with decreasing damping;
 c phase lag of response.

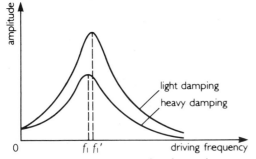

figure 5   Sharpness of resonance; the change in resonant frequency with damping is very small

2   Wave motion:
 a Transverse wave;
 b Longitudinal wave;
 c $v = f\lambda$.

3   Standing waves:
 a On strings;
 b In open pipes;
 c In closed pipes.

4   Subjective aspects:

Quantitative description	Qualitative description
Frequency	Pitch
Amplitude	Loudness
Overtones present	Quality or timbre

Questions

1 Consider the apparatus illustrated in figure 22. The electric motor is fitted with an off-centre drive wheel.

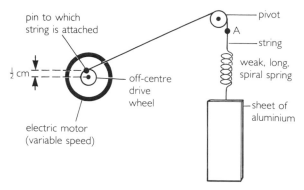

figure 22

a Describe the motion of the aluminium plate when the drive wheel is turned *very slowly* (about 1 rotation per minute).
b Describe the plate's motion when the electric motor is turned off and the aluminium plate is pulled down and released.
c Discuss the physical processes involved which cause the amplitude of the oscillations to become very large when the electric motor rotates at the natural frequency of the plate and spring.
d Sketch a graph to show how the maximum amplitude of the plate's oscillations varies as the driving frequency is gradually increased from zero.
e What would happen to the natural frequency of the plate and spring if a second identical plate were attached to the first one? Explain your answer.

Now consider what happens if the aluminium plate is surrounded by a thick, viscous material such as engine oil.
f In what way will the amplitude of the oscillating plate change due to the presence of the oil? Explain this change.
g If you remove a metal plate from an oil bath it will be coated in a thin film of oil. Therefore, how would you expect the frequency of resonance to change when the plate is moved from air to oil? This is when the plate oscillates completely surrounded by oil, as opposed to being entirely in air. Explain your answer.
h If the plate is disturbed it will try to oscillate at its natural frequency, on the bottom of the spring. The motor will force the top of the spring to move up and down with a frequency equal to the rate of rotation of the motor.

Explain what happens when these two frequencies are slightly different.

2 Consider a metre rule clamped to a bench. When 0.1 kg masses are placed on its free end they cause a displacement, y. The following table gives some typical results obtained by a student who clamped the metre rule at the 10 cm mark.

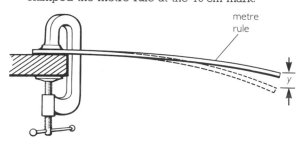

figure 23

Weight/N	1	2	3	4
Displacement y/mm	24	49	74	99

a Plot a graph of the results. Does the graph indicate that the ruler will oscillate with s.h.m. when disturbed? Explain your answer.

The student then attached different masses to the end of the ruler and measured the time for 20 oscillations in each case. He obtained the following results.

Load/kg	0.1	0.2	0.3	0.4
Times for 20 oscillations/s	6.81	8.87	10.59	12.00

b Plot graphs of time-period T against load M, and T^2 against M. Which one is linear?
c If you were to try to measure the time-period of oscillation of the unloaded metre rule you would find it oscillated too quickly for you to measure. Use your graph to determine this frequency.
d How will the natural frequency of oscillation of the metre rule change if the length that is free to oscillate is increased? Explain your answer.
e What would happen to the amplitude of the oscillations if the free end of the metre rule were to be repeatedly hit, at the natural frequency of the rule, by a small hammer? If this process were to be repeated indefinitely, what would be the probable effect on the ruler?
f Modern oil tankers are cheaply constructed. They are basically a deck 400 m long. Explain why they sometimes have to change course when heading into heavy seas. What would

happen if they continued in their original direction?

3 When air flows over a non-streamlined body, vortices occur in the air. This is illustrated in figure 24.

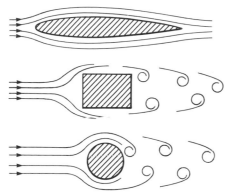

figure 24

The pressure in the gas in the vortex is not quite equal to the pressure in the gas which is flowing smoothly (the Bernoulli effect). Therefore, as vortices break away alternatively from each side of a body, the body will experience alternating forces. If vortices are shed at the natural frequency of the body then resonance occurs, as it did in the Tacoma bridge in 1940. The Tacoma bridge was a large span suspension bridge (similar in appearance to the Severn bridge near Bristol) which underwent torsional oscillations of amplitude more than five metres for over one hour. The steel girders in the structure then failed and the bridge collapsed into the river below.

Small bodies will shed small vortices, and large bodies will shed large vortices (figure 25). Therefore, we should expect the frequency of vortex-shedding to be inversely proportional to the width of the body, i.e. $f \propto 1/d$.

Also the vortex shedding frequency will increase with the speed, v, of the wind, i.e., $f \propto v$.

Experiment shows that $v = 5fd$ for a rod of circular cross-section, when SI units are used.

a What is the Bernoulli effect?
b Explain why it was reasonable to say that $f \propto 1/d$ and $f \propto v$ in the above discussion.

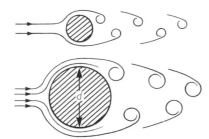

figure 25

c The Forth Road Bridge is suspended from main cables which are 60 cm in diameter. Calculate the frequency at which vortices are shed when the wind blows at a steady 45 km h^{-1}. Explain why there have been no resonance problems with these main cables.
d During construction of the Forth Road Bridge, the main towers oscillated with amplitudes of more than 1 metre in a wind of 36 km h^{-1}. The frequency of oscillation was 0.15 Hz. Use this information to estimate the diameter of the towers.

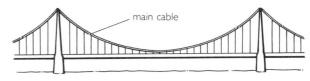

figure 26

e Tall chimney stacks built of light materials can oscillate in the wind. This can seriously damage or weaken them. What would be the effect on the natural frequency of a chimney if the thickness of its walls were doubled? Explain your answer.
f If a system is free from damping, the amplitude of the oscillations will increase indefinitely when the resonant frequency is applied. Explain where the damping occurs in the case of a light metal chimney stack.

4 When car tyres are manufactured they are not always perfectly symmetric. There might be a part with a slightly thick wall. Once the tyre has been bedded-in, by running the car for a hundred miles, the wheel may need to be balanced.
a The BMW 528i has tyres with radius 62 cm. Assume that resonance occurs in the steering wheel assembly when the car is doing 70 km h^{-1}. Calculate the resonant frequency.
b Explain how the wheels are balanced.
c A Fiat 128 has tyres with radius 55 cm. If its unbalanced wheels also produce a resonance in the steering wheel assembly at 70 km h^{-1}, what can you deduce about the two steering wheel systems?

5 a Explain what is meant by a standing wave.
b Explain how it is that when a guitar string is plucked a standing wave results.
c What do you understand by 'fundamental frequency'?
d What is it about a musical note played on one instrument that makes it distinguishable from the same note played on another instrument?
e Explain the difference between 'harmonics' and 'overtones'.

6 In this question, we draw the distinction between the *displacement* of air molecules along a pipe and the changes in *pressure*. The chapter discussed only displacement of air molecules: figures 18 and 20 show nodes and antinodes of displacement, for open and closed pipes respectively.

At the open end of a pipe, there is a displacement antinode; in pressure terms, however, there is a node, because at an open end pressure must always be atmospheric. Conversely, at a closed end there is a displacement node, but a pressure antinode. So displacement diagrams and pressure diagrams look opposite to each other, as figure 27 illustrates.

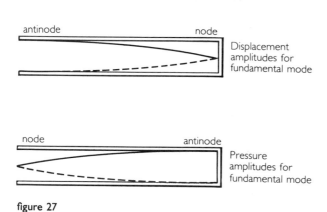

figure 27

a Draw a diagram to show how the amplitude of the *displacement* of air molecules varies along the length of an *open* organ pipe, when it is resonating in its fundamental mode.
b What is the frequency of the note emitted by an open organ pipe 1.94 m long if the speed of sound in air is $330 \, \text{ms}^{-1}$?

Figure 28 shows the smooth tube of a flute. Minute microphones are embedded into the inside walls of the instrument at the positions indicated. The microphones are equispaced. The microphones will respond to the *pressure* variation in the gas at the points where they are located. Each microphone is attached to the Y-plates of a CRO.

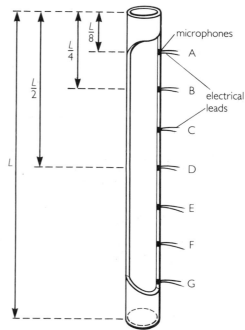

figure 28

c Indicate the relative amplitudes of the signals on the CROs when the flute is blown to resonate in its fundamental mode.
d Explain why blowing harder can cause the flute to resonate in a higher harmonic.
e What would be the relative amplitudes from microphones if the flute were blown so that it resonated at its first overtone?

SECTION B

CIRCUITS

Man-made alternating current . . .

. . . and natural direct current

 9   Amp, volt and ohm   107
10   Resistivity   120
11   Circuits with resistors   129
12   Capacitors   146
13   Square pulses in circuits   159
14   a.c. circuits   165
15   Transistor electronics   177
16   Operational amplifiers   191

CHAPTER 9

Amp, volt and ohm

Introduction

This chapter covers all the key ideas about electric currents in circuits. The basic quantities of current, charge, potential difference and resistance are introduced one at a time. Each one is discussed from first principles: we look at the meaning and usefulness of the quantity, how it may be measured and a suitable unit for it, and how the ideas are applied. We start with current and its effects; then move on to the relationships between current, charge and the flow of individual charged particles summed up by the transport equation. We then explore the issue of energy conversion in a component, and how this leads to the idea of p.d. (electrical potential difference). At this stage we examine the similarity between current and water flow. Finally we discuss resistance, its definition and some useful equations in which it features. A detailed discussion of the nature of resistance in different materials is left to chapter 10.

GCSE knowledge

If you understand and remember everything you learnt for GCSE, then you could in theory skip this chapter. In practice, it would be a good idea to go over this again: though you will have met most of the ideas before, the chapter goes over them very thoroughly, at a depth you may not have explored. And these principles are so important to the rest of the course that you must understand them well. Though the chapter takes the ideas right from the beginning, it will be useful to you to recall what you know on the subjects of current, charge, energy, p.d., power and resistance.

Current

An electric current consists of a movement of charged particles in a particular direction. In most cases, such as when a current passes along a wire, we are not directly aware of charged particles moving – the particles, which in metals are electrons, are much too small for us to see, even with a microscope. We become aware of a current because of the effects it can cause:

☐ magnetic forces;

☐ the conversion of energy into internal (heat) energy and light;

☐ chemical change.

The symbol for current is I. The unit of current is the ampere or amp (A).

The ampere

The ampere is a fundamental SI unit (see chapter 46). All other units involving electrical quantities are derived from it.

The definition of one amp is based on the force of attraction which occurs when currents run in the same direction along two parallel wires as shown in figures 1 and 2.

'One ampere is that constant current which, when passing through two straight parallel conductors of infinite length, of negligible circular cross-section, and placed 1 metre apart in vacuum, would produce between these conductors a force of 2×10^{-7} newton per metre of length.'

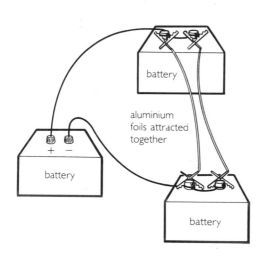

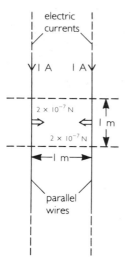

figure I

figure 2 The hypothetical situation used to define one ampere

This is not a situation which can be achieved in practice: and even if it could be, such a small force would be difficult to measure accurately. However, deductions can be made from this definition about the forces in other situations – see chapter 24. Ammeters are calibrated in practice by using one such situation, a current balance.

Measuring current

Ammeters, as their name suggests, are instruments to measure current. An ammeter is connected in series in a circuit, at the point where the value of current is required, as shown in figure 3.

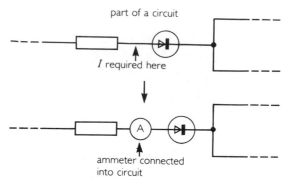

figure 3 Using an ammeter

Two types of ammeter are in common use.

I   *Moving-coil ammeter.* In this, current is passed through a coil which is suspended in a magnetic field. The forces on the coil cause it to rotate, through an angle which is proportional to the current. A pointer attached to the coil moves over a scale, thus giving a reading which can vary smoothly over the complete range available. See figure 4.

figure 4 A moving-coil ammeter

By using suitable adaptors (called 'shunts'), you can usually use one basic meter to cover a whole range of current measurements – from microamps up to kiloamps. See page 130.

2   *Digital ammeters.* These instruments use electronic amplifiers to measure the current, and display it on a digital read-out – one which has to vary in jumps from one value to the next. Again, they are usually made so that one instrument can cover a wide range of currents. See figure 5.

When using any kind of ammeter, it is important to ensure that its insertion in your circuit (figure 3) is not going to affect significantly the current you want to measure. In practice, this means that the electrical resistance of the ammeter must be negligible compared to that of the components in series with it.

In general, any measuring device will put some 'load' on the system being monitored; either the load

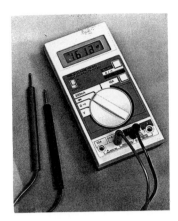

figure 5 A digital ammeter

must be kept negligible, or it must be possible to make allowance for it when assessing the data the device provides.

A third method of measuring current can be useful when the current varies rapidly. That is to insert a small resistor into the circuit, and measure the p.d. across it with a cathode ray oscilloscope.

Current and charge

Consider current I passing a point P in a wire for a time t (figure 6).

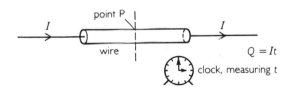

figure 6 To illustrate the relationship between charge and current

How much charge flows past? It is sensible to imagine that the amount of charge Q will be proportional to I and to t. So we say that by definition

$$Q = It$$

The unit of Q is the coulomb (C), where

$$1\,C = 1\,A \times 1\,s$$

Also note that $I = Q/t$; thus current is, in words, rate of flow of charge.

If the current I varies in some way, we might have to evaluate the charge Q which passes in time t by using calculus:

$$Q = \int_0^t I\,dt$$

Alternatively, we could evaluate Q by finding the area under a graph of I against t.

Rules for current

One of the first rules that you learnt by doing experiments with bulbs and cells was that you needed a complete circuit before current would pass at all. This fact sometimes leads to a comparison between charge in wires and water in a system of pipes: you can pump water round a complete circuit of pipes, but if a pipe comes to a dead end then no water can flow through it (figure 7).

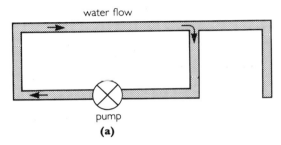

(a)

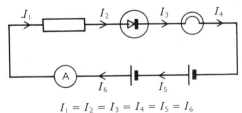

(b)

figure 7 A complete circuit is required for (a) water flow (b) electric current

In both cases, the 'stuff' flowing (water or charge) is conserved: it cannot disappear or be created within the circuit; and it cannot become piled up in one part of the circuit. These ideas lead to two fundamental rules about current, which apply to both the water and the electrical circuits.

☐ The same current passes through all components which are in series. Thus in figure 8 the cells and bulb would have the same current through them.

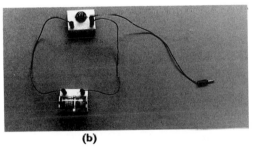

$$I_1 = I_2 = I_3 = I_4 = I_5 = I_6$$

figure 8 The same current passes through all components in series

☐ The total current leaving any point in a circuit is the same as the total current arriving at the point.

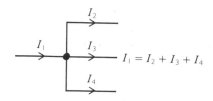

figure 9 Currents leaving a junction add up to current arriving

This second rule is known as Kirchoff's first law. It is particularly relevant in situations where components are connected in parallel.

Note that in the electrical circuit, current I is measured in Cs^{-1}. Flow of water in the water circuit would be measured in m^3s^{-1} (volume of water per unit time).

One other point to note concerns electrons – the charged particles in metals whose movement we call current. Current is always marked on a circuit as passing from the plus pole of a cell to the minus pole – see figure 10. This is sometimes called the 'conventional' current direction. The convention for current direction was established long before individual electrons were discovered (by J. J. Thomson in 1897). This discovery showed that electrons are in fact negative particles, moving round a circuit in the opposite direction to the conventional current. But, even though the early scientists made the wrong guess about the direction of motion of the

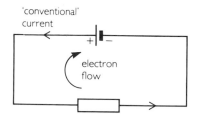

figure 10 Electrons, being negative particles, flow in the opposite direction to the positive (conventional) current

flowing particles in metals, we still retain the convention they established. (Current *is* carried by positive particles in ionic solutions and some semiconductors.)

Thus, for example, in the work which follows we develop the theory as though positive particles are flowing in the conventional direction; only on certain occasions do we need to recall that electrons are negative and flow the other way.

Transport equation

Consider a wire with cross-sectional area A, in which there are n charged particles per unit volume of the material. It would be interesting to examine how the average drift speed v (the average speed with which

the particles move in the direction of the current) is related to the current I.

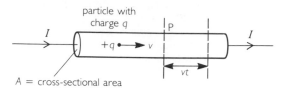

$A = $ cross-sectional area

figure 11 Individual charges moving in a wire

In time t:
  the charges which pass any point, say P, occupy a length $v \times t$ of wire;
  this length has volume $vt \times A$, and contains $vtA \times n$ carriers;
  these carriers carry altogether charge $vtAn \times q$ $(= Q)$.

but
$$I = \frac{Q}{t}$$

$\Rightarrow$
$$I = \frac{vtAnq}{t}$$

$\Rightarrow$
$$I = vAnq$$

We refer to this equation as *the transport equation*.

The *current density J* is sometimes a useful quantity:

$$J = \frac{I}{A} = vnq$$

J is the current per unit area of a conductor and has units of Am^{-2}.

Example Let us consider a sample of germanium semiconductor with dimensions as in figure 12. A current of 50 mA is used. If $n = 1.0 \times 10^{22}\,m^{-3}$, what is the drift velocity of the electrons? (Assume $q = 1.6 \times 10^{-19}\,C$)

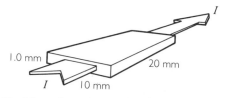

figure 12 Dimensions of a sample of germanium

$$I = nAvq$$

$\Rightarrow$
$$v = \frac{I}{nAq}$$

$$= \frac{50 \times 10^{-3}\,A}{1.0 \times 10^{22}\,m^{-3} \times 1.0 \times 10 \times 10^{-6}\,m^2 \times 1.6 \times 10^{-19}\,C}$$

$\Rightarrow$
$$v = 3.1\,ms^{-1}$$

For comparison, copper has $n = 10^{29}\,\mathrm{m^{-3}}$; so if the same current passed through a sample of copper with the same dimensions, the electrons would be drifting at only $3 \times 10^{-7}\,\mathrm{m\,s^{-1}}$.

In appropriate experimental conditions it is possible to observe the drift of coloured ions in electrolysis. Figure 13 illustrates a possible arrangement.

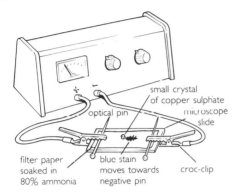

filter paper
soaked in
80% ammonia

blue stain
moves towards
negative pin

croc-clip

small crystal
of copper sulphate

optical pin

microscope
slide

figure 13   Drift of coloured ions

In copper salts the Cu^{2+} ion is blue. Thus with a crystal of copper sulphate on the slide, the blue colour of Cu^{2+} ions would move towards the negative pole of the battery – to the right in figure 13.

Energy and potential difference

Let us consider the simple cell/bulb circuit of figure 14. In this circuit, both the cell and the bulb are converting energy. In the cell, chemical energy is being converted into electrical energy; this enables the cell to apply a force to the charged particles in the wires, pushing them round the circuit. In the bulb, energy is being converted from electrical energy into heat and light.

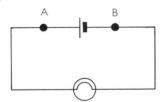

A   B

figure 14   Cell/bulb circuit

As each positively charged particle moves from A through the bulb to B, it transfers some energy into the bulb. The particle thus possesses more energy at point A than at point B. The energy it possesses at A is in a form we call 'electrical potential energy': because when we allow the particle to move to B, by connecting the bulb between A and B, that energy is converted to heat and light.

We thus say that there is an 'electrical potential difference' between points A and B. This is usually

shortened to *potential difference*, or p.d. (symbol V).

In formal terms, the p.d. between two points is the electrical energy converted to another form per unit charge moving from one point to the other. We define V using an equation:

$$V = \frac{W}{Q}$$

where W is the energy converted when charge Q passes from one point to the other.

The unit for a p.d. is one joule/coulomb. This unit is given its own name: one *volt*, V.

$$1\,\mathrm{V} = 1\,\mathrm{J\,C^{-1}}$$

A p.d. is the origin of the driving force causing a current between A and B in figure 14. The p.d. sets up the electric field which causes the forces on the charges.

The electrical energy need not always be converted into heat and light. In an electron gun, for example, the electrons acquire kinetic energy – see figure 15.

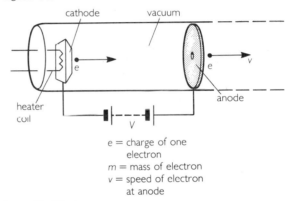

cathode   vacuum

heater
coil

anode

e = charge of one
electron
m = mass of electron
v = speed of electron
at anode

figure 15   Electron gun

In this case, the emerging electrons have kinetic energy E_k given by

$$E_k = W = eV$$

and

$$E_k = \tfrac{1}{2}mv^2$$

$$\Rightarrow \qquad \tfrac{1}{2}mv^2 = eV$$

Finally, consider again the similarity between current and water flow. We have seen already that charge (in coulombs, C) is analogous to volume of water (in $\mathrm{m^3}$). The driving force in a water circuit is created by the pressure difference p maintained by the pump, as shown see figure 16.

As water is pushed round the circuit by the pump, energy is converted in the pipes – to heat, due to the viscosity and turbulence of the water. Thus the pump provides a potential difference: could it be measured using a unit of joules per cubic metre? Let us try:

$$\frac{\mathrm{J}}{\mathrm{m^3}} = \frac{\mathrm{N\,m}}{\mathrm{m^3}} = \frac{\mathrm{N}}{\mathrm{m^2}} = \mathrm{Pa}\ \text{(unit of pressure)}$$

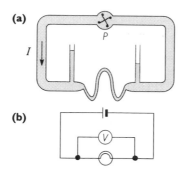

figure 16   (a) Water circuit; (b) electric circuit

Thus the appropriate unit for the p.d. maintained by the pump comes out to be the unit of pressure – just as it should be.

Potential at a point

We have examined what is meant by the statement 'there is a p.d. V between points A and B'. Sometimes, either in circuit work or in electrostatics, it is helpful to regard a specified point as being at 'zero' potential; we can then assign actual values of potential to every other point in the system. For example, following from the above statements, if we also state that B is at zero potential, then A is at potential V (i.e. V above zero).

In circuits, one point is often connected to earth – 'earthed'. That point is taken as being at zero potential. In electrostatics, the usual convention is to regard points infinitely far from the system as being at zero potential.

Measuring p.d.

In principle we can measure a p.d. between two points A and B by finding how much energy is converted (W) between the points when a measured amount of charge (Q) passes: $V = W/Q$. For example, the energy may be converted to heat and used to warm water, as in the arrangement illustrated in figure 17.

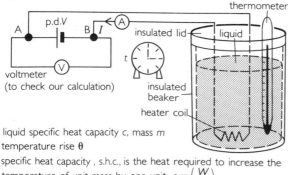

liquid specific heat capacity c, mass m
temperature rise θ
specific heat capacity , s.h.c., is the heat required to increase the temperature of unit mass by one unit. $c = \left(\dfrac{W}{m\theta}\right)$

figure 17   Hypothetical experiment to measure a p.d.

By making suitable measurements (mass of water m, temperature rise θ) we can calculate the amount of heat produced, W; and by recording the current I and the time t for which it flows, we can calculate Q.

$$W = mc\theta$$
$$Q = It$$
$$\Rightarrow \quad V = \frac{W}{Q}$$
$$= \frac{mc\theta}{It}$$

In practice, p.d.s in circuits are usually measured with one of the following:

- ☐ moving-coil voltmeter;
- ☐ cathode ray oscilloscope – see below;
- ☐ potentiometer – see page 132
- ☐ digital voltmeter; as with a digital ammeter, this uses electronic amplification to display the p.d. directly on a read-out.

In all cases the measuring instrument is connected across the p.d. to be measured – as in figure 17. As a practical tip, it is usually a good idea to connect up the main parts of a circuit first, then to connect the p.d.-measuring devices last – in parallel with the appropriate components.

As with measuring current, you must ensure that the load you put on the circuit is negligible. In general, this means ensuring that the current required to drive your voltmeter is a negligible fraction of the current in the component it is connected across: in other words, its resistance is much higher than that of the component.

Using a cathode ray oscilloscope (CRO)

A CRO measures a p.d., and displays it as the calibrated displacement of a spot in the y-direction on a screen. For many applications the spot moves at a constant speed in the x-direction (a *time-base* is applied in the x-direction): this means that the screen actually displays a graph of p.d. against time, as shown in figure 18.

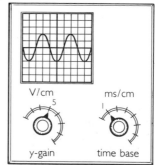

figure 18   CRO-trace, with gain controls

Example Suppose the *y*-gain in figure 18 is set at $5\,\mathrm{V\,cm^{-1}}$, and the time-base is set at $1\,\mathrm{ms\,cm^{-1}}$. What are the amplitude and frequency of the sinusoidally varying p.d. displayed?

$$\text{Amplitude of trace} = 2.0\,\mathrm{cm\ on\ the\ screen}$$
$$= 2.0\,\mathrm{cm} \times 5\,\mathrm{V\,cm^{-1}}$$
$$= 10\,\mathrm{V}$$
$$\text{Time-period of trace} = 4.0\,\mathrm{cm\ on\ the\ screen}$$
$$= 4.0\,\mathrm{cm} \times 1\,\mathrm{ms\,cm^{-1}}$$
$$= 4.0\,\mathrm{ms}$$
$$\Rightarrow \qquad \text{frequency} = 1/\text{time period}$$
$$= \frac{1}{4.0\,\mathrm{ms}}$$
$$= 250\,\mathrm{Hz}$$

A CRO has advantages over other instruments in certain situations, because:

☐ it shows the manner in which a p.d. varies with time;

☐ it responds instantaneously to variations in p.d.;

☐ it draws negligible current compared to moving-coil instruments.

However, its measurements are not as sensitive as those of a moving-coil or digital voltmeter.

Rules for p.d.s in circuits

There are two basic rules for p.d.s in circuits:

1 The p.d. between two points can have only one value at a given instant: so all components in parallel have the same p.d. across them at every instant. See figure 19. (This is equivalent to saying that whatever route is taken by a charge which travels between two points, the same energy will be converted.)

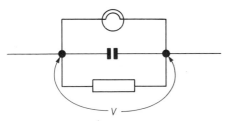

figure 19 All components in parallel have the same p.d. across them

2 For the components X and Y in series, the energy converted as a charge moves through X plus the energy converted as it moves through Y must equal the total energy converted. Thus the p.d.s across components in series add up to give the overall p.d., as shown in figure 20.

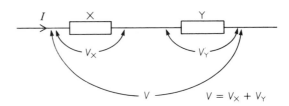

figure 20 The p.d.s across components in series add together

Energy and power

Consider again the basic situation of a component with p.d. *V* across it, and current *I* through it for time *t* (figure 21). A small piece of theory will yield us a rather useful result.

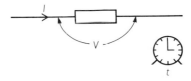

figure 21 Energy and power in a component

The energy converted, *W*, is given by
$$W = VQ,$$
where *Q* is the charge which passes. But $Q = It$,
$$\Rightarrow \qquad W = VIt$$
$$\frac{W}{t} = VI$$

But *W/t* is the energy converted per unit time, in other words, the power *P* being expended in the component.
$$\Rightarrow \qquad P = VI$$

Example A torch bulb has stamped on it '1.5V 0.2A'. What power does it convert when in normal use, and how much energy does it convert in 10 hours (the approximate life of a single cell driving this current)?

$$\text{Power } P = VI$$
$$= 1.5\,\mathrm{V} \times 0.2\,\mathrm{A}$$
$$= 0.3\,\mathrm{W}$$
$$\text{Energy } W = VIt$$
$$= 1.5\,\mathrm{V} \times 0.2\,\mathrm{A} \times 10\,\mathrm{h} \times 3600\,\mathrm{s\,h^{-1}}$$
$$\approx 11\,000\,\mathrm{J}$$

I-against-*V* graphs

We can learn a lot about the electrical behaviour of a component, and the material of which it is made, if we study how the current through it, *I*, varies with the p.d. across it, *V*. In principle we use a circuit such as that shown in figure 22. Certain characteristic graphs emerge – figure 23.

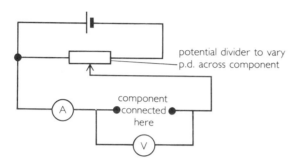

figure 22 Circuit to obtain *I*-against-*V* graph for a component

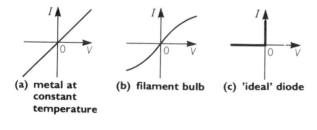

(a) metal at constant temperature   **(b) filament bulb**   **(c) 'ideal' diode**

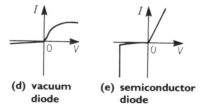

(d) vacuum diode   **(e) semiconductor diode**

figure 23 Typical *I*-against-*V* graphs

Resistance

The word resistance suggests that a component with a high resistance would be hard to push current through. Since the 'pushing' comes from the applied p.d., it is reasonable to define the resistance *R* of a component by the equation

$$R = \frac{V}{I}$$

Thus *I* is small when *R* is high, and *I* is large when *R* is small. The unit of *R* is thus one volt/amp. This unit is given its own name, one ohm (Ω).

$$1\,\Omega = \frac{1\,\text{V}}{1\,\text{A}}$$

Figure 23 shows that the resistance of many components varies with both the value and direction of the p.d. across them.

For the special case where the *I*-against-*V* graph is a straight line through the origin (figure 23(a)) the component is said to obey Ohm's law, which states:

'For certain materials, the resistance of a sample remains constant, provided physical conditions (such as temperature) remain constant.'

A large number of materials obey Ohm's law over a small range of p.d. and current. Metals obey it over a wide range. If the words 'conductor' or 'resistor' are applied to a component, this usually implies that the component has constant resistance.

The instantaneous rate at which energy is converted to heat inside a resistor is given by $P = VI$ (see above). Using the equation defining *R* in its other forms:

$$V = IR \text{ and } I = \frac{V}{R}$$

we may write down two more expressions for the power being converted in a resistor:

$$P = I^2R \text{ and } P = \frac{V^2}{R}$$

Example Consider a standard light bulb, rated at 60 W to operate from a 240 V mains supply. What is its resistance and what current does it take?

Using $\qquad\qquad P = \frac{V^2}{R}$,

$$60\,\text{W} = \frac{240\,\text{V} \times 240\,\text{V}}{R}$$

$\Rightarrow \qquad\qquad R = 960\,\Omega$

Using $\qquad\qquad P = VI$,

$$60\,\text{W} = 240\,\text{V} \times I$$

$\Rightarrow \qquad\qquad I = 0.25\,\text{A}$

Summary

1 Current is a flow of charged particles.

2 Current may give rise to magnetic effects, chemical change and conversion of energy into heat and other forms.

3 The unit of current – one amp – is a fundamental SI unit.

4 An ammeter is connected in series in a circuit, and ideally has negligible resistance.

5 Charge: $Q = It$

$$1\,C = 1\,A \times 1\,s$$

6 Rules for currents in circuits:
    components in series have the same current; the currents leaving a junction add up to the current arriving.

7 Electrons are negative, and flow in the opposite direction to the (conventional) current.

8 Transport equation:

$$I = vAnq$$

9 Current density:

$$J = vnq$$

10 Electrical potential-energy difference (p.d.):

$$V = \frac{W}{Q}$$

$$1\,V = 1\,JC^{-1}$$

11 For an electron gun:

$$\tfrac{1}{2}mv^2 = eV$$

12 An earthed point in a circuit is at zero potential. The potential of any other point in the circuit is equal to the potential difference between it and the earthed point.

13 Analogy between current and water flow:

$$I\,(Cs^{-1}) \equiv I\,(m^3 s^{-1})$$

$$V\,(JC^{-1}) \equiv P\,(Jm^{-3} \text{ or } Nm^{-2})$$

14 A voltmeter is connected across the p.d. to be measured, and ideally has very high resistance.

15 A cathode ray oscilloscope is a voltmeter which displays a p.d.-against-time graph.

16 Rules for p.d.s in circuits:
    parallel components all have the same p.d. across them;
    the p.d.s across components in series add up to the total p.d.

17 Energy and power:

$$P = VI$$

$$W = VIt$$

18 Different components give quite different graphs of I against V.

19 Resistance:

$$R = \frac{V}{I}$$

$$1\,\Omega = 1\,V/A$$

20 Other useful equations:

$$V = IR \quad I = \frac{V}{R} \quad P = I^2 R \quad P = \frac{V^2}{R}$$

Questions

1 A current of 2 A passes in a metal wire with cross-section 0.5 mm^2.
   a How many electrons pass a point in the wire each second?
   b What is the current density in the wire?

2 A simple model of a hydrogen atom is to imagine an electron moving in a circular orbit round a proton. If the electron orbits at $6 \times 10^{15}\,s^{-1}$, what is the current at a point on the orbit?

3 The belt of a Van der Graaff generator moves at 5 ms^{-1}; it is 10 cm wide, and carries charge at a density of $2 \times 10^{-6}\,Cm^{-2}$ up to the dome of the generator. Calculate the current moving to the dome.

4 In an article about the large-scale electrolysis of sodium chloride to produce chlorine gas, one part reads: 'Forty-eight electrolytic cells are operated in series. The current used to produce the large-scale output of chlorine (5 kg per second) is 300 000 A.'
   Check whether the figures in this passage are consistent, showing clearly your reasoning. The following data may be useful:
   $e = 1.6 \times 10^{-19}\,C$
   $L = 6.0 \times 10^{23}\,mol^{-1}$
   1 mole of chlorine atoms has mass 35.5 g.
   The chlorine ion is Cl$^-$.
(Hints: If you are not sure how to proceed, you might try calculating these quantities (for one

second in all cases): charge reaching each anode, number of electron charges reaching each anode, amount (number of moles) of chlorine reaching each anode, mass of chlorine reaching each anode, mass of chlorine reaching all anodes.)

5 A typical electric light bulb is suspended by copper flex with cross-section $2\,\text{mm}^2$, and carries $\frac{1}{4}$ A.

If copper has electrons, charge e, with number density $n = 10^{29}\,\text{m}^{-3}$, calculate the drift speed of the electrons in the wire.

6 Consider an alternating current in a wire, using the following parameters:
maximum I 2.0 A
cross-section A 1.0 mm^2
number density n $1.0 \times 10^{29}\,\text{m}^{-3}$
electron charge e $1.6 \times 10^{-19}\,\text{C}$
frequency f 50 Hz.
a Calculate
(i) the maximum drift velocity v of the electrons;
(ii) the amplitude of the electrons' travel, assuming they oscillate with s.h.m. about a fixed position, without making collisions with atoms.
b Comment on the likelihood of the assumptions in a(ii) being valid, in view of the answer you obtain.

7 In the summer of 1985, the AA reported that one tailback on the M4 into London was 56 miles long. Assume that traffic filled all three lanes, and was averaging 5 miles per hour. By making suitable intelligent guesses, use the transport equation to estimate how many cars per hour were entering London along this road.

8 This question is about applying the transport equation to one lane of a motorway. The *Highway Code* has this advice for drivers: 'Keep one car's length between you and the car in front for every 10 mph $(5\,\text{ms}^{-1})$ of speed.'
Let us use the following nomenclature:
v = speed of cars (ms^{-1})
x = average car length (m)
I = flow of cars along lane of motorway (cars s^{-1})
a Write down an expression for the distance from the front of one car to the front of the next, assuming they are all travelling at the same speed and are observing the *Highway Code* rule.
b We will use the transport equation $I = nAvq$ to write down an expression for I in terms of v and x. (n must be in carsm^{-1}; $A = 1$; $q = 1$). First write down n.
c Now show that I must be given by

$$I = \frac{5v}{x(5+v)}$$

d Sketch on a graph how I changes with changing v.
e I tends to a limiting value as v becomes very large. Using a suitable value for x, estimate this limiting value for I. Does this seem to be a sensible result?

9 A cell electrolysing a molar copper sulphate solution between copper electrodes has the dimensions shown in figure 24.

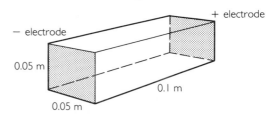

figure 24

Assuming that all the current of 5.0 A is carried by Cu^{2+} ions, calculate
a the current density, J;
b the number density, n, of the Cu^{2+} ions;
c the drift speed of the ions.

($L = 6.0 \times 10^{23}\,\text{mol}^{-1}$; $e = 1.6 \times 10^{-19}\,\text{C}$)

10 A radioactive source fixed inside an evacuated enclosure emits α-particles (charge 3.2×10^{-19} C) uniformly in all directions. It emits on average 10^{10} particles per second.
a Calculate the current density, J, at a distance of 0.1 m from the source.
b If the speed of the emerging α-particles is $2 \times 10^7\,\text{ms}^{-1}$, calculate the number density (number/cubic metre) at any moment at a distance of 0.1 m from the source.

11 On a particular campsite, you have to put a 10p coin in a slot for a six-minute hot shower. The shower fitting says on it '3 kW', and you know that one unit of electrical energy costs 6p. Is your shower good value? (Justify your answer quantitatively.)

12 In many washrooms managements have decided to install electric hand-driers in place of paper-towel dispensers. On the basis of the costs and data below, advise a manager about his best course of action.
An electric drier costs £100 to install initially.
It gives a 30-second blast of air each time it is operated.
It dissipates 2.5 kW.
One unit of electricity costs 6p.
The lifetime of the drier is roughly 2000 operating hours.
In an average day 100 people wash their hands in this washroom.
A pack of 100 disposable towels costs £1.00.

13 The graph of figure 25 represents half a cycle of
an alternating current.

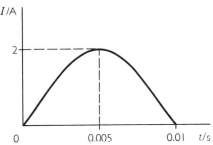

figure 25

   a Estimate how much charge flows past one
   point during this half-cycle.
   b During the next half-cycle, this same amount of
   charge flows back in the opposite direction.
   Thus in one complete cycle, no net charge
   flows past any point in the circuit. How, then, is
   energy transferred by an alternating current?

14 A particular battery is specified by the
manufacturers as follows:
   '12 V 24 Ah at a 20 hr discharge rate'.
   a According to this, what current would the
   battery be capable of driving for 20 hours
   before becoming discharged?
   b What charge would flow altogether?
   c How much energy does the battery store when
   fully charged?

   (Hint: Ah means 'ampere-hour'.)

15 In the circuit of figure 26, a steady current of
0.50 A is flowing.
   (For electrons, $e = 1.6 \times 10^{-19}$ C;
   $m = 9.1 \times 10^{-31}$ kg.)
   a What are the charge-carrying particles in each
   of the three components A, B and C?
   b Is the value of the charge carrier drift speed
   the same in the tungsten filament as in the
   thicker tungsten wires leading to it?
   Justify your answer.
   c How many electrons pass a point P in one of
   the connecting wires each second?

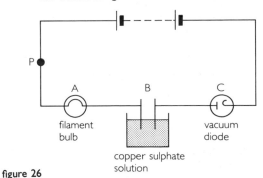

figure 26

   d Calculate approximately the speed reached by
   one of the charge carriers in the diode if the
   p.d. across it is 100 V.

16 In many school laboratory experiments, electrons
are accelerated through a p.d. of 5000 V, and
then strike a screen. The beam current may be
about 2.0 mA. Calculate (assuming
$e = 1.6 \times 10^{-19}$ C, $m = 9.1 \times 10^{-31}$ kg)
   a the energy acquired by each electron;
   b the speed of each electron;
   c the number of electrons hitting the screen
   each second;
   d the number of electrons in flight at any one
   time, if the tube is 0.10 m long;
   e the rate of conversion of energy at the screen.
   Comment on this answer in relation to the
   brightness that you observe when doing this
   experiment, compared to, say, a bulb of the
   same power.

17 Manufacturers sell resistors covering a wide
range of values; also the resistors are specified
according to the power they can tolerate.
   a Explain why, for a given construction of carbon
   resistor, the 2 W resistors are (i) all of similar
   size, (ii) bigger than the 1 W resistors.
   b One of the available values in the 1 W range is
   10 Ω. What are the maximum values this
   resistor should have for (i) the current through
   it (ii) the p.d. across it?

18 This question is about the various electrical
quantities and units.
   a What quantity would the unit $1\,CA^{-1}$ measure?
   b How would you usually express this quantity:
   '300 CA^{-1}'?
   c What is $1\,VA^{-1}$ usually called?
   d Convert the quantity 1 kWh into an SI quantity.
   e What simple quantity is the same as $5\,WsV^{-1}$?
   f Convert the unit $1\,VJ^{-1}s^{-1}$ into a unit
   containing only fundamental SI units.

19 A diode has the following characteristic graph
(figure 27).

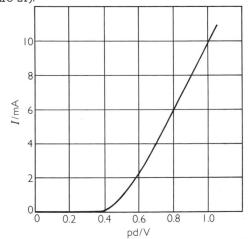

figure 27

Sketch the graph of I against V which would result if such a diode were placed in the circuits below. Show as much quantitative information as possible, in the range indicated.

(a)

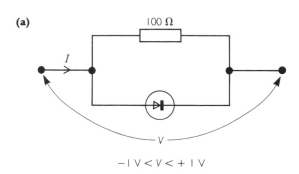

$-1V < V < +1V$

(b)

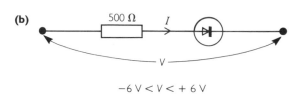

$-6V < V < +6V$

(c)

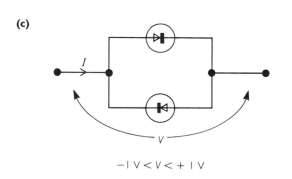

$-1V < V < +1V$

(d)

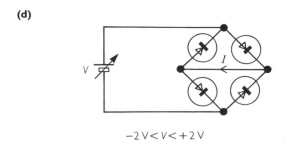

$-2V < V < +2V$

figure 28

20 Two circuit components A and B have I-against-V graphs as shown in figure 29.

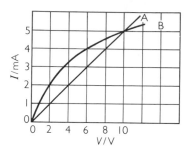

figure 29

a Describe in words how the resistance of each component changes with V.
b Suggest what B might be. For the suggestion you have made, say how you would expect the graph to continue beyond $V = 10$ V, up to say $V = 40$ V, and explain your reasoning.
c The two components are connected in series with a supply (as in figure 30), and a current of 3 mA flows through them. What is the p.d. across the supply terminals?

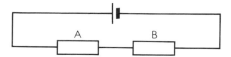

figure 30

d With the components connected as in (c), the supply voltage is changed to 6 V. What current passes through them now?
e A and B are connected in parallel with a supply (as in figure 31).
If the supply voltage is 8 V, what current comes from it?
f With A and B still connected as in figure 31, the current from the supply is adjusted to 7 mA. To what voltage has the supply been set?
g If A and B are connected in series as in figure 30, sketch two graphs, on the same axes, to show how the power expended in each component varies with the supply voltage, up to a supply voltage of 20 V.

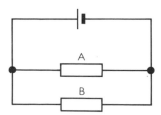

figure 31

21 You have several bulbs, with the following manufacturer's specifications.

1. 240 V   60 W
2. 1.25 V   0.25 A
3. 6.5 V   0.15 A
4. 3.5 V   0.3 A
5. 6.5 V   2 W
6. 6.5 V   0.06 A

In each of the situations below (figure 32), describe carefully what you think you would observe as the p.d. is raised slowly from zero to the value indicated.

(a)

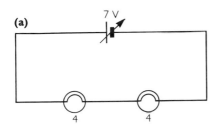

(b)

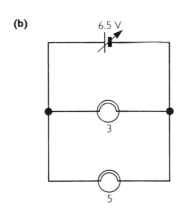

(c)

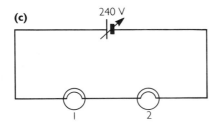

(d)

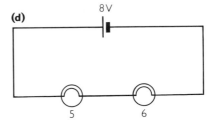

(e)

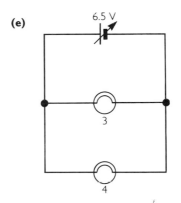

(f)

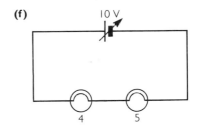

(g)

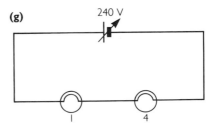

figure 32

CHAPTER 10

Resistivity

Introduction

This chapter deals in quite a detailed way with the electrical conduction properties of the variety of materials we know: metals, semiconductors, insulators. We start by considering experiments leading to the concept of resistivity: a quantity which is a property of each material, giving a measure of its resistive nature. Then we develop a model for conduction in a metal, and show how this accords with experiment – in particular, how the effect of temperature change on resistivity may be predicted. We go on to discuss semiconductors, intrinsic, n- and p-type; and finish with a brief look at insulators.

GCSE knowledge

For this chapter you will need a clear understanding of resistance, either from GCSE or from chapter 9. Other ideas which come in briefly are about force, acceleration and velocity. It will help to know that some electrons are attached to atoms, others are free to conduct. An idea about electrons in shells round atoms would help at one stage, when we look at semiconductors. Most of this chapter, though, does not require previous knowledge from GCSE.

Resistance of wires

You should at some stage use the usual resistance-measuring circuit of figure 1 to investigate how the resistance R of a piece of wire depends on its dimensions – length ℓ and cross-sectional area A.

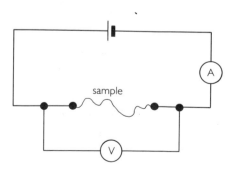

figure 1   Circuit to measure resistance

For any particular material, you should find this (provided the wire doesn't get too warm):

$$\text{resistance } R \propto \text{length } \ell \qquad \text{(i)}$$

(if you double the length of wire, you double the resistance);

$$\text{resistance } R \propto \frac{1}{\text{area } A} \qquad \text{(ii)}$$

(if you double the cross-sectional area you halve the resistance).

Doubling the length of wire is the same as adding a second length in series with the first; so this is compatible with the rule that resistors in series add up.

Doubling the cross-section of the wire is like adding a second wire in parallel; and the result that the resistance is halved is compatible with the rule for combining resistors in parallel.

Example A metre of copper wire with diameter 0.10 mm has resistance 2.2 Ω. What would be the resistance of a 5.0 m length of copper with diameter 0.20 mm?

Let us consider changing one factor at a time, length first. Length is increased by a factor 5, so new resistance would be

$$2.2\,\Omega \times 5 = 11\,\Omega$$

if only the length altered. Area is increased by a factor 4 (since diameter is doubled); so new resistance is

$$11\,\Omega \div 4$$
$$= 2.8\,\Omega$$

Resistivity

The treatment offered in the above example is rather cumbersome, and becomes difficult if the numbers involved are awkward. We can do better by developing an equation.

From equations (i) and (ii) above, we can obtain

$$R \propto \frac{\ell}{A}$$

We insert a constant of proportionality, ρ, a property of the material which we call its *resistivity*. We then obtain

$$R = \frac{\rho\ell}{A}$$

We can make ρ the subject of the equation to obtain

$$\rho = \frac{RA}{\ell}$$

Thus the unit for ρ is $\Omega\,m^2/m$, or $\Omega\,m$.

Example Calculate ρ for copper, using the data above.

$$R = 2.2\,\Omega$$
$$\ell = 1.0\,m$$
$$A = \pi r^2$$
$$= \pi \times \left(\frac{0.1 \times 10^{-3}}{2}\right)^2 m^2$$
$$= 7.9 \times 10^{-9}\,m^2$$
$$\rho = \frac{RA}{\ell}$$
$$= \frac{2.2\,\Omega \times 7.9 \times 10^{-9}\,m^2}{1.0\,m}$$
$$= 1.7 \times 10^{-8}\,\Omega\,m$$

The resistivities of different materials vary very widely. Materials are divided, slightly arbitrarily, into three categories: conductors (low resistivity), semiconductors (medium) and insulators (high resistivity). The resistivity of every material is affected by its temperature: the resistivities of all metals increase with temperature, whereas those of other materials usually decrease. (These effects are discussed further later in the chapter.) A table showing some resistivity values is given below.

Conductivity

Sometimes physicists find it useful to refer to the *conductivity* of a material, as opposed to its resistivity. This quantity allows us to make a direct analogy with thermal conduction by materials. As you may have guessed, the electrical conductivity σ of a material is the inverse of its resistivity:

$$\sigma = \frac{1}{\rho}$$

The units for σ are $\Omega^{-1}m^{-1}$.

Material	Resistivity at 20°C/$\Omega\,m$	Change in resistivity with rising temperature	Classification
silver	1.6×10^{-8}	increases	conductor
copper	1.7×10^{-8}	increases	conductor
lead	2.1×10^{-7}	increases	conductor
graphite	8.0×10^{-6}	deceases	semiconductor
germanium	5.0	deceases	semiconductor
silicon	2.5×10^3	decreases	semiconductor
glass	10^{12}	decreases	insulator

Metallic conductors

In most metals, current is carried by free electrons: electrons which are not bound at all to their parent atoms, but are free to roam within the volume of the piece of metal. Such electrons have rapid, random motion within the metal, rather like the molecules of a gas; typical electron speeds at room temperature are around $10^5 \, \mathrm{ms}^{-1}$.

When a p.d. is placed across the ends of a metal sample, each electron experiences a force in the same direction. Hence a systematic drift of electrons in one direction is superimposed on their random motion. On page 111 we calculated the drift speed of electrons carrying a typical current in a copper sample as $3 \times 10^{-7} \, \mathrm{ms}^{-1}$: very much slower than the random speeds at which they move.

A model for resistivity of metals

A piece of theory at this stage gives some interesting results. Consider a length of metal carrying a current, with the parameters shown in figure 2.

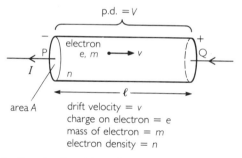

figure 2   Sample of metal conductor

Let us assume that an electron moves something like this. The force on the electron accelerates it to the right, and it gains drift velocity; but after a short time it collides with an atom, and loses its drift velocity completely. The process then repeats itself. See figure 3.

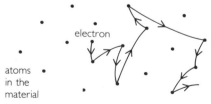

figure 3   Random path of an electron, with drift velocity to the right

There is a force F pulling the electron towards the right. The work done by this force in taking the electron from P to Q is $F \times \ell$. The work done is equal to the loss in electrical potential energy

$$\Rightarrow \qquad F \times \ell = eV$$

$$\Rightarrow \qquad F = \frac{eV}{\ell}$$

and so the acceleration a of the electron is given by

$$a = \frac{eV}{\ell m}$$

Now let us say that on average there is a time t between each collision the electron makes with an atom. Then the drift velocity v' acquired before the next collision will be:

$$v' = \text{acceleration} \times \text{time}$$

$$\Rightarrow \qquad v' = \frac{Ve}{\ell m} . t$$

The average drift velocity over the whole time t will be just half of this, that is,

$$v \, (\text{drift}) = \frac{Vet}{2\ell m}$$

Now the resistivity of the material is given by

$$\rho = \frac{RA}{\ell}$$

But

$$R = \frac{V}{I}$$

so

$$\rho = \frac{VA}{I\ell}$$

But the transport equation gives I in terms of n, A, v, e:

$$I = nAve$$

Thus

$$\rho = \frac{VA}{nAve . \ell}$$

$$= \frac{V}{ne\ell} . \frac{1}{v}$$

Using the equation above that we have derived for v, we obtain

$$\rho = \frac{V}{ne\ell} . \frac{2\ell m}{Vet}$$

$$= \frac{2m}{ne^2 t}$$

The interesting features of this expression are the n and t on the bottom line.

The n factor predicts that materials with a high density of free electrons will have low resistivity – as we would expect. It is thought that for a given metal n stays almost constant, whatever the temperature.

The t factor predicts that when the electrons are colliding frequently with metal atoms (i.e. t is low), the resistivity will be high. This will occur when the random speeds of electrons are high, that is, at higher temperatures. In fact, metal resistivities do all rise with temperature; they also tend to become very small at low temperatures, when random speeds would be very small and the t factor would be very large. See figure 4.

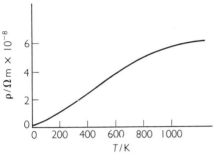

figure 4 Resistivity ρ of copper at varying temperatures

Pure semiconductors

In an intrinsic, or pure, semiconductor, such as germanium or silicon, charge is carried partly by free (conduction) electrons, and partly by 'holes'; a hole is a site in the outer electron shell of an atom where an electron is missing because it has become free. When a p.d. is applied across the material, (valence) electrons attached to atoms can move into nearby holes, without needing to become free; the effect is that the hole moves in the direction of the p.d. as though it were a positive charge. See figure 5.

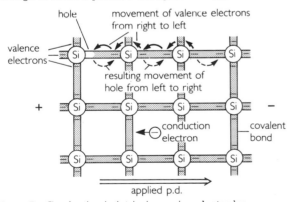

figure 5 Conduction in intrinsic semiconductor by electrons and holes

It is reasonable to expect a theory similar to the one we derived for metals (page 122) to apply to intrinsic semiconductors: in which case we would expect

$$\rho \propto \frac{1}{nt}$$

n will be the density of conduction electrons, and also the density of holes. n increases very rapidly with temperature, as more atoms acquire enough energy to release an electron. The increase in n swamps the

decrease in t that occurs in the same way as with a metal; and the net result is that the resistivity of semiconductors falls rapidly as temperature rises.

n- and p-type semiconductors

The conduction properties of an intrinsic semiconductor can be radically altered by adding quite small traces of impurity materials.

Figure 5 illustrates a pure four-valent material behaving as an intrinsic semiconductor. Both silicon and germanium are materials of this type. The addition of a small proportion of five-valent impurity, such as antimony or arsenic, adds extra conductor electrons: each impurity atom behaves within the crystal structure like its four-valent neighbours, meaning that the spare fifth electron in its outer shell is released for conduction. See figure 6. The majority of the charge carriers are thus negative electrons; and the material is called an n-type semiconductor.

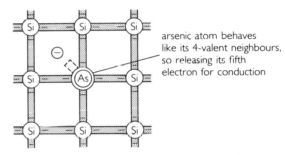

figure 6 Five-valent impurity atom creating an n-type semiconductor

If a three-valent impurity, such as indium or gallium, is added to a four-valent intrinsic semiconductor, the impurity atoms again behave in the lattice as though they were four-valent. In this case, though, the effect is to add extra holes. See figure 7.

The material now has a majority of positive holes as charge-carriers, and is called a p-type semiconductor.

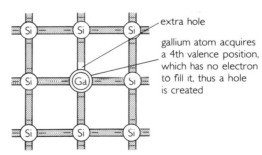

figure 7 Three-valent impurity atom creating a p-type semiconductor

Insulators

We can consider insulators as an extreme case of intrinsic semiconductors, in which the energy required to liberate conduction electrons is very great. At normal temperatures the materials contain very small numbers of conduction electrons indeed, and any conduction which occurs is due to impurities or stray ionisation by, for example, background radiation.

Conduction can occur in insulators in two possible ways.

1 With a large p.d., any stray charged particles in the material may acquire enough kinetic energy between collisions with atoms to ionise the atoms they hit. This then causes an avalanche of further charged particles, leading to electrical breakdown of the insulator, and conduction. This occurs, for example, during a lightning flash through air.

2 If the material becomes hot, a few atoms may acquire enough energy to release conduction electrons, allowing the material to conduct. As with intrinsic semiconductors, the resistivity decreases with rising temperature.

Summary

1 For samples of material,
$$R \propto \ell$$
$$R \propto \frac{1}{A}$$

2 $R = \dfrac{\rho \ell}{A}$ where ρ = resistivity; unit: $\Omega\,\text{m}$.

3 Conductivity $\sigma = \dfrac{1}{\rho}$

4 Materials are classified as conductors, semiconductors or insulators.

5 a Metals conduct because of free electrons.
  b The density of free electrons in a metal stays constant with temperature; but because electrons make collisions more often as the metal gets hotter, the resistivity rises with temperature.

6 a Pure (intrinsic) semiconductors conduct because of conduction (free) electrons and holes.
  b At higher temperatures there are more conduction electrons and holes, and resistivity goes down.
  c n-type semiconductors have more electrons than holes.
  d p-type semiconductors have more holes than electrons.

7 a Insulators have very few conduction electrons indeed.
  b Insulators may break down and conduct with high electric fields across them, or at high temperatures.

Questions

1 Calculate the resistances of the following specimens of mild steel. The resistivity for this material is $15 \times 10^{-8}\,\Omega\,\text{m}$.
  a a wire 2.0 m long, with cross-sectional area $1.0 \times 10^{-4}\,\text{m}^2$;
  b a wire 2.0 km long, with radius 0.50 mm;
  c a rod 0.40 m long, with diameter 2.0 cm;
  d a wire 5.0 km long, with diameter 0.20 mm.

2 A square $1\,\text{cm} \times 1\,\text{cm}$ of conducting paper has good electrical contact made between opposite sides, as in the two parts of figure 8. The resistance between opposite sides is then $1000\,\Omega$.

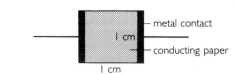

figure 8(a)

  a Estimate, using reasoned arguments, the resistance of pieces of similar paper in the following situations:

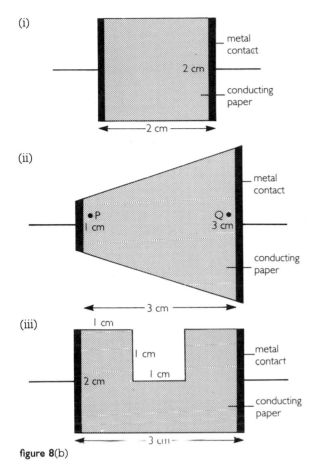

(i)

metal contact

2 cm

conducting paper

2 cm

(ii)

metal contact

•P
1 cm

Q•
3 cm

conducting paper

3 cm

(iii)

1 cm

1 cm

1 cm

metal contact

2 cm

conducting paper

3 cm

figure 8(b)

b Which of your answers would you regard as being the most accurate? Explain.
c Suppose that in piece (ii) of part (b) a certain current passes. What is the ratio of the drift speeds of charge carriers at P to those at Q?

3 Estimate the resistance of a pencil lead.
ρ for graphite $= 3.5 \times 10^{-5}\,\Omega\,\text{m}$.

4 Wood is thought to have a resistivity of about $10^{10}\,\Omega\,\text{m}$.
a Estimate the resistance from end to end of a laboratory metre rule.
b Would you expect to be able to detect current passing through this ruler if it were connected to a 1.5 V cell, using equipment easily available in your own lab? Explain your answer, and, if yes, specify the meter you would use.

5 The quantity resistivity (ρ) sums up the electrical behaviour of a material using one parameter. But usually engineers find it more convenient to work with tables showing the resistance per metre, R_l, for different diameters of wire, of various materials.
Complete the spaces in the following table,

which shows values for resistance per metre, in $\Omega\,\text{m}^{-1}$, for three different diameters of three different materials.

Material	$\rho/10^{-8}\,\Omega\,\text{m}$	R_l for different diameters of wire/$\Omega\,\text{m}^{-1}$		
		0.020 mm	0.10 mm	(a) mm
Cu	(b)	55	(c)	0.14
Al	2.6	(d)	(e)	(f)
Constantan	47	(g)	(h)	(i)

6 Some car owners carry 'jump leads', for connecting their starter motor to a friendly neighbour's battery when their own is flat. Suppose the specification for the leads is as follows:
the leads must consist of two wires each 4 m long;
they will be made from copper ($\rho = 1.8 \times 10^{-8}\,\Omega\,\text{m}$);
they must be able to carry 100 A, and cause a drop in voltage of no more than 1 V.

What is the minimum diameter for the wires?

7 A student sets up an experiment to try to determine the resistivity of plastic bag material. He sandwiches a single thickness of the bag between a pair of conducting plates which measure 25 cm × 25 cm. With 12 V between the plates, he detects a current of 300 pA. He measures the thickness of the plastic as 0.3 mm.
a Calculate a value for the resistivity of the material.
b He finds in a book that the resistivity of some plastic materials is about $10^{15}\,\Omega\,\text{m}$. This makes him suspicious about the results of his experiment. Suggest a possible non-trivial hypothesis to explain his results, and outline a procedure he could use for testing your hypothesis.

8 a What is the unit of conductivity, σ, (i) in terms of ohms, (ii) in terms of fundamental SI units?
b What are the conductivities of (i) aluminium ($\rho = 2.6 \times 10^{-8}\,\Omega\,\text{m}$), (ii) constantan ($\rho = 4.7 \times 10^{-7}\,\Omega\,\text{m}$)?

9 This question is about strain gauges, which are widely used throughout the engineering world to sense changes in length of, for example, mechanical machine parts and struts within structures. A typical gauge consists of a zig-zag pattern of thin conducting foil (often a copper–nickel alloy) printed on a base. The base is glued firmly onto the component being tested, so that both base and foil change length with the sample. See figure 9.

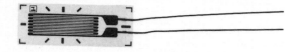

figure 9

When the foil changes length, its area of cross-section also changes, so that its volume stays roughly constant. If ℓ_0 and A_0 refer to its initial length and area, then its initial resistance R_0 is given by

$$R_0 = \frac{\rho \ell_0}{A_0}$$

For one particular gauge, the following data apply:
R_0 $120\,\Omega$
max. permissible strain   4%
total length of foil   50 mm
average width of foil   0.040 mm
resistivity of foil materials $4.0 \times 10^{-7}\,\Omega\,\text{m}$.

a   Calculate the thickness of the foil.
b   If the length of the foil increases by 4%, by what factor roughly will its area change?
c   By what factor will its resistance change? What assumption are you making?
d   What will be its new resistance?

Note that these changes are the maximum possible. In many uses the gauge will be sensing strains very much smaller than in the above calculations.

10   Read the passage below, then answer the questions which follow it.

p–n junction diode
If p- and n-type regions of the same crystal of germanium or silicon are in contact, the resulting component acts as a diode – it has quite different resistances in different directions. Figure 10 helps to explain this.

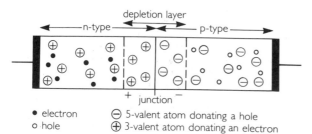

figure 10

'Throughout the material, holes and electrons have random motion ("thermal motion"), like the molecules in a gas. At the junction, this motion causes some holes and electrons to stray to the opposite side; a hole straying into the n-type region will neutralise one of the conduction electrons there, and vice versa. A small region round the junction thus has many of its charge-carriers neutralised, and becomes effectively insulating – it is known as the "depletion layer". Overall, n- and p-type materials are of course usually neutral; but because of this migration of electrons and holes, the n-type side of the depletion layer is now positive, while the p-type side is negative. These regions of charge create an electric field in the layer (from left to right in figure 10), and this field acts to discourage more holes and electrons from crossing the junction.

'Now suppose a cell is connected as in figure 11. This cell acts to *increase* the field across the junction. This discourages holes and electrons even more from crossing; thus the diode has high resistance in this direction. Another way of looking at it is to say that the positive end of the battery draws electrons in the n-type material away from the depletion layer – and similarly the

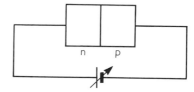

figure 11

negative end of the battery draws holes away on the p-type side – making the layer even wider and more insulating.

'On the other hand, the cell of figure 12 acts to decrease the left-to-right field which occurs naturally across the junction. Holes and electrons are both pushed towards the junction from their respective sides. At a certain voltage in this direction the depletion layer disappears altogether, and the diode conducts easily.

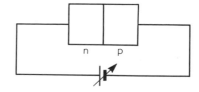

figure 12

'The graph of figure 13 shows the graph of current against p.d. for a typical silicon diode in the forward direction. In the reverse direction (figure 11) there may be a current of order 10^{-8} A.

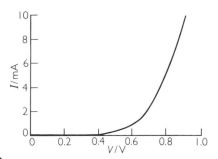

figure 13

n–p–n junction transistor
'This device consists of two p–n junctions back to back. See figure 14.

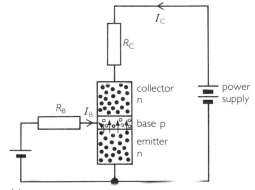

figure 14

'If the lower junction has a small current passing through it in the forward direction, then many electrons from the n-type emitter cross this junction. The base layer is very thin, and a large proportion of these electrons pass right across the upper junction to the n-type collector, which is made positive by the power supply. They then constitute I_C. Furthermore, small changes in I_B cause large changes in I_C; and so we have a device which amplifies currents. By connecting suitable resistors R_B and R_C, we can make it amplify p.d.s as well.'

a Spell out in words the sentence which 'vice versa' in the second paragraph implies.

b Why is n-type material 'usually neutral', even though there are far more conduction electrons in it than holes?

c For the silicon diode whose behaviour is illustrated in figure 13, what p.d. in the forward direction is needed to reduce the width of the depletion layer to zero?

d What is the resistance of the diode in the forward (conducting) direction when (i) the current through it is 10 mA, (ii) the p.d. across it is 0.7 V?

e Some transistors are made p–n–p instead of the n–p–n type described. Draw a circuit showing the directions of currents and cell connections which you think would be appropriate for a p–n–p transistor.

11 An ordinary laboratory lead is made of stranded copper. There are 32 strands, each of copper wire of diameter 0.20 mm. (ρ for copper = $1.7 \times 10^{-8}\,\Omega\,\text{m}$.)

a What is the resistance of a lead 1.5 m long?

b This lead is rated at 6 A maximum. What power would be generated in it at that current?

c Why is it better to use stranded wire than to make the lead from solid copper?

12 An engineer is considering the design of a high voltage power cable, to transmit up to 1 GW at 475 000 V. One idea is for six 1 cm diameter strands of aluminium to be grouped round a single 1 cm diameter strand of steel. (ρ for Al = $2.7 \times 10^{-8}\,\Omega\,\text{m}$; ρ for steel = $9.0 \times 10^{-8}\,\Omega\,\text{m}$.)

a What maximum current will the cable carry?

b What resistance will one metre of this cable have?

c How much energy will be lost as heat in each metre of cable when the cable transmits a nominal 1 GW?

d Up to what approximate distance will this cable be able to transmit a nominal 1 GW and lose less than 1% of the input power?

13 Read the article below, then answer the questions which follow it.
(from 'The development of ion implantation technology in the U.K. semiconductor industry', K. E. Dickson, *Physics in Technology*, vol. 16, (Institute of Physics, 1985)).

Ion implantation

Ion implantation is one of several techniques for introducing electrically charged particles (ions) into solid materials thus enabling the material to acquire various stable regions with differing electrical properties. This latter phenomenon is the basis of all semiconductor technology, so that all semiconductor devices, from the earliest transistors to the latest microelectronic circuits, have required 'doping' (as the process of ion introduction is called) during their manufacture. Initially and during much of the history of semiconductor production, thermal diffusion techniques satisfied this need but more recently, as more exacting requirements in semiconductor manufacture have developed, they have been replaced by ion implantation techniques which are able to provide a more controlled doping pattern.

Ion implantation not only permits greater control over the number of 'impurity' ions introduced into the targeted material but also, and more importantly, it also produces more accurate distribution of the dopant within the targeted material, as indicated in figure 1. In semiconductor devices this sharper profile is crucial as it permits much finer circuit designs to be manufactured. The method of introducing ions into solids requires sophisticated high-voltage equipment that can electrically accelerate the ions and direct them onto the targeted materials which in the semiconductor industry would normally be wafers of silicon.

Essentially therefore, an ion implanter incorporates an ion source, high-voltage power supply, analysing magnets (to purify the ion beam), accelerator structure, target chamber and some form of scanning system, along with suitable controls and monitoring instruments. A simplified, schematic layout of an ion implanter is shown in figure 2. Typically, the acceleration energy available in today's machines will range from 20 to 200 keV with the ion beam currents ranging from around 1 mA to over 10 mA depending on the type of machine and the ions being accelerated.

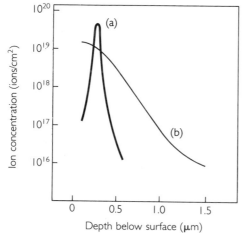

figure 1   Typical concentration distributions of impurity ions in silicon brought about by (a) ion implantation and (b) thermal diffusion techniques (after Dearnaley and Freeman, 1969)

a   What is the purpose of introducing impurity atoms into silicon?

b   Why is it desirable to be able to manufacture much finer circuit designs?

c   Although the impurity atoms are introduced to the silicon in the form of ions, they are unlikely

to remain ions for long after they have been implanted. Why is this?

d   Roughly, the conductivity of the silicon is proportional to the number density n (usually measured in units of m^{-3}) of the impurity atoms.

(i) Explain this.

(ii) Compare the conductivity of a piece of silicon doped by ion implantation at depths of 0.25 μm and 0.5 μm.

(iii) Compare the conductivity of a piece of silicon doped by thermal diffusion at the same depths.

e   What do you think 'thermal diffusion' means in the context of introducing impurity atoms?

f   What power would be incident on the specimen at the highest values of p.d. and current mentioned?

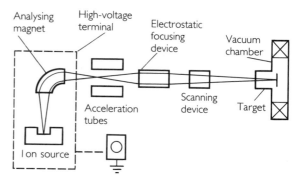

figure 2   Typical commercial implanter (from Ryssel and Glawischnig, 1982)

g   What is the purpose, in the apparatus of figure 2 in the article, of (i) the analysing magnets, (ii) the scanning device?

CHAPTER 11

Circuits with resistors

Introduction

This chapter covers all the ideas you will need about resistors within circuits. Resistance was defined in chapter 9. Now we see how the effects of several different resistors in one circuit can be calculated: first in series and parallel, then in more complex situations. We apply the ideas to the practical problem of converting meters to measure other ranges of current and p.d. We look at the potential divider circuit, and the potentiometer method of comparing p.d.s. The Wheatstone bridge circuit provides a useful control technique, while Kirchoff's laws offer us a way of analysing a circuit theoretically. Also we consider the effects that the internal resistance of a supply can have.

GCSE knowledge

To follow this chapter you will need a thorough understanding of the concepts of current and p.d. You will have met these during your GCSE course; but they are also thoroughly recapped and expanded in chapter 9.

Resistors in series

It is useful to be able to calculate the combined resistance of several resistors connected in series. We can derive a simple rule for this, by applying some of the circuit rules discussed in chapter 9. Consider the situation of figure 1.

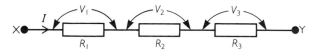

figure 1   Three resistors in series

The relevant circuit rules are that the current is the same through each resistor; and that the p.d. between X and Y is the sum of the p.d.s across each resistor.

Using the latter rule, V_{XY}, the p.d. between X and Y, is given by

$$V_{XY} = V_1 + V_2 + V_3$$
$$= IR_1 + IR_2 + IR_3$$

But if R_T is the effective resistance between X and Y, then

$$V_{XY} = IR_T$$
$$\Rightarrow \qquad R_T = R_1 + R_2 + R_3$$

Resistors in parallel

As with resistors in series, we can apply circuit rules to derive a rule for calculating the combined resistance of two or more resistors in parallel. Consider figure 2.

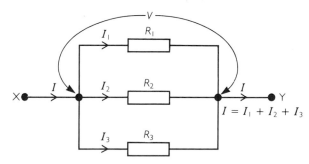

figure 2   Three resistors in parallel

The appropriate circuit rules in this case are that each resistor has the same p.d. across it (V); and that the total current is the sum of the currents through each resistor.

The current in R_1 is V/R_1; and similarly for R_2 and R_3.

Thus
$$I = \frac{V}{R_1} + \frac{V}{R_2} + \frac{V}{R_3}$$

If R_T represents the combined resistance between X and Y, then

$$I = \frac{V}{R_T}$$

$$\Rightarrow \qquad \frac{V}{R_T} = \frac{V}{R_1} + \frac{V}{R_2} + \frac{V}{R_3}$$

$$\Rightarrow \qquad \frac{1}{R_T} = \frac{1}{R_1} + \frac{1}{R_2} + \frac{1}{R_3}$$

This is a rather cumbersome formula to apply in practice. But in many cases we are dealing with only two resistors; and in that case by making R_T the subject of the equation we can reach a much more manageable formula:

$$R_T = \frac{R_1 R_2}{R_1 + R_2}$$

As an example, let us calculate the combined resistance of $1.0\ \Omega$ in parallel with $100\ \Omega$.

$$R_T = \frac{1.0\ \Omega \times 100\ \Omega}{1.0\ \Omega + 100\ \Omega}$$

$$= \frac{100}{101}\ \Omega$$

$$= 0.99\ \Omega$$

Note that when two resistors which are very different in value are in parallel, the combined value of resistance is just a little lower than the smaller individual one.

Adapting a moving-coil meter

A moving-coil meter (see figure 4 on page 108) consists of a coil of wire positioned in a magnetic field. When current passes through the coil, it rotates through an angle which depends on the current, and comes to rest there.

The basic meter can be adapted to read any range of currents or voltages you like (provided these are above the full-scale-reading values: see page 263). To perform the necessary calculations (which provide good examples of applying circuit ideas), we need to know two features of the meter: the resistance of its coil, and the current it needs to give a full-scale reading.

In this photograph you can see thousands of lights. Are they all connected in series or in parallel?

Example We will do some calculations on a meter with coil resistance 1000 Ω, and full-scale-reading current of 100 μA. The p.d. for full-scale reading for this basic meter is given by

$$V = IR$$

$$= 100 \,\mu\text{A} \times 1000 \,\Omega$$

$$= 0.10 \,\text{V}$$

1 *To convert the meter into an ammeter to read higher currents,* we need to connect a 'shunt' resistor in parallel with it, so that some of the excess current bypasses the meter itself. What value of shunt resistor would we need to make it read currents up to 1 A? See figure 3.

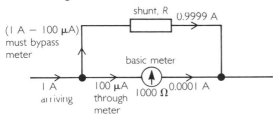

figure 3 Adapting a basic meter to read a higher current

When 1 A arrives at the junction, 100 μA must be made to pass through the meter, so that it will read full-scale.
Therefore 1.0000 A – 100 μA, or 0.9999 A, must pass through the shunt.

p.d. across shunt = p.d. across meter

$$\Rightarrow \qquad R \times 0.9999 \,\text{A} = 1000 \,\Omega \times 100 \times 10^{-6} \,\text{A}$$

$$\Rightarrow \qquad R = 0.10001 \,\Omega$$

$$\approx 0.1 \,\Omega$$

The complete ammeter to read up to 1 A thus consists of the basic meter (resistance 1000 Ω), with the shunt resistor of about 0.1 Ω in parallel with it.
The total resistance of the meter is 0.1 Ω, since there is a p.d. of 0.1 volt across it and a total of 1 A through it. This is likely to be negligible in many circuits where there are currents of order 1 A; and therefore this ammeter will generally fulfil the requirement that its resistance be negligible.

2 *To adapt the same basic meter to read p.d.s higher than 0.1 V,* we need to connect a 'multiplier' resistor in series with it, so that some of the higher p.d. can be across the multiplier. What value of multiplier would we need to make it read p.d.s up to 1 V?
Figure 4 shows the situation in which the meter will be intended to read full-scale.

We know the p.d. across the meter must be 0.1 V.

$$\Rightarrow \quad \text{p.d. across multiplier} = 1.0 \,\text{V} - 0.1 \,\text{V}$$

$$= 0.9 \,\text{V}.$$

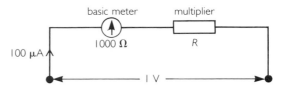

figure 4 Adapting a basic meter to read a higher p.d.

$$\Rightarrow \qquad R = \frac{0.9 \,\text{V}}{100 \,\mu\text{A}}$$

$$= 9000 \,\Omega$$

Note that it gives the complete voltmeter a resistance of 10 000 Ω. For many circuits this will be sufficiently high for the voltmeter not to load the circuit significantly; this is equivalent to saying that the current to drive the meter, 100 μA, is negligible compared to the other currents in the circuit.
Note also a rather useful piece of knowledge about resistors in series: the division of the total p.d. of 1 V between them is in the same ratio as their resistances.

$$\frac{0.9 \,\text{V across multiplier}}{0.1 \,\text{V across meter}} = \frac{9000 \,\Omega \text{ for multiplier}}{1000 \,\Omega \text{ for meter}}$$

Circuit-loading by meters

Let us consider briefly how ammeters and voltmeters might affect the currents and p.d.s they are intended to measure. Consider this simple circuit (figure 5), with a single cell and two resistors.

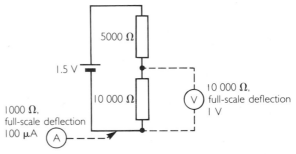

figure 5 To illustrate the effect on a circuit of an ammeter and a voltmeter

Without meters in the circuit, the current I_0 should be given by

$$I_0 = \frac{1.5 \,\text{V}}{10\,000 \,\Omega + 5000 \,\Omega}$$

$$= 1.0 \times 10^{-4} \,\text{A}$$

$$= 100 \,\mu\text{A}$$

In principle, therefore, if we insert the basic meter from the last section as an ammeter into the circuit as shown, it should read full-scale (100 μA).
In fact, however, inserting this meter adds a resistance of 1000 Ω in series in the circuit. So the

actual current, I_m, is given by

$$I_m = \frac{1.5\ \text{V}}{10\,000\ \Omega + 5000\ \Omega + 1000\ \Omega}$$

$$= 94\ \mu\text{A}$$

Thus the ammeter, which reads I_m, 94 μA, reads a different current to that which we set out to measure (100 μA).

What about the p.d.s in the circuit? Without any meters, the p.d. V_0 across the 10 000 Ω resistor is given by

$$V_0 = 100\ \mu\text{A} \times 10\,000\ \Omega$$

$$= 1.0\ \text{V}$$

Therefore, connecting the adapted meter from the last section as shown in figure 6 should make it read full-scale. However, we now have two resistances of 10 000 Ω in parallel; these combine to make 5000 Ω; so we now effectively have 5000 Ω and 5000 Ω in series, which will make the p.d. across each of them 0.75 V.

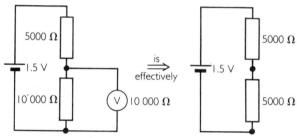

figure 6   To illustrate the effect of a voltmeter

The voltmeter, which reads the p.d. across its ends, thus reads 0.75 V, instead of the value of 1 V which we set out to measure.

The effect of inserting both ammeter and voltmeter into the circuit together is left to you to work out in question 7.

Note that this example was deliberately set using quite high resistances in the circuit; for many purposes, the meters used in the calculations above would introduce negligible amounts of error.

Potential divider circuit

Let us pursue the idea raised earlier about the division of p.d. across two resistors in series, since the principle of potential division is widely used. Consider the part of a circuit shown in figure 7.

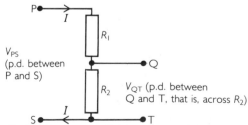

figure 7   To illustrate the division of potential

Assume for a moment that no current passes through the contacts at Q and T. Then I is given by

$$I = \frac{V_{PS}}{R_1 + R_2}$$

and

$$V_{QT} = I \times R_2$$

$$= \frac{V_{PS}}{R_1 + R_2} \times R_2$$

$$= V_{PS} \times \frac{R_2}{R_1 + R_2}$$

(This is equivalent to saying that the p.d.s across R_1 and R_2 are in proportion to their values, as we saw before.)

If R_1 and R_2 are two parts of a rheostat or potentiometer, and Q is the variable contact, then the arrangement of figure 7 allows V_{QT} to be adjusted to any value between zero and V_{PS}. (This statement is true even if current passes from Q to T.)

This circuit arrangement is very useful. Two applications are illustrated in figures 8 and 9.

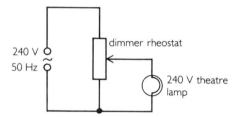

figure 8   The brightness of the lamp can be varied continuously from fully on to completely off

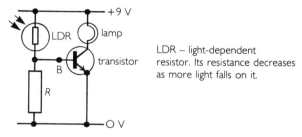

figure 9   The 9 V is divided between the LDR and the fixed resistor R. When light alters the resistance of the LDR, the p.d. input to the transistor changes. (The lamp comes on when the light on the LDR increases.)

Comparing p.d.s with a potentiometer

The idea of the potential divider circuit can be adapted to provide a method of comparing p.d.s. The principle is illustrated in figure 10.

The cell or other source of p.d. V is connected to QT through a meter, as shown. Then if the p.d. across R_2, V_{QT}, is equal to V, no current will pass through the meter.

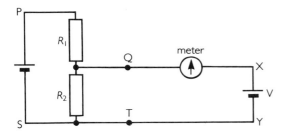

figure 10 To illustrate the principle of potentiometer measurements

To operate this principle, we make R_1 and R_2 from a single potentiometer, with Q the variable connection. Then we adjust it until the meter reads zero.

Now since $V = V_{QT}$, and V_{QT} is proportional to R_2, we have

$$V \propto R_2$$

If the potentiometer is linear, then the value of R_2 is proportional to the dial setting on it. Alternatively, R_1 and R_2 could be a continuous length of wire, with a sliding connection. In this case, R_2, and thus V, is proportional to the length of wire between S and the sliding contact. See figures 11 and 12.

figure 11 A dial potentiometer

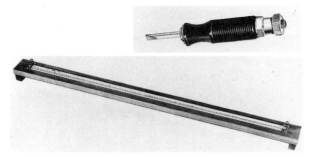

figure 12 A linear potentiometer

To compare two p.d.s, we find the value of R_2, or the length of wire ℓ_1, corresponding to no meter

current for p.d. V_1; then we repeat the process with the second p.d., V_2, between X and Y, to obtain a second length ℓ_2.

Since

$$V \propto \ell$$

we obtain

$$\frac{V_2}{V_1} = \frac{\ell_2}{\ell_1}$$

This may seem a laborious procedure, when we could use simple voltmeters for V_1 and V_2. But it can be useful in some circumstances. Since there is no current through the meter when we measure a p.d., the process places no load on the p.d.s being observed; the technique can be made extremely accurate, more accurate than moving-coil meters; and finally, no calibrated meter is needed, only one which can indicate, as sensitively as possible, when no current is passing.

The Wheatstone bridge circuit

This circuit is widely used to monitor and control situations of all kinds. Before we can discuss how this is done, we must do a little analysis of the circuit itself.

Consider the arrangement of four resistors shown in figure 13. We need to choose the values of the four resistors so that the p.d. between P and Q is roughly zero. We must work out the condition for this.

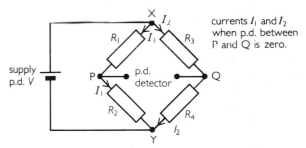

figure 13 The Wheatstone bridge circuit

If there is no p.d. between P and Q, then

potential at P, V_P = potential at Q, V_Q

$\Rightarrow$ p.d. between P and X,   p.d. between Q and X,
$$V_{PX} = V_{QX}$$

$\Rightarrow$
$$R_1 I_1 = R_3 I_2 \qquad \text{(i)}$$

Similarly
$$V_{PY} = V_{QY}$$

$\Rightarrow$
$$R_2 I_1 = R_4 I_2 \qquad \text{(ii)}$$

Dividing equation (i) by equation (ii),

$$\frac{R_1 I_1}{R_2 I_1} = \frac{R_3 I_2}{R_4 I_2}$$

$\Rightarrow$
$$\frac{R_1}{R_2} = \frac{R_3}{R_4}$$

We can use the circuit to operate a control function like this. Consider an example in which we want to maintain an oven at a fixed temperature which we can set. For simplicity, we will choose R_1 and R_2 to be equal. R_3 will be a resistance which increases with rising temperature: we will place this inside the oven. R_4 will be a variable resistor with a dial calibrated with temperatures on the control panel. See figure 14.

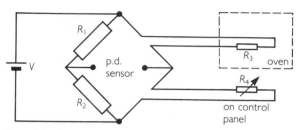

figure 14 Control of temperature using a Wheatstone bridge circuit

The setting of R_4 for a particular temperature required is equal to the value of R_3 at that temperature. So if the oven is at the right temperature, $R_3 = R_4$, and no p.d. is detected by the sensor.

If, however, the oven temperature falls, R_3 becomes less than R_4. A p.d. then appears across the sensor, which can be used to switch on the heating mechanism to return the oven to the correct temperature. Furthermore, this p.d. is roughly proportional to $(R_4 - R_3)$; so by amplifying the p.d. suitably, you can control the supply to the heater to provide more power if the oven is a long way from the correct temperature.

Sources of electrical energy: e.m.f.

To initiate an electric current, we require some device which is capable of supplying energy and setting up the necessary electric field to exert force on the electrons. Such a device must convert energy from some other form into electrical energy. Examples are:

☐ cells (converting chemical energy),

☐ turbines and dynamos (converting kinetic energy),

☐ thermocouples (converting heat).

We define the e.m.f. (electromotive force) E of a supply of electrical energy using this word equation:

e.m.f. of a source

$$= \frac{\text{energy converted into electrical form } (W)}{\text{charge passing through source } (Q)}$$

or
$$E = \frac{W}{Q}$$

Note that this is a complementary definition to that of p.d., both of them being energy converted per unit charge passing. The difference is that e.m.f. involves energy being converted into electrical energy, whereas p.d. involves it being converted from electrical energy into some other form. Thus the unit of e.m.f. is one volt.

Note also that the rate P at which the supply converts energy is given by

$$P = EI$$

Internal resistance

A source of e.m.f. cannot be 100% efficient; that is, some of the energy converted from another form is lost inside the supply. Thus the supply is said to have an 'internal resistance', r.

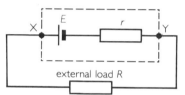

source of e.m.f. with external connections X and Y

external load R

figure 15 To illustrate the internal resistance of a source of e.m.f.

Look at the circuit in figure 15. Thinking about where the energy goes in this circuit in time t leads us to the following equation:

(energy converted into electrical energy by source)
$= $ (energy lost in r) $+$ (energy converted in R)

$$\Rightarrow \qquad EIt = I^2rt + I^2Rt$$

$$\Rightarrow \qquad E = Ir + IR \qquad \text{(i)}$$

Also
$$I = \frac{E}{r + R}$$

Note also that the p.d. V across the source, and across R, is given by

$$V = IR = \frac{R}{R+r}.E \qquad \text{(ii)}$$

It is interesting to examine how the p.d. V across a supply with internal resistance varies as more current I is drawn from it. You can do this using the circuit of figure 16.

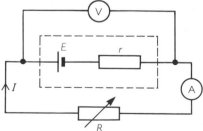

figure 16 Circuit to investigate p.d. across a supply and current

From equations (i) and (ii) we obtain a further equation to describe this situation:

$$E = Ir + IR$$
$$= Ir + V$$
$$\Rightarrow \qquad V = E - Ir$$

This equation predicts that the experiment will give a graph of V against I for the supply like figure 17.

Note qualitatively what happens: when the supply is not delivering current, the p.d. across its terminals

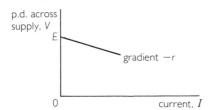

figure 17   Predicted p.d.-against-current graph for a supply

is equal to its e.m.f. E; but as you draw more current the p.d. falls.

Different sources have quite different values of internal resistance. As an example, a car battery has a small r, around $0.01\ \Omega$; but you may notice its effect when the starter motor operates, causing the lights to dim. This occurs because the starter motor takes a current of something like 200 A.

Thus V across terminals $= E - Ir$

$$= 12\ \text{V} - 200\ \text{A} \times 0.01\ \Omega$$
$$= 10\ \text{V}$$

Thus the p.d. across the battery falls far enough to be noticeable.

The mains supply to your house has an extremely small internal resistance, so falls in p.d. are barely noticeable: though if you switch on something taking a big current like an electric fire you may just notice the lights change in brightness.

By contrast, the internal resistance of a 1.5 V cell is about $0.5\ \Omega$; so when you add several bulbs in parallel, one at a time, each one draws a bit more current and makes the ones already there dimmer.

Using the data from the experiment already described above, you could also plot a graph of the power which the supply delivers to $R(= VI)$ against the value of $R(= V/I)$. The result is something like figure 18.

There is thus a maximum amount of power which you can obtain from a supply. This maximum occurs when the load resistance R equals the internal resistance r.

This fact has important applications. For example, stereo systems are designed so that the resistance of the speakers matches the internal resistance of the amplifier system: in this way the maximum amount of energy can be converted to sound. Typically speakers and amplifier both have resistance of $8\ \Omega$.

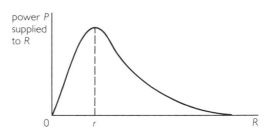

figure 18   To illustrate the maximum power that a source can supply

Analysis of circuits using Kirchoff's laws

The analysis of circuits we have carried out so far has been based on the four rules for circuits already discussed. To recap, for currents, components in series have the same current through them, while currents leaving a junction add up to those arriving; for p.d.s, all components in parallel have the same p.d. across them, while the p.d.s across components in series add up to the total p.d.

Just occasionally you may meet a circuit problem which cannot be solved using the four rules. Kirchoff's laws may help in this case – or, to be precise, his second law, since his first is the same as the currents-at-a-junction rule restated above. This second law is: 'Round any closed circuit the sum of the e.m.f.s is equal to the sum of the p.d.s.'

Algebraically this can be stated as

$$\Sigma E = \Sigma\ \text{p.d.} \qquad \text{for a closed loop}$$

This is really just an extension of the energy equation used in connection with internal resistance.

We have already used this law in simple situations, without having it stated formally. For example, it can be used to calculate the current in the circuit of figure 19.

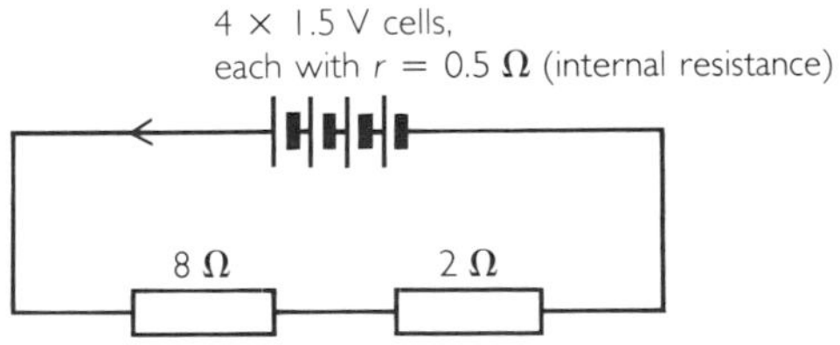

figure 19   A simple application of Kirchoff's second law

$$\Sigma E = 4 \times 1.5\ \text{V} = 6\ \text{V}$$
$$\Sigma\ \text{p.d.} = \Sigma IR$$
$$= I \times 4 \times 0.5\ \Omega + I \times 8\ \Omega + I \times 2\ \Omega$$
$$= I \times 12\ \Omega$$
$$\Rightarrow \qquad 6\ \text{V} = I \times 12\ \Omega$$
$$\Rightarrow \qquad I = 0.5\ \text{A}$$

In more complicated circuits, using this law often leads to difficult sets of equations; you should only use it if there is more than one e.m.f. source in a circuit, and other methods and ideas have not worked.

As an example, consider the circuit of figure 20; this is roughly what may happen in a car electrical system when the generator is driven by the engine. The problem is to find I_1 and I_2.

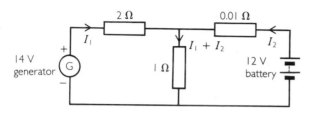

figure 20   Simplified car electrical system

Using the currents-at-a-junction rule (Kirchoff's first law):

$$\text{current in } 1\,\Omega \text{ resistor} = I_1 + I_2$$

Using Kirchoff's second law on the left hand circuit (clockwise):

$$14\,\text{V} = I_1 \times 2\,\Omega + (I_1 + I_2) \times 1\,\Omega$$
$$\Rightarrow \qquad 14\,\text{A} = 3I_1 + I_2 \qquad\qquad (\text{i})$$

Using Kirchoff's second law on the right hand circuit (anticlockwise):

$$12\,\text{V} = I_2 \times 0.01\,\Omega + (I_1 + I_2) \times 1\,\Omega$$
$$\Rightarrow \qquad 12\,\text{A} = I_1 + 1.01\,I_2 \qquad\qquad (\text{ii})$$

To solve these simultaneous equations, we can substitute an expression for I_2 from equation (i) into equation (ii):

$$12\,\text{A} = I_1 + 1.01\,(14\,\text{A} - 3I_1)$$
$$\Rightarrow \qquad 12\,\text{A} = I_1 + 14.14\,\text{A} - 3.03I_1$$
$$\Rightarrow \qquad 3.03I_1 - I_1 = 14.14\,\text{A} - 12\,\text{A}$$
$$\Rightarrow \qquad I_1 = 1.05\,\text{A}$$

Substituting this value into equation (i) gives:

$$14\,\text{A} = 3 \times 1.05\,\text{A} + I_2$$
$$\Rightarrow \qquad I_2 = 14\,\text{A} - 3.15\,\text{A}$$
$$= 10.85\,\text{A}$$

Summary

I   For resistors in series:

$$R_T = R_1 + R_2 + \ldots$$

2   For resistors in parallel:

$$\frac{1}{R_T} = \frac{1}{R_1} + \frac{1}{R_2} + \ldots$$

3   For two resistors in parallel:

$$R_T = \frac{R_1 R_2}{R_1 + R_2}$$

4   For two very different resistors in parallel, the combined resistance is just less than the smaller one.

5   To adapt a basic meter to read higher currents, connect a shunt resistor of suitable value in parallel with it.

6   To adapt a basic meter to read higher p.d.s, connect a multiplier resistor of suitable value in series with it.

7   The division of p.d. between two resistors in series is in proportion to their resistances.

8   If you insert an ammeter with significant resistance into a circuit, it will reduce the current you are trying to measure.

9   If you use a voltmeter which draws significant current to measure a p.d., it will reduce the p.d. you are trying to measure.

10   You can use a potential divider circuit to obtain a continuously-variable supply ranging from zero up to your supply voltage. See figure 8.

11   You can use a potentiometer to compare two p.d.s:

$$\frac{V_2}{V_1} = \frac{\ell_2}{\ell_1}$$

12   When a Wheatstone bridge circuit is balanced:

$$\frac{R_1}{R_2} = \frac{R_3}{R_4}$$

See figure 13.

13   A Wheatstone bridge circuit is useful for control and measurement applications because the p.d. obtained is proportional to $(R_4 - R_3)$.

14 Sources of e.m.f. convert energy into electrical potential energy from some other form.

15 Internal resistance: $E = Ir + IR$

$$V = IR$$

$$I = \frac{E}{R+r}$$

16 Internal resistance causes the p.d. of a supply to fall when you draw current from it.

17 For maximum power from a supply, $R = r$.

18 Kirchoff's laws:
(i) at a junction Σ currents in $= \Sigma$ currents out
(ii) round a closed loop $\Sigma E = \Sigma$ p.d.

Questions

1 Find a value for the resistance between points X and Y in each of the combinations of resistors shown in figure 21.

(a)

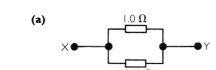

(b)

(c)   **(d)**

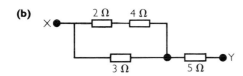

figure 21

2 In the circuits shown in figure 22, find any values of current, p.d. or resistance labelled I, V or R.

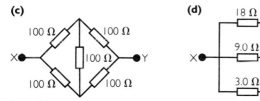

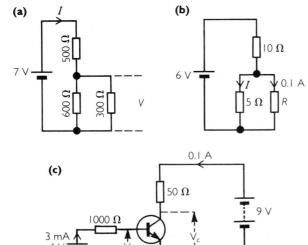

figure 22

3 Sketch a graph to show how the resistance R between points A and B of the circuit of figure 23(a) varies with the position x of the slider of the potentiometer.

(a)

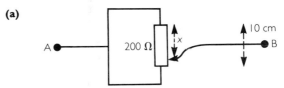

(b) R between A and B/Ω

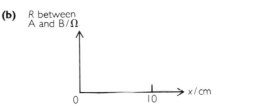

figure 23

4 A student is building circuits using a lot of individual $1\ \Omega$ resistors.
   a She first connects three as shown in figure 24. What is the resistance between A and B?

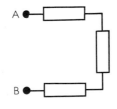

figure 24

   b She then adds three more resistors to obtain the circuit of figure 25. Calculate the resistance between C and D now.

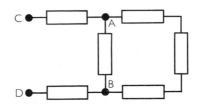

figure 25

c She then adds another three to make the circuit of figure 26. Find a value for the resistance between E and F.

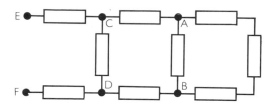

figure 26

d If she goes on adding sections, the resistance between the open ends eventually reaches a limiting value – you can probably see that this is happening from your answers so far. Call this limiting value R. Then imagine adding one more section of three resistors to a chain whose value is already R – see figure 27.

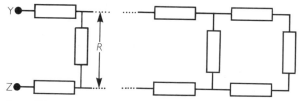

figure 27

Write down a value in terms of R for the resistance between Y and Z. But this extra section must still leave the total value equal to R: hence find R.

5 You are doing experiments with two wires of the same material and a 12 V battery. The wires are the same length, but wire X has twice the diameter of wire Y.

a When you connect them in series to the 12 V battery, which one gets hot first? Explain.

b When you connect them in parallel, which one gets hot first? Explain.

6 The following three questions refer to tests carried out, in sequence, on a four-terminal 'puzzle' box, such as that illustrated in figure 28.

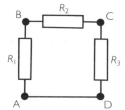

figure 28

Terminals A and D are connected by a thick wire; there may or may not be resistors connected in the arms AB, BC, CD. Explain what deduction you can make after each test. (Assume the voltmeters have high resistance, and the battery pack has negligible internal resistance.)

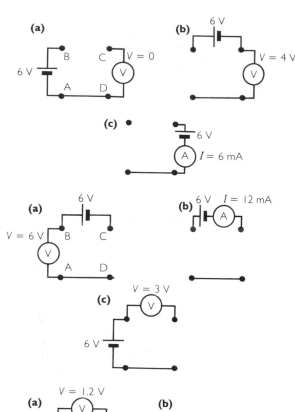

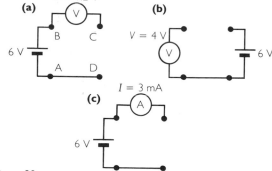

figure 28

7 This question follows on from the worked examples on page 131 about the load placed on circuits by adding meters to them. The original problem was to measure the current through and p.d. across the 10000 Ω resistor in figure 29.

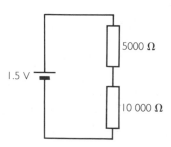

figure 29

The meters available are:
voltmeter: full-scale deflection 1 V; resistance 10 000 Ω,
ammeter: full-scale deflection 100 μA; resistance 1000 Ω.
The worked example showed the effects of connecting into the circuit each meter on its own.
a What will each meter read if connected in the circuits of (i) figure 30, (ii) figure 31?

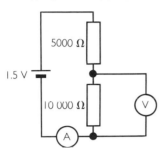

figure 30

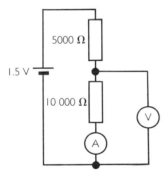

figure 31

b Which of the methods of connecting the ammeters would you regard as preferable in this particular example? Justify your answer.

8

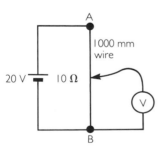

figure 32

In figure 32, AB is a uniform wire, 1000 mm long, of resistance 10 Ω. Neglect the resistances of connecting wires. The cell has e.m.f. 20 V, with negligible internal resistance.
a If V is an infinite resistance voltmeter, what would it read if the variable contact was positioned 30 cm from B?
b If V is a 1000 Ω voltmeter, where would you have to put the contact for the meter to read exactly 10 V?

9 a How would you convert a galvanometer with a 1000 Ω, 100 μA movement into a 10 V voltmeter?

It is common practice to quote voltmeter resistances as a number of ohms/volt. This unconverted meter has a resistance of 10 000 Ω/V.
b What is the corresponding quantity for the converted meter?
c Is it desirable for a voltmeter to have a high or a low number of ohms/volt?
d How does the quantity being measured in Ω/V relate to the f.s.d. (full-scale deflection) current?

10 a How would you convert a meter with resistance 100 Ω and f.s.d. current 100 μA into a 1 A ammeter?
b In practice the shunt can be adjusted to precisely the correct value like this. The meter plus roughly correct shunt are placed in series with a standard meter, and the current is adjusted to 1.00 A according to the standard meter. Suppose that when this is done the combination being adjusted reads slightly less than 1.00 A. Must the shunt resistance be increased or decreased? Explain.

11 What will the voltmeter in this circuit read if it has a resistance of (i) 2000 Ω, (ii) ∞?

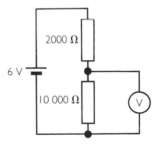

figure 33

12 A student is doing experiments using the following items:
a cell with e.m.f. 1.5 V and internal resistance 0.5 Ω;
a bulb marked 1.5 V 0.2 A;
an ammeter marked 0–5 A (resistance 0.02 Ω);
a voltmeter marked 0–5 V (resistance 50 kΩ).
He connects up, one after the other, the two circuits in figure 34.

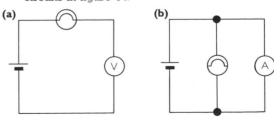

figure 34

In figure 34(a)
a will the bulb light?
b approximately what will the voltmeter read?
In figure 34(b),
c will the bulb light?
d roughly what will the ammeter read?

13 a What is the *power* which is transferred to the
fixed 20 Ω resistor in the circuit in figure 35
when the variable resistor is set at (i) 0,
(ii) 20 Ω?

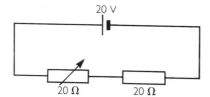

figure 35

b Sketch a different circuit, using the same
components, which will allow you to vary the
p.d. across the fixed resistor between 0 and
20 V.

14 Calculate the p.d.s marked *V* in the two circuits
in figure 36, in which the cells have zero internal
resistance.

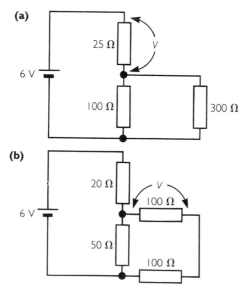

figure 36

c How would your answer to (a) be affected if
the 6 V cell was replaced by a 12 V power
supply having an internal resistance of 100 Ω?

15 A student is attempting to use a 1 kΩ potential
divider to light a 3.0 V 0.25 A lamp. She first
connects it like this to a 6 V supply:

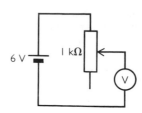

figure 37

When she varies the slider of the
potentiometer from one end to the other, the
voltmeter readings only vary between 6 V and
5 V.
a At which end would the slider be for the
reading of 5 V?
b Calculate the resistance of the voltmeter.

She now alters the circuit to this:

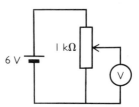

figure 38

She finds the setting of the potentiometer
which gives 3 V on the meter. She then replaces
the voltmeter by the lamp.
c Explain why the slider is now set just over
half-way up the potentiometer.
d Assume for simplicity that it is set exactly half-
way up. Calculate the approximate current in
the lamp. Will it light?

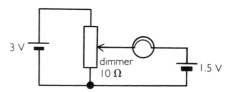

figure 39

16 The bulb in figure 39 requires 1.0 V across it to
light it normally; its resistance when lit normally
is 5 Ω. The dimmer is 20 cm long, and its
resistance is 10 Ω.
a Where must the dimmer be set so that no
current passes through the bulb at all?
b Calculate the two positions at which it can be
set to light the bulb normally. Assume that the
cells have no internal resistance.

17 You are using a metre-wire potentiometer to compare the e.m.f.s of two cells, each with nominal e.m.f. 1.5 V. Figure 40 shows the circuit.

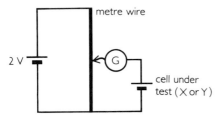

metre wire

2 V

G

cell under test (X or Y)

figure 40

a Cells X and Y give balance lengths of 764 mm and 746 mm respectively. What is the ratio of their e.m.f.s (e.m.f. of X ÷ e.m.f. of Y)?
b Approximately what difference would you obtain in balance lengths if two such cells were compared which differed in e.m.f. by 1%?
c Assume the metre-wire potentiometer can be balanced correct to the nearest millimetre. What percentage difference in e.m.f. between two such cells would be detectable?
d You want to make the arrangement even more sensitive to small differences between the two test cells. If you could alter the supply voltage to the metre wire, would you increase it or decrease it? Explain.

18 You are designing a stage-lighting system. You plan to use a potential-divider dimming arrangement. To begin with, you draw the simplified circuit of figure 41.

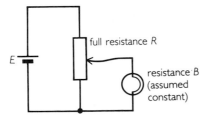

full resistance R

E

resistance B (assumed constant)

figure 41

In general, you have no control over B; for a particular power of bulb, B will be fixed, since at brightest setting the bulb will have p.d. E across it. What you need to do is select the value of R.
a Where is the slider setting which gives full brightness?
b Your first thought is that R can have any value, since in all cases you will get a full range of values of p.d. across B, varying from off to full brightness; therefore you will have R very large. Why is this unsatisfactory?

c Why is it also unsatisfactory to have R very small?
d What sort of value do you think would be best for R? Justify your answer as fully as you can.

19

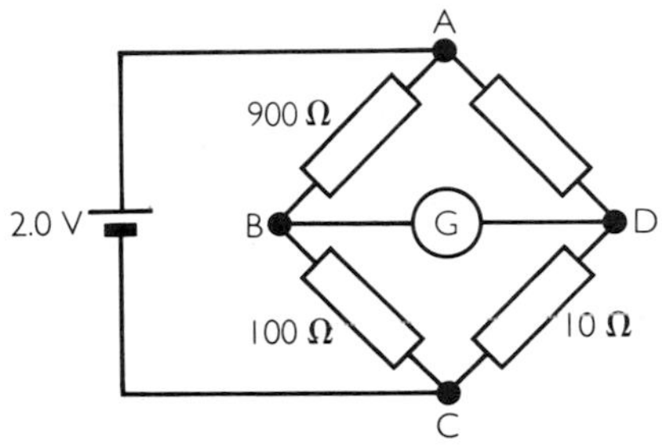

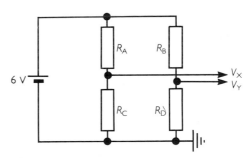

2.0 V

900 Ω

A

B   G   D

100 Ω   10 Ω

C

figure 42

a In the circuit of figure 42, what resistor in arm AD is needed to balance the Wheatstone bridge?
b With the bridge balanced, what is the p.d. across (i) AB, (ii) BC, (iii) AD, (iv) CD?

20 This question is about the detailed properties of a Wheatstone bridge circuit. Consider the circuit in figure 43.

R_A R_B

6 V

V_X
V_Y

R_C R_D

figure 43

Suppose $R_A = R_B = R_C = R_D = 1000\ \Omega$.
a Explain why $V_Y - V_X = 0$. (V_Y and V_X are the potentials at Y and X respectively.)
b Suppose R_A increases by 1% (to 1010 Ω). Calculate $V_Y - V_X$ now.
c Calculate $V_Y - V_X$ also in these cases:
(i) R_A and R_B both become 1010 Ω;
(ii) R_A and R_D both become 1010 Ω.

Now consider this general situation. You can select R_A, R_B, R_C, and R_D to be any values you like; during the experiment, R_A will increase by up to 1%.
d Can you find any combination of values for R_A, R_B, R_C and R_D which will give you a value for $V_Y - V_X$ greater than that in part (b)? Or is the choice such that $R_A = R_B = R_C = R_D$ the best for obtaining a large value of $V_Y - V_X$?

21 Read the passage below, then answer the questions about it which follow.

The corrosometer

Problems

At an oil refinery or chemical factory, corrosive liquids have to be pumped through pipes. How much corrosion is taking place in the pipes and when do they need replacing? It is not practicable to look inside the pipes, so how can the amount of corrosion be quantified?

Principles

(i) Experiments show that the electrical resistance R of a given length of material varies inversely with the cross-sectional area A of the material

$$\text{i.e. } R \propto \frac{1}{A}$$

(ii) As a metal corrodes, the corroded layer on the surface of the metal tends to flake off, reducing the cross-sectional area of the conducting metal. The resistance of the material therefore increases.

(iii) A change in resistance can be detected by a Wheatstone bridge arrangement.

Practice

A piece of wire made from the material to be tested is bent into a U shape (see below). This piece of wire is called the '*measuring* element'. The measuring element is connected to a probe as shown.

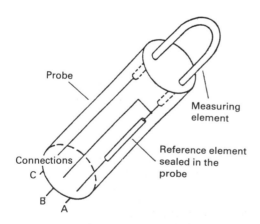

The probe contains another piece of wire of the same material as the measuring element. This piece of wire is sealed into the probe and therefore not subject to corrosion. It is called the '*reference* element'.

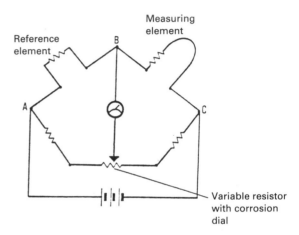

The probe containing the measuring and reference elements is connected to a unit with two fixed resistors (see above), and a galvanometer, to form a Wheatstone bridge arrangement. Before the measuring element is exposed to the corrosive environment a variable resistor is adjusted so that there is no current in the galvanometer. The probe is then left in the corrosive environment for a period of time and finally connected to the meter unit to complete the Wheatstone bridge when a corrosion reading is required. As corrosion proceeds, the resistance of the measuring element increases, while that of the reference element remains constant. This change in resistance of the measuring element upsets the balance of the bridge and a current is produced in the galvanometer. The variable resistor is now adjusted so that the meter read zero again.

A scale is attached to the variable resistor and the extent of the corrosion is read directly from this scale.

Reproduced by kind permission from BP Educational Service.

a How does the resistance of a conductor depend on its cross-sectional area (lines 4, 5)?

b Suggest two reasons why the reference element is sealed into the probe, rather than kept in the remainder of the circuit.

Now suppose that initially the reference element and the measuring element have exactly the same resistance; and that each is a wire $1000 \, \mu m$ in diameter. Also suppose that the part of the circuit between A and C consists of a single dial potentiometer, with linear markings of zero (slider at A) to 100 units (slider at C). See figure 44.

c What reading on the dial would you expect initially?

d Suppose corrosion has now proceeded to a depth of $100 \, \mu m$ all round the measuring element. Calculate:
(i) the new cross-sectional area of the measuring element;
(ii) the ratio

$$\frac{\text{cross-sectional area of reference element}}{\text{cross-sectional area of measuring element}}$$

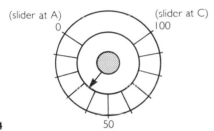

figure 44

(iii) the new setting on the potentiometer for balance of the bridge.

e calculate some other points on the dial (readings of the dial for different depths of corrosion). Sketch the graph of dial reading against depth of corrosion. How linear is the scale?

22 This question is about the use of strain gauges to measure the tensile load on mechanical components. A slightly flexible ring is fixed in series with the tensile load to be measured, and four strain gauges are fixed onto it as shown in figure 45.

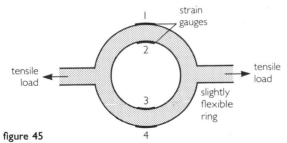

figure 45

a When the tensile load is applied, the ring tends to flex, so two gauges are stretched while two are compressed. Explain which gauges are stretched and which ones are compressed.

b Suppose that each gauge has initial resistance R; but that when this stretching occurs, the resistance of each one changes by ΔR. It is usual to wire the four gauges into a Wheatstone bridge circuit as in figure 46.

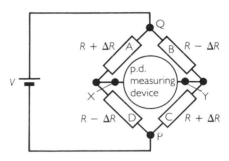

figure 46

How will gauges 1, 2, 3, 4 be identified with the resistors A–D in the circuit?

c Assuming no current passes into the p.d. measuring device, calculate in terms of R, ΔR and V the p.d.s (i) across QX, (ii) across QY, (iii) across XY.

d The value of the p.d. across XY should be directly proportional to ΔR. Will the p.d. therefore be proportional to the average strain on the four gauges? Explain.

23 When a battery known to have internal resistance $1000 \, \Omega$ is connected to a digital (high-resistance) voltmeter, the meter reads 1.5 V. When the same battery is connected instead to another (moving-coil) voltmeter, this meter reads 1.2 V.

a What is the resistance of the moving coil voltmeter?

b What would each meter read if they were both connected simultaneously in parallel across the battery? Explain.

24 A 5000 V supply for use in a school laboratory has a large 'internal resistance' built into it for safety. Its voltmeter, which is across the external terminals, reads 5000 V when the unit is first switched on, with nothing connected to it; but the reading falls to zero when the unit is short-circuited across a milliammeter. The milliammeter then reads 2.5 mA. See figure 47.

a What is the internal resistance of the unit?

b Why does the voltmeter reading fall to zero as described?

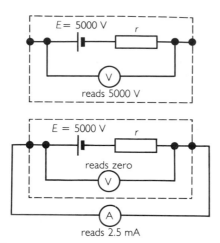

figure 47

25 What is the maximum power you can obtain in an external load from:
(i) a 12 V car battery with internal resistance 0.05 Ω;
(ii) a 1.5 V dry cell with internal resistance 0.5 Ω?

26 In the circuit network in figure 48, find the value of the current marked *I*.

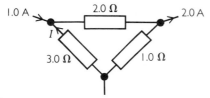

figure 48

27 Consider the hypothetical circuit of figure 49.

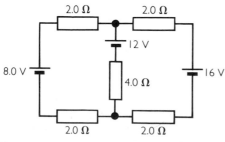

figure 49

Find the current in each battery (size and direction). Assume the batteries have zero internal resistance.

29 A 1.2 V cell with internal resistance 2.0 Ω and a 2.0 V cell with internal resistance 3.0 Ω are connected in parallel (figure 50). If a 3.0 Ω resistor is connected in parallel with both of them, what power is dissipated in it?

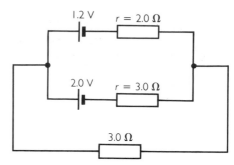

figure 50

29 Figure 51 shows data obtained from a solar cell. For three different levels of illumination, graphs are plotted of the p.d. across the cell and the current through it.

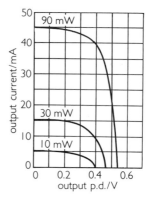

figure 51

a Draw a diagram of a circuit which could have been used to obtain these measurements.
b Estimate the maximum power obtainable from the cell for an incident power of 90 mW. How efficient is the cell at this power output?
c Estimate the maximum efficiencies of the cell with incident powers of (i) 30 mW, (ii) 10 mW.
d Suggest a reason for the efficiency changing as it does.
e For 90 mW incident power, what is the load resistance when the output current is (i) 40 mA, (ii) 25 mA?

30 Sketch the CRO traces that would be seen in each of the following cases. In every case the a.c. is at 50 Hz, with peak value 1 V, and the spot takes $\frac{1}{25}$ s to cross the screen. Indicate the amplitude of p.d.s, in volts, on each trace.

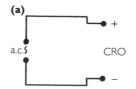

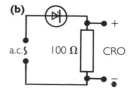

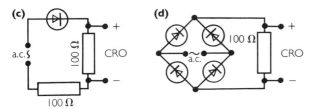

figure 52

31 This question is about the pumped-storage hydroelectric power system at Dinorwig in North Wales. Schematically, the system is illustrated in figure 53.

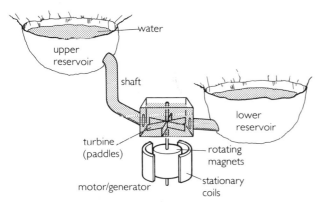

figure 53

Unlike nuclear or coal-burning power stations, a hydroelectric power system takes almost no time to start up or shut down: the Dinorwig system can be brought up from zero load to supplying 1320 MW in just 10 seconds. So it is the best sort of system to use for 'topping up' the grid at times of peak demand for power – see figure 54.

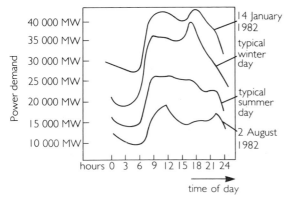

figure 54

The modern coal and nuclear stations on the grid are run 24 hours a day, so that there is no waste as they are started up or shut down. At

times when little power is needed by consumers (midnight to 6 a.m.), some of the power these stations produce is used at Dinorwig to pump water back up from the lower reservoir to the higher one; then during the day the water is used to generate power as required.

In each generator/turbine unit, there is a turbine, where water is directed onto blades fixed to a vertical rotating shaft. Lower down the same shaft is the generator assembly. The unit is reversible; when spun clockwise by water coming down hill it acts as a generator. But when supplied with power it turns anticlockwise, pumping water through the turbine back up the hill. Neither turbine nor generator is 100% efficient, and each has a different efficiency.

Use the data below to answer the questions which follow.

For each unit:

maximum electrical output	313.5 MW
maximum power consumption when pumping	283 MW
generating voltage	18 kV
efficiency of motor/generator	0.95
efficiency of turbine when generating	0.925
efficiency of turbine when pumping	0.917

For system as a whole (six generator/turbine units):

working volume of water	6.7×10^6 m^3
normal pumping period	6 h
full output generating period	5 h
effective head of water	535.8 m
overall efficiency	0.75

For shaft (supplying all six units):

maximum flow rate (generating)	420 m^{3}s^{-1}
maximum flow rate (pumping)	348 m^{3}s^{-1}

Shaft diameter   10 m
Density of water   1000 kg m^{-3}

a How much gravitational potential energy does the system store when the top reservoir is full? Answer in joules and kilowatt hours.

b Show how the efficiencies of each stage in the generating/pumping process enable you to predict the overall efficiency. Suggest reasons for any discrepancy.

c Starting from the power consumption of each unit when pumping, work out the rate at which water can be pumped uphill. Show how this ties up with the time quoted as 'normal pumping' period.

d Starting from the maximum output of each unit, work out the flow rate needed if the system is on maximum load. Is this compatible with (i) the maximum permissible flow rate, (ii) the full output generating period?

e At the maximum permitted flow rate (generating), what is the speed of the water in the shaft?

f When giving its maximum electrical output, what is the current flowing from each unit?

CHAPTER 12

Capacitors

Introduction

This chapter is all about a circuit component which is probably new to you: a capacitor. By discussing the results of simple experiments with a capacitor, we piece together a picture of how it behaves. We see that it can, in one sense, store charge, and that when charged it stores energy. We define quantitatively the property of capacitance, and see how we may measure it. We develop rules for finding the result of combining several capacitors. Finally, we examine how charge decays from a capacitor through a resistor in an exponential way.

GCSE knowledge

This chapter requires no previous knowledge to be transferred directly from your GCSE course. You must, though, be thoroughly confident about the quantities charge, current, p.d. and resistance, which were developed fully in chapter 9. It may be that you already know something about capacitors from GCSE electronics, physics or technology courses: any such knowledge will of course be a useful starting point.

What does a capacitor do?

A capacitor is a circuit component which behaves quite differently from a resistor or any of the components whose current-against-p.d. graphs you met earlier.

Let us consider some experiments you may have done, and what each one tells us about a capacitor's behaviour.

1   Using the circuit of figure 1, touching the flying lead to X causes a short pulse of current clockwise through the ammeter. After that the ammeter continues to read zero, even if you remove the flying lead from X and then retouch it.

Conclusions:
a A capacitor doesn't allow steady current through itself. In fact a capacitor is made with insulation between its two sides, so this isn't surprising.
b Once current has stopped, the p.d. across the capacitor must be equal to the full p.d., *V*.

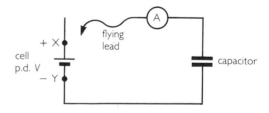

figure 1   Initial experiments with a capacitor

c Since no further current passes even after you remove and replace the flying lead, the capacitor must retain this p.d. across itself in some way when disconnected.

2   If you now touch the flying lead to Y, a pulse of current passes anticlockwise round the circuit. It appears to be the same size of pulse (i.e. same amount of charge) as that which passed clockwise initially.

Conclusions:

d Once a p.d. is established across the capacitor, it is capable of passing charge round a circuit: that is, it acts like a source of e.m.f., with stored energy.

e It passes back the same amount of charge as it received initially.

f The charge must be returning from one side of the capacitor to the other, since the only other item in the circuit is the ammeter.

3 You can confirm conclusion (f) by repeating experiments 1 and 2 with a second ammeter in the circuit – figure 2. A_1 and A_2 always read exactly the same. This confirms that when one plate of the capacitor gains charge, the other plate loses an identical amount of charge.

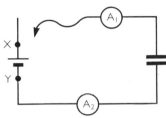

figure 2   A second ammeter in the circuit

Thus we say that a capacitor with a p.d. across it is 'charged': there is charge on one plate ready to be sent back to the other when a suitable connection is made. However, the lower plate cannot just be missing charge. It was neutral to start with; taking away charge from a neutral object makes it negative. So in fact when charge $+Q$ has passed round the circuit to the top plate, the bottom plate has charge $-Q$.

4 The next experiment uses the more complex circuit of figure 3.

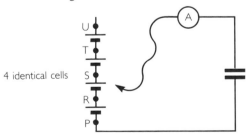

4 identical cells

figure 3   Experiments using four cells

If, starting with the capacitor uncharged, you touch the flying lead successively to R, then S, then T, then U, the ammeter shows a succession of clockwise pulses of equal size.

Conclusion:

g Each time the p.d. across the capacitor changes by a fixed amount (call it ΔV), the charge on the

capacitor changes by a fixed amount (call it ΔQ). So when the p.d. is ΔV (flying lead at R) the charge on the capacitor is ΔQ; when the p.d. is $2\Delta V$ the charge is $2\Delta Q$, and so on. In other words, if the charge on the capacitor is in general Q, and the p.d. is V, we have

$$Q \propto V$$

or
$$\frac{Q}{V} = \text{constant}$$

By definition, that constant is called the *capacitance* C of the capacitor:

$$C = \frac{Q}{V}$$

The unit of capacitance is therefore one coulomb/volt. This is given a special name: one farad (F).

$$1\,F = 1\,CV^{-1}$$

N.B. Do not confuse C for capacitance with C for coulomb. The former will appear in an equation in *italic* type and the latter after a number as a unit in roman type.

5 Using just these simple pieces of equipment we can make an estimate of C, the capacitance value of the capacitor we are using. Each cell may have p.d. 1.5 V. The pulse of current may be 10 μA, lasting for 0.3 s.

Then
$$Q = It$$
$$= 10\,\mu A \times 0.3\,s$$
$$= 3\,\mu C$$

$$\Rightarrow \qquad C = \frac{Q}{V}$$
$$= \frac{3\,\mu C}{1.5\,V}$$
$$= 2\,\mu F$$

Note that a capacitance of one farad is a very large capacitance. So values in microfarads, nanofarads and picofarads are much more usual.

In answer to the question posed at the beginning of the section, 'What does a capacitor do?', many people would answer 'It stores charge'. We can now see that the answer is a little more subtle: one plate loses charge, to become negative; the other gains the charge from the first, to become positive.

There is a useful analogy which we can develop by comparing a capacitor to a tank of water. The tank has a stretchable rubber membrane across the middle of it and is connected to the pump as shown in figure 4(a). The essential point of the analogy is brought out in figure 4(b): here the pump has been turned on and water has been taken from one side of the membrane and put in the other. There has been no change to the total amount of water stored in the tank; however the membrane has been stretched and energy is stored in it. When the pump is removed from the circuit the membrane is capable of driving the water through a pipe to equalise the amounts of water in each side of the tank. We can now see the similarity between the water tank in figure 4(c) and the capacitor in figure 4(f): in this diagram the capacitor has $+Q$ on one plate and $-Q$ on the other. The total charge on the capacitor is still zero, but the capacitor has stored energy which is capable of driving the charge back round the circuit.

figure 5   Reed switch (removed from its magnetizing coil)

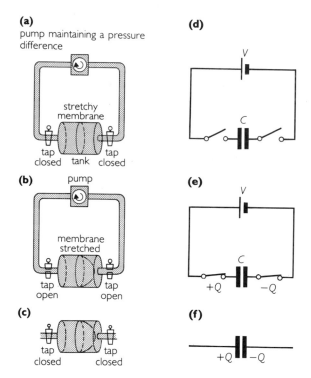

figure 4   Analogy between capacitor and tank with stretchy membrane: (a) tank and pump; (b) some water has moved round the circuit; (c) membrane storing energy; (d) capacitor and cell; (e) charge has moved round the circuit; (f) capacitor storing energy

figure 6   A circuit to measure capacitance

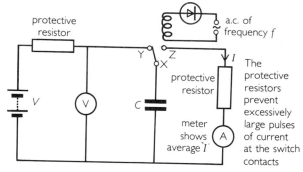

figure 7   Circuit for measuring capacitance using a reed switch

Measuring capacitance

The standard laboratory method is to use a reed switch as shown in figures 5, 6 and 7.

The current in the coil of the switch rises then falls to zero at the frequency of the a.c. supply. (A diode prevents current in the reverse direction.) Each time the current rises a magnetic field forms inside the coil; this magnetises contacts X and Z, which attract and touch. When the current falls, the natural springiness of the materials takes X back to Y.

If the frequency of the a.c. supplied to the coil is f,

then the reed will move through one complete cycle (from Z to Y and back to Z) in time T given by

$$T = \frac{1}{f}$$

Each time the capacitor is charged, it receives charge $Q = CV$. This amount of charge then passes through the ammeter.

$$\Rightarrow \quad \text{average } I = \frac{CV}{T} = CVf$$

Although this current passes in a series of pulses, provided f is reasonably high then the ammeter, because of its inertia, will give a steady reading equal to the average I.

Then:

$$C = \frac{I}{Vf}$$

Typically, I might be 2×10^{-3} A, for $V = 5$ V and $f = 200$ Hz. Then

$$C = \frac{2 \times 10^{-3} \text{ A}}{5 \text{ V} \times 200 \text{ Hz}}$$
$$= 2 \times 10^{-6} \text{ F}$$
$$= 2 \, \mu\text{F}$$

It is possible to measure capacitance more directly, using a coulombmeter. This is an electronic device which takes the charge to be measured onto a capacitor of known capacitance, amplifies the p.d. produced, and gives a calibrated read-out of the charge. A suitable circuit is shown in figure 8.

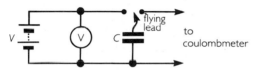

figure 8 Circuit for measuring capacitance using a coulombmeter

To obtain the value of C, we simply apply the equation $C = Q/V$.

Capacitors in parallel

Like all components, capacitors in parallel must all have the same p.d. across them. See figure 9.

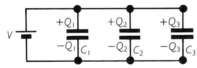

figure 9 Three capacitors in parallel

Each of the capacitors C_1, C_2 and C_3 is governed by the equation $Q = CV$.

$$\Rightarrow \qquad Q_1 = C_1 V$$
$$Q_2 = C_2 V$$
$$Q_3 = C_3 V$$

The total charge Q_T which has passed out from the positive pole of the battery must be equal to the sum of the charges that have accumulated on the top plates of the capacitors.

$$\Rightarrow \qquad Q_T = Q_1 + Q_2 + Q_3$$
$$= C_1 V + C_2 V + C_3 V$$
$$= (C_1 + C_2 + C_3) V$$

But if C_T is the total combined capacitance we can write:

$$Q_T = C_T V$$
$$\Rightarrow \qquad C_T = C_1 + C_2 + C_3$$

So we can say that when we have several capacitors in parallel, the total capacitance is simply the sum of the individual capacitances.

Capacitors in series

In this case each capacitor carries the same charge. This is because, being in series, the same current passes to each one of them for the same length of time. Alternatively, think of the dotted region of figure 10: the total charge in this region must remain zero, since there is insulator at each end of it.

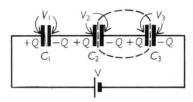

figure 10 Three capacitors in series

The p.d. V across the composite capacitor C_T is given by

$$V = V_1 + V_2 + V_3$$

but

$$V = \frac{Q}{C_T}$$

and

$$V_1 = \frac{Q}{C_1}; \; V_2 = \frac{Q}{C_2}; \; V_3 = \frac{Q}{C_3}$$

$$\Rightarrow \qquad \frac{Q}{C_T} = \frac{Q}{C_1} + \frac{Q}{C_2} + \frac{Q}{C_3}$$

$$\Rightarrow \qquad \frac{1}{C_T} = \frac{1}{C_1} + \frac{1}{C_2} + \frac{1}{C_3}$$

Example What is the total capacitance between points X and Z in figure 11?

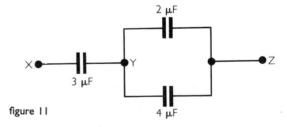

figure 11

The capacitance between Y and Z is given by:

$$C = C_1 + C_2$$
$$\Rightarrow \qquad C = 2 \, \mu\text{F} + 4 \, \mu\text{F}$$
$$= 6 \, \mu\text{F}$$

Thus the capacitance between X and Z is given by:

$$\frac{1}{C_T} = \frac{1}{3\,\mu F} + \frac{1}{6\,\mu F}$$

$$\frac{1}{C_T} = \frac{1}{2\,\mu F}$$

$$C_T = 2\,\mu F$$

Energy stored in a capacitor

A charged capacitor stores energy, since it is able to drive current, through a resistor, for example. See figure 12.

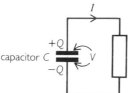

figure 12   Charged capacitor driving current through a resistor

To calculate the amount of energy stored, consider the graph of p.d. against charge for a particular capacitance C (figure 13).

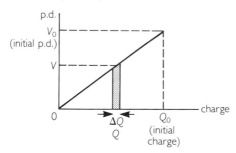

figure 13   p.d.-against-current graph for a capacitor discharge

At some stage in the discharging process, the capacitor has charge Q, and p.d. V. When the capacitor drives the next very small charge ΔQ through the resistor, the electrical energy ΔW it gives the charge is given by

$$\Delta W = V \times \Delta Q$$

$V\Delta Q$ is roughly equal to the shaded area under the graph in figure 13. We can think of the whole discharging process as a large number of such small movements of charge; and the total energy delivered from the capacitor into the resistor, which is equal to the stored energy, is the whole area under the graph.

$$\Rightarrow \qquad W = \tfrac{1}{2} Q_0 V_0$$

By substituting $Q_0 = CV_0$ into this equation, we obtain

$$W = \tfrac{1}{2} C V_0^{\,2}$$

And by substituting $V_0 = Q_0/C$ we obtain

$$W = \tfrac{1}{2}\frac{Q_0^{\,2}}{C}$$

The energy-stored formula can also be shown by integration:

$$\Delta W = V\Delta Q$$

$$= \frac{Q}{C}\Delta Q$$

$$\Rightarrow \qquad W = \sum \frac{Q}{C}\Delta Q$$

$$= \int_0^{Q_0} \frac{Q}{C}\,dQ$$

$$= \tfrac{1}{2}\frac{Q_0^{\,2}}{C}$$

N.B. In the above work we have used Q and V to denote variable values, and Q_0 and V_0 to denote final values. The various capacitor formulae are usually quoted using simply Q and V, when there is no ambiguity.

Discharge of a capacitor through a resistor

In the last section we examined the energy aspect of a capacitor discharging through a resistor. It is also interesting to see the manner in which this discharge proceeds with time. You may have made some measurements about this, using a circuit like figure 14.

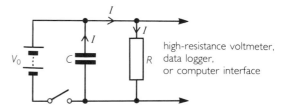

figure 14   Circuit for observing the discharge of a capacitor through a resistor

The p.d. across C falls with time in the qualitative manner illustrated in figure 15.

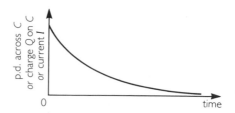

figure 15   Discharge of a capacitor through a resistor

Since the instantaneous charge Q on the capacitor is CV, then Q also falls in this manner; and since the instantaneous current is given by $I = V/R$, I falls in a similar way too.

Figure 16 shows a graph of some computer-logged data from such an experiment; the computer acts as a voltmeter and then displays its recorded data graphically.

If you examine the graph of figure 16, you will be able to see the interesting property that in any chosen time interval the value of V always falls by the same fraction. For example, the value of V initially is 8.8 V, and after 10 seconds it is 7.3 V. Thus it has fallen to a fraction P of its starting value given by

$$P = \frac{7.3\ \text{V}}{8.8\ \text{V}} = 0.83$$

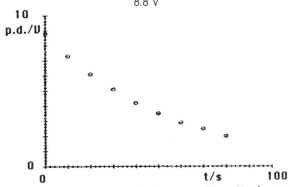

p.d. against time graph for capacitor discharge
figure 16

If we then read V at 50 s and 60 s, we obtain values of 3.6 V and 3.0 V respectively. In this case P is given by

$$P = \frac{3.0\ \text{V}}{3.6\ \text{V}} = 0.83$$

In other words, within experimental error the fractional change in the two 10-second periods is the same.

A related feature of this sort of change is that V (or Q, or I) always takes the same time to halve. In the graph of figure 16, that time is about 38 s. This is the same property as we find with radioactive decay. Any process with this property is said to change *exponentially*.

Analysis of exponential capacitor discharge

We can learn more about this exponential decay of charge on a capacitor by developing some theory about it.

Figure 17 shows a particular instant in the discharge process. For a short time interval Δt, we can take I as being approximately constant. Then

$$\Delta Q = -I\Delta t$$

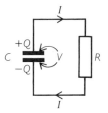

figure 17   An instant in the discharge process

(ΔQ is negative because the current is resulting in the value of Q becoming smaller.)

Now $$I = \frac{V}{R} = \frac{Q}{RC}$$

$\Rightarrow$ $$\Delta Q = -\frac{Q}{RC}\Delta t \qquad \text{(i)}$$

Also $$\frac{\Delta Q}{\Delta t} = -\frac{Q}{RC} \qquad \text{(ii)}$$

Equation (ii) is the characteristic equation shown by all exponentially decaying quantities. In words, the rate of decay of charge at any moment is proportional to the amount of charge remaining.

Using equation (i), you can construct a theoretical graph for the way in which Q changes with time t. To do this you calculate the loss of charge ΔQ in each succeeding period Δt, and then subtract that ΔQ from the earlier value of Q before repeating the calculation.

For example, take $R = 10^5\ \Omega$, $C = 100\ \mu\text{F}$, $\Delta t = 1$ s.

Then $$\Delta Q = -\frac{Q \times 1\ \text{s}}{10^5\ \Omega \times 10^{-4}\ \text{F}}$$

$$= -Q \times 0.1$$

If the initial p.d. is 6 V, then the initial Q is given by

$$Q = CV$$
$$= 100\ \mu\text{F} \times 6\ \text{V}$$
$$= 600\ \mu\text{C}.$$

The first value of ΔQ is $-600\ \mu\text{C} \times 0.1 = -60\ \mu\text{C}$.

Therefore the next value of Q is $600\ \mu\text{C} - 60\ \mu\text{C} = 540\ \mu\text{C}$; the next ΔQ is $-54\ \mu\text{C}$; and so on.

Plotting these calculated values out on a graph gives a curve in pretty good agreement with figure 15. However, it is a fairly tedious process and it is much more fun to get a computer to do it. Below is shown a program in BASIC that will do this job for us. You can see that the key lines in the program are lines 140 and 150. If you do decide to try the program out for yourself, make sure that you choose an interval dt that is small in comparison with RC. (We will leave you to work out why that is important.)

```
10 REM C-R PROGRAM
20 MODE 7
30 PRINT ' " * R–C CIRCUIT *"
40 INPUT "Value of capacitance (μF) ",C
50 C=C*1E–6
60 INPUT "Value of resistance (ohms) ",R
70 INPUT "Time interval (s) ",dt
80 INPUT "How much initial charge (μC) ",Q0
90 REM Start of loop
100 Q=Q0
110 PRINT TAB(5)"Interval";TAB(15)"Charge (μC) "
120 PRINT0;TAB(18)Q0
130 FOR T=1 TO 15
140 dQ=−Q*dt/(R*C)
150 Q=Q+dQ
160 PRINT T;TAB(18)INT(Q)
170 NEXT T
180 END
```

The capacitor would discharge in a time RC if it carried on discharging at its initial rate; this is illustrated in figure 18. We can justify this statement using equation (ii).

The initial rate of discharge is given by

$$\frac{\Delta Q}{\Delta t} = -\frac{Q_0}{RC}$$

where Q_0 is the original charge. If the capacitor were to continue to discharge at this rate it would be totally discharged when $\Delta Q = -Q_0$.

$$\Rightarrow \quad \text{time for discharge } \Delta t = RC$$

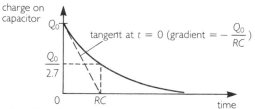

figure 18 The significance of the time constant RC

Because of the fall in rate as the p.d. across it falls, it actually takes much longer to discharge than that. (In fact, it is never theoretically discharged.) But the time RC gives a good guide to the discharge time. Three ideas are worth remembering.

- ☐ RC is the time for the charge (or p.d. or current) to fall to 0.37 (roughly $\frac{1}{3}$) of its initial value. (We justify this below mathematically).

- ☐ A capacitor is fully discharged (as good as) after about 5RC.

- ☐ RC gives the order of magnitude of the discharge time.

RC is called the *time constant* of the circuit. Qualitatively, you can see why making either R or C larger will increase the time for discharge:

- ☐ a larger R will result in a smaller current, so it will take longer for the charge to flow off;

- ☐ a larger C will mean there will be more charge on the capacitor to flow off, even though the current is no larger.

- ☐ You should note one final point about a capacitor–resistor circuit. When the capacitor is being charged, in the circuit of figure 19, the graph of charge against time looks like figure 20. This is exactly the same graph as the discharging process, but upside down. The charge value approaches its final value CV as in the discharging process.

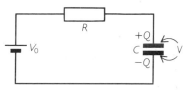

figure 19 Circuit for charging a capacitor through a resistor

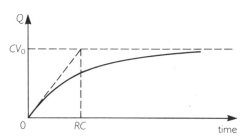

figure 20 Graph to illustrate charging a capacitor through a resistor

The mathematics of *RC* circuits

If we start from equation (ii) on page 151, we can proceed using calculus from there.

$$\frac{\Delta Q}{\Delta t} = -\frac{Q}{RC}$$

for charge ΔQ leaving a capacitor C through a resistor R in time Δt;

$$\Rightarrow \quad \text{instantaneously } \frac{dQ}{dt} = -\frac{Q}{RC}$$

$$\Rightarrow \quad \int_{Q_0}^{Q}\frac{dQ}{Q} = -\int_{0}^{t}\frac{dt}{RC}$$

$$\Rightarrow \quad \ln\frac{Q}{Q_0} = -\frac{t}{RC}$$

$$\Rightarrow \quad Q = Q_0 e^{-t/RC}$$

This is the mathematical description of the decay of charge from a capacitor. It fulfils all the properties of the decay process discussed earlier.

From the equation above we can easily find the fraction of charge left after a time equal to one time constant RC.

When $\qquad t = RC$

$\Rightarrow \qquad Q = Q_0 e^{-1}$

$\Rightarrow \qquad Q = 0.37 Q_0$

We can also derive the mathematical description of the charging process, starting from some moment during the process (figure 21).

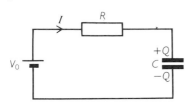

figure 21 A moment during the charging of a capacitor

Kirchoff's second law for the circuit gives:

$$V_0 = IR + \frac{Q}{C}$$

Differentiation of this equation with respect to time gives:

$$0 = R\frac{dI}{dt} + \frac{I}{C} \qquad \text{since } I = \frac{dQ}{dt}$$

$$\Rightarrow \frac{dI}{I} = -\frac{dt}{RC}$$

so the current decays exponentially

$\Rightarrow \quad I = I_0 e^{-t/RC}$

$\qquad = \frac{V_0}{R} e^{-t/RC}$

since the maximum current $= \dfrac{V_0}{R}$

So the p.d. across the resistor is:

$$V_R = V_0 e^{-t/RC}$$

The p.d. across the capacitor can be calculated using:

$$V_0 = V_R + V_C$$

$\Rightarrow \qquad V_C = V_0 - V_R$

$\qquad\qquad = V_0(1 - e^{-t/RC})$

and the charge on this capacitor is:

$$Q = CV_0(1 - e^{-t/RC})$$

Summary

1 A capacitor stores charge $+Q$ on one plate, and $-Q$ on the other plate.

2 A capacitor will not allow steady current through itself.

3 $$C = \frac{Q}{V}$$

$$1\,\text{F} = \frac{1\,\text{C}}{1\,\text{V}}$$

4 You can measure capacitance using a reed-switch circuit, or a coulombmeter.

5 For capacitors in parallel,
$$C_T = C_1 + C_2 + \dots$$

6 For capacitors in series,
$$\frac{1}{C_T} = \frac{1}{C_1} + \frac{1}{C_2} + \dots$$

7 Energy stored in a capacitor:

$$W = \tfrac{1}{2}QV = \tfrac{1}{2}CV^2 = \tfrac{1}{2}\frac{Q^2}{C}$$

8 a The decay of charge from a capacitor through a resistor is exponential: a fixed fraction of the charge leaves in a certain time interval.
   b V and I also fall exponentially.
   c The time constant of the decay is RC.
   d You can construct a theoretical graph of the decay using a step-by-step calculation method.
   e The equation describing the decay is

$$Q = Q_0 e^{-t/RC}$$

9 The equation describing the charging of a capacitor through a resistor is

$$Q = CV_0(1 - e^{-t/RC})$$

Questions

I Complete the spaces in the following table, which refers to the charge Q stored on various capacitors C when there is p.d. V across them.

C	Q	V
1000 μF	(a)	5.0 V
(b)	50 μC	2.0 V
500 μF	200 μC	(c)
5000 pF	20 μC	(d)
(e)	100 nC	500 mV
50 pF	(f)	200 mV

2 You are experimenting with the circuit in figure 22. Whatever you do with the flying lead, the two ammeters show the same deflections and readings at all times.
   a What do you conclude from this observation?
   b How do you reconcile this finding with the idea that the capacitor stores charge?

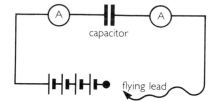
figure 22

3 Capacitors are being discussed. Angela says they store e.m.f. Ben says they store current, since current passes into them. Carol says they store energy. Comment on these three points of view.

4 You are 'spooning' charge onto a coulombmeter, using the arrangement shown in figure 23. You touch the tablespoon carefully onto the +5000 V terminal of the EHT (extra-high tension) high voltage supply, then onto the coulombmeter. At each transfer, the coulombmeter shows an additional 15 nC deposited.
   a What is the capacitance of the spoon?
   b What charge would be carried at each transfer if the EHT supply was set at 2000 V?
   c Guess what capacitance a teaspoon would have.

5 Figure 24 provides a simple laboratory method of measuring capacitance, C. The reed switch moves backwards and forwards at frequency f.

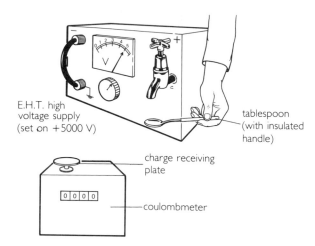
figure 23

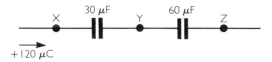
figure 24

   a Sketch a graph of the actual value of I against time, for a few cycles of the reed switch.
   b Why does the ammeter show a steady value?
   c What assumptions must be made if C is to be calculated from values of V, I and f?
   d If the ammeter reads 15 μA, and $f = 300$ Hz, how much charge is transferred by each cycle of the reed switch?
   e What is the value of C?

6 Two capacitors are arranged in series as shown in figure 25.

figure 25

   Suppose charge 120 μC moves onto the left plate of the 30 μF capacitor, as shown.
   a What charge must pass (i) point X, (ii) point Y?
   b What now is the p.d. (i) across the 30 μF capacitor, (ii) across the 60 μF capacitor?
   c What is the p.d. between X and Z?
   d What is the combined capacitance of these two capacitors in series?

7 Four identical capacitors are connected up into the two circuits shown in figure 26.

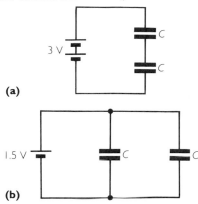

(a)

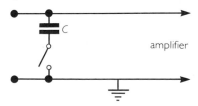

(b)

figure 26

a In which circuit will the capacitors have more charge on them? Explain.
b In which circuit will the capacitors store more energy? Explain.

8 You have a supply of 10 μF capacitors rated at 25 V. How many would you need to make up a composite capacitor of 15 μF rated at 100 V?

9 A widely-used digital coulombmeter/voltmeter has switchable ranges like this:

figure 27

The left knob switches in an internal capacitor of capacitance C for the coulombmeter function: the charge transfers onto the capacitor, and the voltmeter circuitry then measures the p.d. across it. See figure 28.

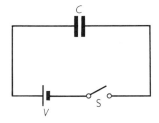

internal wiring

figure 28

a What is the value of C in this meter?
b What assumption is made about C in relation to the capacitance of the object from which the charge is being transferred?

10 The graph in figure 29 shows how the charge q on one capacitor varies with the p.d. V across it.

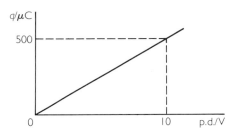

figure 29

a Calculate the size of the capacitance.
b How much energy does it store when charged to a p.d. of 10 V?
c How would the line for a capacitor with twice the capacitance appear on these axes?

11 Two capacitors are being used, of values 1 μF and 1000 μF.
a What is their combined capacitance if they are arranged (i) in series, (ii) in parallel?
b If they are in parallel, what is the ratio

$$\frac{\text{energy stored in 1000 μF capacitor}}{\text{energy stored in 1 μF capacitor}}$$

when a p.d. is put across the combination?
c What is this ratio if they are in series?

12 If you close the switch S in the circuit in figure 30, charge q flows through the cell, and charges the capacitor C.

figure 30

a How much energy is converted by the cell as charge q passes through potential difference V?
b How much energy ends up stored on the capacitor?
c Can you explain the loss of half of the energy converted by the cell?

13 You are doing experiments with a capacitor. You put 5 V across it. You then disconnect it, and connect across it a voltmeter of resistance 45 000 Ω. The meter reads 5 V to start with, but after 10 seconds the reading has dropped to about 4 V. Estimate, with clear reasoning,
a the average current which passed in those 10 seconds;

b how much charge flowed off the capacitor in those 10 seconds;
c the original charge on the capacitor;
d the capacitance of the capacitor.

14 The graph shows how the initial charge of 10 μC on a 4 μF capacitor discharges with time through a resistor.

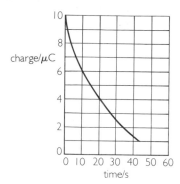

figure 31

a Estimate from the graph the initial current through the resistor.
b Work out the resistance of the resistor.

15 A student is using the discharging of a capacitor through a resistor as a way of measuring the speed of a falling ball bearing. The circuit is shown in figure 32.

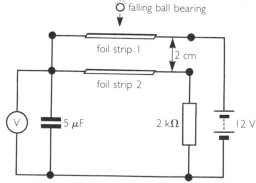

figure 32

a What will the voltmeter read before the ball breaks foil strip 1?
b What happens once strip 1 is broken?

The student observes that when the ball has fallen through both strips, the voltmeter reads 8 V.
c How much charge has flowed off the capacitor?
d What, approximately, was the average current through the resistor?
e Find a value for the time the ball takes to fall from strip 1 to strip 2.

16 A capacitor of unknown capacitance C is charged to a p.d. of 5 V, then discharged through a resistor of unknown resistance R. The current at the capacitor is measured, and is shown on figure 33.

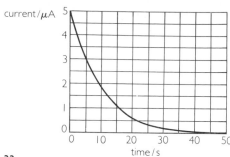

figure 33

Using the available information, calculate
a the resistance R,
b the capacitance C.

17 In this circuit, the ammeter at one moment reads 100 μA.

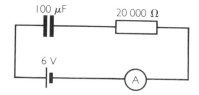

figure 34

At that moment
a what is the p.d. across the resistor?
b what is the p.d. across the capacitor?
c how much charge is stored in the capacitor?
d what would the charging current have been at the beginning of the charging process?
e Make an estimate of the time for which the charging has been in progress, showing your reasoning.

18

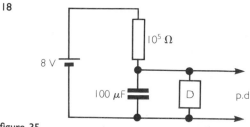

figure 35

In this circuit, D is a special device: it has a very large resistance until the p.d. across it rises to 4.0 V; at that point its resistance becomes very small, until the p.d. across it falls to zero, when its resistance again becomes very high.

a Sketch a graph of the p.d. across D against time.
b Estimate the periodic time of this varying p.d. (the 'repeat time').

19 The following graphs (figure 36) could refer to various aspects of capacitors storing charge. Pick the appropriate graph for each of the situations described below.

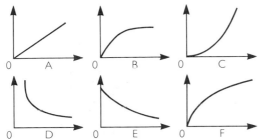

figure 36

a For a fixed value of capacitance, the charge stored against the p.d.
b For a fixed value of capacitance, the energy stored against the p.d.
c for a fixed value of capacitance, the energy stored against the charge stored.
d For a fixed value of charge stored, the capacitance required against the p.d.
e For a fixed value of charge stored, the energy stored against the capacitance of the capacitor.
f The amount of charge left on a capacitor which is discharging through a resistor, against time.
g The amount of charge which has flowed off a capacitor discharging through a resistor, against time.

20 One particular electrolytic capacitor in a manufacturer's catalogue is described like this:

Nominal capacitance $10\,000\,\mu\text{F}$
Rated voltage $100\,\text{V}$

In addition, the complete range of capacitors in the catalogue are specified as follows:

Tolerance -10%, $+50\%$
Leakage current $0.003\,\mu\text{A}$ per μFV.

a Within what range of values should the capacitance of this component actually be?

Suppose this capacitor actually has it specified nominal capacitance, and is charged to its rated voltage.
b What charge is on it?
c How much energy does it store?
d Why would this capacitor in this state be extremely dangerous if left lying about the lab?

The leakage current is the tiny current which crosses from one plate to the other, if there is a p.d. between them, through the insulating

material. (You need to know, for these questions, that the capacitance C of a capacitor is proportional to the area A of the two plates.)

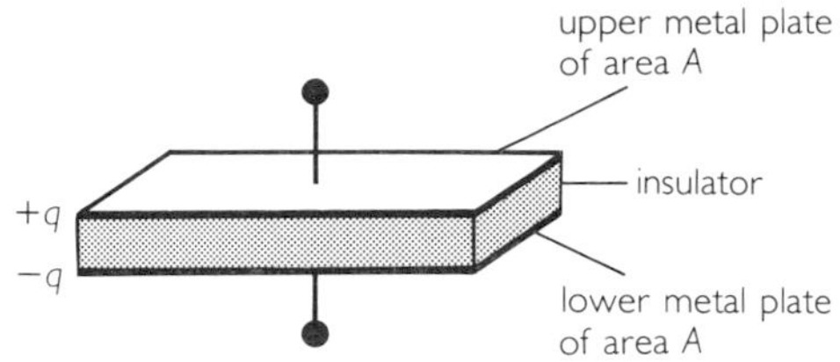

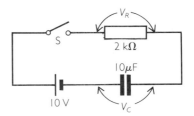

figure 37

e Explain why the way in which the leakage current I is specified implies that I is proportional to the value of the capacitor, C, and the p.d. across it, V.
f Give physical reasons why $I \propto C$ and $I \propto V$.
g Show that the unit in which the leakage current is specified, μA per μFV, can be simplified to s^{-1}.
h If this capacitor is left charged but isolated, it gradually becomes discharged due to the leakage current. Use the value given for the leakage current to find the time constant for this discharge. (Hint: Write down the equation which describes the discharge of a capacitor through a resistor, and relate it to the way the leakage current is specified.)

21 In the circuit in figure 38, the capacitor is initially uncharged (at time $t = 0$).

figure 38

The table below represents the first three lines in a step-by-step calculation to show how the capacitor charges with time when the switch S is closed.

t/ms	$q/\mu\text{C}$	V_C/V	V_R/V	I/A	$\Delta q/\mu\text{C}$
0	0	0	.... ①	.... ②	
1		.... ④	.... ⑤	.... ⑥	.... ⑦	.... ③
2		.... ⑨	.... ⑩	.... ⑪	.... ⑫	.... ⑧

a The key step in the calculation is to work out how much charge Δq moves onto the capacitor in the time interval Δt, selected to be 1 ms, assuming the current I remains constant for that time interval. Comment on whether this assumption is reasonable, for $t = 1$ ms and the component values in this circuit.

b Write down the equations for calculating the following:
(i) V_R, if you know V_C;
(ii) I, if you know V_R;
(iii) Δq, if you know I;
(iv) the new q value;
(v) the new V_C value.

c Calculate values to go in the spaces in the table, in the order indicated in the table.

d Write a computer program to plot out values of q against t, using this method of calculation.

22 Imagine a situation where a capacitor C is discharging through a non-ohmic circuit component. This component passes current I related to the p.d. across it, V, by the expression $I = 10^{-6} V^2$

Consider a moment when the charge on the capacitor is q.

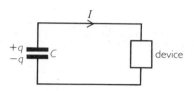

figure 39

a For a time interval Δt, which is short enough for us to assume q hardly changes, how much charge Δq leaves the capacitor
(i) in terms of I and Δt,
(ii) in terms of V and Δt,
(iii) in terms of q, C and Δt?

b Suppose $C = 2\ \mu F$, and it starts charged to a p.d. of 100 V. Find how long it will take to discharge to a p.d. of 50 V.
(You can do this from equation a(iii) *either* by solving the differential equation, *or* by writing a suitable computer program.)

CHAPTER 13

Square pulses in *RC* and *RL* circuits

Introduction

When we say a square pulse has been fed into a circuit we mean that the potential difference *between the two wires feeding into* the circuit has changed in a manner similar to that illustrated here. The time it takes the p.d. to change from 0 to V is very small (10^{-9} s), compared with the time spent at the higher value.

The purpose of this chapter is to investigate what happens when square pulses are fed into circuits containing resistors and capacitors (*RC* circuits) or resistors and inductors (*RL* circuits). The circuits we choose may be used as filters, which will allow some frequencies to pass but not others. In addition we meet the idea of integrating and differentiating circuits.

GCSE knowledge

We assume that you know the current through a resistor is proportional to the potential difference between its ends.

Other than this, most of the basic physics for this chapter is also to be found elsewhere in the book. *RC* circuits were first introduced in the last chapter and the inductor is encountered in chapter 27 as an important example of the application of the theory of electromagnetic induction. This chapter contains some quite difficult mathematics, which you can omit at a first reading.

Square pulses into *RC* circuits

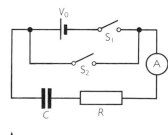

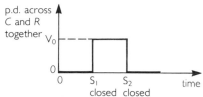

figure 1

Figure 1 shows a circuit in which it is possible to charge a capacitor by closing S_1 and then discharge the capacitor by opening S_1 then closing S_2.

First we will consider the discharge. When S_2 is closed the capacitor has a charge Q. You may recall from chapter 12, page 152, that the decay of this charge from the capacitor is exponential; and that mathematically this is described by the equation

$$Q = Q_0 e^{-t/RC}$$

The p.d. across the capacitor falls exponentially as well, also with time constant *RC*. This is illustrated graphically by figure 2(a).

When S_1 is closed and S_2 opened the capacitor begins to charge up. The equation that describes the charging process is

$$Q = CV_0(1 - e^{-t/RC})$$

(a) Decay of charge, p.d.

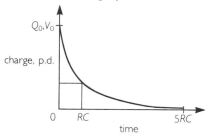

(b) Growth of charge, p.d.

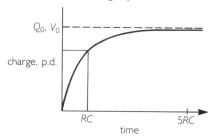

figure 2

(see page 153). The form of the p.d. rise across the capacitor (which is proportional to Q) is shown in figure 2(b).

The two results above may now be used to analyse the output of the circuits shown in figure 3 and figure 4. To the input of each of these circuits a square wave p.d. is applied which switches between 0 and V_0, remaining at V_0 for a time T.

In each circuit, when the input p.d. rises to V_0 the capacitor starts to charge up, the p.d. across it varying as shown in figure 2(b); providing that the time duration T of the pulse of p.d. at potential V_0 is about five times bigger than the time constant of the circuit, RC, the capacitor will become charged to a potential very nearly equal to V_0. After time T the

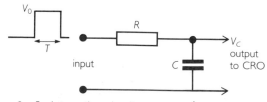

figure 3   An integrating circuit

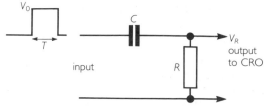

figure 4   A differentiating circuit

input p.d. drops to zero and the capacitor begins to discharge. The p.d. across the capacitor now varies as shown in figure 2(a). During the charging up process the current is greatest at time $t = 0$, because charge is being transferred at the greatest rate. The current is again large at $t = T$ just as the discharge of the capacitor commences. Now the current is in the opposite sense and the p.d. across the resistor is now negative. Figure 5(a) shows V_{in}, V_R and V_C for our two circuits for $5RC \approx T$. It should be noted that at all times

$$V_{in} = V_C + V_R$$

Figure 5(b) shows V_{in}, V_R and V_C for the case $T \gg 5RC$. In this situation the capacitor charges and discharges very rapidly in comparison to the duration of the pulses. Sharp p.d. spikes are observed across the resistor as the current passes only briefly to charge or discharge the capacitor. Figure 5(c) shows the p.d. variations in the situation where $T \ll 5RC$. In

(a) $T \approx 5RC$

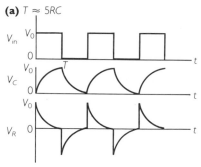

(b) $T \gg 5RC$

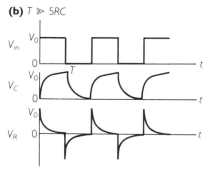

(c) $T \ll 5RC$

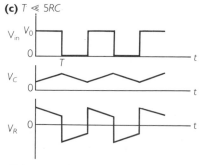

figure 5   Voltage-against-time graphs for the circuits of figures 3 and 4

this case there is insufficient time for the capacitor to become fully charged before the input p.d. reverts to zero. Thus V_C increases and decreases by only a small amount.

Figure 4 is known as a *differentiating circuit*. With $T \gg 5RC$, V_{out} is roughly proportional to dV_{in}/dt, or the gradient of the input p.d. This circuit may also be used as a *high-pass filter*. At high frequencies, i.e. $T \ll 5RC$, the capacitor hardly charges up at all and the output p.d. is of the same form as the input; so high frequencies pass through. On the other hand, at low frequencies, i.e. $T \gg 5RC$, the capacitor charges up very rapidly and so the output p.d. has a very different form to the input.

Figure 3 is known as an *integrating circuit*. When $T \ll 5RC$, the output p.d. is roughly equal to the integral with respect to time of the input pulse. (The integral with respect to time is equal to the area under a p.d.-against-time graph.) This circuit can also be used as a *low-pass filter*. At low frequencies, i.e. $T \gg 5RC$, the p.d. across the capacitor resembles that of the input p.d., since the capacitor has ample time to charge up; thus the circuit in figure 3 passes low frequencies.

Both the high- and low-pass filters may also be explained, perhaps more easily, using the idea of reactance which you will meet in the next chapter.

Square pulses into *LR* circuits

In case you do not know or have forgotten what an inductor is, we will give a brief resumé of the discussion in chapter 27. An inductor is a coil of wire, which is often filled with iron. When the current changes through the coil, there is an e.m.f. induced across its ends which is proportional to the rate of change of current. The size of the induced e.m.f. is $L \, (dI/dt)$ where L is thus a constant for the particular coil. L is known as the *self inductance* of the coil; the unit of self inductance is the henry (H). By Lenz's law of electromagnetic induction, the e.m.f. acts in such a direction that it opposes the change of magnetic flux through the coil. Thus in figure 6, when S_1 is closed there is an e.m.f. across L as well as a p.d. across R.

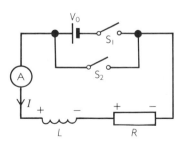

figure 6

We can apply Kirchoff's law for p.d.s round a circuit, to write down this equation:

$$V_0 = L\frac{dI}{dt} + IR$$

This equation can be solved mathematically, so that the growth of the current I with time t is described by this equation:

$$I = \frac{V_0}{R}\left(1 - e^{-Rt/L}\right)$$

A graph of I against t is shown in figure 7(a).

Let us now consider the case when we switch off the current through the inductor. Suppose a current has been passing through L, driven by the cell, then simultaneously we open S_1 and close S_2. The p.d.s are now related by the equation:

$$L\frac{dI}{dt} + IR = 0$$

whence we get

$$\frac{dI}{I} = -\frac{R}{L}dt$$

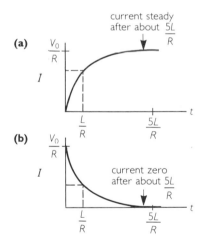

figure 7

This is a familiar equation showing that I decreases by a constant fraction per unit time, i.e. an exponential decrease. The solution is thus in the form: $I = I_0 e^{-Rt/L}$, where the maximum current $I_0 = V_0/R$ and L/R is the time constant of the decay. The decrease in current through L and R is shown in figure 7(b).

We may now use these two results to analyse the outputs of the two circuits shown in figures 8 and 9, when a square wave p.d. is applied to the input.

In both circuits, when the input p.d. rises to V_0 the current grows in the way shown in figure 7(a). When the input p.d. drops back to zero the current decays as shown in figure 7(b). Thus, provided that the

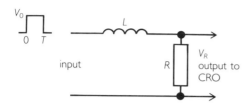

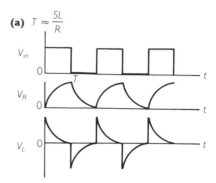

figure 8

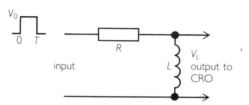

figure 9

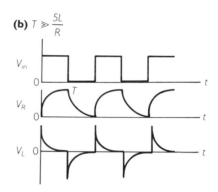

figure 10

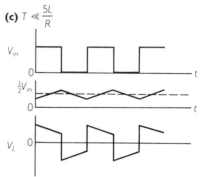

duration of the pulse T is about $5L/R$, the variation of V_R will be as shown in figure 10(a). The p.d. across the inductor, V_L, will be biggest at $t = 0$, just after the input p.d. reaches V_0, because then the current is growing at its greatest rate. The current also changes most rapidly, but in the opposite sense, just after the input p.d. drops to zero after time $t = T$. At this instant the p.d. V_L will be large and negative. Note that at all times $V_{in} = V_R + V_L$.

Figure 10(b) shows V_{in}, V_R and V_L when $T \gg 5L/R$. In this case there is a lot of time available for the growth of current; an oscilloscope will show V_R having nearly the same form as the input p.d., and V_L will appear as a series of sharp spikes. Under these conditions, figure 9 can be used as a differentiating circuit, as V_L is nearly proportional to the gradient of the input p.d.

Figure 10(c) shows V_{in}, V_L and V_R when $T \ll 5L/R$. Under these circumstances there is insufficient time for the current to grow to its maximum value of V_{in}/R. As a result, V_R varies only by small amounts about the average value of V_{in}. V_L approximates closely to the square wave form of V_{in}, but the average value of V_L is zero.

Thus when $T \ll 5L/R$, figure 8 may be used as an integrating circuit, as the output V_R is nearly proportional to the integral of V_{in} with respect to time.

The circuits discussed above may also be used as high- or low-pass filters. Figure 8 may be used as a low-pass filter because at low frequencies, i.e. $T \gg 5L/R$, the p.d. across R is nearly equal to the input p.d. At high frequencies, on the other hand, the p.d. across the output of figure 8 is nearly constant. Figure 9 will allow high frequencies to pass through, since when $T \ll 5L/R$ most of the voltage appears across the inductor.

Summary

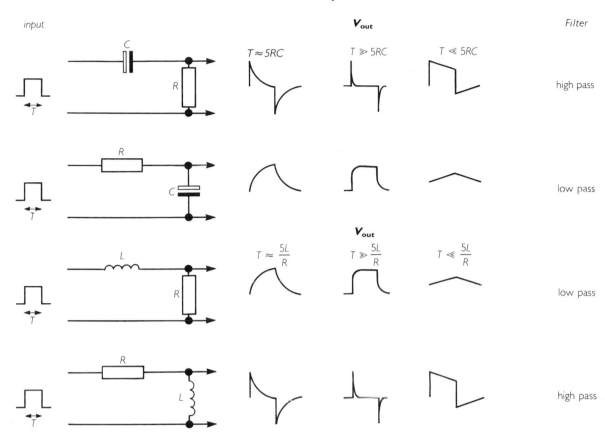

figure 11

Questions

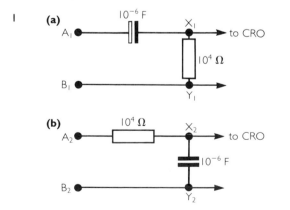

I (a)

(b)

(c)

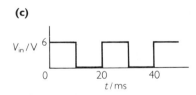

figure 12

a A square waveform p.d. as shown in figure 12(c) is applied across each of the pairs of terminals A_1B_1 and A_2B_2. Sketch carefully and explain the form of the p.d. that would be displayed on an oscilloscope connected across (i) X_1Y_1 and then (ii) X_2Y_2.

b Sketch further diagrams of the CRO traces across X_1Y_1 and X_2Y_2 that would be seen if the frequency of the applied square wave signal were (i) increased by a factor of 10, (ii) decreased by a factor of 10.

2 A p.d. of the form shown in figure 13 is applied in turn to each of the circuits shown in question 1.

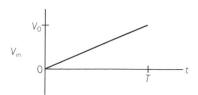

figure 13

a Show that the maximum p.d. at any time across the resistor R is

$$V_0 \cdot \frac{RC}{T}.$$

b Sketch graphs of the p.d. waveform appearing across $X_1 Y_1$ and $X_2 Y_2$ for $T = 10RC$, between the times 0 and T.

3

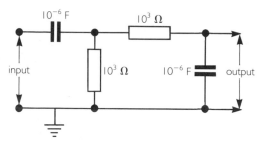

figure 14

Figure 15 shows a simple filter using capacitors and resistors. Explain why a relatively large fraction of a 1000 Hz signal applied to the input will reach the output, whereas a much smaller fraction of a 100 Hz or 10000 Hz signal will reach the output. How does such a filter help to improve the 'signal to noise' ratio?

4

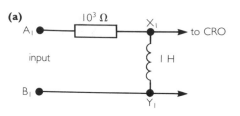

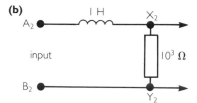

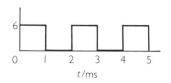

figure 15

a A square wave alternating p.d. as shown in figure 14(c) is applied across each of the terminals $A_1 B_1$ and $A_2 B_2$. Sketch carefully and explain the form of the p.d. that would be displayed on an oscilloscope connected across (i) $X_1 Y_1$ and (ii) $X_2 Y_2$.

b Sketch further diagrams of the p.d. waveforms across $X_1 Y_1$ and $X_2 Y_2$ that would be seen if the frequency of the applied p.d. were: (i) increased by a factor of 10, (ii) decreased by a factor of 10.

c Explain which of the circuits shown in this question could be used as a high-pass filter and which could be used as a low-pass filter.

a.c. circuits

Introduction

In this chapter we examine the behaviour of the resistor, the capacitor and the inductor when they are subjected to an alternating current (a.c.) supply. The resistor behaves in exactly the same way as it does in a d.c. circuit. However, the inductor and capacitor behave in a surprising way. Both of these components have a 'resistance' that depends on the frequency of the supply, but neither of them uses any power. We also see how a circuit using a capacitor and an inductor can help to detect radio signals.

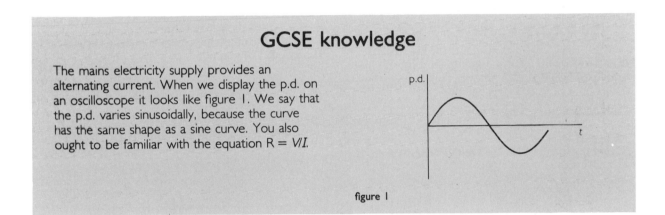

GCSE knowledge

The mains electricity supply provides an alternating current. When we display the p.d. on an oscilloscope it looks like figure 1. We say that the p.d. varies sinusoidally, because the curve has the same shape as a sine curve. You also ought to be familiar with the equation $R = V/I$.

figure 1

Mains electricity supply

We have an a.c. mains electricity supply for the simple reason that it is considerably cheaper to distribute electricity with an a.c. supply than it is with a d.c. supply. It is possible, with transformers, to step up the p.d. of the supply to 400 000 V so that the current in the overhead power lines is as small as possible. In this way enormous power losses are avoided. The variation of our mains supply with time is sinusoidal, as shown in figure 2.

The value of the supply p.d. varies between +339 V and −339 V; the average value of the p.d. is zero. The time for one complete cycle is 0.02 s and so the frequency of the supply is: 1/0.02 s = 50 Hz. The variation of the mains supply may be described mathematically by:

$$V = 339\,\text{V} \times \sin(2\pi \times 50\,\text{Hz} \times t)$$

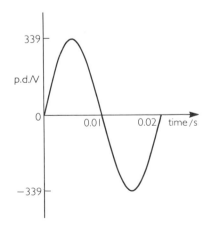

figure 2   Mains supply p.d. on the live wire

In general a sinusoidal supply may be described by:

$$V = V_0 \sin 2\pi f t$$

where V_0 is the peak p.d. and f the frequency of the supply.

Root mean square value of a.c.

Our mains supply is rated at 240 V; this is called its *root mean square* (r.m.s.) value. The significance of the r.m.s. value of an a.c. supply is this: if a resistor is attached to the mains supply it dissipates exactly the same power that it would have done had it been connected to a 240 V d.c. supply. Consequently the average power dissipated in an a.c. circuit is:

$$P = I_{rms}^2 \times R$$

However, at any instant the power dissipated is $I^2 R$, and thus the average power dissipated is $R \times$ average value of (I^2).

Thus $\quad I_{rms}^2 \times R = $ average value of $(I^2) \times R$

and $\quad\quad I_{rms}^2 = $ average value of (I^2).

The average value of I^2 is called the *mean square current*. Figure 3 shows a sinusoidally varying current together with a graph of I^2 as a function of time. It can be seen from the symmetry of the I^2 graph that its average value is $I_0^2/2$, where I_0 is the peak value of the current.

Thus $\quad\quad I_{rms}^2 = I_0^2/2$

and $\quad\quad I_{rms} = I_0/\sqrt 2$

Using this relationship for r.m.s. values for a sinusoidal supply, we can see that the mains supply with a peak value of 339 V has an r.m.s. voltage of $339/\sqrt 2$ V = 240 V.

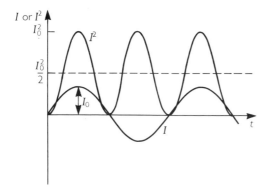

figure 3   A sinusoidally varying current

It is important to realise that the relationship derived above between r.m.s. and peak values of a current or p.d. holds true only for a sinusoidal supply. Figure 4 shows a square wave p.d. varying between $+V_0$ and $-V_0$. The mean square p.d. is V_0^2 and the r.m.s. value is V_0.

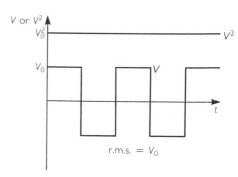

figure 4   A square wave p.d.

Capacitative circuits

When a capacitor is placed in a circuit as shown in figure 5, connected to a cell, no current flows. However, at the moment that the switch S is closed a small surge of current occurs as the capacitor charges up. There is no flow of charge actually through the capacitor itself: charge flows off one plate and onto the other. In the same way, when a capacitor is connected to an a.c. supply, charge flows continuously onto and off the plates. Thus an a.c. ammeter placed in series with a capacitor connected to an a.c. supply will register a current.

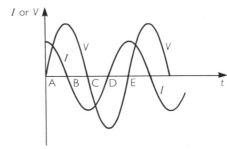

figure 5

The behaviour of a capacitor in an alternating current circuit is somewhat different to that of a resistor. The r.m.s. current through a capacitor is also proportional to the r.m.s. p.d., but the current is 90° out of phase with the applied p.d. (figure 6).

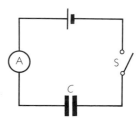

figure 6

We can begin to understand this phase difference between p.d. and current when we consider how the capacitor charges and discharges. This process is illustrated in figure 7. Between times A and B on the graph the capacitor is charging, so a current passes (say in a positive sense). By time B the capacitor is fully charged as the p.d. across it is now a maximum. The current is now zero and the capacitor is just about to start discharging. Between times B and C the capacitor still has charge on it, but the charge is leaving the capacitor so the current is now a negative one. At time C the p.d. across the capacitor starts to become negative so the capacitor starts to charge up but with the charge on its plates reversed, so that the left plate in figure 7(e) is now becoming negatively charged. By time D the capacitor is again fully charged and the current is momentarily zero before the current reverses direction and discharge begins again.

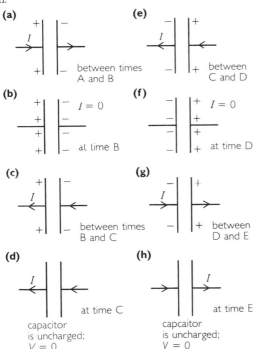

figure 7 The charging and discharging of the capacitor in figure 5

Alternatively, this phase relationship may be understood by a more direct mathematical line of argument. For a capacitor we may write:

$$Q = CV$$
$$\Rightarrow \quad \frac{dQ}{dt} = I = C\frac{dV}{dt}$$

or in words, 'the rate of flow of charge (or current) is equal to the capacitance times the rate of change of p.d.'. Thus in the graph in figure 6, at time A where the p.d. is changing at its greatest rate, charge flows onto the capacitor at its greatest rate and the current

is a maximum. At time B the p.d. has reached its maximum value and so the capacitor is fully charged and is about to begin discharging. At this instant dV/dt is zero and so the current is also zero. At time C, the p.d. is again changing at its greatest rate but in a negative sense, and so the current is a maximum but in the opposite direction to the current at time A.

For a particular r.m.s. p.d., the r.m.s. current through a resistor does not depend on the frequency, f, of the supply. This is not the case for a capacitor; the r.m.s. current (for a given r.m.s. p.d.) is proportional to the capacitance and proportional to the frequency. The charge stored on a capacitor is proportional to the capacitance C, and so a large capacitor will allow more charge to flow per second than a smaller capacitor working at the same frequency. At a high frequency a given capacitor is charged and discharged more times in a second than at a low frequency, thus the current is increased as the frequency increases. We may summarise this paragraph by:

$$I_{rms} \propto fC$$

We define the *reactance* of a capacitor by the equation:

$$\text{reactance} = \frac{V_{rms}}{I_{rms}}$$

It follows from the argument above that the reactance of a capacitor is $\propto 1/fC$. For a sinusoidal supply:

$$\frac{V_{rms}}{I_{rms}} = \frac{1}{2\pi fC}$$

(see below for proof).
Reactance is defined in a similar way to resistance and has the same unit, the ohm. However, the two should not be confused. In a resistor the p.d. and current are in phase and heat is dissipated; in a reactance the p.d. and current are 90° out of phase and no heat is dissipated, as shown below.

Reactance of a capacitor to a sinusoidal current
Let the p.d. applied to a capacitor be $V = V_0 \sin 2\pi ft$.

Thus $$Q = CV = CV_0 \sin 2\pi ft$$
$$\Rightarrow \quad I = \frac{dQ}{dt}$$
$$= CV_0 . 2\pi f \cos 2\pi ft$$
$$\Rightarrow \quad \frac{V_0}{I_0} = \frac{V_0}{2\pi fCV_0} = \frac{1}{2\pi fC}$$
$$\Rightarrow \quad \text{reactance} = \frac{V_{rms}}{I_{rms}}$$
$$= \frac{V_0}{I_0}$$
$$= \frac{1}{2\pi fC}$$

Power calculations for capacitors

The power being supplied to a capacitor at any instant may be calculated by multiplying together the values of p.d. and current for that particular instant. Figure 8 shows the variation of power over one cycle of the applied p.d. For half the cycle the power being supplied is negative, and the average value of power is zero. The idea of negative power is perhaps surprising, but what is happening is this: over the period AB, charge is flowing onto the capacitor until at B the maximum p.d. is across it. At B the maximum energy is stored in the capacitor's electric field. Over the region BC the capacitor's stored energy is fed back to the power supply as the capacitor discharges (the current is now in the opposite sense). Over the period CD power is again taken from the supply as the capacitor is charged up once more (although the p.d. across it is reversed), and over the next quarter of a cycle the stored energy is returned to the supply as it was over the period BC.

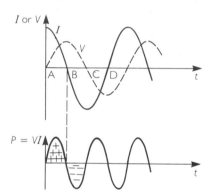

figure 8   Variation of power supply to a capacitor

One has to be wary when calculating the power dissipated in a circuit containing a resistor and a capacitor. The power dissipated in the circuit of figure 9 is not given by $V_s \times I$, but by $P = V_R \times I$, where I is the r.m.s. current and V_R the r.m.s. p.d. across the resistor. The point is that the p.d. across the resistor is in phase with the current, thereby dissipating power. The p.d. across the capacitor and the current are 90° out of phase, so power is not dissipated.

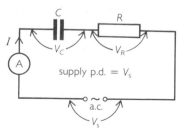

figure 9

An additional puzzling point about the circuit in figure 9 is that if one were to *measure* the p.d.s V_s, V_C and V_R *with a meter*, it would be found that $V_s \neq V_C + V_R$. In fact, at any instant $V_C + V_R$ must be equal to the supply p.d., but the p.d.s are out of phase so we cannot simply add up the meter readings that we see. Figure 10 illustrates this idea: the supply p.d. V_s reaches its maximum somewhere in between the times when V_R and V_C are maximum. Thus the maximum value of V_s is less than the sum of the maximum values of V_R and V_C.

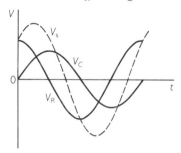

figure 10

The *impedance* of the circuit shown in figure 9 may be defined as follows:

$$\text{impedance} = \frac{\text{r.m.s. supply p.d.}}{\text{r.m.s. current}}$$

Impedance, like resistance and reactance, has units of ohms. Impedance is the term that we use to describe a circuit that contains both resistive and reactive components.

Inductive circuits

In this section we discuss the response of an inductor to a sinusoidally varying supply. Before you read this section it is helpful if you have first read chapter 27, in which the nature of an inductor is explained more fully. For this section it is important that you remember that when the current through an inductor is constant there is no p.d. across it. However, when the current is changing at a rate dI/dt, the p.d. across the inductor is:

$$V_L = L\frac{\mathrm{d}I}{\mathrm{d}t}$$

where L is the inductance of the inductor. Thus, in figure 11 immediately after the switch S is closed there will be a p.d. across the inductor as the current is changing. However, after some while the current will have reached a steady value, $I = V_0/R$, and the p.d. across the inductor will be zero.

If the cell in figure 11 were replaced by a sinusoidal a.c. supply, the behaviour of the inductor would be somewhat different. The current would be switching backwards and forwards continuously, and so there would be a p.d. across the inductor all the

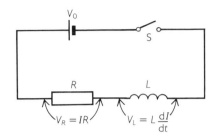

figure 11

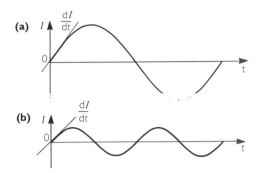

(a)

(b)

figure 13 Effect on current in an inductor of doubling frequency

time, not just immediately after the supply is turned on.

Experiments show that the r.m.s. current through an inductor is proportional to the r.m.s. p.d. applied to it, but the p.d. and current are 90° out of phase as shown in figure 12.

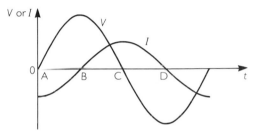

figure 12 R.m.s. voltage and current for an inductor

However, the phase relationship between p.d. and current differs from that of a capacitor. For a capacitor in an alternating current circuit, the current rises to a maximum before the p.d., but for an inductor the current rises to a maximum after the p.d. We may explain this behaviour as follows. The p.d. across the inductor is given by:

$$V = L\frac{dI}{dt}$$

At times A and C the rate of change of current is instantaneously zero and so the applied p.d. at those points must be zero. At time B the rate of change of current dI/dt is a maximum, and so therefore the p.d. must be at its maximum value. At time D, dI/dt is again a maximum but now in the opposite sense, and so the applied p.d. has its maximum negative value.

This phase relationship between p.d. and current is consistent with our experience of inductors in the last chapter. There we learnt that when a steady p.d. is applied to a circuit with inductance, there is a delay between applying the p.d. and the current rising to its maximum value. In the a.c. circuit the current through the inductor also lags behind the applied p.d.

Because the p.d. and current for the inductor are 90° out of phase in an a.c. circuit, no power is dissipated (see below). The ratio V_{rms}/I_{rms} is called the *reactance* of the inductor. The reactance of the inductor is proportional to the inductance L and the frequency of the supply f. Figure 13(a) shows the variation of current with time through a particular

inductance L when a sinusoidal alternating p.d. is applied across it.

Suppose now that the frequency of the supply is doubled but the maximum p.d. of the supply remains constant. This means that the maximum rate of change of current must also remain constant, since $dI/dt = V/L$. But the act of doubling the frequency has halved the time available for the current to grow and thus the maximum current is halved as shown in figure 13(b).

Suppose now that the frequency is kept constant but that the inductance is doubled for a particular applied p.d. (figure 14). Then because $dI/dt = V/L$, it follows that the maximum value of dI/dt must be halved. Thus the maximum value of current is halved (figure 14(b)).

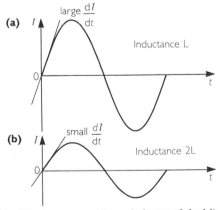

figure 14 Effect on current in an inductor of doubling inductance

These results show us that for a particular applied p.d.:

$$I_{rms} \propto \frac{1}{fL}$$

$$\Rightarrow \quad \text{reactance} = \frac{V_{rms}}{I_{rms}} \propto fL.$$

Below we show that for a sinusoidal supply the reactance of an inductor is $2\pi fL$.

Reactance of an inductor For an inductor the p.d. across it is given by:

$$V = L\frac{dI}{dt}$$

Suppose that the current is: $I = I_0 \sin 2\pi ft$.

Then
$$\frac{dI}{dt} = 2\pi f I_0 \cos 2\pi ft$$

$$\Rightarrow \quad V = L\frac{dI}{dt}$$

$$= 2\pi f L I_0 \cos 2\pi ft.$$

Now the maximum value of p.d. $V_0 = 2\pi f L I_0$

Thus
$$\text{reactance} = \frac{V_{rms}}{I_{rms}}$$

$$= \frac{V_0}{I_0}$$

$$= 2\pi f L.$$

Energy dissipation in an inductor

The average power used in a pure inductor is zero. The arguments used to show this are similar to those applied to the case of the capacitor. The power being used at any instant by the inductor may be calculated by multiplying together the instantaneous values of voltage and current.

Figure 15 shows the variation of power over one cycle of the applied p.d. For half the cycle the power is negative, and its average value is zero.

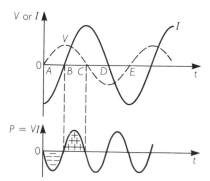

figure 15 Variation of power in an inductor with time

The energy stored in the inductor's magnetic field is greatest when the current is at a maximum. Thus in the interval AB, as the current decreases the field decreases and energy is actually given back to the power supply. Over the interval BC, the current increases from zero to its maximum value; power is drawn from the supply as energy is used to build up the magnetic field.

Electrical resonance in *LC* parallel circuits

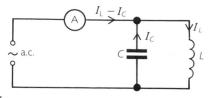

figure 16

Figure 16 shows a capacitor and an inductor in parallel, connected to a sinusoidal a.c. supply. The phase relationship between p.d. and current for each of these components has already been discussed; the current through the capacitor, I_C, leads the applied p.d. by 90°, while the current through the inductor, I_L, lags behind the applied p.d. by 90°. Figure 17 summarises the relative phases of current and p.d. for each of the two components; the important point to notice is that I_L and I_C are always exactly out of phase.

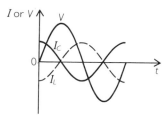

figure 17

The net current through ammeter A will vary considerably as a function of frequency, and at one special frequency the current will be zero. The reactance of the capacitor varies in inverse proportion to the frequency, while that of the inductor varies in direct proportion to the frequency (figure 18). Thus at one frequency, f_0, the reactance of the capacitor is exactly equal to that of the inductor, and so $I_L = I_C$. Therefore the current flowing through the ammeter will be zero at that point, due to their phase difference of 180°.

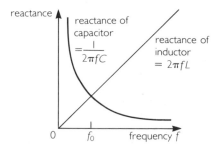

figure 18

f_0 is known as the *resonant frequency* of the circuit, and may be calculated as follows. At the resonant frequency, the reactances of the inductor and capacitor are equal; thus we may write

$$\frac{1}{2\pi f_0 C} = 2\pi f_0 L$$

$$\Rightarrow \quad 4\pi f_0^2 LC = 1$$

$$\Rightarrow \quad f_0 = \frac{1}{2\pi}\sqrt{\frac{1}{LC}}$$

Figure 19 shows the variation of current through the ammeter, I_A, as a function of frequency. At very low frequencies the current through the inductor is high and the current through the capacitor low. At very high frequencies the situation is reversed; but in both cases the net current through the ammeter is high. Only near f_0 does the current become small.

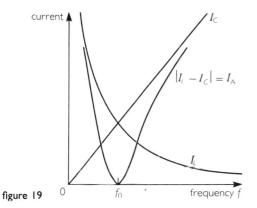

figure 19

This parallel resonance circuit may be well illustrated using the circuit shown in figure 20.

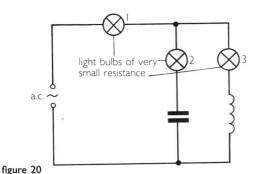

figure 20

At high frequencies bulbs 1 and 2 light; at low frequencies bulbs 1 and 3 light; near f_0 bulbs 2 and 3 light and bulb 1 is out. This is because near f_0 the current flowing through bulb 1 is too small to make it light; the currents through bulbs 2 and 3 are relatively large but since they are out of phase and nearly the same size they add up to give only a small current flowing through bulb 1.

Resonance in a series *L, C, R* circuit

Figure 21 shows an inductor, a capacitor and a resistor in series. In this case all the components have the same current, but the p.d.s across each component will have different phases, as shown in figure 22. (You can check these phases using arguments similar to those used earlier.) In this circuit the p.d. across the inductor, V_L, and the p.d. across the capacitor, V_C, are always out of phase. At the resonant frequency of the circuit, given by

$$f_0 = \frac{1}{2\pi}\sqrt{\frac{1}{LC}},$$

the reactances of the inductor and capacitor are equal, which means that the p.d.s across each are equal in size but in antiphase. Thus the net p.d. across the inductor and capacitor taken together is zero and therefore the impedance of the circuit has its minimum possible value. Figure 23 shows the variation of current with frequency.

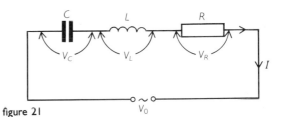

figure 21

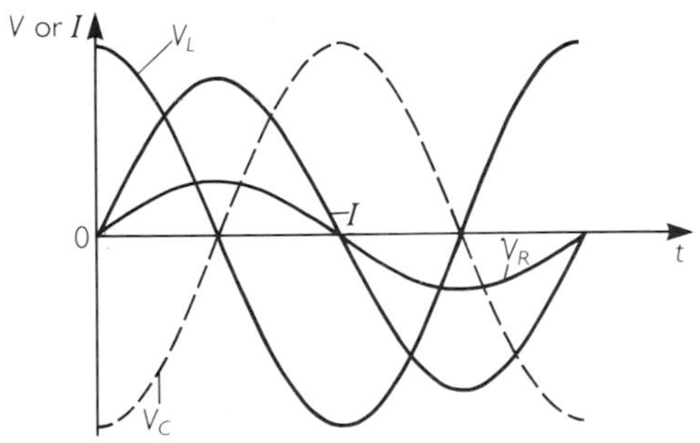

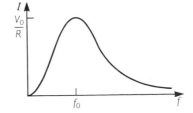

figure 22   Current and p.d.s at resonant frequency. (The *phases* are related like this at *all* frequencies; at f_0, V_C and V_L are equal in size

figure 23

This circuit may be used to generate p.d.s well in excess of the applied p.d. V_0. At resonance, the p.d. across the resistor is V_0.

Thus
$$I = \frac{V_0}{R}$$

$\Rightarrow$
$$V_L = I \times 2\pi f L$$

$$= V_0 \times \frac{2\pi f L}{R}$$

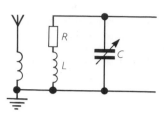

figure 24 Radio tuning circuit

so if the reactances of the inductor and capacitor are much greater than the resistance R the voltages V_L and V_C will be much in excess of V_0.

The series resonant circuit is of particular use for the detection of radiowaves. Radiowaves of frequency f emitted from a radio station induce a varying p.d. in the aerial, which in turn induces a p.d., by mutual induction, across the coil of the tuning circuit. When the capacitance is altered the resonant frequency of the circuit changes. If the resonant frequency is f, a large current will pass, causing a large p.d. to be induced across C; this makes it easy to amplify the radio signal. Radiowaves of other frequencies will cause only very small currents in the tuning circuit and so will not be detected.

Summary

1 For a sinusoidal a.c. supply:

$$\text{r.m.s. p.d.} = \frac{\text{peak p.d.}}{\sqrt{2}}$$

2 Power dissipated in a resistor $= V_{rms} \times I_{rms}$

3 Reactance of capacitor $= \dfrac{V_{rms}}{I_{rms}} = \dfrac{1}{2\pi f C}$

4 Reactance of inductor $= \dfrac{V_{rms}}{I_{rms}} = 2\pi f L$

5 The phase relationship between p.d. and current for a capacitor and inductor are shown in figure 25. Neither a capacitor nor a pure inductor use any power.

6 The resonant frequency of an LC circuit is

$$f = \frac{1}{2\pi}\sqrt{\frac{1}{LC}}$$

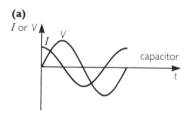

(a)

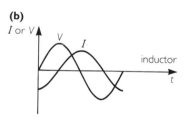

(b)

figure 25 The phase relationship between p.d. and current for (a) a capacitor; (b) an inductor

Questions

1 A laboratory power pack supplies 12 V a.c. (r.m.s.). What is the peak p.d. across the terminals?

2 A certain torch bulb is designed to operate with a steady p.d. of 1.5 V across it. Suppose you now want to operate it from an a.c. supply.
  a What r.m.s. value should the a.c. supply have? (Assume the supply is sinusoidal.)
  b What peak value should the supply have?

3 time base: 10 ms/division

y-gain 0.5 V/division

figure 26

From the CRO trace shown in figure 26, find
a the frequency of the supply;
b the peak p.d. of the supply;
c the r.m.s. p.d. of the supply.

4 A sinusoidal a.c. supply of peak p.d. 10 V is applied across a 20 Ω resistor. What average power is dissipated in the resistor?

5 Two identical light bulbs are made to light to exactly the same brightness using the two circuits shown in figure 27.
   The first light bulb is connected to a 3 V d.c. supply and the second to a low voltage 100 Hz a.c. supply. The p.d.s across each bulb are measured using a dual beam oscilloscope. The

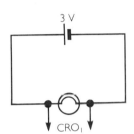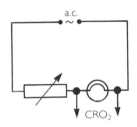

figure 27

Y-input on each channel is switched to 'direct', and the Y-gain control is set to $1\,\text{V cm}^{-1}$. The time base is adjusted to $2\,\text{ms cm}^{-1}$. The oscilloscope screen measures 10 cm × 10 cm.
   Draw, with as much detail as is possible, what you would see on the screen of the oscilloscope. Justify your answer.

6 In Britain the electricity mains supply is alternating and sinusoidal, with a root mean square value of 240 V and frequency 50 Hz.
   a Explain the term 'root mean square'.
   b Suppose that you lived in a part of the country so remote that there is no mains electricity supply, and it is thus necessary to install a d.c. generator. What value of p.d. would you choose so that heaters, designed to operate on a mains r.m.s. supply of 240 V, would work normally?
   d What is the peak p.d. provided by the mains supply?
   e What is the peak current supplied by the mains that would pass through a resistor of resistance 120 Ω?
   f What power would be dissipated in the 120 Ω resistor?

7 A signal generator produces an e.m.f. whose value at any instant is given by
   $E = 10\,\text{V}\sin(100\pi\,\text{Hz})t$.
   a What is the peak value of this e.m.f.?
   b What is the r.m.s. value of this e.m.f.?

c What is the frequency of this signal generator?
d What average power would this generator deliver to a 20 Ω resistor?
e What e.m.f. is produced at the following times: $t = 0$; $t = 1.67\,\text{ms}$; $t = 5.0\,\text{ms}$; $t = 10\,\text{ms}$?

8 (a)

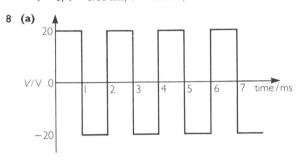

(b)

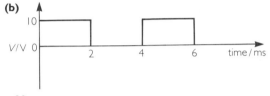

figure 28

a Figure 28 shows the output from two signal generators. For each of the oscillators, what are the values of: (i) the mean p.d., (ii) the mean square p.d. (iii) the r.m.s. p.d.?
b If the two generators were connected together in series, what then would be the r.m.s. p.d. of the combined supply?

9 Show that the r.m.s. p.d. of the waveform shown in figure 29 is $V_0/\sqrt{3}$.

figure 29

10 At a frequency of 100 Hz a certain capacitor has a reactance of 20 Ω. An a.c. p.d. of 15 V r.m.s. is applied across it.
   a What r.m.s. current flows in the capacitor?
   b What is the power dissipated by the capacitor?
   c What current will flow in the capacitor if the frequency of the supply is changed to 200 Hz?

11 A 300 μF capacitor is connected to a variable frequency sinusoidal a.c. supply.
   a What is the reactance of the capacitor at (i) 10 Hz, (ii) 100 Hz, (iii) 1000 Hz, (iv) 10 000 Hz?
   b What reactance does the capacitor have when connected to a dry cell?

12 When a sinusoidal alternating p.d. is applied across a capacitor, the current is found to be 90° out of phase with the p.d. Give a clear explanation of this, illustrating your answer with appropriate graphs.

13 The reactance of a capacitor is given by the equation:

$$\text{reactance} = \frac{1}{2\pi f C}$$

a Explain what f and C represent, and say under what circumstances the equation is true.
b Without using calculus, explain why
(i) reactance $\propto 1/f$, (ii) reactance $\propto 1/C$

14 A sinusoidal a.c. supply of peak value V_0 is applied across a capacitor of value C. Figure 30 shows the variation with time of current and p.d. What charge flows around the circuit over the interval of time AB?

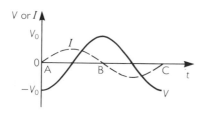

figure 30

15 In turn a resistor and a capacitor are connected to a sinusoidal a.c. supply of frequency 100 Hz. Figure 31(a) shows the variation of p.d. across, and current through, the resistor as a function of time; Figure 31(b) displays the same quantities for the capacitor.
a What is the resistance of the resistor?
b What is the reactance of the capacitor?
c Explain carefully why energy is dissipated in the resistor but not in the capacitor.

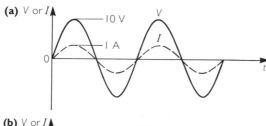

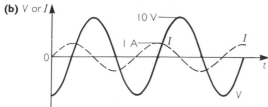

figure 31

d Give one similarity and one difference between a reactance of 10 Ω and a resistance of 10 Ω.
e The capacitor and resistance are now connected together in series so that a peak current of 1 A passes through them. Sketch graphs of the p.d. across the two components, V, and the current through them, I.

16 A 200 µF capacitor and a 1 kΩ resistor are placed in series with a sinusoidal a.c. supply. An oscilloscope is used to measure the peak-to-peak p.d.s V_{XY}, V_{YZ}, V_{XZ}. Peak-to-peak values are obtained as follows: $V_{XY} = 6$ V; $V_{YZ} = 8$ V; $V_{XZ} = 10$ V.

figure 32

a Explain why $V_{XY} + V_{YZ} \neq V_{XZ}$
b Calculate the power dissipated in the circuit.

17 A pure inductor is connected to an alternating current supply. Figure 33 shows part of the p.d. and current graphs for the inductor. The peak p.d. is 4 V and the peak current is 2 mA.
a Use the graph to estimate the maximum rate of change of current in the inductor.
b Hence calculate the value of the inductance.
c Calculate the inductor's reactance.
d Calculate the frequency of the supply.

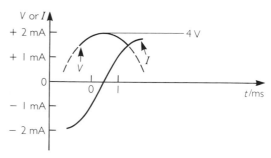

figure 33

18 When a sinusoidal alternating p.d. is applied across an inductor, the current is found to be 90° out of phase with the applied voltage. Explain.

19 The reactance of an inductor is given by the equation: reactance $= 2\pi f L$.
a Explain what f and L represent, and say under what circumstances the equation is true. (Hint: What p.d. waveform should be used?)
b Give a physical explanation (without using calculus) of why (i) reactance $\propto f$, (ii) reactance $\propto L$.

20 Explain why a pure inductor dissipates no power when it is connected to an alternating current supply.

21 Figure 34 shows an inductor and a capacitor connected in parallel to a variable frequency sinusoidal a.c. supply.

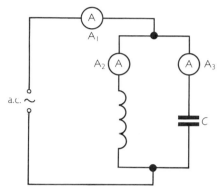

figure 34

a Explain carefully why, at all frequencies, the currents measured by A_2 and A_3 are always exactly 180° out of phase. You should illustrate your answer with suitable sketched graphs.
b Explain why, if the frequency is adjusted to a particular value, the current in ammeter A_2 is the same size as that in ammeter A_3, thus making that in ammeter A_1 very small.
c At the resonant frequency (described in part (b)) there is a continual interchange of energy between the inductor and capacitor. (i) In what form is the energy stored in the inductor and capacitor? (ii) Sketch graphs, using the same axes, to show the variation with time of the applied p.d., the energy stored in the capacitor, and the energy stored in the inductor.
d Describe the energy changes that occur in a chosen mechanical resonant system. Is it possible to draw an analogy between the electrical and mechanical resonant systems?

22

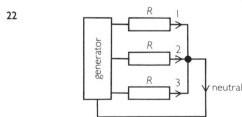

figure 35

Figure 35 shows a three-phase electricity generating system. Three loads are fed independently with currents of a different phase from the same generator. The currents then return to the generator along the same neutral

line. The currents vary with time as follows:
$I_1 = I_0 \sin \omega t$, $I_2 = I_0 \sin(\omega t + 2\pi/3)$,
$I_3 = I_0 \sin(\omega t - 2\pi/3)$.
a Show that there is no current in the neutral wire.
b Explain how the use of a three-phase system enables the Electricity Generating Board to cut both capital and running costs.
c As there is no current in the neutral line, why is it necessary to have one at all?

23

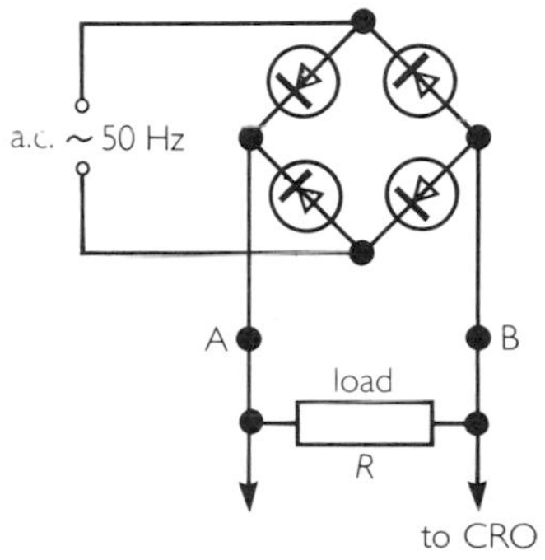

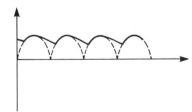

figure 36

a Figure 36 shows a full wave bridge rectifier. Sketch the p.d. waveform that would appear across the load resistor R, if a 50 Hz sinusoidal a.c. supply is connected to the rectifier.
b Suppose now that the load resistor is disconnected and that a 500 μF capacitor is connected across AB instead. Explain why a constant p.d. would now be recorded across the capacitor, once it has been charged up.
c If a load resistance of 100 Ω is placed across the capacitor a waveform as shown in figure 37 is obtained.

figure 37

Explain in as much detail as possible the form of the p.d.
d Sketch a graph of current as a function of time.
e How would the waveform be altered if the load resistance were (i) 1000 Ω, (ii) 10 Ω?
f Explain why the waveform would be smoother if a very large capacitor were used. What in practice, however, limits the size of the smoothing capacitor? (Hint: Think what happens when the power supply is initially turned on.)

24 The rectified unsmoothed p.d. V_1 that was produced by the circuit in the last question can be considered as made up of two parts. There is a steady d.c. component, V_2, and a 50 Hz a.c. ripple, V_3; see figure 38.

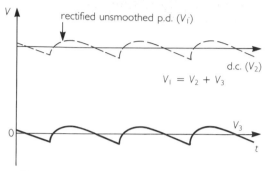

figure 38

a Approximately what is the reactance of a 15 H inductor to (i) the steady d.c. p.d. V_2, (ii) the 50 Hz a.c. ripple p.d. V_3?

b What is the reactance of a 100 μF capacitor to (i) the steady p.d. V_2, (ii) the 50 Hz a.c. p.d. V_3?

c If the p.d. V_1 is placed across the inductor and capacitor as shown in figure 39, what is the p.d. across (i) the inductor, (ii) the capacitor?

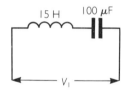

figure 39

d Draw a circuit diagram to show how you would convert a 240 V r.m.s. a.c. supply into a d.c. supply capable of delivering a constant p.d. of up to 5000 V.

25 In question 21 you may have considered why at a certain (resonant) frequency the current going into a circuit with an inductor and capacitor in parallel is zero. We now examine what happens if each of L and C has a certain resistance in series with it, as in figure 40.

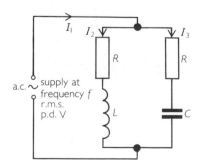

figure 40

In the questions that follow, the resonant frequency f_0 is defined as the frequency when the reactance of the capacitor is the same size as that of the inductor.

a Figure 41 shows how the r.m.s. current I_1 varies with frequency when the resistances R are *very small* in comparison with the reactances of L and C at the resonant frequency. Use an energy argument to explain why I_1 cannot equal zero at the resonant frequency.

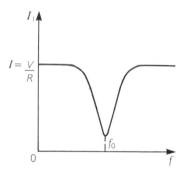

figure 41

Figure 42 shows how I_1 varies with frequency when the resistances R are *very big* in comparison with the reactances of L and C at the resonant frequency. In fact at f_0 the current I_1 is about twice the current at very high or very low frequencies.

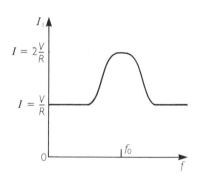

figure 42

b Explain why at resonance $I_1 = 2V/R$, where V is the r.m.s. supply p.d.
(Remember the reactances are small.)

c (i) Explain why for a *very low* frequency $I_3 = 0$ and $I_2 = V/R$. (ii) Explain why for a *very high* frequency $I_2 = 0$ and $I_3 = V/R$.

d Now explain why the current I_1 rises rather than falls in figure 42.

CHAPTER 15

Transistor electronics

Introduction

This chapter revolves around the use of the inverter. We explain how inverters may be used to construct logic gates, astables, bistables and amplifiers. At the beginning of the chapter we explain briefly how an inverter is constructed using a transistor. The transistor is the building brick of all electronics; however it is not our intention to explain the inner workings of the transistor. (To some extent this has been done in chapter 10, question 10.) Electronics is a branch of science that has expanded so rapidly that we feel it is far more important that you learn about the applications of this area rather than the detailed action of one device.

GCSE knowledge

We do not assume any prior knowledge of electronics, although in practice many of you will have gained considerable experience of this area through your GCSE courses.

The transistor

When viewed from the outside, the transistor is an unimpressive plastic or metal cylindrical casing out of which protrude three 'legs'. These 'legs' are the connecting wires to the transistor's three terminals, which are known as the base, collector and emitter. Transistors are made out of specially manufactured germanium or silicon.

The most important points about the behaviour of a transistor may be illustrated with reference to figure 2. In this circuit the collector C has been connected to the positive terminal of a 6 V battery and the base B has been connected to the positive terminal of a power supply that can vary between 0 and 1 V. The emitter E has been set to a potential of zero and is connected to the negative terminal of each supply. The table below illustrates the way in which the collector current I_C, the base current I_B and the emitter current I_E vary with the potential difference V_{BE} between the base and the emitter.

figure 1

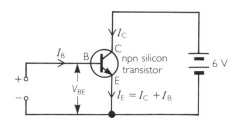

figure 2

V_{BE}/V	I_B/mA	I_C/mA	I_E/mA
0	0	0	0
0.5	0	0	0
0.55	0.01	1.0	1.01
0.6	0.06	6.0	6.06

The table shows us these points:

☐ Until the base is about 0.6 V above the potential of the emitter no base current will flow.

☐ Until there is a base current, no current flows from the collector to the emitter.

☐ As expected $I_E = I_C + I_B$. However I_C and I_E are about 100 times the size of I_B.

It is these properties of a transistor that allow it to perform two very important operations. Firstly, a very small change in base current causes a very big change in collector current; thus the transistor can be used to amplify currents. Secondly, a small change in the p.d. between base and emitter will switch the transistor on or off. These are the two key functions of the transistor: a switch and an amplifier.

The inverter

Figure 3 shows a transistor connected in a circuit to make an inverter. The collector is connected to the positive terminal of a 6 V supply through a 2 kΩ resistor. The emitter is connected to the other terminal of the supply and is set at zero potential.

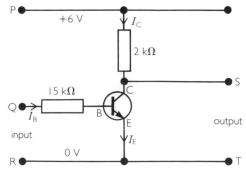

figure 3   An inverter

A p.d. may be applied to the input of the inverter (between Q and R) and this will determine the p.d. across the output which is between S and T. If the input has no p.d. across it then no base current I_B will flow. As a result no current will flow from the collector to the base. If I_C is zero then there can be no potential drop across the 2 kΩ resistor and as a result the potential at S is 6 V. On the other hand, if Q is made to have a high potential (say 2 V) then the p.d. between base and emitter will be above 0.6 V and current will now flow from collector to emitter. Suppose the collector current is 3 mA: then the p.d. across the 2 kΩ resistor is 6 V and so the potential at S is now zero. You can now see why this is called an inverter: if Q is at a high potential S is at a low potential, and if Q is at a low potential then S is at a high potential.

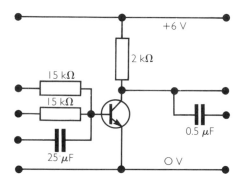

figure 4   Circuit for a 'basic unit'

It is possible to use such an inverter, with a few modifications, as a basic building block for an electronics course. Such a circuit is shown in figure 4; this is sometimes known as a 'basic unit'. A manufactured basic unit is shown in the photograph, figure 5. The obvious advantage of such a ready-made kit is that we are spared the chore of soldering capacitors or resistors as required for different experiments.

figure 5   A 'basic unit'

The symbol for an inverter is shown in figure 6. On the top of the basic unit this symbol has been used instead of the symbol for the transistor. This is a very useful shorthand which avoids the necessity of drawing in the 6 V and 0 V supply lines. We will often use this terminology in future, though of course all inverters must be connected to a power supply.

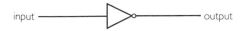

figure 6 Symbol for an inverter

Logic gates

(a) NOT-gate or inverter
This is the most basic of the logic gates. We have already seen the transistor circuit for the inverter in figure 3. Earlier we explained that its behaviour was such that when the input was 'high' the output was 'low' and vice versa. These properties may be summarised in a 'truth table' where the digit 1 represents high and the digit 0 represents low.

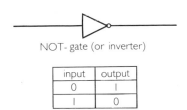

NOT- gate (or inverter)

input	output
0	I
I	0

figure 7

The output of a NOT-gate is only high when the input is *not* high.

(b) NOR-gate

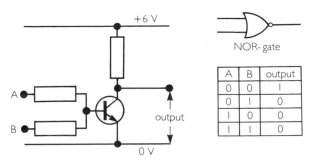

NOR- gate

A	B	output
0	0	I
0	I	0
I	0	0
I	I	0

figure 8

A NOR-gate can be constructed by using a NOT-gate with two inputs. The basic unit described earlier would be suitable for this purpose; such a circuit is shown again in figure 8. If either of the inputs A or B is high then a current will flow through the transistor and the output will be low. If A and B are both low then the transistor will be switched off and the output

will be high. The symbol for the NOR gate, and a truth table summarising its behaviour, are shown in figure 8.

The output of the NOR-gate is only high if neither A *nor* B is high.

(c) OR-gate

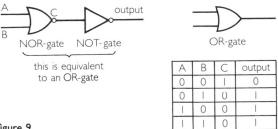

NOR-gate   NOT- gate

this is equivalent to an OR-gate

OR-gate

A	B	C	output
0	0	I	0
0	I	U	I
I	0	0	I
I	I	0	I

figure 9

Figure 9 shows how an OR-gate can be constructed using a NOR-gate and a NOT-gate. The output of the OR-gate is high if either one input *or* the other is high. The NOT-gate inverts the truth table of the NOR-gate.

(d) AND-gate

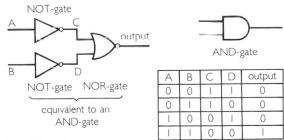

NOT-gate   NOR-gate

equivalent to an AND-gate

AND-gate

A	B	C	D	output
0	0	I	I	0
0	I	I	0	0
I	0	0	I	0
I	I	0	0	I

figure 10

Figure 10 shows how an AND-gate can be constructed using two NOT-gates and a NOR-gate. The output of an AND-gate is only high if one input *and* the other input are high. The output of the NOR-gate will only be high if C and D are low; C and D will be low if A and B are high.

(e) NAND-gate

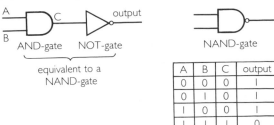

AND-gate   NOT-gate

equivalent to a NAND-gate

NAND-gate

A	B	C	output
0	0	0	I
0	I	0	I
I	0	0	I
I	I	I	0

figure I I

Figure 11 shows how a NAND-gate is made using an AND-gate and a NOT-gate. ('NOT-AND' is abbreviated to NAND.) The NOT-gate inverts the truth-table of the AND-gate so that the output of the NAND-gate is high when either A or B or both are low.

While in practice one may buy all these gates ready made up, it is in fact possible to construct all of them from either NAND-gates or NOR-gates.

Example Use four NAND-gates to make a NOR-gate.

Figure 12 shows how this can be done. The first thing to realise is that a NAND-gate can be made into a NOT-gate by joining the two input terminals together. You can check for yourself the validity of the truth table: the columns A and B and the output give the same truth table as a NOR-gate.

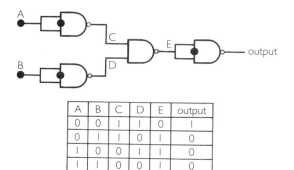

A	B	C	D	E	output
0	0	I	I	0	I
0	I	I	0	I	0
I	0	0	I	I	0
I	I	0	0	I	0

figure 12

The astable

Figure 13 shows an inverter which has a resistor R and a capacitor C connected to its input; both these components are connected to the positive side of the 6 V power supply. At present the input of the inverter is high and so the output is low; the easiest way for the output state of an inverter to be registered is by using an indicator lamp which will be on when the output is high. (The indicators which are used in most electronics kits are modules which themselves contain transistors which are switched on and off by drawing only a small current from the output of the inverter.)

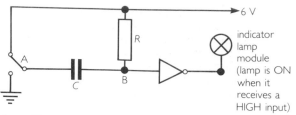

figure 13

Now if, in figure 13, point A is connected to earth, the light can be turned on for a small interval. Here is the reason why. When A is connected to earth both sides of the capacitor are reduced to zero potential. The result is that the input of the inverter, point B in the diagram, is low and the output is therefore high and the lamp is on. However, this state of affairs does not last because the capacitor begins to charge, and soon it has enough p.d. across it to switch the inverter: once again the input is high and the lamp is off.

You know from previous work, chapter 12, that the time taken for a capacitor to charge up is of the order of RC. Thus in figure 13, if R were 20 kΩ and C were 25 μF, then the time the light would be on for is roughly:

$$RC = (2 \times 10^4 \, \Omega) \times (25 \times 10^{-6} \, \text{F}) = 0.5 \, \text{s}$$

Clearly we can control the time of the pulse by adjusting the values of R and C.

We may use the ideas above to make a circuit to switch lamps on and off continuously. Such an arrangement is shown in figure 14; this is called an *astable circuit*.

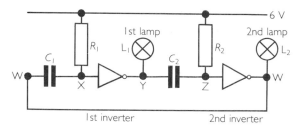

figure 14   An astable circuit

We may explain the action of this circuit with the help of the table below. We have taken the time of switching for each inverter to be about 1 s.

Time/s	W	X	Y	Z	C_1	C_2	L_1	L_2
0	low	low	high	high	charging up	zero p.d. across it	ON	OFF
1	high	high	low	low	zero p.d. across it	charging up	OFF	ON
2	low	low	high	high	charging up	zero p.d. across it	ON	OFF
	etc.							

At the beginning of the sequence shown in the table, the input to inverter 1 is low and its output is high; inverter 2 has a high input and a low output. However this is not a stable state as after about 1 second C_1 has been charged by current flow through R_1 and X has become high. This causes Y and Z to go low and the output of inverter 2 goes high; thus L_1 goes off and L_2 goes on. Again this second state is unstable, as C_2 is now charging up due to current flowing through R_2. Thus the two inverters switch between their two states and the lamps flash on and off.

We can control the frequency of these flashes by adjusting the values of R_1, R_2, C_1 and C_2. If these values are made very small then we can make a fast astable which can produce pulses in the audio-frequency range.

Bistable circuit (or flip-flop)

Unlike the astable circuit, which has no stable position, the bistable circuit has two stable states. Figure 15(a) shows two inverters connected to form a bistable. The inputs A and B are not connected permanently to the 6 V line. Let us now consider what happens if A is connected to the 6 V supply. X goes high, Y and Z go low and W goes high; L_1 is off and L_2 is on. If A is now disconnected, L_1 remains off and L_2 remains on. The state is stable because the high output of the second inverter keeps the input of the first high. The only way that the state can be changed is by making input B high; now Z is high, W and X low, and Y high. In the second state L_2 is off and L_1 is on.

The bistable circuit is extremely important: it is the basis of computer memories. The circuit we have just described remembered which of the two inputs A or B was last high. We have thus stored one bit of information.

Often bistable circuits are simply incorporated into

a single module, figure 15(b). The two outputs are labelled Q and $\bar{Q}$. These are called complementary outputs; when Q is high, $\bar{Q}$ is low, and vice versa. Q is set to high by making S high and Q is set to low, or cleared, by making C high.

University Challenge

Figure 16 shows a very elementary 'University Challenge' circuit. The idea is that a light will show which team got a finger on the button first, but of course the other team's light must not come on. In figure 16 both NOR-gates, N_1 and N_2 are initially connected to the 6 V line, and both lamps are off. Each NOR-gate has one input high and the other low.

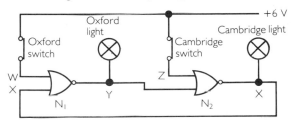

figure 16   Elementary 'University Challenge' circuit

If the Oxford team presses its switch (which disconnects W from the 6 V line, making it low), then both inputs to N_1 are low, and the Oxford lamp lights. It does not matter if the slower Cambridge team now presses its switch, since Y is now high so the output of N_2 is low, no matter what state Z is.

Binary counter – the triggered bistable

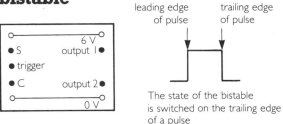

figure 17   Triggered bistable module

The operation of a binary counter depends on the triggered bistable. Figure 17 shows a triggered bistable module. The state of the bistable can be switched in the usual way by connecting either of the inputs S or C to the 6 V rail. Alternatively, the bistable may be switched by a series of pulses which are fed into the trigger. The module then directs each of these pulses in turn to each of the inputs so that the state of the bistable is switched. The vital point in this module is that the switch of the bistable state occurs as the voltage pulse dies away: see figure 17. We say that the bistable triggers on the trailing edge of the pulse.

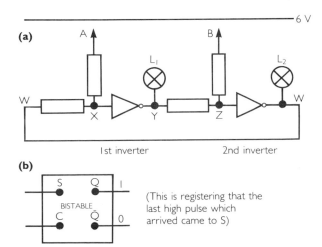

(a)

1st inverter   2nd inverter

(b)

(This is registering that the last high pulse which arrived came to S)

figure 15   (a) A bistable circuit; (b) a bistable module

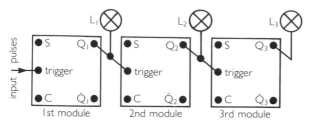

figure 18   A binary counter

Figure 18 shows how three such triggered bistable modules may be used to count up to 7. The counting operation may be explained with the aid of figure 19. Initially the bistables are set so that each of the outputs Q_1 and Q_2 and Q_3 are low; $\bar{Q}_1$, $\bar{Q}_2$ and $\bar{Q}_3$ will all be high. Then the pulses are applied. On the trailing edge of the first pulse, $X_1 X_1$, the first module will trigger and Q_1 becomes high.

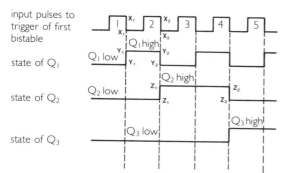

figure 19

However, the second module does not trigger since there is now the leading edge of a pulse, $Y_1 Y_1$, going into its trigger. So after one pulse L_1 is on, L_2 is off and L_3 is off. This corresponds to binary state 001. After the second pulse the first module triggers again on the trailing edge $X_2 X_2$. Now Q_1 drops back to a low state but the second module is triggered by the trailing edge, $Y_2 Y_2$, of the pulse from Q_1, and Q_2 therefore goes high. The binary state after 2 pulses is 010. We will leave you to check that the counter provides the results tabulated below.

No. of pulses	L_1 units	L_2 twos	L_3 fours	Binary state
0	OFF	OFF	OFF	000
1	ON	OFF	OFF	001
2	OFF	ON	OFF	010
3	ON	ON	OFF	011
4	OFF	OFF	ON	100
5	ON	OFF	ON	101
6	OFF	ON	ON	110
7	ON	ON	ON	111

Such a counter, with many more bistable modules, could be put to use to count pulses caused in a Geiger–Müller tube by radiation. Alternatively, the counter could be used as a millisecond timer by counting the number of millisecond pulses that enter it.

Input/output characteristics of an inverter

So far we have dealt with the inverter as a switch. We have only been interested in its output as a high (1) or low (0) state. By knowing a little bit more about how the inverter switches from one state to the other we can put it to greater use.

We need to know how the output p.d., V_{out}, of the inverter varies as the input p.d., V_{in}, is changed. Figure 20 shows a simple way in which we can measure this.

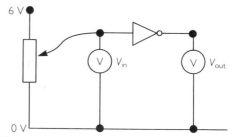

figure 20

The input p.d. can be varied between 0 and 6 V by using a potential divider, and V_{in} and V_{out} are simply recorded using voltmeters. We may then use our values of V_{in} and V_{out} to plot a graph to show the characteristics of a particular inverter.

Alternatively we could use a dual beam oscilloscope to plot our graph for us, as shown in figure 21. Here an a.c. p.d. is applied to the input of the inverter; a diode ensures that only a positive p.d. is applied. A dual beam oscilloscope switched to XY mode is then connected to the input and output as shown. This enables the output potential (Y-plates) to be recorded as the input potential (X-plates) is varied.

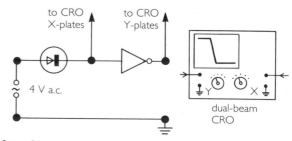

figure 21

Figure 22 shows input/output characteristics for two inverters. For the inverter using a single transistor, the switch of the output from its high state to its low state occurs gradually, while the input potential changes from about 1 V to 1.5 V. On the other hand, the switch for the integrated circuit inverter is very abrupt.

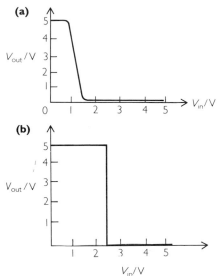

figure 22 Input/output characteristics for (a) an inverter using a single transistor, (b) an inverter based on an integrated circuit

Controlling a light

It is now a fairly simple matter to use an inverter to control the turning on and turning off of a light. Figure 23 shows how this aim can be achieved.

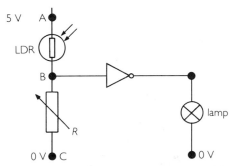

figure 23 Circuit to switch a light on as darkness falls

The input potential of the inverter (at point B) is controlled by a variable resistor R and a light dependent resistor (LDR). The resistance of an LDR, as its name suggests, depends on the intensity of light falling on it; at very low intensities of light its resistance may be $10 \text{ k}\Omega$ and yet in bright sunlight its resistance may only be $200 \, \Omega$. We would need to decide in advance what level of light intensity we consider dark; this might occur when the resistance of the LDR is $2.5 \text{ k}\Omega$. If we were using an inverter whose output switches from high to low at an input

p.d. of 2.5 V (as shown in figure 22(b)), we would then adjust the variable resistor to $2.5 \text{ k}\Omega$. The control of the lamp works as follows.

In bright light the resistance of the LDR is very low, so most of the 5 V p.d. across AC will appear across BC and so the potential at B is high. The output of the inverter is low therefore, and the lamp is off. When the resistance of the LDR reaches $2.5 \text{ k}\Omega$ the p.d. across AC is now split evenly across AB and BC and the input potential is 2.5 V. If the resistance of the LDR increases a fraction more, as the light intensity drops further, the p.d. across AB rises a little more and the potential at B drops below 2.5 V. At this point the input of the inverter is effectively low and the output of the inverter switches to high and the lamp is now on.

A latched burglar alarm – an example of an electronic system

We have now met quite a few electronic devices and we are in a position to put some of them together to make an electronic system, for example, a burglar alarm. Before starting to design the alarm it is of course important to decide what we want it to do. We want our alarm to be triggered when the burglar walks through a light beam that falls onto a light dependent resistor. Once the burglar has walked through the beam the alarm must of course stay on. For this alarm we do not intend to make a great noise since it will frighten the burglar away and we want to catch him. So instead, the intrusion of a burglar will set a light flashing at the police station.

Figure 24(a) shows a diagram of a system that will fulfil this aim. The diagram does not include the minute details of the circuit, but simply shows parts of the system and allows us to describe their function. The first part of the system is the light sensor whose

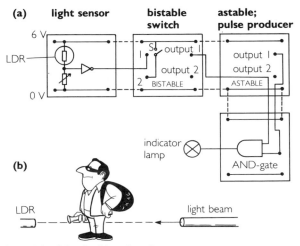

figure 24 A latched burglar alarm

operation we described before. When the burglar passes through the light beam (figure 24(b)), the output of the light sensor goes high. As soon as the burglar has finished walking through the beam the output of the light sensor goes low again. However the output of the sensor is connected to a bistable unit. Before the burglar arrived, output 1 of this module was low and output 2 was high. The pulse entering input 1 of the bistable causes its outputs to reverse, so output 1 becomes high; the bistable remains in this state even after the burglar has passed by the light beam. Once we have caught the burglar the alarm can be switched off and reset by closing the switch S; this second input of the bistable causes its outputs to revert to their original state.

There are two more parts to the burglar alarm. The function of the astable is to produce pulses; these can be controlled in length by a suitable choice of resistors and capacitors. Finally, the AND-gate operates the lamp. For the lamp to be on, both inputs of the AND-gate must be high. Before the burglar triggered the light sensor one input of the AND-gate was low and so the pulses from the astable were blocked. Once the alarm has been activated the output of the AND-gate switches from high to low as the output from the astable switches.

The amplifier

In this section we show how an inverter may be used to make a simple amplifier. The circuit for such an amplifier is shown in figure 25(a).

First, however, let us refer to the input/output characteristics of the inverter, figure 25(b). Over the regions AB and CD of the graph, the output p.d. does not change with the input p.d. However, over the region BC a 0.6 V change in the input causes 6 V change in output. In this region we can say that:

$$\frac{\Delta V_{\text{out}}}{\Delta V_{\text{in}}} = 10$$

This factor is called the *gain* of the amplifier. If we want to amplify an a.c. signal of peak-to-peak p.d. 0.1 V, we cannot simply feed it in to the input of an inverter since this would switch the input between +0.05 V and −0.05 V. As you can see from figure 25(b), this would leave the output constant at 6 V. So what we do is to *bias* the inverter. This is done by using two resistors so that the input is centred round 1.5 V. For our particular inverter 1.5 V is in the middle of the region where the output changes rapidly with the input. We have chosen resistors of 4.5 kΩ and 1.5 kΩ because with a voltage supply of 6 V the p.d. across them will be 4.5 V and 1.5 V respectively.

We can now apply our signal p.d. to the input, but it is fed in to the input of the inverter through a capacitor. The purpose of the capacitor is to isolate the inverter from the source of our a.c. signal, so that

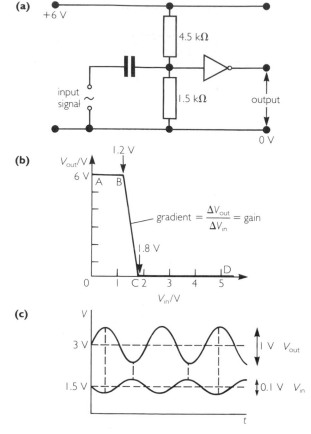

figure 25 (a) An inverter as an amplifier; (b) input/output characteristics of the inverter; (c) input and output p.d.s for the amplifier

the biasing is not affected; a capacitor passes an a.c. p.d. but blocks a d.c. p.d. Now the small signal of 0.1 V peak-to-peak is turned into a signal of 1 V peak-to-peak. The input and output p.ds are out of phase because when the input of the inverter rises the output falls: figure 25(c).

Negative feedback

One problem with the simple single-stage amplifier shown in figure 25(a) is that its gain is likely to be subject to fluctuations. The gain of the amplifier would increase if the temperature rose, and may well drop off at high or low frequencies.

Figure 26 shows an amplifier with a feedback loop. R_1 and R_2 are chosen so that a small fraction of the output, V_{out}/x, is fed back to the input. Since the output is out of phase with the input, the gain of the amplifier is reduced, but as we show below the gain is now independent of the inner workings of the inverter, and is determined only by the values of the resistors R_1 and R_2.

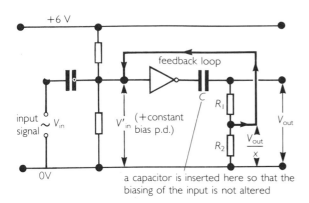

figure 26 Amplifier with feedback loop

Suppose the intrinsic gain of the inverter is m; this quantity is equal to the slope of the output/input characteristic graph (see figure 25(b)).

Then $\quad\quad\quad\quad\quad V_{out} = mV'_{in}$ $\quad\quad$ (i)

where V'_{in} is the net input voltage:

$$V'_{in} = V_{in} - \frac{V_{out}}{x} \quad\quad\quad (ii)$$

V_{in} is our signal to be amplified but it is reduced in size due to the feedback from the output.

From equations (i) and (ii) it follows that:

$$V_{out} = m\left(V_{in} - \frac{V_{out}}{x}\right)$$

$$\Rightarrow \quad\quad V_{out} + \frac{mV_{out}}{x} = mV_{in}$$

$$\Rightarrow \quad\quad V_{out}\left(1 + \frac{m}{x}\right) = mV_{in}$$

If $m \gg x$, then we can ignore the 1 in the bracket,

$$\Rightarrow \quad\quad V_{out} \times \frac{m}{x} \approx mV_{in}$$

$$\Rightarrow \quad\quad V_{out} \approx xV_{in}$$

Typically m is about 100 and x could be 10, so this approximation is a fair one to make. The quantity x is determined simply by the values of R_1 and R_2. For example, if we wanted x to be 10, then the ratio of R_1/R_2 would have to be 9, since then 1/10 of the output voltage would appear across R_2.

Summary

I   Inverter

input	output
1	0
0	1

```
input          output
  A ——[>o—— B
```

figure 27

2   NOR-gate

A	B	output
0	0	1
0	1	0
1	0	0
1	1	0

```
A ——\
      )o—— output
B ——/
```

figure 28

3   OR-gate

A	B	output
0	0	0
0	1	1
1	0	1
1	1	1

```
A ——\
      )—— output
B ——/
```

figure 29

4   AND-gate

A	B	output
0	0	0
0	1	0
1	0	0
1	1	1

```
A ——\
      )—— output
B ——/
```

figure 30

5   NAND-gate

A	B	output
0	0	1
0	1	1
1	0	1
1	1	0

```
A ——\
      )o—— output
B ——/
```

figure 31

6   Bistable

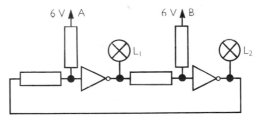

figure 32

L_2 stays on and L_1 off after A has been made high; the state of the bistable may be reversed (i.e. L_2 off and L_1 on) only by making B high and not by making A low.

7   Astable

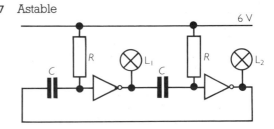

figure 33

L_1 and L_2 switch states continuously; when L_1 is on L_2 is off and vice versa. The length of the pulses is controlled by choosing R and C.

8   Amplifier

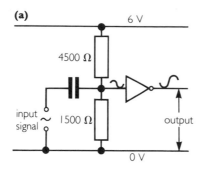

(b)
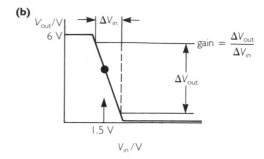

figure 34

The input is biased so that the output changes rapidly with small changes in the input. Thus small a.c. signals can be amplified.

Questions

1

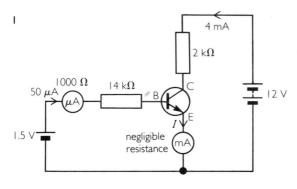

figure 35

This question is about a circuit with a transistor in it, but you do not have to know how a transistor works to be able to answer it.

a What is the current that flows through the milliameter?

b What is the p.d. between B and E?

c What is the p.d. between C and E?

d Show that the largest current that could flow from the 12 V battery in this circuit is 6 mA.

2   What are the logic states of X, Y and Z in figure 36?

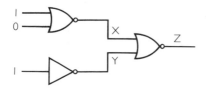

figure 36

3   What is the logic state of P in figure 37?

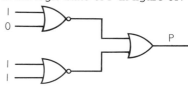

figure 37

4   What logic states do A, B and C have to have to make D low in figure 38?

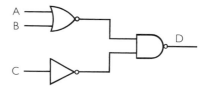

figure 38

5   In each of the four cases shown in figure 39, draw up a truth table to show the logic states (1 or 0) for each of the labelled points in the circuit.

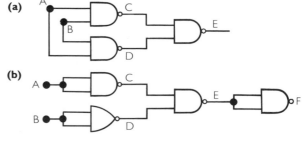

figure 39   (a) and (b)

(c)

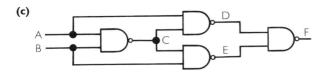

(d)

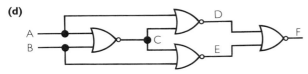

figure 39 (c) and (d)

6 Figure 40 shows two inverters and an OR-gate. Which single logic gate could be used instead?

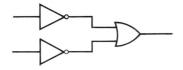

figure 40

7 For a 2-input NOR-gate to have a high output, neither one input nor the other must be high. Explain why figure 41 shows a 4-input NOR-gate.

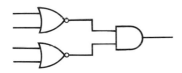

figure 41

a How would you make an 8-input NOR-gate?
b How would you make a 4-input AND-gate?

8 a An exclusive OR-gate is a logic gate for which the output is high only if either of its inputs is high but not both. Figure 42(a) shows how to make such a gate. Verify its action by drawing up a truth table. The symbol for an exclusive OR-gate is shown in figure 42(b).

(a)

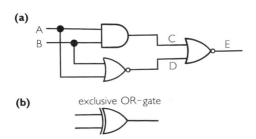

(b) exclusive OR–gate

figure 42

b A Parity-gate is a logic gate whose output is high if both inputs have the same logic state, but low if the logic states are different. Draw the appropriate truth table, and then design such a circuit. Can you suggest a suitable symbol for the Parity-gate?

9 *Half adder and full adder*
One of the most important functions carried out by a computer is the task of binary addition; since we only have two logic states, 0 and 1, we have to use a binary scale. The rules are simple: 0 and 0 make 0, 1 and 0 make 1 and 1 and 1 make 10. In the last addition 0 is called the *sum* and 1 is called the *carry*; we have of course met this idea in our everyday arithmetic – '8 + 7 make 5 carry 1'.
The addition of two digits can be carried out by a *half adder*, figure 43.

(a)

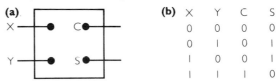

(b)

X	Y	C	S
0	0	0	0
0	1	0	1
1	0	0	1
1	1	1	0

figure 43

Show that the half adder may be constructed from the circuit shown in figure 44.

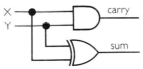

figure 44 A half adder

10 The circuit in figure 45 contains two full adders and a half adder. Explain how this arrangement will help us to do the sum

$$\begin{array}{r} X_3\ X_2\ X_1 \\ +\ Y_3\ Y_2\ Y_1 \\ \hline Z_4\ Z_3\ Z_2\ Z_1 \end{array}$$

Check that our circuit gives the right answer to the sum 110 + 101.

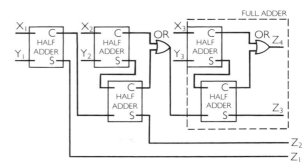

figure 45

11 This question refers to figure 46. Y is always held in a low state; initially X and Q are also in a low state. Then X is made high with a voltage pulse of duration 1 μs. Draw a graph to show the states of P and Q from a time just before X was made high until a time 1 ms later.

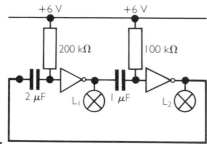

figure 46

12 Describe the behaviour of the light bulbs shown in figure 47. Give as much detail as possible.

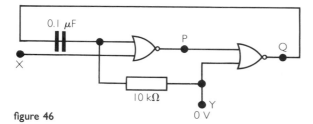

figure 47

13 Approximately how many times will the lamp flash in figure 48 when the switch is thrown from B to A?

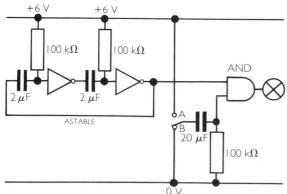

figure 48

14 Figure 49 shows two NAND-gates and two switches. Describe what happens to the outputs of the two gates when first S_1 and then S_2 is pressed. What name is usually given to such a circuit? (The inputs to these gates are high unless connected to earth.)

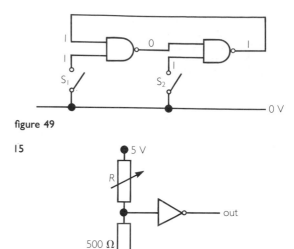

figure 49

15

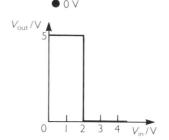

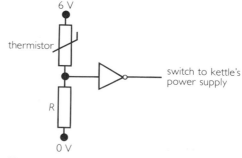

figure 50

Figure 50(b) shows the output/input characteristic for the inverter shown in figure 50(a). What will the output p.d. be when the variable resistor has a value of (i) 2000 Ω, (ii) 1000 Ω, (iii) 500 Ω, (iv) 250 Ω?
Explain your answers.

16 Figure 51 shows an electronic switch whose function is to turn off the power supply to an electric kettle once the steam has warmed up a thermistor placed near the top of the kettle.

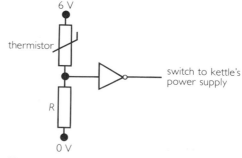

figure 51

a Explain how the device works.
b Calculate what value we need to choose for R, given that the output of the inverter is high if

its input is at a potential lower than 1 V. The resistance of the thermistor is given by $R = R_0 \exp(A/T)$; $A = 960$ K, T is the Kelvin temperature of the thermistor, $R_0 = 120$ Ω.

c What is the resistance of the thermistor at (i) very low temperatures, (ii) very high temperatures?

17 Figure 52 shows how the output p.d. of an inverter varies with its input. Figure 53 shows three waveforms that are to be applied to the input of the unbiased inverter. For each case, make an accurate copy of the input p.d. on graph paper, and then draw in the output.

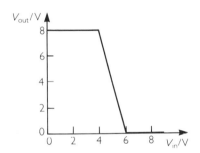

figure 52

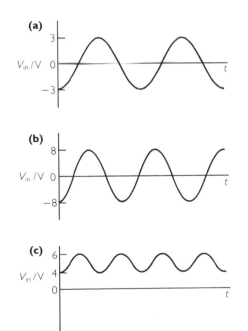

figure 53

18 Draw a fully labelled circuit diagram to explain how you would make a two-stage amplifier; its function will be to amplify sounds picked up by a microphone so that they can be heard in an earphone. Your circuit should contain a volume control.

19 In figure 54, the amplifier has a gain A; it has an extremely large input resistance so we may assume a negligible current goes into it. The circuit is connected so that V_2/V_1 is negative.

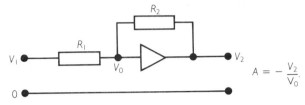

$$A = -\frac{V_2}{V_0}.$$

figure 54

a What can you say about the current flowing through R_2 and R_1?

b Show that

$$\frac{V_0 - V_2}{V_1 - V_0} = \frac{R_2}{R_1}$$

c Use the equation above and the information at the start of the question to show that if $A \gg 1$ then

$$\frac{V_2}{V_1} = \frac{-R_2}{R_1}$$

Explain why this is a useful result.

20 Very often we want to use a logic gate to help us to control the operation of another device. This device may be a motor or heater or electric bell, which is going to need a lot of current. Our logic gates only operate on currents of a few milliamps. Figure 55 shows how we may use a relay to overcome the problem.

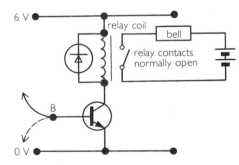

figure 55

Explain how the circuit works. Why do we need a diode in the circuit? (Hint: the relay contains a coil in which large e.m.f.s may be induced.)

In this case we have included the full circuit diagram of the transistor, since an understanding of how the transistor switches on and off will help you to explain this problem.

21

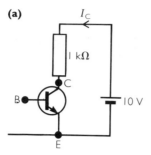

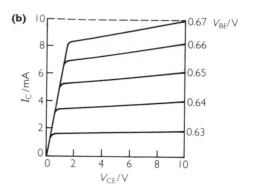

figure 56

a   A transistor with the characteristics shown in figure 56(b) is connected as shown in figure 56(a) with a p.d. of 0.65 V between B and E. What is the p.d. between C and E?

b   An additional peak-to-peak voltage of 0.01 V is applied at B; what variation is observed in V_{CE}?

CHAPTER 16

Operational amplifiers

Introduction

In this chapter we meet an electronic chip, called an operational amplifier (op. amp.), which has a wide variety of uses: among others, as amplifier, adder, digital-to-analogue converter, integrator and control device. Throughout the chapter we consider the device as a 'black box' – we are not concerned with its internal workings, but only with its characteristics and properties. First of all we examine its general features; then we look in detail at a number of circuits in which the op. amp. can be made to perform some of the functions listed above.

GCSE knowledge

You need for this chapter a very thorough understanding of currents, p.d.s, resistance and capacitance. You met these ideas in your GCSE course, and they were discussed again in greater detail in chapters 9 to 12. In particular, you should be able to handle with confidence the idea of the potential divider circuit.

You need to understand simple binary numbers; and you need to be familiar with certain other circuit components: light dependent resistors (LDRs), diodes, relays.

Operational amplifiers

An operational amplifier (or op. amp.) is an integrated circuit which can perform a number of different electronic tasks. Originally, in the 1960s, op. amps. were developed as amplifiers to carry out specific mathematical operations in analogue computing – hence the name. To start with they were developed using combinations of discrete transistors, resistors, capacitors and diodes; now they are fabricated in one piece as integrated circuits, or chips. See figures 1 and 2.

We shall consider an op. amp. purely as a 'black box' – a piece of equipment that gives predictable and repeatable outputs when you put certain inputs into it. We need not know anything about the internal working of the chip itself. We simply use it to do a series of jobs. And, as we shall see, the range of uses for op. amps. within electronic systems is now considerable, even though the chip itself has only seven working connections to it, of which we actively use only three – two inputs and one output.

figure 1 An op. amp. chip

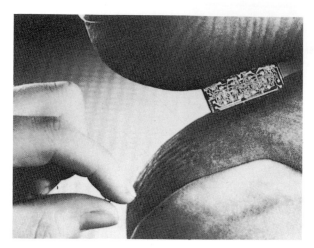

figure 2   Magnified view of a tiny integrated circuit

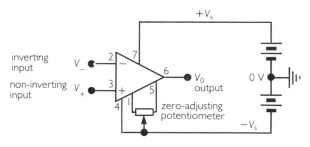

figure 3   Connections to an op. amp. chip

Figure 3 shows the connections to the chip. (The triangular shape is a symbol for an amplifier.) The chip must receive energy from an external power supply, which must go $+V_s$ and $-V_s$ (specified by the manufacturers) of a zero reference point: $+V_s$ and $-V_s$ are connected to pins 7 and 4 respectively. Pins 1 and 5 are connected via a potentiometer to pin 4; this provides a zero adjustment, to allow you to set the output (pin 6) to zero when the two inputs (pins 2 and 3) are both zero.

Now that we have seen the connections made to the power supply and zeroing potentiometer, we will omit these from future circuit diagrams, for simplicity – but remember that they always need to be connected for the circuit to work. Thus our basic op. amp. has the connections of figure 4.

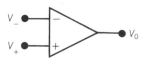

figure 4   Input and output connections to an op. amp.

Characteristics of a basic op. amp.

The basic op. amp. amplifies the difference between V_+ and V_-, by a factor called the *open-loop voltage gain*, A. Thus

$$V_0 = A(V_+ - V_-)$$

A is very large – typically about 2×10^5. Thus if V_- is connected to 0 V (earth), the op. amp. just acts as an amplifier with a very high gain, with output the same sign as the input; similarly if V_+ is connected to 0 V, we have a high gain amplifier with an inverted output. See figure 5.

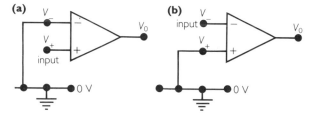

figure 5   Basic amplifiers: (a) non-inverting; (b) inverting

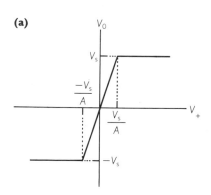

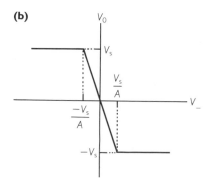

figure 6   Characteristics of basic op. amp.: (a) non-inverting; (b) inverting

The characteristic graphs of the op. amp. in these two modes of operation are shown in figure 6. Notice that the output is only proportional to the input, i.e. analogue, when V_0 lies between $+V_s$ and $-V_s$, where V_s is the supply voltage.

In fact, we never use an op. amp. just like this. The gain A is too high to be useful, and it is very sensitive to changes in temperature and frequency of signal. In practice, we use external feedback arrangements of resistors and perhaps capacitors to control the effective gain. There is one important point arising from the large A, however, provided we are in the analogue range, $V_0 \leqslant V_s$. Typically, $V_s = 10$ V. This means that $V_0 \leqslant 10$ V; therefore

$$A(V_+ - V_-) \leqslant 10 \text{ V}$$

$$\Rightarrow \qquad V_+ - V_- \leqslant \frac{10 \text{ V}}{2 \times 10^5}$$

$$\Rightarrow \qquad V_+ - V_- \leqslant 50 \text{ } \mu\text{V}$$

This is a very tiny p.d., which means that the difference between V_+ and V_- is virtually zero. If one of them is connected to 0 V, then the other is virtually at 0 V also – in fact, we say it is at 'virtual earth'. For the analysis that follows, we will assume that

$$V_+ - V_- = 0$$

for analogue uses of an op. amp.

One other point needs mention before we see what an op. amp. can do. Its input resistance is extremely high. This means that virtually no current passes into the device at either input. We will assume that zero current passes into the device; this is approximately true for all uses.

An inverting amplifier

You can use an op. amp. as an ordinary audio amplifier in, for example, a stereo system. The usual circuit for this is shown in figure 7.

Two resistors are used: an input resistor R_{in}, and a feedback resistor R_f. (*Feedback*, that is, making a connection from the output back to an input, is more usually taken to the inverting input of an op. amp. This is called *negative feedback*; it stabilises the amplifier against fluctuations, because small rapid changes are inverted as they are amplified, and come back to the input inverted – partially cancelling the initial change. *Positive feedback*, on the other hand, amplifies the initial change, and may lead to instability.)

Since V_+ is at 0 V, V_- remains at 0 V as well. No current enters the op. amp., so the same current I which enters R_{in} for some particular value of V_{in} passes round through R_f and joins the current coming out of the op. amp.

Using
$$I = \frac{V}{R}$$

$$I = \frac{V_{in}}{R_{in}}$$

$$= \frac{-V_{out}}{R_f}$$

$$\Rightarrow \qquad \frac{V_{out}}{V_{in}} = -\frac{R_f}{R_{in}}$$

In other words, the voltage gain of the complete system just depends on the values of R_f and R_{in}. (It is inverted, but this doesn't matter for audio-amplification.)

For example, to obtain a gain of 10, you can choose $R_{in} = 10$ kΩ and $R_f = 100$ kΩ.

Notice that the gain does not depend on the open-loop voltage gain A of the op. amp. itself. Thus A can vary with frequency or temperature without affecting the gain of the system.

An integrator

If you replace R_f in figure 7 by a capacitor, the circuit becomes an integrator: one of the mathematical operations for which op. amps. were originally designed. Such an integrator is shown in figure 8.

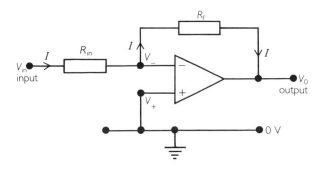

figure 7   Using an op. amp. as a practical inverting amplifier

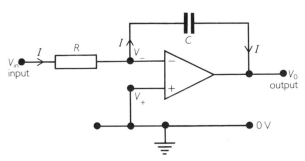

figure 8   Using an op. amp. as an integrator

As before,
$$I = \frac{V_{in}}{R}$$

Also, $I = dq/dt$ for the capacitor, since the effect of the current is to charge the capacitor. V_{out}, which will be on the negative side of the capacitor, will thus also change, so that

$$I = \frac{dq}{dt} = -C\frac{dV_{out}}{dt}$$

$$\Rightarrow \qquad \frac{V_{in}}{R} = -C\frac{dV_{out}}{dt}$$

$$\Rightarrow \qquad V_{out} = -\frac{1}{RC}\int V_{in}dt$$

and also
$$\frac{dV_{out}}{dt} = -\frac{V_{in}}{RC}$$

Thus, for example, a square wave input will result in a saw tooth output. See figure 9.

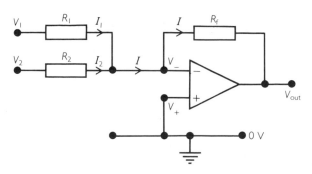

figure 9   An integrator converts a square wave input to a sawtooth output

Summing amplifier

A variation of the inverting amplifier situation of figure 7 enables us to use an op. amp. to do addition. We connect extra input resistors in parallel with the first as shown in figure 10.

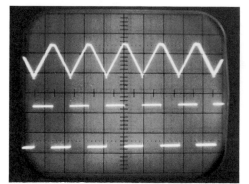

figure 10   Using an op. amp. as a summing amplifier

Since
$$I = I_1 + I_2$$

$$\Rightarrow \qquad \frac{V_{out}}{R_f} = -\left(\frac{V_1}{R_1} + \frac{V_2}{R_2}\right)$$

A particular use of this circuit is in making a digital-to-analogue converter. This is a system which takes a binary number, consisting of an array of 1s and 0s, and turns it into a p.d. proportional to the actual number.

The circuit of figure 10 will handle a 2-digit binary number. Suppose we feed the binary digit p.d.s (which are either on or off) as follows:

1s digit to R_1,
2s digit to R_2.

Thus we want V_2, if on, to give twice as much contribution to V_{out} as V_1 when it is on. So we need to choose resistors such that

$$R_2 = \tfrac{1}{2}R_1.$$

For example, $R_1 = 8\,k\Omega$, $R_2 = 4\,k\Omega$.

Now, whatever 2-digit binary combination comes into R_1 and R_2, the value of V_{out} will be in proportion to the binary number being signalled.

Non-inverting amplifier

The most useful form of this circuit is shown in figure 11.

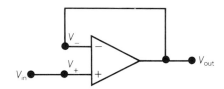

figure 11   Using an op. amp. as a buffer

Recall that in an op. amp., $V_+ = V_-$ always (provided $V_0 \leqslant V_s$).
In the circuit above, $V_+ = V_{in}$; $V_- = V_{out}$.
Therefore $V_{out} = V_{in}$.

This would seem to be a rather useless device. What does it do? The answer is that it is a 'buffer', or p.d. follower. Remember that op. amps. have very high input resistance; so you can, for example, effectively increase the resistance of a moving-coil voltmeter: see figure 12.

The voltmeter still receives the p.d. to be measured, but the op. amp. offers a very high resistance. So you can, for example, measure the p.d. across a capacitor without it discharging. This is the way in which your laboratory coulombmeter works.

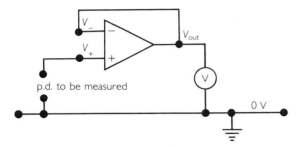

figure 12 Using a buffer to increase the effective resistance of a voltmeter

Follower circuit with variable gain

If we modify the circuit of figure 12 by adding two resistors to form a potential divider, and thus change the feedback, we can change the gain of the follower circuit. (The single buffer circuit had a gain of 1.) The circuit of figure 13 does this.

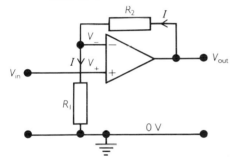

figure 13 Using an op. amp. as a voltage follower with variable gain

Now current I flows from V_{out} through R_2 through R_1 to the 0 V rail.

$$I = \frac{V_-}{R_1} = \frac{V_{out}}{R_1 + R_2}$$

But

$$V_+ = V_- = V_{in}$$

$$\Rightarrow \qquad \frac{V_{in}}{R_1} = \frac{V_{out}}{R_1 + R_2}$$

$$\Rightarrow \qquad V_{out} = V_{in}\left(1 + \frac{R_2}{R_1}\right)$$

This circuit is useful if you want to amplify a voltage to a voltmeter as well as buffer it; or if you want to perform the mathematical function of multiplication.

Comparator

In this application, we use the ability of the op. amp. to detect and amplify the difference between its two inputs:

$$V_0 = A(V_+ - V_-)$$

To control a system, for example, you can arrange that when the system is correctly set, $V_+ = V_-$. But when the system moves away from the setting you want, a voltage change will occur on one of the inputs; thus $V_+ - V_-$ becomes finite, and a voltage appears at V_0 which you can use to operate a correcting mechanism.

As an example, consider the circuit in figure 14. You can use this circuit to switch on the lamp L if the amount of light falling on the LDR from elsewhere becomes less than a preset level.

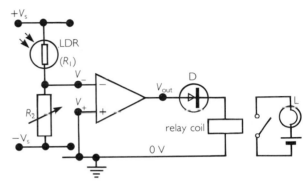

figure 14 Using an op. amp. as a comparator

You adjust the variable resistor R_2 so that at the required critical level of light R_2 equals the resistance of the LDR, R_1.

If the light level is higher than this, $R_2 > R_1$, and therefore $V_- > 0$; since $V_{out} \propto (V_+ - V_-)$, V_{out} is negative. In this case, the diode D prevents current flowing in the relay, and L remains off.

If the light level on the LDR falls, then at the critical level we reach the point $R_1 > R_2$. Then $V_- < 0$, V_{out} becomes positive, current flows in the relay, and L is switched on.

Notice that this use of the op. amp. is a digital, not an analogue, use. We simply want current in the relay to be on or off; it doesn't matter if the amplifier 'saturates', that is, the value of V_{out} reaches V_s.

Simple comparator/amplifier

In the circuit of figure 14, we might wish to control the level of brightness of L to compensate progressively as the external light level falls. To achieve this, we need to use the op. amp. in an analogue mode, which requires feedback and input resistors. One possible circuit would be that of figure 15.

With this circuit, as with that of figure 14, if the external brightness is above our critical level, then $V_{in} > 0$, $V_{out} < 0$, and the diode prevents current flowing to the left.

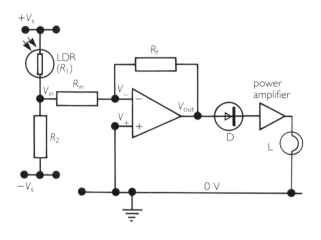

If the brightness falls below critical, then $R_1 > R_2$, and $V_{in} < 0$. The op. amp. is in its inverting amplifier mode, so

$$V_{out} = -\frac{R_f}{R_{in}} \cdot V_{in}$$

So V_{out} is proportional to $-V_{in}$; thus L will be lit, via the power amplifier, to a level dependent on how far V_{in} has fallen below zero; that is, the less light falls on the LDR, the brighter L will be.

If you place the LDR so that light from L falls upon it, then, with suitable values of R_f and R_{in}, the circuit will always reach a balance such that the total intensity of light on the LDR stays roughly constant – independent of the external light level.

Though we have only looked at one example, the use of an op. amp. for control purposes is widespread. For example, robots can be made to follow white lines, production lines keep their feed rates constant, and dish antennae on the earth track satellite transmissions.

figure 15   Using an op. amp. to control a lamp

Summary

I   The name 'operational amplifier' (op. amp.) refers to an integrated circuit amplifier designed to perform mathematical operations.

2   By suitable circuit design, we can use an op. amp. to perform a variety of tasks.

3   An op. amp. requires a power supply: $+V_s$, 0, $-V_s$.

4   Most op. amp. circuits with feedback use negative feedback, to the inverting input.

5   Below saturation:
   a $V_{out} = A(V_+ - V_-)$
   b $V_+ \approx V_-$
   c I_{in} (either input) ≈ 0

6   For an inverting amplifier:

$$V_{out} = -\frac{R_f}{R_{in}} \cdot V_{in}$$

7   For an integrator:

$$V_{out} = -\frac{1}{RC}\int V_{in} dt$$

8   For a summing amplifier:

$$\frac{V_{out}}{R_f} = -\left(\frac{V_1}{R_1} + \frac{V_2}{R_2} + \dots\right)$$

9   For a non-inverting buffer:

$$V_{out} = V_{in}$$

10   For a non-inverting amplifier (follower) with variable gain:

$$V_{out} = V_{in}\left(1 + \frac{R_2}{R_1}\right)$$

11   For a simple differential amplifier:

$$V_{out} = -\frac{R_f}{R_{in}} \cdot V_{in}$$

$V_{in} \propto$ displacement of system from required setting

Questions

1 The 741 op. amp. has an open-loop voltage gain of 2×10^5, and an input resistance of 2 MΩ.
a Explain what these two quantities mean.

Suppose the non-inverting input is at zero potential, and the output is at +2 V.
b What is the potential at the inverting input?
c What is the current at the inverting input?

2 An op. amp. with appropriate power supplies is connected in this circuit:

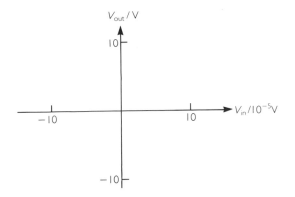

figure 16

a Copy and complete the graph shown in figure 17, if the open-loop gain A of the op. amp. has value 2×10^5.
b How would your answer change if the connections to the two inputs were changed over?

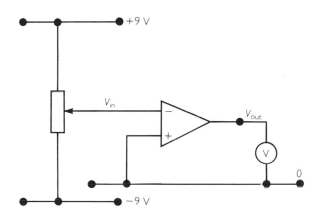

figure 17

3

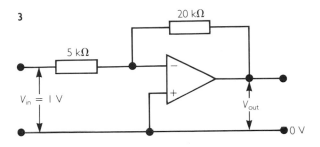

figure 18

Consider the circuit shown in figure 18.
a What is the potential at the inverting input?
b What is the current through the 5 kΩ resistor?
c What current passes into the inverting input?
d What current passes through the 20 kΩ resistor?
e What is the value of V_{out}?

4 An op. amp. with appropriate power supplies is connected in this circuit:

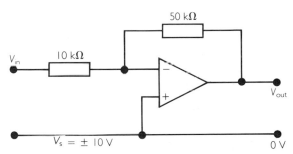

figure 19

a What assumption do we make (i) about the potential at the inverting input, (ii) about the current into the inverting input?
b Copy the axes shown in figure 20, and draw onto them the graph you would obtain if you varied V_{in} between + 10 V and − 10 V.

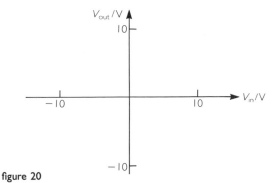

figure 20

c Copy the graphs below, and draw onto them, in a different colour, the outputs that would result from each input waveform.

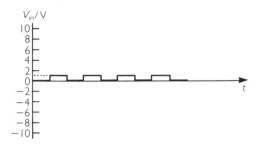

figure 21

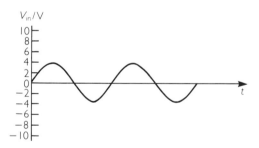

figure 22

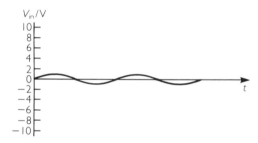

figure 23

5 This question is about using an op. amp. as an adder.

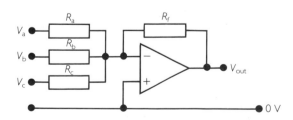

figure 24

Consider the circuit in figure 24. Assume the potential at the inverting input, and the current into it, are both zero.
a What current passes (i) in resistor R_a, (ii) in R_b, (iii) in R_c?
b What is the current in R_f?
c Write down an expression for V_{out} in terms of the other p.d.s and resistances in the circuit.
d Suppose $R_a = R_b = R_c = R_f = 10\,k\Omega$. How does V_{out} now relate to the voltages V_a, V_b, and V_c?

Now suppose we want to modify this circuit to use it as a digital-to-analogue converter; this means that a three-digit binary number will arrive at the inputs, and the output voltage must be proportional to the complete binary number. See figure 25.

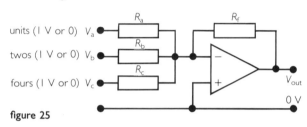

figure 25

Suppose we want an output of $-1\,V$ to represent the binary number 1 (001), $-2\,V$ to represent the number 2 (010), etc.
e Suggest values of R_a and R_f which will give the required output for the binary number 001.
f For the values given in (e), what values of R_b and R_c will be required?
g Suppose you want positive voltages as the output, instead of negative ones. Draw a further circuit using a second op. amp. which will invert the output from the circuit above.

6 The circuit in figure 26 is known as an integrator.

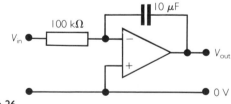

figure 26

Suppose V_{in} is 2 V.
a What is the current through the $100\,k\Omega$ resistor?
b Assuming all this current flows onto the $10\,\mu F$ capacitor, what happens to V_{out}?
c The graphs in figure 27 show how two inputs at V_{in} vary with time. Draw on copies of the same axes how V_{out} will vary in each case.

(i)

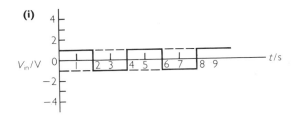

(ii)

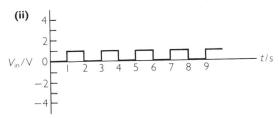

figure 27

7   This question is about using an op. amp. as an integrator to model exponential decay. Consider the circuit in figure 28.

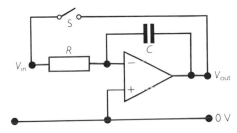

figure 28

a   Write down the relationship between V_{in} and V_{out} when the switch S is open, in the form
$V_{out} = \ldots\ldots\ldots\ldots$
b   Show that the relationship in (a) can be rewritten as

$$V_{in} = -RC\frac{dV_{out}}{dt}$$

If switch S is closed, then $V_{in} = V_{out}$. Thus

$$V_{out} = -RC\frac{dV_{out}}{dt},$$

or

$$\frac{dV_{out}}{dt} = -\frac{1}{RC}V_{out}$$

You have met equations of this sort before: in radioactive decay, where V_{out} would be replaced by N, the number of nuclides remaining; and in the decay of charge on a capacitor. The situation in the above circuit is of course exactly that: the charge on the capacitor flowing away through the resistor.

The mathematical solution to the above equations can be expressed like this:

$$V_{out} = V_0\,e^{-t/RC}$$

where $V_{out} = V_0$ at $t = 0$.
c   Suppose you want to model an exponential decay of radio nuclides with a half-life of 60 s. Suggest suitable values of R and C. (Hint: Recall that in radioactive decay, with decay constant λ and half-life $t_{1/2}$, $\lambda \times t_{1/2} = 0.69$. See page 388.)
d   Suppose you also want to model the growth of the daughter nuclide (which is stable). Suggest how you could use a second op. amp. to give a p.d. proportional to the number of daughter nuclides produced. Draw a circuit diagram, and indicate where the p.d. required appears.

8   This question is about using an op. amp. as a follower or multiplier circuit. The general arrangement is this:

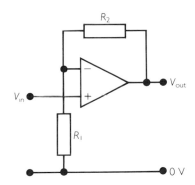

figure 29

V_{out} is related to V_{in} by the equation

$$V_{out} = V_{in}\left(1 + \frac{R_2}{R_1}\right)$$

a   If $R_2 = 0$ (short circuit) and there is no connection at R_1 ($R_1 = \infty$), then how are V_{out} and V_{in} related?
b   For what purpose is the arrangement of (a) useful?
c   This circuit can be used to perform multiplication. Suggest values for R_1 and R_2 so that the circuit multiplies by 10.
d   Can you suggest an arrangement of two op. amps. which together divide an input by 10 (keeping it the same sign)?

9   This question is about the use of op. amps. in analogue computing: that is, making electrical circuits to model various other kinds of system. Consider the circuit in figure 30.

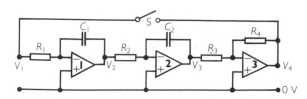

figure 30

This circuit is a simple analogue computer. We will consider two ways in which it can be used to model mechanical systems. With S open, the circuit models free fall; while if S is closed the circuit models simple harmonic motion.

We consider the free-fall model first, with S open. At first ($t = 0$), all the p.d.s (V_1 to V_4) are zero.

If V_1 is suddenly made equal to 1 V, then V_2 changes according to the equation

$$V_2 = -\frac{1}{R_1 C_1} \int V_1 \, dt \qquad \text{(see page 194)}$$

$$= -\frac{t}{R_1 C_1} \times 1 \, V$$

Now suppose the fixed value of 1 V at V_1 represents the fixed acceleration of $10 \, ms^{-2}$ which is the property of an object in free fall.

a Explain why the numerical value of V_2 represents the speed of the object.

b If we want a p.d. of 1 V at V_2 to represent a speed of $20 \, ms^{-1}$, suggest suitable values of R_1 and C_1. Explain your reasoning. (The model is to operate using the same time-scale as a real falling object.)

Op. amp. 2 behaves in the same way as op. amp. 1:

$$V_3 = -\frac{1}{R_2 C_2} \int V_2 \, dt$$

c Explain why V_3 gives the distance travelled by the falling object.

d If a p.d. of 1 V at V_3 is to represent a distance of 100 m, suggest suitable value of R_2 and C_2. Explain your reasoning.

We have now seen how the circuit can calculate the displacement (at V_3) for a fixed acceleration (at V_1). We will next modify this circuit to make it model s.h.m.

The important condition for s.h.m. is that

$$a \propto -s \qquad \text{(see page 83).}$$

If we can invert the displacement V_3, and feed back a suitable fraction of it as the acceleration V_1, then we will be modelling s.h.m. Op. amp. 3 performs just this function, if the switch S is closed. We will analyse this circuit from the right hand end, at V_4.

e What is V_3 in terms of V_4, R_3 and R_4?

f Explain why V_2 is given by

$$V_2 = -R_2 C_2 \frac{dV_3}{dt} \qquad \text{(i)}$$

g Show from (e) and (f) that V_2 is given by

$$V_2 = \frac{R_2 R_3 C_2}{R_4} \cdot \frac{dV_4}{dt}$$

h Using equation (i), write down an expression for V_1 in terms of V_2, R_1 and C_1.

i Show from (g) and (h) that V_1 is given by

$$V_1 = -\frac{R_1 R_2 R_3 C_1 C_2}{R_4} \cdot \frac{d^2 V_4}{dt^2}$$

But if S is closed, $V_1 = V_4$. Hence we have an equation for V_4 of this form:

$$\frac{d^2 V_4}{dt^2} = -\frac{R_4}{R_1 R_2 R_3 C_1 C_2} \cdot V_4$$

You should recognise this as an equation for oscillation, in which V_4 oscillates. It is analogous to $a = -\omega^2 s$ in s.h.m.

j Show that $R_4/R_1 R_2 R_3 C_1 C_2$ and ω^2 both have the same units.

k If $R_1 = R_2 = R_3 = R_4 = 1 \, M\Omega$ and $C_1 = C_2 = 1 \, \mu F$, calculate the natural frequency of the oscillations.

l Which p.d. in figure 30 corresponds to (i) displacement, (ii) acceleration?

m One of the p.d.s corresponds to the negative value of the velocity. Which one? Why is it the negative value?

n Suppose you wanted to model forced oscillations of the oscillatory system. Explain how you would modify the circuit.

o What features of the circuit would limit the amplitude which these forced oscillations would be able to reach?

10 This question is about using an op. amp. as a differential amplifier. A circuit for this application is given below:

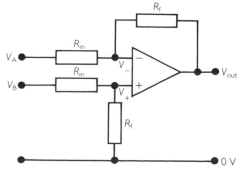

figure 31

We will analyse this circuit to see how it works.
a Find an expression for V_+ in terms of V_B, R_{in} and R_f.
b Find an expression for V_- in terms of V_A, R_{in}, R_f and V_{out}.
c So long as we are in the linear range of the amplifier (that is, $V_{out} < V_s$), we can use the approximation $V_+ = V_-$. Hence find an expression for V_{out} in terms of V_A, V_B, R_{in} and R_f.

One application of this circuit is to amplify the voltage difference between two points in a Wheatstone's bridge which is out of balance. See figure 32.

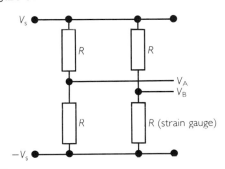

figure 32

d The strain gauge has a resistance which increases with strain. This unbalances the bridge. Is the expression $(V_B - V_A)$ then greater or less than zero? Explain.

The differential amplifier can be used to amplify this difference by any desired factor, perhaps to operate a recording device or correcting mechanism.

11 Design a circuit using one or more op. amps. to fulfil the following tasks.
a Switch on a refrigerator motor when the temperature of a sensor rises above a level which can be set.
b Increase the effective input resistance of a moving-coil voltmeter.
c Multiply a voltage input by (i) +4, (ii) −3.
d Amplify a microphone input by a factor 5.
e From two inputs V_A and V_B, give out a voltage
  (i) of value $(2V_A + 3V_B)$
  (ii) of value $5(V_A - V_B)$.

SECTION C

FIELDS

Force fields are mysterious, forces act at a distance. We can write equations to describe the size of these forces, but we still do not really understand how force fields work. Challenger is pulled down by Earth's gravitational field...

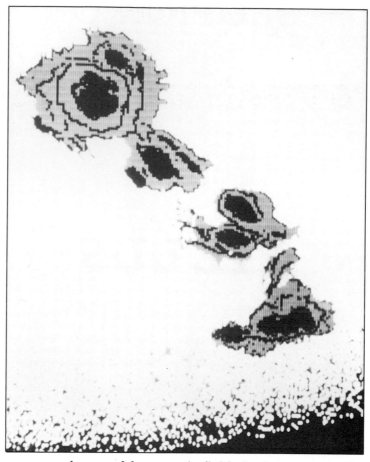

. . . and powerful magnetic fields in the sun control these sunspots

17 Gravitational fields   205
18 Gravitational potential   212
19 Orbits   222
20 Electric field   226
21 Electric potential   233
22 The parallel-plate capacitor   243
23 Electric flux   247
24 Magnetic fields near
    currents   251
25 Magnetic force on charged
    particles   264
26 Electromagnetic induction   271
27 The applications of electromagnetic
    induction   280

CHAPTER 17

Gravitational fields

Introduction

We begin by discussing Newton's law of gravitation and Kepler's laws of planetary motion. The effect of gravitation on a spacecraft is investigated, and an experiment to measure the gravitational constant G is described. The idea of a gravitational field is then introduced.

GCSE knowledge

We assume you know that:

☐ A falling object close to the surface of the Earth accelerates towards the ground at a rate of $10\,\mathrm{ms}^{-2}$.

☐ Mass is a measure of how hard it is to accelerate something: a given force will accelerate a large object less rapidly than a small object. Mass is a constant and is only determined by the amount of matter in an object; the mass of an object is the same anywhere in the universe.

☐ The weight of an object is the force that the Earth exerts on it due to its gravitational pull. We say that the pull of the Earth is 10 N on each kilogram, or that the Earth's gravitational field strength is $10\,\mathrm{Nkg}^{-1}$ at its surface. Thus the weight of a 20 kg object is 200 N.

In addition, it will help you with this chapter if you are familiar with the principle of the conservation of momentum, and Newton's third law which says that 'whenever one object exerts a force on a second object, the second also exerts an equal and opposite force on the first.'

Newton's law of gravitation

In this chapter we shall be discussing a law of Physics which is both elegant and simple, but which reaches far in its consequences. This is Newton's law of gravitation, which states that every object in the universe attracts every other object with a force that is proportional to the mass of each of them and is inversely proportional to the square of the distance between them. Mathematically this statement is expressed by the equation

$$F = \frac{Gm_1 m_2}{r^2}$$

G ('big gee') is the universal constant of gravitation.

The reason for the importance of the law is that if we add to it the principle that a resultant force will accelerate an object, as described by the familiar equation

$$F = ma$$

it is then perfectly possible, with the help of some mathematics, to deduce all the consequences of the two principles. It is indeed our purpose to apply the two principles to describe recent space flights and to investigate several twentieth century astronomical discoveries.

However, only to mention Newton and his laws would be to slight the many men who have devoted their lives to the pursuit of understanding astronomy, throughout the previous two or three thousand years. The story begins with the ancient philosophers who argued about the nature of the physical world in which we live and the motion of the heavens. They invented models to describe the movements of the stars and planets. The most famous of these is the Ptolemaic System of epicycles (Alexandria, AD 120) which survived until the sixteenth century as the most accurate means of predicting the position of the planets.

The fifteenth century witnessed debates as to whether the planets really went round the sun, as championed by Copernicus, or whether the epicyclic motions of Ptolemy were correct. A complete change in the nature of the subject we now call physics occurred when Tycho Brahe suggested that to sit back and argue philosophically about the universe would resolve nothing. His idea was that we could only understand the nature of the planets and stars if we knew their positions very accurately, to which end he spent a great deal of time. Brahe's data enabled Kepler to deduce his three laws of planetary motion, which in turn provided Newton with the opportunity to tie together a lot of apparently disjointed facts with his theory of gravitation.

Kepler's laws

First Kepler discovered that planets moved in elliptical orbits, with the Sun at one focus: figure 1.

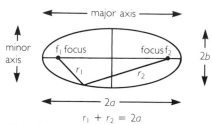

figure 1   An ellipse

Secondly, Kepler observed that the planets do not go round the Sun at a uniform rate, their speed being greater when they are near the Sun and slower when further away. The precise way in which the speed varies may be described with the aid of figure 2.

Suppose that when a planet is near to the Sun it travels from A to B in a month; when it is in a different part of its orbit it might travel from C to D in the same time. The shaded areas will be the same in each case. Kepler's second law can be summarised as 'An imaginary line joining the Sun to the planet sweeps out equal areas in equal times'.

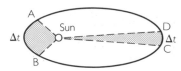

figure 2   Kepler's second law

The third law discovered by Kepler was that the ratio T^2/a^3 was a constant for all planets; T is the time taken for a planet to complete one orbit and a is half of the major axis of the ellipse. In cases where planets described circular orbits, a would be equal to the radius of the orbit.

The inverse square law of gravitation

As we have seen earlier, the equation $F = Gm_1m_2/r^2$ is a mathematical description of Newton's law of gravitation. To the question 'Why do forces obey such a law?' we could answer 'that is just the way things are', because a satisfactory mechanism for the action of gravity has yet to be discovered. So we could just take it as a law that works and leave it at that. However, we can at least convince ourselves that such a law is reasonable if we call on our experience of physics.

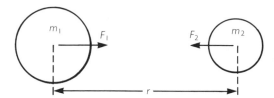

figure 3   Gravitational attraction

1 $F_1 \propto m_1$: we know from experience that the gravitational acceleration (on Earth) of two objects of different masses is the same (provided air resistance is negligible). Since $a = F/m$ it follows that the force of attraction F_1 towards m_2, experienced by a body of mass m_1, must be proportional to m_1.

2 $F_1 \propto m_2$: in the world in which we live we have learned to expect a certain symmetry. It would be odd if in figure 3 F_1, the force that m_1 experiences, were any different from F_2, the force that m_2 experiences. Suppose the two masses were released from rest in space, and that F_1 were greater than F_2, then upon collision the two bodies (as we look at them) would continue moving to the right. Our experience with trolleys tethered together by elastic bands tells us that, no matter what the mass of the trolleys, they will not be moving after such a collision – figure 4.

(Such considerations lead us to the generalisation that 'momentum is conserved'.)

Returning to our two masses we may argue thus:

$F_1 \propto m_1$ and $F_2 \propto m_2$.
If $F_1 = F_2$ then $F_1 \propto m_2$
So $F_1 \propto m_1m_2$.

You will no doubt recognise Newton's third law in the argument above.

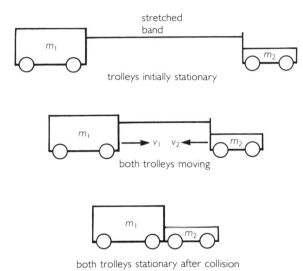

figure 4   Conservation of momentum illustrated by colliding trolleys

3 $F \propto 1/r^2$. You may be familiar with the inverse square law for light intensities. A possible argument goes like this. If a card were held at a distance x from a small light bulb (figure 5), it would receive energy at a certain rate W. If the card were removed, that energy would illuminate four similar cards at a distance $2x$; so the rate of energy reception by these cards would be $W/4$. Although we cannot see 'gravitational waves' spreading outwards from planets, it is reasonable to suppose that the gravitational influence of a planet might weaken with distance in the same way as the intensity of light decreases with distance, i.e. inversely proportional to the (distance)^2.

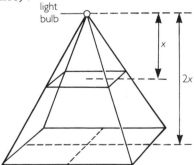

figure 5   Light bulb analogy

Recent space flights have provided a most convincing proof of the truth of the inverse square law. Such an example is shown below.

Example In figure 6, points 1 and 2 are very close together at an average distance of 6.3×10^7 m from the centre of the Earth. At point 1 a spacecraft is drifting almost directly towards the Earth at a speed

of 508 m s^{-1}; at 2, 10 minutes later, its speed is 568 m s^{-1}. During this time the motors were switched off. Show that these data are consistent with the inverse square law.

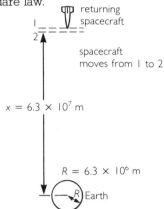

figure 6   Spacecraft returning to Earth

During the 10 minutes from 1 to 2, the craft's acceleration is:

$$a = \frac{v - u}{t}$$

$$= \frac{568\,\text{m s}^{-1} - 508\,\text{m s}^{-1}}{600\,\text{s}}$$

$$= 0.10\,\text{m s}^{-2}$$

On the surface of the Earth $g \approx 10$ N kg^{-1}.

From figure 6 we can see that the spacecraft is ten times further away from the Earth's centre than we are on the Earth's surface. From the inverse square law the gravitational field in the region near the spaceship will be smaller than our field by a factor of $(1/10)^2$. Thus the field near the spaceship will be

$$g_1 = (1/10)^2 \times 10\,\text{N kg}^{-1}$$

$$= 0.10\,\text{N kg}^{-1}$$

This prediction is consistent with the measured acceleration.

It is interesting to note that even before space flights, people had placed immense faith in Newton's law. After the discovery of Uranus in 1781 its path was carefully plotted. By 1845 its position was not as expected; the discrepancy could not be explained even when perturbations from Saturn and Jupiter were taken into account. Rather than discard Newton's law, scientists predicted that another planet must lie outside Uranus – and indeed Neptune was discovered in 1846.

Measurement of *G*

Figure 7 illustrates the apparatus used by Boys in 1895 to determine G. Two small gold balls of mass 2.5×10^{-3} kg were suspended on a very fine quartz fibre. Two lead spheres of mass 7.5 kg were

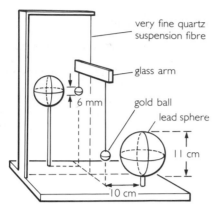

figure 7 Boys' apparatus

suspended so as to exert a gravitational force on the balls and thus twist the quartz fibre through a small angle. The pairs of spheres were suspended at different heights in order to reduce the effect of each lead sphere on the more distant gold ball.

The angle through which the arm was rotated when the lead spheres were put in place was measured by the reflection of a light ray from its glass surface. The observations of the light ray were carried out by a telescope placed some seven metres away. The position of the telescope not only provided a large magnification of the movement, but had the added advantage of keeping the observer at a distance from the delicate quartz fibres.

In addition to the minimising of vibrations, it was also necessary to keep the suspensions draught-free and this was achieved by enclosing the apparatus in a large container. Finally, it is important to realise that any electrical forces could easily have outweighed those gravitational ones that the experimenter was trying to measure, and so it was essential to maintain electrical neutrality by earthing the spheres and balls before starting the experiment (see below).

Example to calculate G   Data for an
experiment similar to Boys':
- mass of a lead sphere   7.5 kg
- mass of a gold ball 2.5×10^{-3} kg
- length of suspension rod x 5.0 cm
- separation of centres of a pair of lead and gold spheres R 10 cm.

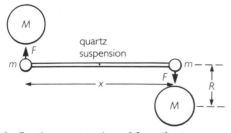

figure 8   Boys' apparatus viewed from the top

It was also discovered by a separate experiment that a torque C of 1.0×10^{-7} Nm was required to turn the quartz fibre through 1.0 radian.

Finally it was observed that the telescope, at a distance of 7.0 m from the apparatus, had to be moved sideways a distance of 0.88 mm when the lead spheres were placed next to the suspended gold balls.

The torque turning the balls from gravitational forces is:

$$F.x = \frac{GMm}{R^2}.x$$

The torque required to turn the quartz suspension through an angle θ is $C\theta$.

So

$$C\theta = \frac{GMm}{R^2}.x$$

or

$$G = \frac{C\theta R^2}{Mmx}$$

From figure 9 we may see that when the quartz suspension turns through an angle θ, the light beam is turned through 2θ.

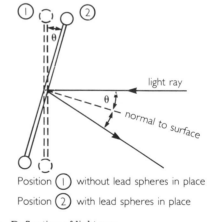

Position ① without lead spheres in place

Position ② with lead spheres in place

figure 9   Deflection of light ray

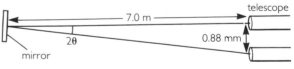

figure 10   Angle of deflection

From figure 10 we can see that

$$2\theta \approx \tan 2\theta = \frac{0.88}{7000}$$

or

$$\theta = 6.3 \times 10^{-5} \text{ rad}$$

Thus

$$G = \frac{C\theta R^2}{Mmx}$$

$$= \frac{(1.0 \times 10^{-7}\,\mathrm{Nmrad^{-1}})(6.3 \times 10^{-6}\,\mathrm{rad})(0.10\,\mathrm{m})^2}{(7.5\,\mathrm{kg})(2.5 \times 10^{-3}\,\mathrm{kg})(0.050\,\mathrm{m})}$$

$$= 6.7 \times 10^{-11}\,\mathrm{Nm^2\,kg^{-2}}$$

Digression To what p.d. would one of the small gold balls have to be charged, for electrostatic forces to outweigh gravitational forces?

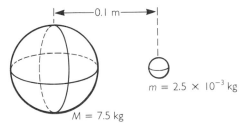

figure 11

Gravitational force between a pair of spheres is

$$F = \frac{GMm}{R^2}$$

$$= \frac{(6.7 \times 10^{-11}\,\mathrm{Nm^2kg^{-2}}) \times (7.5\,\mathrm{kg}) \times (2.5 \times 10^{-3}\,\mathrm{kg})}{(0.10\,\mathrm{m})^2}$$

$$= 1.26 \times 10^{-10}\,\mathrm{N}$$

If we assume that each sphere carries an identical charge Q, then we can work out what size Q has to be for electrostatic forces to be of the same size as the gravitational ones that are to be measured.

$$F = \frac{Q^2}{4\pi\epsilon_0 R^2}$$

$$\Rightarrow \quad Q^2 = 4\pi\epsilon_0 FR^2 \qquad \text{(see page 229)}$$

$$\Rightarrow \quad Q = \sqrt{4\pi\epsilon_0 FR^2}$$

$$= \sqrt{4\pi \times (8.85 \times 10^{-12}\,\mathrm{Fm^{-1}}) \times (1.26 \times 10^{-10}\,\mathrm{N}) \times (0.10\,\mathrm{m})^2}$$

$$= 1.2 \times 10^{-11}\,\text{coulomb}$$

The radius of a small gold ball of mass $2.5 \times 10^{-3}\,\mathrm{kg}$ will be about 3.0 mm. Using this we may thus estimate the potential of such a small ball that carries the charge calculated above.

$$V = \frac{Q}{4\pi\epsilon_0 r} \qquad \text{(see page 235)}$$

$$= \frac{1.2 \times 10^{-11}\,\mathrm{C}}{4\pi \times (8.85 \times 10^{-12}\,\mathrm{Fm^{-1}})(3.0 \times 10^{-3}\,\mathrm{m})}$$

$$= 36\,\mathrm{V}$$

Since potentials of several thousand volts may be induced by friction, it is clear that great care must be taken to ensure electrical neutrality. A sphere even at a potential of 1 V would introduce serious errors into a measurement of G.

Gravitational fields

From Newton's law of gravitation it follows that in the region close to any lump of matter, an object will experience a force. We all experience such a force on the surface of the Earth, and we say that in this region a gravitational field acts. The strength of a gravitational field, g-field, is measured by the force exerted on a unit mass. This may be defined formally by the equation

$$g = \frac{F}{m}$$

showing that the units of field are $\mathrm{Nkg^{-1}}$

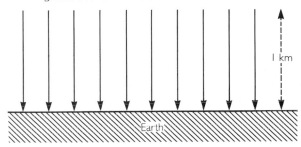

figure 12   Uniform gravitational field near the Earth's surface

Close to the Earth's surface, figure 12, the g-field is approximately uniform: over a height of several hundred metres we do not notice any change. However, at large distances the field strength diminishes, figure 13.

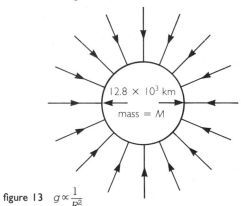

figure 13 $g \propto \dfrac{1}{R^2}$

From the definition of gravitational field and Newton's law, we find that

$$F = mg = \frac{GMm}{r^2}$$

$$\Rightarrow \qquad g = \frac{GM}{r^2}$$

where M is the mass of the planet and r the distance from its centre.

Example 1 Calculate the mass of the Earth given that its radius is 6400 km and the g-field near its surface is approximately $10\,\mathrm{N\,kg^{-1}}$.

$$g = \frac{GM}{r^2}$$

$$\Rightarrow \qquad M = \frac{gr^2}{G}$$

$$= \frac{(10\,\mathrm{N\,kg^{-1}})(6.4 \times 10^6\,\mathrm{m})^2}{6.7 \times 10^{-11}\,\mathrm{N\,m^2\,kg^{-2}}}$$

$$= 6.0 \times 10^{24}\,\mathrm{kg}$$

Example 2 Calculate the gravitational field strength on the surface of Jupiter, given that its mean density is 0.24 times that of the Earth and that its radius is 11 times that of the Earth.

$$g = \frac{GM}{r^2}$$

but $M = \frac{4}{3}\pi\rho r^3$ where ρ is the density of Jupiter.

$$\Rightarrow \qquad g = \frac{4}{3}\pi G\rho r$$

Thus the g-field on the surface of any planet is proportional to its mean density and radius. For Jupiter,

$$g = (10\,\mathrm{N\,kg^{-1}}) \times 0.24 \times 11 = 26\,\mathrm{N\,kg^{-1}}$$

Summary

1 $F = \dfrac{Gm_1 m_2}{r^2}$

where F is the gravitational attraction between two objects with masses m_1 and m_2, r is the separation of the centres of mass, and G is the universal gravitational constant.

2 Kepler's laws describing the motion of planets:
(i) The planets move in elliptical orbits, with the sun at one focus.
(ii) An imaginary line joining the sun to a particular planet sweeps out equal areas in equal times.
(iii) The expression T^2/a^3 is the same for all planets, where T is the time for one revolution, and a is half the major axis of the ellipse.

3 G may be measured using a carefully constructed experiment in a laboratory. For example, Boys' experiment (1895) measured the force between two pairs of spheres made from dense metal.

4 Gravitational field g is defined by the equation

$$g = \frac{F}{m}$$

where F is the force on a mass m placed in the field.

5 The field at distance r from a mass M is given by

$$g = \frac{GM}{r^2}$$

6 At the surface of a spherical planet of uniform density ρ and radius R, the gravitational field strength g is given by

$$g = \frac{4}{3}\pi G\rho R$$

Questions

In the following questions,
$G = 6.7 \times 10^{-11}\,\mathrm{N\,kg^{-2}\,m^2}$.

1 The gravitational field strength on the surface of a planet of radius $1.0 \times 10^6\,\mathrm{m}$ is $4.0\,\mathrm{N\,kg^{-1}}$. What is the field strength at a distance of $2.0 \times 10^6\,\mathrm{m}$ from its centre?

2 Astronomers think that black holes are formed when large stars collapse rapidly at the end of their lives. A black hole has a mass of $2.0 \times 10^{33}\,\mathrm{kg}$ and a radius of 1.0 km. Calculate the gravitational field strength near to its surface (its 'event horizon').

3 An astronaut of mass 100 kg stands on a planet of mass $3.0 \times 10^{25}\,\mathrm{kg}$ with a radius $2.0 \times 10^7\,\mathrm{m}$. What is his weight?

4 Two spherical masses of 8.0 kg are placed so that the distance between their centres is 1.0 m. Calculate the gravitational force of attraction between them.

5 Calculate the force acting on a woman of mass 70 kg standing on the surface of the Earth, using Newton's law of gravitation. Assume that the mass of the Earth is $6.0 \times 10^{24}\,\mathrm{kg}$ and that the radius of the Earth is $6.4 \times 10^6\,\mathrm{m}$. Comment on your answer, bearing in mind that the gravitational

field strength on the surface of the Earth is
$9.8 \, N \, kg^{-1}$.

6 The gravitational field strength on the surface of a
planet is $20 \, N \, kg^{-1}$; the planet has a radius of
$1.0 \times 10^4 \, km$. What will be the gravitational pull
due to the planet on a mass of 100 kg at a
distance of $1.0 \times 10^5 \, km$ from the centre of the
planet?

7 Define *gravitational field strength*.

8

A spacecraft is travelling away from the earth
with speed v when the engines are switched off.
The speed of the craft is insufficient to escape
from the earth. Sketch graphs to show the
variation with time of (i) the distance of the craft
from earth; (ii) the velocity of the craft.

9 Reports are just coming in that two Russian
cosmonauts landed recently on Mars. Their
landing was slightly harder than expected and
they wondered whether on Mars, G, the
universal constant of gravitation, might be
different from on Earth; this discrepancy could
have explained their unfortunate landing.

They had landed next to the extinct volcano,
Olympus Mons, which towers 20 km above the
surrounding plains and has a base diameter of
80 km. They planned to use a plumbline and
observe its deflection away from the vertical
when placed near the mountain. Fortunately, they
discovered that at regular time intervals a star
appeared directly overhead, when viewed from
the plains. They decided that by taking sightings
with a telescope on this star they would be able
to determine the deflection of the plumbline.

a Would the cosmonauts be able to make
measurements on their plumbline at any time
during the day?

b It was decided that the experiment should be
carried out about 5 km above the base of the
mountain, since this would then make them
about on a level with the centre of mass of the
mountain. Explain whether or not you think this
is a good idea.

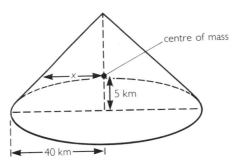

figure 15 Olympus Mons: assumed conical

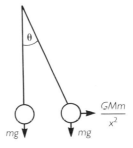

figure 16

c The mountain is approximately like a cone; the
density of the surface rocks was found to be
$3000 \, kg \, m^{-3}$. Calculate the mass of the
mountain. (Volume of a cone $= \frac{1}{3} \times$ base
area $\times$ height.)

d Show that the sideways deflection of the
plumbline is given by:

$$\tan \theta = \frac{GM}{x^2 g}$$

θ is measured in radians, M is the mass of the
mountain, g is the gravitational acceleration on
Mars. In a separate experiment g was
determined to be $3.7 \, ms^{-2}$.

e Calculate x and hence calculate G, given that
the cosmonauts measured a deflection in their
plumbline of $(1.8 \pm 0.2) \times 10^{-3} \, rad$. Comment on
your answer. (It might help to know that for
$\theta \ll 1$, $\tan \theta \approx \theta$, provided θ is in radians.)

10 Newton's third law is sometimes stated like this:
'To every force there is an equal and opposite
force.' When a book is at rest on a desk, the
Earth exerts a gravitational force on it; in this
case what is the equal and opposite force? (Hint:
It is *not* the desk pushing back!)

CHAPTER 18

Gravitational potential

Introduction

We start by considering the potential energy gained by an object as it is raised from the floor, which we then extend to the case of a rocket leaving the Earth. This leads on to the idea of escape velocity. We show how gravitational field strength can be equated to gravitational potential gradient. Finally we investigate how the Earth's gravitational field strength varies with distance from the centre of the Earth.

GCSE knowledge

Although the mathematics in this chapter complicates things a little, there are only two fundamental pieces of physics that you ought to know to start with.

☐ Work done = force × distance
   ⇒ gain in gravitational potential energy = weight × height
   $$= mgh$$
   This is true provided that g does not change over a distance h.

☐ As an object falls it loses gravitational potential energy but gains kinetic energy. This can be summarised by the equation

$$mgh = \tfrac{1}{2}mv^2$$

Energy changes in gravitational fields

If we wish to calculate the change in potential energy ΔE when a body of mass m is lifted through a vertical distance Δx we write

$$\Delta E = mg.\Delta x$$

provided that Δx is sufficiently small for the field to be considered uniform over that distance. This enables us to define the change in gravitational potential ΔV (which is analogous to electric potential, see chapter 21) as

$$\Delta V = \frac{\Delta E}{m} = g.\Delta x$$

whence

$$g = \frac{\Delta V}{\Delta x}$$

ΔV is an energy change per unit mass and has units $J\,kg^{-1}$.

Example Calculate the potential change between the top and bottom of a 100 m high building.

$$\Delta V = g.\Delta x$$
$$\Delta V = (10\,N kg^{-1}) \times (100\,m)$$
$$= 1000\,J kg^{-1}$$

If we wish to calculate the change in potential over a large distance (g not constant), we may use the same equation as in the example above, but we have to use it many times. We are already familiar with the idea of using the area under a force-against-extension graph to calculate the work done in stretching a piece of wire; we use the same principle here. Figure 1 shows how the gravitational field varies near to an imaginary planet of mass M_p. A thin strip of thickness Δx is shown; the area of this strip is $g.\Delta x$, which is of course the change in potential in moving a distance Δx. If we wish to calculate the change in potential in going from point 1 to point 2, we add up all the strips between them. This, of course, is the area under the graph.

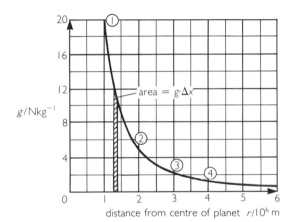

figure I Variation of gravitational field strength

It will be shown later that the change in potential in going from a distance r_1 to r_2 is

$$\Delta V = GM_p\left\{\frac{1}{r_1} - \frac{1}{r_2}\right\} \qquad (i)$$

For the time being we will use this equation without rigorous proof, but we will check its validity.

Example Change in potential in going from point 1 to point 2 = area under graph = ΔV_{12}. The number of squares under the graph between points 1 and 2 is 10. Each square represents a potential change of

$$(2\,\text{Nkg}^{-1}) \times (0.5 \times 10^6\,\text{m}) = 1.0 \times 10^6\,\text{Jkg}^{-1}$$

Thus the potential change

$$\Delta V_{12} = 10 \times 1.0 \times 10^6\,\text{Jkg}^{-1} = 1.0 \times 10^7\,\text{Jkg}^{-1}$$

We can check this result by using the equation above, but first we need to know GM_p. At any distance r from the planet,

$$g = \frac{GM_p}{r^2}$$

$$\Rightarrow \qquad GM_p = gr^2,$$

Using the values at point 1 on the graph,

$$GM_p = 20\,\text{Nkg}^{-1} \times (1.0 \times 10^6\,\text{m})^2$$
$$= 2.0 \times 10^{13}\,\text{Nm}^2\text{kg}^{-1}$$

Thus $\quad \Delta V = GM_p\left\{\frac{1}{r_1} - \frac{1}{r_2}\right\}$

$$= 2.0 \times 10^{13}\,\text{Nm}^2\text{kg}^{-1}\left\{\frac{1}{1.0 \times 10^6\,\text{m}} - \frac{1}{2.0 \times 10^6\,\text{m}}\right\}$$
$$= 1.0 \times 10^7\,\text{Jkg}^{-1}$$

You should now check the equation using the area under the graph between points 2 and 3 or 3 and 4.

Potential

So far we have only talked about *changes* in potential. How can potential be defined at any point? When we are dealing with problems near the Earth's surface, it is convenient to define the Earth's surface as zero potential. Thus a point 20 m above the Earth's surface has potential of 200 Jkg^{-1}. When we are dealing with situations a long way from the Earth's surface it is sensible to define a point infinitely far from the Earth's centre as zero potential. Equation (i) shows us that the potential difference between a point at a distance r from the Earth's centre and infinity is

$$\Delta V = \frac{GM_E}{r}$$

From this definition it follows that the potential V is negative at any point r which is closer to the Earth than infinity. The potential is zero at infinity, and so, since potential energy is lost as an object moves towards the Earth, its potential has to be negative; the potential becomes increasingly negative the closer the object moves towards the Earth. Thus at a point r away from the Earth the gravitational potential is:

$$V = -\frac{GM_E}{r}$$

Example Calculate the minimum speed that a rocket must have at the Earth's surface to escape completely from the Earth's gravitational field. (The rocket engines only burn for a short burst at take-off, so that its maximum speed is reached very close to the Earth's surface.)

Gain in gravitational potential energy = loss of kinetic energy. If the rocket starts off at the escape speed V_e, and finishes with zero speed at infinity, then, for a rocket of mass m,

$$\frac{GM_E}{r}.m = \tfrac{1}{2}mV_e^2$$

$$\Rightarrow \qquad V_e^2 = \frac{2GM_E}{r}$$

$$= \frac{2 \times 6.7 \times 10^{-11}\,\mathrm{Nm^2kg^{-2}} \times 6.0 \times 10^{24}\,\mathrm{kg}}{6.4 \times 10^6\,\mathrm{m}}$$

$$= 1.26 \times 10^8\,\mathrm{m^2s^{-2}}$$

$$\Rightarrow \qquad V_e = 1.1 \times 10^4\,\mathrm{ms^{-1}}$$

This is often called the Earth's *escape velocity*.

It is interesting to note that the average speed of hydrogen molecules at room temperature is only $2000\,\mathrm{ms^{-1}}$, which is why our atmosphere can contain a few traces of hydrogen. The escape velocity on the Moon is about $2100\,\mathrm{ms^{-1}}$. When the solar system was formed, some 4500 million years ago, the temperature on the Earth's and Moon's surfaces was a lot higher than it is now. Consequently gas molecules would have been moving a lot faster; the gravitational pull of the Moon was too weak to hold onto its atmosphere but the Earth's was strong enough.

Example   Suppose that we have a binary star, the two components of which are separated by 7 Astronomical Units (A.U.) (7 times the Earth–Sun distance). The mass of star A is twice that of its companion star B, but each star has a radius of 0.5 A.U. (Each star would have to be a red giant to have such a radius.) Calculate the distance along a line connecting the centres of A and B where a gravitational neutral point will occur.

This problem can be solved graphically, by plotting potential as a function of distance from one of the stars. Now, we have not got enough information to calculate potential in $\mathrm{J\,kg^{-1}}$, but in this case it will suffice to work in arbitrary units. We will define a new system of units in which $G = 1\,\mathrm{u}$, $M_A = 2\,\mathrm{u}$, $M_B = 1\,\mathrm{u}$ and 1 Astronomical Unit $= 1\,\mathrm{u}$. Thus potential near the surface of star A (not including any effects from star B) is

$$V = -\frac{GM}{r}$$

$$= -\frac{1 \times 2}{\frac{1}{2}}\,\mathrm{u}$$

$$= -4\,\mathrm{u}$$

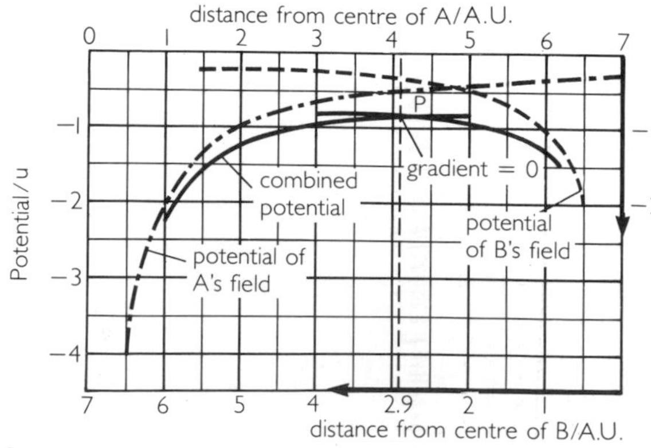

distance from centre of A/A.U.

Potential/u

figure 2   Variation of potential between two stars

distance from centre of B/A.U.

Figure 2 shows three curves plotted by calculating V at different distances from A and B. Two show how the potential would vary in the region of each star in isolation and the third curve shows how the potential varies in the region between the two stars together. The third curve is simply obtained by summing the first two; potential is a *scalar* quantity because we are adding energies.

The combined potential graph has a gradient of zero at P, a distance of 4.1 units from A. Using the relation

$$g = \frac{\Delta V}{\Delta r}$$

we deduce that the field is zero at P.

This answer may now be checked by calculating the field strengths at P due to the two stars. We will define a field acting towards B as positive. We will still work in our arbitrary system of units.

Thus at P

$$g_A = -\frac{GM_A}{(4.1)^2}\,\mathrm{u}$$

$$= \frac{2}{(4.1)^2}\,\mathrm{u}$$

$$= -0.12\,\mathrm{u}$$

$$g_B = \frac{GM_B}{(2.9)^2}\,\mathrm{u}$$

$$= \frac{1}{(2.9)^2}\,\mathrm{u}$$

$$= +0.12\,\mathrm{u}$$

The combined field is

$$g = g_A + g_B$$

$$= -0.12\,\mathrm{u} + 0.12\,\mathrm{u}$$

$$= 0$$

Note that fields add as *vectors* since they are directional.

Digression: a more mathematical treatment of potential. In general, if an object moves a small distance $\mathrm{d}x$ in the direction of an applied force F the potential energy decreases by an amount $\mathrm{d}E$. So we may write:

$$\mathrm{d}E = -F\,\mathrm{d}x$$

figure 3

It follows from the definition of gravitational field and potential that

$$dV = -g\,dx$$

where dV is the decrease in potential as an object approaches a planet. Thus as an object goes away from the planet we may write:

$$dV = g\,dx$$

⇒ the potential difference between r_1 and r_2 is given by:

$$V = \int_{r_1}^{r_2} \frac{GM}{x^2}\,dx$$

$$= \left[\frac{-GM}{x}\right]_{r_1}^{r_2}$$

$$= GM\left(\frac{1}{r_1} - \frac{1}{r_2}\right)$$

A further example on field and potential

How does the gravitational field vary both inside and outside a solid planet of uniform density? How can we work out the gravitational potential at the centre of such a planet?

To answer these questions we use two theorems:

1 The resultant gravitational force outside a thin, uniform spherical shell is as if a mass identical to that of the shell were placed at its centre point.

2 The resultant gravitational force inside a thin, uniform spherical shell is zero.

These results can be proved with the aid of integral calculus, but we can arrive at them using symmetry arguments and the inverse square law.

Strictly speaking, Newton's law of gravitation as discussed in chapter 17 holds only for *point* masses; it should not be used for any other masses without some justification. However, it is perfectly reasonable to argue that if a planet is spherical then the field associated with it must share its spherical symmetry. Thus gravitational field lines will be everywhere perpendicular to the shell surface, and so above the planet's surface it appears as if these lines have diverged from the centre of the shell – figure 4. If the planet had been a cube, then such an approximation could not have been made. It is interesting to note that the Earth is not a perfect sphere and any satellite in orbit around it (including the Moon) is affected by the Earth's equatorial bulge.

The second theorem can be justified as follows. In figure 5, O is the apex of two imaginary cones. The cones are similar but one has four times the linear dimensions of the other; the angle at the top of the cone is very small indeed. O is surrounded by a thin uniform shell, of mass μ per unit area.

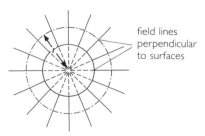

figure 4   Spherical symmetry

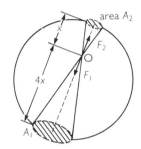

figure 5   Spherical shell

The masses enclosed in the base of each cone would each exert a force on a small mass m at O, but in opposite directions. These forces are:

$$F_1 = \frac{GA_1\mu.m}{(4x)^2}$$

$$F_2 = \frac{GA_2\mu.m}{x^2}$$

but $A_1 \times 16A_2$

⇒ $$F_1 = F_2$$

Thus the net force from the two areas of mass on m is zero. It is now easy to generalise this result to show that the net field anywhere inside the shell is zero.

We may now answer the original question. Outside the planet,

$$g \propto \frac{1}{r^2} \quad \text{(theorem 1)}$$

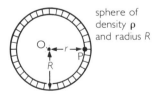

figure 6   Spherical shell

Inside the planet at point P we now know that the contribution of the shaded area is zero (theorem 2). We also know that the field at P is due to the matter enclosed in the sphere of radius r, and it is as if all that matter is concentrated at the centre, O.

Thus the field at P is:

$$g = \frac{G\frac{4}{3}\pi\rho r^3}{r^2}$$

$$= \frac{4}{3}\pi\rho Gr$$

So inside the sphere, $g \propto r$.

These two results are summarised in figure 7.

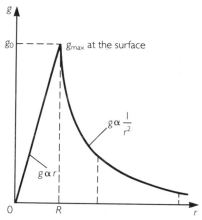

figure 7   Variation of g

We may now use the graph to evaluate the potential at O. First we know that the potential change between infinity and the planet surface (radius R) is GM/R

Then the potential difference between R and O is

$$\Delta V = \frac{1}{2} g_0 R$$

using the area under the graph.

But $$g_0 = \frac{GM}{R^2}$$

$$\Rightarrow \qquad \Delta V = \frac{GM}{2R}$$

so the total potential difference between O and infinity is

$$\frac{3}{2} \frac{GM}{R}$$

and the potential at the centre of the planet is therefore

$$V = -\frac{3}{2}\left(\frac{GM}{R}\right)$$

Summary

1   A gravitational potential difference between two points, A and B, is defined as the work done on a unit mass in taking it from A to B.

$$\Delta V = \frac{W}{M}; \text{ units } \text{J}\,\text{kg}^{-1}$$

2   We define the zero of potential as a point a long way from any planet, and therefore all potentials at points close to planets are negative, since the potential has to become more positive as we go away from a planet.

3   The potential at a distance r from a mass M is given by:

$$V = -\frac{GM}{r}$$

4   The field strength may be related to the potential gradient by the equation:

$$g = \frac{\Delta V}{\Delta x}$$

Questions

1   Define gravitational potential.

2   It was once suggested that we ought to have a new unit called the gravitational volt or gravolt. How do you think the gravolt should be defined?

3   Calculate the gravitational potential difference between the top of Mount Everest (height 9000 m) and sea level.

4   Figure 8 shows the variation of g-field near to a planet.

a   Use a simple numerical test to show that the variation of field with distance obeys an inverse square law.

b   An object of mass 2 kg is at a distance of 8×10^6 m from the centre of the planet. How much work must be done to move the object 10 m further away from the planet?

c   How much work must be done to move the same object from a distance of 4×10^6 m to a distance 12×10^6 m from the centre of the planet?

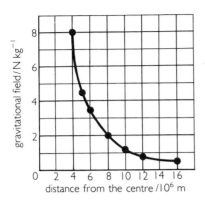

figure 8

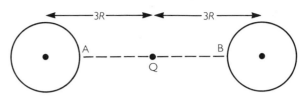

figure 11

5   Figure 9 shows lines of equal gravitational potential near to the surface of a small asteroid.

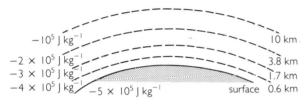

figure 9

a   Why are all the values of potential negative?
b   Why does the spacing of the lines get bigger as the distance from the asteroid increases?
c   Estimate the gravitational field strength near to the surface.

6   A planet A of radius R has a potential at its surface of $-6.0 \times 10^7 \, \mathrm{J\,kg^{-1}}$, and a gravitational field strength of $9 \, \mathrm{N\,kg^{-1}}$.

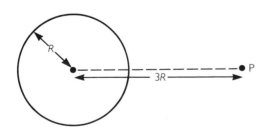

figure 10

a   Explain the significance of the minus sign in the value of the potential at the planet's surface.
b   Calculate (i) the value of potential at P, (ii) the value of field at P.

c   Suppose that the planet is actually part of a binary system, each component being of the same mass and radius. (The effects of B on the initial data given were ignored.)
     Calculate (i) the value of potential at Q, (ii) the value of field strength at Q.
d   Sketch graphs to show how field and potential vary along the line AB.

7   An astronaut stands on an asteroid of radius 3 km and density $3000 \, \mathrm{kg\,m^{-3}}$.
a   Calculate the gravitational field strength on the surface of the asteroid.
b   Calculate how fast the astronaut would have to jump to leave the asteroid.
c   If the astronaut were to jump with the same speed on Earth, how high would she jump?

Approach to Jupiter by Voyager 1 – data handling questions

Voyager 1 was launched on 5 September 1977. After a journey of nearly two years it reached Jupiter. On 30 January 1979 *Voyager 1* was 35.1 million km away from Jupiter. The data below show how *Voyager* approached Jupiter during the next month until passing close by Jupiter on 5 March. The times recorded are from 00.00 h GMT on 30 January 1979.

figure 12

	Distance from Jupiter's centre $R/10^8$ m	$(1/R^2)$/m^{-2}	Velocity v/ms^{-1}	Time from 00.00 hr GMT 30 Jan. 1969				Date
				days	hrs	min	sec	
1	351.0	8.11×10^{-22}	10 900	0	00 :	00 :	00	Jan 30
1A	347.1		10 904	0	10 :	00 :	00	
2	280.8	1.28×10^{-21}	11 000	7	10 :	00 :	00	Feb 6
2A	276.8		11 006	7	20 :	00 :	00	
3	210.6	2.25×10^{-21}	11 135	14	18 :	00 :	00	Feb 13
3A	206.6		11 145	15	4 :	00 :	00	
4	140.4	5.10×10^{-21}	11 402	21	23 :	00 :	00	Feb 20
4A	138.8		11 411	22	3 :	00 :	00	
5	70.2	2.03×10^{-20}	12 165	28	21 :	00 :	00	Feb 27
5A	69.3		12 184	28	23 :	00 :	00	
6	35.1	8.12×10^{-20}	13 564	32	1 :	00 :	00	Mar 3
6A	34.6		13 601	32	2 :	00 :	00	
7	28.1	1.28×10^{-19}	14 212	32	15 :	00 :	00	Mar 3
7A	27.6		14 271	32	16 :	00 :	00	

8 *Gravitational field*
During all the approach to Jupiter the motors on *Voyager* were closed down. The data in points 1–5 cover a period when the spacecraft was travelling more or less directly towards the centre of Jupiter.
  a The pair of values 3 and 3A are 10 hours apart. During that time the spacecraft increased its speed, relative to Jupiter, by $10\,\mathrm{ms}^{-1}$. Calculate the mean acceleration over this interval.
  b Write down an estimate for the gravitational field strength at a point 2.086×10^{10} m from the centre of Jupiter, that being the average distance of points 3 and 3A.
  c Use the pairs of points 1–5 to plot a graph which will test whether or not the gravitational field due to Jupiter varies as the inverse square of the distance.
  d Given that the radius of Jupiter is 70 020 km, calculate the gravitational field strength on the surface of the planet near to the poles, using your answer to (b) together with the inverse square law.

9 *Potential changes*
  a Since points 3 and 3A are very close together we may assume that the gravitational field is uniform over this relatively small distance. Using your answer to question 8(b) calculate the work done per unit mass by the field in taking the craft from 3 to 3A.

  b Write down the change in gravitational potential of the craft. Has the potential increased or decreased?
  c Calculate the change in kinetic energy per unit mass of the craft in going from 3 to 3A. Comment on your answer.
  d Show also that when the craft goes from point 6 to point 7 the amount of potential energy lost by the craft is equal to the kinetic energy gained by it. (Hint: Is the field uniform over this distance? $GM = 1.28 \times 10^{17}\ \mathrm{Nkg}^{-1}\mathrm{m}^2$ for Jupiter.)
  e Check the calculation in part (d) for several other sets of points, e.g. 1 and 7, 2 and 5.

10 *Inverse square law for light*
A cepheid is a star whose luminosity is variable; over a period of days or weeks the brightness of a cepheid might change by a factor of two or three. By observing large numbers of cepheids in the Magellanic Clouds (two galaxies at a distance of about 150 000 light years away from us) it has been discovered that the average luminosity of such a cepheid is related to its time-period. This is called the cepheid period–luminosity law and is illustrated in figure 13.
  Thus by observing the time-period of a cepheid in a distant galaxy, we can calculate its luminosity; the fact that a distant cepheid looks duller than a similar one in the Magellanic Clouds gives us a measure of distance.

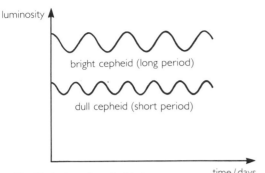

figure 13 Variation of cepheid star

Figure 14 shows how the apparent average luminosity of cepheids in the Magellanic Clouds and the Andromeda Galaxy depend on their time-periods.

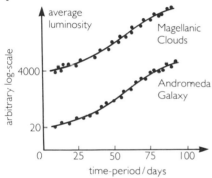

figure 14 Time-periods of cepheid stars

a Suggest suitable units for the quantity 'luminosity'.

b Explain carefully why the apparent luminosity of a star should vary as the inverse square of our distance from it.

c Given that the average distance of the Magellanic Clouds from us is 150 000 light years, use the inverse square law for light intensity, together with the information in figure 14, to calculate our distance from the Andromeda Galaxy.

11 *An expanding universe? – a comprehension question*

At the beginning of the twentieth century astronomers were unaware of the existence of anything beyond our own galaxy, the Milky Way. They had seen various, hazy, nebulous clouds which had been given the collective name *nebulae*. It was only with the arrival of the modern telescope that it was eventually realised that the term nebulae embraced all external galaxies. We now believe that galaxies are scattered throughout a vast universe, whose bounds stretch 15 000 million light years* from us.

* A light year is the distance that light travels in 1 year. Light travels at $3 \times 10^8 \, \mathrm{ms^{-1}}$ and there are $3.16 \times 10^7 \, \mathrm{s}$ in a year. So 1 light year = $(3.00 \times 10^8 \, \mathrm{ms^{-1}}) \times (3.16 \times 10^7 \, \mathrm{s}) = 9.48 \times 10^{15} \, \mathrm{m}$.

Galaxies exist in groups; our own local group of galaxies contains some 15 galaxies spread over a distance of about 2 million light years – figure 15.

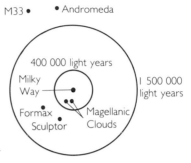

figure 15 This shows our own local group of galaxies projected onto the plane of the Milky Way

Further out into space we find other groups of galaxies – figure 16. The photograph (figure 17) shows the cluster in Corona Borealis; the small round spots and objects with spikes are stars in our own galaxy. Other structures are galaxies at a distance of about 600 million light years.

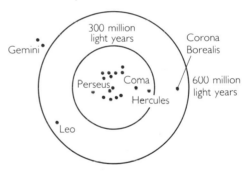

figure 16 On this scale our own local group is reduced to a small dot in the centre of the diagram. Each dot represents a cluster of 50 galaxies or so

figure 17

Edwin Hubble was the first man to fathom the depths of the universe; he determined the distance to nearer galaxies by means of the cepheid period–luminosity law (see question 10). In the case of galaxies too distant for individual stars to be resolved, he argued along similar lines. He made the bold assumption that galaxies of a chosen shape all have approximately the same absolute luminosity. The further the galaxy is, the smaller its apparent luminosity, so it should be possible to calculate its distance from its estimated absolute and apparent luminosities, by employing the inverse square law for light intensity.

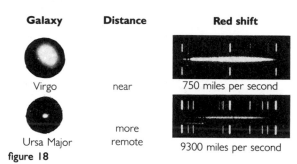

Galaxy	Distance	Red shift
Virgo	near	750 miles per second
Ursa Major	more remote	9300 miles per second

figure 18

Hubble photographed the spectra of galaxies and noticed that certain characteristic absorption lines were shifted towards the red end of the spectrum; the conclusion that he drew was that these galaxies are receding rapidly from us, the shift in the spectral lines being caused by a Doppler effect (see chapter 28, question 8, page 303).

The photographs on the right hand side of figure 18 show the extent of the shifts, the photographs on the left hand side show the apparent size of the galaxy in question. It is clear that the smaller the apparent size of the galaxy (and hence the greater the distance), the greater is the Doppler shift (and hence the receding velocity). Hubble was able to show that the velocity of recession of a group of galaxies, v, is proportional to its distance from us, R: figure 19. This can be summarised mathematically as $v = HR$; H is the Hubble constant.

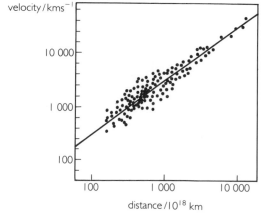

figure 19

a The majority of astronomers now believe that the universe originated with a 'Big Bang' some 20×10^9 years ago. How does Hubble's discovery support this idea?

b Hubble's constant is quoted as $15.3\,\mathrm{km\,s^{-1}}$ per million light years. What does this mean? Show that the Hubble constant $H = 1.6 \times 10^{-18}\,\mathrm{s^{-1}}$.

c Explain why the fact that galaxies are getting further apart and slowing down means that the Hubble constant must be getting less as time passes.

d Comment on Hubble's 'bold' assumption that galaxies of a chosen shape all have approximately the same absolute luminosity. What other assumption is implicit in his calculations about the light reaching us?

Some cosmologists occupy themselves by contemplating whether or not the universe will carry on expanding for ever. Figure 20 illustrates the universe with a couple of spiral galaxies near the edge travelling away from the centre.

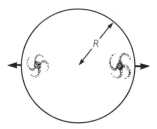

figure 20 Our universe

These galaxies are attracted back towards the centre of the universe by all the matter within; thus they have a certain amount of potential energy $-E_\mathrm{p}$. They also have kinetic energy E_k. They will only escape from the universe, so allowing the universe to expand for ever, if $E_\mathrm{k} > E_\mathrm{p}$, i.e. the total energy of the galaxy must be positive. If the universe were to expand for ever, it would be called open; if the universe were to expand and then contract again under its own gravitational forces it would be called closed. Figure 21 illustrates the separation of typical galaxies in open and closed universes, as a function of time.

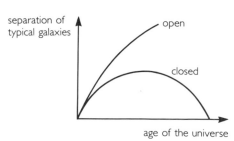

figure 21

e Sketch graphs to show how (i) the kinetic energy, (ii) the potential energy, (iii) the total energy of a typical galaxy changes with time in both an open and a closed universe.

Since we know the Hubble constant, we are in a position to calculate whether we live in an open or a closed universe. If the kinetic energy E_k of a galaxy exactly equals its potential energy E_p, then the universe is 'critical', i.e. the gravitational pull of the universe is just sufficient to prevent it from expanding indefinitely. We will now calculate the critical density, ρ_c, for our universe.

f If the radius of our universe is R, what is the potential energy of a galaxy of mass m at a distance R from the centre? Express your answer in terms of ρ_c and R.

g What is the speed of recession of this galaxy?

h What is the kinetic energy of this galaxy?

i Now use your answers to (f) and (h) to show that the critical density of the universe is:

$$\rho_c = \frac{3H^2}{8\pi G}$$

Use your answer to (b) to calculate ρ_c.

j The average cosmological density is quoted as 3 hydrogen atoms per cubic metre. Given that a hydrogen atom has a mass of 1.7×10^{-27} kg, state whether or not we live in a closed universe. Comment on the certainty of your answer.

k Assuming that the universe is of the critical density, cosmologists have shown that the age of the universe is $2/3H$. Use the answer to part (b) to calculate the age of the universe.

l Prove the result quoted in part (k) [optional].

CHAPTER 19

Orbits

Introduction

In this very short chapter we deal with orbits. This is an interesting area in which we can apply our knowledge of gravitation. We deal mostly with circular orbits, although one example is included which requires some knowledge of elliptical orbits and Kepler's laws.

The physics of circular orbits revolves around one simple idea: in such an orbit the gravitational pull of the planet provides the necessary centripetal force to keep the satellite in its circular path.

GCSE knowledge

We assume you know that
Force = mass × acceleration;

$$\text{average speed} = \frac{\text{distance}}{\text{time}};$$

the circumference of a circle = $2\pi r$

Other than these points the basic physics for this chapter is given in chapter 4 (Circular motion) and chapter 17 (Gravitational fields).

Circular motion and gravitation: Kepler's third law

In chapter 17 we discussed Newton's law of gravitation, and suggested some arguments to show that it at least appeared a plausible theory. Here we see how Newton tested his theory and made it fit existing experimental data.

Newton began by considering a planet in a circular orbit around the Sun. He realised that the gravitational attraction between the planet and the Sun provided the centripetal force necessary to keep the planet in its circular path.

Assuming Newton's law we may write:

$$\frac{mv^2}{r} = \frac{GMm}{r^2} \qquad \text{(i)}$$

but

$$v = \frac{2\pi r}{T}$$

T is the time taken for the planet to complete one orbit of the Sun.

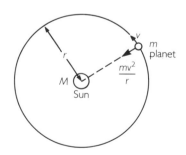

figure 1   Planetary orbit about the Sun (*not to scale*)

This gives

$$\frac{4\pi^2 r^2}{T^2} = \frac{GM}{r}$$

and

$$\frac{r^3}{T^2} = \frac{GM}{4\pi^2} \qquad \text{(ii)}$$

This result is consistent with Kepler's third law and confirmed for Newton his idea of an inverse square law. (It should be noted that the equation above still holds for elliptical orbits provided r is taken as half of the major axis of the ellipse.)

The Moon test

Newton tested the validity of his law by considering the centripetal acceleration of the Moon towards the Earth, and comparing it with the known gravitational acceleration on the Earth's surface. We will use modern data to illustrate his idea.

Moon–Earth distance 3.82×10^8 m = 60 Earth radii
Moon goes round the Earth in 27.32 days.

Assuming the Moon moves in a circular orbit around us we may calculate its centripetal acceleration towards the Earth:

$$a = \frac{v^2}{r}$$

$$= \frac{(2\pi r)^2}{rT^2}$$

$$= \frac{4\pi^2 \times (3.82 \times 10^8 \, \text{m})}{(27.32 \times 24 \times 3600 \, \text{s})^2}$$

$$= 2.7 \times 10^{-3} \, \text{ms}^{-2}$$

The gravitational acceleration on the Earth's surface is $9.81 \, \text{ms}^{-2}$. Since the Moon is 60 times further away from the centre of the Earth than we are, from the inverse square law, it ought to experience an acceleration of

$$a = 9.81 \times \left(\frac{1}{60}\right)^2 \, \text{ms}^{-2}$$

$$= 2.7 \times 10^{-3} \, \text{ms}^{-2}$$

Thus the Moon provided the first practical proof of Newton's law.

Mass of the Sun

Calculate the mass of the Sun, given that the Earth's average distance from the Sun is 1.5×10^{11} m, and that we take 365.25 days to orbit the Sun once.

From equation (ii) above:

$$\frac{r^3}{T^2} = \frac{GM}{4\pi^2}$$

$$\Rightarrow \qquad M = \frac{4\pi^2 r^3}{GT^2}$$

$$= \frac{4\pi^2 \times (1.5 \times 10^{11} \, \text{m})^3}{(6.7 \times 10^{-11} \, \text{Nm}^2\text{kg}^{-2}) \times (365.25 \times 24 \times 3600 \, \text{s})^2}$$

$$= 2.0 \times 10^{30} \, \text{kg}$$

Conclusion

Chapter 17 began on a historical note to illustrate man's understanding of the motion of heavenly bodies at the time of Newton's birth. It should now be clear that Newton's theory elevated the study of the night sky from an awesome mystery to an exact science. That the law of gravitation could provide a rational explanation for Kepler's obscure third law and the motion of the Moon was a great triumph. Since Newton's day we have been able to explain our tides, the precession of the equinoxes, why planetary orbits tend to be elliptical rather than circular, the formation of stars and galaxies; most impressive of all we have transported men to the Moon (and back) and directed space probes beyond the limit of the solar system.

Summary

1 For a planet of mass m in a circular orbit round a sun of mass M:

$$\frac{GMm}{r^2} = \frac{mv^2}{r}$$

2 $$\frac{r^3}{T^2} = \frac{GM}{4\pi^2}$$

3 Similar equations also apply for moons and satellites orbiting a planet.

Questions

1 A spy satellite is in an orbit 100 km above the Earth's surface. How long does it take to complete one orbit? Take G to be $9.8 \, \text{Nkg}^{-1}$ and R_E to be 6400 km.

2 When a satellite is in a circular orbit around the Earth there is always a force acting on it. Explain why this force does no work on the satellite, and therefore its kinetic energy remains constant during its orbit.

3 Why does an astronaut in an orbiting spacecraft experience the sensation of weightlessness? Is he really weightless?

4 a Calculate the force of attraction between the Sun and the Earth.
 b Calculate the gravitational field strength, due to the Sun's influence, in the vicinity of the Earth.
 c Calculate the centripetal acceleration of the Earth towards the Sun, using the relation $a = v^2/r$.
 d Explain, with the aid of a diagram, why the Earth does not get significantly closer to the Sun as time passes.

$M_E = 6 \times 10^{24}$ kg; $M_s = 2 \times 10^{30}$ kg; Earth–Sun distance = 1.5×10^8 km.

5 a Calculate how high above the surface of the Earth you would have to place a satellite so that it would rotate around the Earth once every 24 hours. ($G = 6.7 \times 10^{-11} \, \text{Nm}^2 \, \text{kg}^{-2}$; $M_E = 6 \times 10^{24}$ kg; radius of the Earth = 6400 km)
 b Such an orbit is called 'geosynchronous'. Why? Could a geosynchronous orbit take a satellite over the poles?

6 The work required to remove an object of mass m from a planet of mass M and radius R is GMm/R. Show that if the same object is in orbit just above the surface of the planet the work required to remove the object is now only $GMm/2R$.

7 A spacecraft is in a circular orbit around a planet; the radius of the orbit is 3000 km and the time-period of one orbit is 2 h. The craft fires its rockets for a short interval, in a direction tangential to its orbit, and increases its speed to $3000 \, \text{ms}^{-1}$; the craft now goes into a new orbit, its furthest distance from the planet being 9000 km.
 a Sketch the new orbit, marking on it the point at which the rockets were fired.
 b Calculate the slowest speed that the craft has in its new orbit.
 c Calculate the time-period of the new orbit.

8 a Mars has two moons: Phobos and Deimos. Phobos' mean radius of orbit is 9300 km; Deimos' mean radius of orbit is 23 360 km. Calculate the time-period of orbit for each moon. Mass of Mars = 6.48×10^{23} kg.
 b Mars itself rotates on its own axis in a time of 24 h 37 min; Phobos and Deimos rotate in the same direction as Mars. Use your answers to part (a) to discuss how the motion of these moons would appear to a Martian.

9 Below is tabulated some data about the five moons of Uranus.

Satellite system of Uranus

	Mean distance from Uranus	Period of revolution
Miranda	130 000 km	1 d 10 h
Ariel	192 000 km	2 d 12 h 29 min
Umbriel	266 000 km	4 h 3 h 28 min
Titania	436 000 km	8 d 16 h 56 min
Oberon	582 000 km	13 d 11 h 7 min

Check the validity of Kepler's third law for the Uranus system.

Further questions on Jupiter and Voyager

10 a Jupiter rotates once on its own axis in 9 h
       50 min; calculate the centripetal acceleration of
       an object near the equator of Jupiter.
    b Near to the poles of Jupiter the weight of a
       1 kg mass is about 25 N. Calculate the apparent
       weight of the same 1 kg mass near the equator.
       Explain carefully why the mass should appear
       to weigh less. (Radius of Jupiter = 70 000 km).

11 After *Voyager 1* had left Jupiter it continued
    towards Saturn.
    Explain carefully why Voyager's velocity, relative
    to Saturn, was increased substantially by its
    encounter with Jupiter. (Hint: Jupiter's average
    orbital speed around the Sun is about $13\,\mathrm{km s^{-1}}$.)

12 Pictures sent back by *Voyager* of Saturn's rings
    showed them to be made up of small ice
    particles, each individually in orbit around the
    planet. However, photographs also revealed
    radial spokes which suggested that the rings, in
    places, rotate as a corporate mass. Could
    gravitational forces alone produce such radial
    structures in the rings? Explain your answer.

figure 2   Part of the ring system of Saturn

CHAPTER 20

Electric fields

Introduction

In this chapter we consider the effects associated with electric charge which is stationary, or static. The forces between static charges lead to the idea of an electric field; the fields due to various arrangements of charge are considered from both practical and theoretical viewpoints; and the comparison between electric (E) and gravitational (g) fields is brought out by reference to results already considered in chapter 17.

GCSE knowledge

We expect that from GCSE knowledge you will already be able to:

☐ recall that static electric charges can be created by rubbing suitable insulating materials together; and that the charges formed can be of two distinct types, labelled arbitrarily positive (+) and negative (−);

☐ discuss how the formation of such static charges involves the separation of + and − charges on insulating materials which were initially neutral; and how equal amounts of + and − charge recombine to make material neutral again;

☐ recall that two objects carrying like charge repel each other, while two objects carrying unlike charges attract each other;

☐ remember that we measure charge in coulombs.

Static charge

The study of static electric charges, or electrostatics, is a branch of physics that dates back several hundred years. Nowadays we are used to producing static electric charge at the flick of a switch, from a Van de Graaff generator or high voltage power supply. In earlier times charges were produced by rubbing together two different materials.

Using modern ideas we may explain this method of producing charges by friction as follows. Each atom is electrically neutral, having a certain number of positively charged protons in its nucleus (as well as some uncharged neutrons) and an equal number of negatively charged electrons outside the nucleus. The charges on a proton and an electron are exactly equal in size, but opposite in sign. When, for

example, a polythene rod is rubbed by a duster electrons are removed from the duster and transferred to the polythene. Thus the duster is left with a positive charge, and the polythene with an equal negative charge.

It is interesting to compare the magnitudes of charges and potential differences that we encounter in static and current electricity. The charge on an object acquired by friction will be very small, and yet the resulting potential differences will be high. For example, if we wished to demonstrate electrostatic repulsion between two small positively charged balls, we would have to charge them up to a p.d. of about $10\,\text{kV}$, and yet each ball would probably only carry about $10^{-9}\,\text{C}$ of charge. Even in a lightning flash the charge discharged to Earth from the base of the thundercloud is only about 10 C. Yet a 3 V battery will quickly pass 10 C of charge through a 1 W torch bulb. (Prove to yourself that the time taken would be 30 s.)

This vast discrepancy between the charges involved in static and current electricity causes us to think again about what we mean by an insulator. For instance, a wooden ruler may have a resistance of $10^{12}\,\Omega$. We will calculate what effect the ruler will have if we touch it to a gold-leaf electroscope which is charged to a potential of 1000 V above earth, and carries a charge of 10^{-10} C; see figure 1.

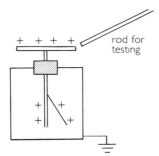

figure 1

Current will pass through the ruler back to earth, causing the electroscope to discharge. During discharge the average p.d. will be 500 V. Thus the average discharge current will be given approximately by:

$$I = \frac{V}{R}$$
$$= \frac{500\,\text{V}}{10^{12}\,\Omega}$$
$$= 5 \times 10^{-10}\,\text{A}$$

but
$$Q = It$$

$\Rightarrow$
$$t = \frac{Q}{I}$$
$$= \frac{10^{-10}\,\text{C}}{5 \times 10^{-10}\,\text{A}}$$
$$= 2 \times 10^{-1}\,\text{s}$$

Thus in a few tenths of a second the electroscope will have discharged.

A person's resistance is typically $10^5\,\Omega$, so if we touch an electroscope it will discharge even more rapidly. Satisfactory insulators for the purposes of electrostatics must have resisitivities of the order of $10^{12}\,\Omega$m; modern plastics and glass are suitable.

Forces on charges

We will use a Maltese Cross apparatus (figure 2) to illustrate some basic electrostatic principles. A 6 V

supply delivers current to a filament, making it hot enough for electrons to be emitted from its surface. The filament is connected to the negative terminal of an EHT supply, providing say 4 kV; the positive terminal of the supply is attached to a hollow cylindrical anode. The negatively charged electrons are thus repelled from the filament and attracted towards the positively charged anode. The electrons pass through the anode with a large kinetic energy; this energy is converted to light when the electrons strike a fluorescent screen. A metal cross lies in the path of the electrons and causes a shadow to appear on the screen; see figure 2(b).

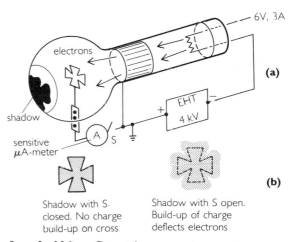

Shadow with S closed. No charge build-up on cross

Shadow with S open. Build-up of charge deflects electrons

figure 2   Maltese Cross tube

In the region between the filament and anode the charged electron experiences a force which accelerates it. The electron is affected in a similar way to a falling object being accelerated by the force of gravity. The falling object is being affected by a *gravitational field*, in which any *massive* object experiences a force. We say that the electron is being affected by an *electric field*, in which any object carrying *charge* experiences a force.

We may use the familiar concept of energy conservation to calculate the speed of the electrons when they leave the anode. At the filament the electrons have electrical potential energy. Due to the action of the accelerating force on each electron this potential energy is converted into kinetic energy. A potential difference of 1 volt between filament and anode would cause 1 joule of kinetic energy to be imparted to each coulomb of charge that was accelerated (using the definition of the volt.) Thus an electron (charge e) accelerated through a p.d. of value V would gain kinetic energy eV.

Thus to calculate the speed of the electrons at the anode we may write:

$$eV = \tfrac{1}{2}\,mv^2$$

Substitution of the values of charge and mass of an electron and the filament–anode p.d. gives:

$$1.6 \times 10^{-19}\,C \times 4000\,V = \tfrac{1}{2} \times 9.1 \times 10^{-31}\,kg \times v^2$$

$$\Rightarrow \qquad v^2 = \frac{1.6 \times 10^{-19} \times 4000}{\tfrac{1}{2} \times 9.1 \times 10^{-31}}\,m^2s^{-2}$$

$$= 1.4 \times 10^{15}\,m^2s^{-2}$$

$$v = 3.7 \times 10^7\,ms^{-1}.$$

Electric field strength

In chapter 17 the strength of a gravitational field g was defined as the force acting on unit mass:

$$g = \frac{F}{m}$$

In the same way we say that the strength of an electric field E (or its intensity) is defined as the force that acts on unit charge:

$$E = \frac{F}{q}$$

The unit of E is thus $1\,NC^{-1}$. (On page 233 we see that this is the same as $1\,Vm^{-1}$.)

The *direction* of an electric field is the direction of the force acting on a *positively* charged particle placed in it. The field in the region round a charged object may be represented by a pattern of lines called field lines, or *lines of force*; these lines show the direction of the force on a positively charged particle at each part of the region. On such a pattern parallel lines represent constant field strength, while lines which converge in a particular direction indicate an increasing field strength. This is similar to the representation of magnetic fields using magnetic field lines (see page 251). Note that electric field is a vector quantity, having both strength and direction.

Mapping electric fields

Magnetic field lines are often mapped using iron filings. Using a similar method, electric field patterns may be mapped in two dimensions by floating small particles of an insulating material, such as chopped hair, on a layer of a suitable liquid, such as castor oil. An electric field is created by applying a high p.d. to electrodes which are dipped into the castor oil (figure 3); the small particles then orientate themselves along the direction of the electric field.

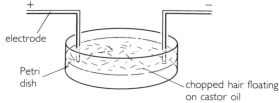

figure 3

The mechanism by which the particles respond is as follows. The charge on the positive electrode (figure 4) will attract electrons to one end of a piece of hair, while the negative charge on plate B will repel electrons from the end nearest to it. Consequently at each end of the piece of hair there will be induced charges. This process is called *polarisation*. As a consequence, each end of the hair experiences a force toward the electrode nearer to it, and the hair aligns itself along the direction of the field at that point. In the same way an iron filing aligns itself along the direction of a magnetic field.

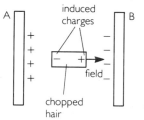

figure 4

Figures 5(a) and (b) show the patterns of electric field lines produced by a pair of parallel electrodes and by a point surrounded by a ring of charge. Between the parallel electrodes (figure 5(a)) the lines of chopped hair are parallel and evenly spaced (except near the ends of the electrodes); this means that the direction and strength of the field are the same at all points. Such a field is called *uniform*. The point electrode (figure 5(b)) produces a radial field which has radial symmetry. This field is not uniform; the lines of force are most concentrated near the point.

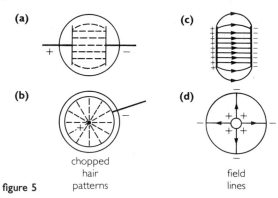

figure 5

We may understand these fields more easily by recalling that charges occur in pairs. Thus there is as much positive charge on one electrode as there is negative charge on the other. With parallel electrodes (figure 5(c)) the charges are spread evenly on each electrode, giving rise to the uniform field. In figure 5(d) the positive charges are all concentrated on a point. The field lines leave a small point but spread out to the negative electrode which is much bigger. So there is a strong field near the point and a small field near the surrounding ring.

Coulomb's law

The starting point of our work on gravitation was the statement of Newton's law, which governs the forces that act between two point masses. Similarly, in electrostatics we start with Coulomb's law, which states that the force F between two point charges q_1 and q_2 is given by

$$F = \frac{kq_1q_2}{r^2}$$

where r is the distance between the charges and k is a constant of proportionality.

Unlike the gravitational case, the electrostatic force is not always attractive. If q_1 and q_2 are like charges the force is repulsive, and if q_1 and q_2 are unlike charges then the force is attractive. Strictly speaking, this law applies only to point charges. However, much of the later work of the chapter concerns charged spheres, and for our purposes it is satisfactory to assume that the electric field near to a charged sphere is the same as that produced by a point charge placed at its centre.

Experimental test of Coulomb's law

Coulomb's law may be tested experimentally using the apparatus shown in figure 6. Two small polystyrene spheres coated in a conducting paint are charged from a high voltage source. It is important that all measurements are made immediately after charging the spheres, since the charge leaks away quite quickly. Sphere A is held in a fixed position, and sphere B is suspended using a thin nylon thread.

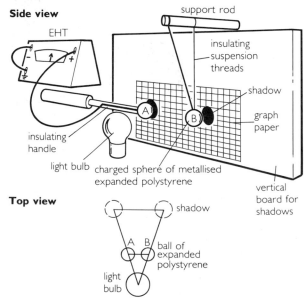

Side view

Top view

figure 6 Experimental test of Coulomb's law

Thus B is free to move, and will be repelled by the similar charge on A; the distance that B moves may be deduced with reasonable accuracy from the movement of its shadow projected onto a screen.

Figure 7 shows the forces acting on sphere B, if its mass is M; Mg is the ball's weight, F the electrostatic repulsive force and T the pull from the string.

From the diagram we can see that

$$F = Mg\tan\theta$$

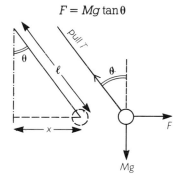

figure 7

Figure 7 shows the displacement, x, of B from its original position. The length of the string is ℓ. Thus $\sin\theta = x/\ell$. However, if the angle of deflection θ is very small, then $\sin\theta \simeq \tan\theta$, and thus we may deduce that the repulsive electrostatic force F is given by:

$$F = Mg\frac{x}{\ell}$$

A test of the relationship $F \propto 1/r^2$ may now be carried out by taking a series of measurements of x and r, the separations of the two spheres. A graph of x against $1/r^2$ should be linear, confirming the relationship. This relationship, $F \propto 1/r^2$, is sometimes called an 'inverse square' relationship.

To make an estimate of the constant of proportionality k in our original statement of Coulomb's law, we also need to measure the charge q on each sphere. This can be done by using a coulombmeter, which gives a direct reading of charge.

A knowledge of values for F, q, and r now enables us to work out a value for k.

$$F = \frac{kq_1q_2}{r^2}$$

$$\Rightarrow \qquad k = \frac{Fr^2}{q_1q_2}$$

We now substitute the values from the experimental data in questions 2 and 3 at the end of the chapter, where $F = 5 \times 10^{-5}$ N, $r = 0.02$ m, $q_1 = q_2 = 1.5 \times 10^{-9}$ C.

Thus
$$k = \frac{5 \times 10^{-5}\,\text{N} \times (0.02\,\text{m})^2}{(1.5 \times 10^{-9}\,\text{C})^2}$$

$$= 9 \times 10^9\ \text{Nm}^2\text{C}^{-2}.$$

Thus Coulomb's inverse square law of force may be written as

$$F = (9 \times 10^9 \, \mathrm{Nm^2C^{-2}}) \frac{q_1 q_2}{r^2}$$

Note that the above value of k was measured for charges separated in air. If the charges are separated by a different material, k will have a different value. Note also, however, that if the charges have *no* material between them, that is, they are in free space, then k has a value virtually identical with that measured in air.

Example What is the force between a proton and an electron separated by a distance of 10^{-10} m? How does this compare with the gravitational force between them?

The electrostatic force is:

$$F = (9 \times 10^9 \, \mathrm{Nm^2C^{-2}}) \times \frac{q_1 q_2}{r^2}$$

$$= \frac{(9 \times 10^9 \, \mathrm{Nm^2C^{-2}})(1.6 \times 10^{-19} \, \mathrm{C})(1.6 \times 10^{-19} \, \mathrm{C})}{(1 \times 10^{-10} \, \mathrm{m})(1 \times 10^{-10} \, \mathrm{m})}$$

$$= 2 \times 10^{-8} \, \mathrm{N}$$

The gravitational force is:

$$F = \frac{G m_1 m_2}{r^2}$$

$$= \frac{(6.67 \times 10^{-11} \, \mathrm{Nm^2kg^{-2}})(1.7 \times 10^{-27} \, \mathrm{kg})(9.1 \times 10^{-31} \, \mathrm{kg})}{(1 \times 10^{-10} \, \mathrm{m})(1 \times 10^{-10} \, \mathrm{m})}$$

$$= 1 \times 10^{-47} \, \mathrm{N}$$

Thus the electrostatic force is about 10^{39} times bigger than the gravitational force.

Electric field strength near a point charge

We may now determine the electric field strength at a distance r from a point charge q situated in air. In figure 8 the field near to q is investigated using a small charge q'.

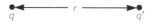

figure 8

From Coulomb's law, the force between the two charges is given by:

$$F = (9 \times 10^9 \, \mathrm{Nm^2 \, C^{-2}}) \frac{qq'}{r^2}$$

But the field strength E near q is given by $E = F/q'$. From the two equations above it follows that

$$E = \frac{(9 \times 10^9 \, \mathrm{Nm^2C^{-2}})q}{r^2}$$

This equation may be used to calculate the electric field strength at any point in space due to a single point charge, or due to any specified arrangement of point charges.

Example 1 Calculate the electric field at a distance of 0.1 m from a small charge of $+10^{-9}$ C.

$$E = \frac{(9 \times 10^9 \, \mathrm{Nm^2C^{-2}})q}{r^2}$$

$$= \frac{9 \times 10^9 \times 10^{-9}}{(0.1)^2} \, \mathrm{NC^{-1}}$$

$$= 900 \, \mathrm{NC^{-1}}$$

Example 2 Calculate the electric field strength at the point P, shown in figure 9, where A and B are both small charges of $+10^{-9}$ C. In which direction does the field act?

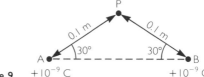

figure 9   A $+10^{-9}$ C   B $+10^{-9}$ C

This problem is considerably harder than the first, because we have to add together two electric fields. Since electric field is a vector quantity, we must add them as vectors.

In example 1 it was shown that the electric field strength at a distance of 0.1 m from a charge of 10^{-9} C was $900 \, \mathrm{NC^{-1}}$. Thus at P the two fields E_A and E_B, are as shown in figure 10(a). These two vectors may be resolved into components parallel and perpendicular to the line AB. The component parallel to the line AB has a magnitude of $(900 \, \mathrm{NC^{-1}}) \cos 30° = 780 \, \mathrm{NC^{-1}}$; the component perpendicular to the line AB has a magnitude of $(900 \, \mathrm{NC^{-1}}) \sin 30° = 450 \, \mathrm{NC^{-1}}$. From figure 10(b) it may be seen that the components of E_A and E_B parallel to AB cancel out, but the two components perpendicular to AB are in the same direction and so add up to give a field of strength $900 \, \mathrm{NC^{-1}}$. The direction of the field is towards the top of the paper.

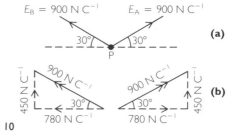

figure 10

Summary

1. By comparison with the p.d.s and charges typical in circuit electricity, static p.d.s tend to be very large, and amounts of charge very small.

2. An electric field causes a charged object to experience a force.

3. $E = \dfrac{F}{q}$

4. Unit of E: $1\,NC^{-1}$ or $1\,Vm^{-1}$

5. Electric fields may be mapped using electric field lines (lines of force).

6. The force between two point charges is:

$$F = \frac{kq_1 q_2}{r^2}$$

(Coulomb's law)

7. Coulomb's law may be tested by experiment, and a value for k obtained.

$$k \approx 9 \times 10^9\,NC^{-2}m^2$$

8. Field near a point charge:

$$E - \frac{kq}{r^2}$$

Questions

1. A ping pong ball covered in conducting paint is placed between two charged plates as shown in figure 11.

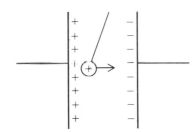

figure 11

Explain why the ball oscillates backwards and forwards between the plates once the ball has been charged. Sketch a graph to show how the force on the ball varies with time.

2.

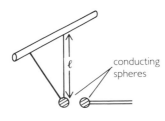

figure 12

The apparatus shown in figure 12 was used to test Coulomb's law which is expressed mathematically in the form:

$$F = \frac{kq_1 q_2}{r^2}$$

where q_1 and q_2 are two point charges separated by a distance r. The idea of the experiment is that both balls receive a charge from a Van de Graaff generator. When the balls are placed close together they will then experience a repulsive force. The sideways deflection of the suspended ball can be measured and this will be proportional to the force between the two balls.

a Explain carefully why the sideways deflection of the suspended ball is proportional to the force acting on it.

b The table below shows the results for five different separations of the balls. Between each set of measurements the balls were recharged to ensure that the charge on each ball was constant throughout the experiment.

Distance between centres of balls/mm	Deflection of hanging ball/mm
100	1
50	5
40	8
30	14
20	30

Discuss, with the aid of a suitable graph, whether these results confirm that the force between two charged spheres varies as the inverse square of their separation.

3. In the experiment described in question 2, a student wished to confirm that the constant of proportionality in Coulomb's law really is $9 \times 10^9\,Nm^2C^{-2}$.

a First it is necessary to calculate the force that acts on each ball. The mass of the suspended ball is 0.1 g. Hence show that the force of repulsion between the balls F marked in figure 13(b), is 5×10^{-5} N.

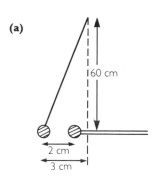

(a)

60 cm

2 cm

3 cm

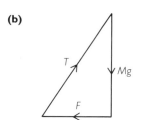

(b)

T

Mg

F

figure 13

b Secondly, the student needed to measure the charge on a ball. (The balls were identical, and therefore, since they were connected to the same p.d. source, they carried the same charge.) He did this by transferring the charge from a ball onto a capacitor of value 6×10^{-9} F; he then measured the p.d. across the capacitor as 0.25 V. Calculate the charge on the ball, stating clearly any assumptions that you make in your calculation.

c Use your answers to (a) and (b) to determine the value of the constant, k, in the equation:

$$F = k.\frac{q_1 q_2}{r^2}$$

Comment on your answer.

4 The radius of a proton is about 10^{-15} m. Estimate the force of repulsion between two touching protons in a nucleus.

5 In figure 14, x is 0.1 m and q is 10^{-6} C. Calculate the field at P.

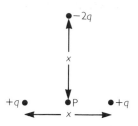

figure 14

6 Two positive point charges A and B are of equal magnitudes (figure 15).

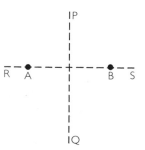

figure 15

a Sketch diagrams to show how the electric field varies along the two lines RS and PQ.
b Repeat part (a) for the case when A is a positive charge and B is a negative charge of equal magnitude.

7 At O, in figure 16, there is a positive point charge. At A the electric field strength is $100\ \text{NC}^{-1}$. Calculate the field strengths at points B, C and D.

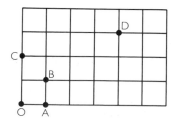

figure 16

CHAPTER 21

Electric potential

Introduction

In this chapter we develop the idea of electrical potential energy as the energy gained when work is done moving an object through an electric field.

GCSE knowledge

You should know that work done = force × distance, and that when work is done on an object energy is transferred. For instance, hitting an object with a hammer generates some heat, and lifting an object up increases its gravitational potential energy.

Electric potential energy

All of us know that it is very hard work to climb up a mountain. This is because, as we climb, we are increasing our gravitational potential energy by working against the Earth's gravitational field. In the same way, if a charge is moved in an electric field then work has to be done to move it, if we are to increase its electrical potential energy.

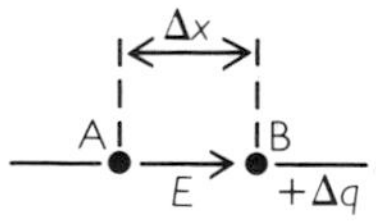

figure 1

In figure 1, a small charge Δq is to be moved a small distance Δx, from A to B, away from a large positively charged object. In the region round A and B the electric field has strength E, which is constant if Δx is small enough. The force F acting on the small charge is given by $F = E\Delta q$. The work done by the electric field in moving Δq the small distance Δx is thus given by

$$F\Delta x = E\Delta q\Delta x$$

The charge Δq has thus lost an amount of potential energy equal to $E\Delta q.\Delta x$. Its change in potential energy, ΔW, is thus given by

$$\Delta W = E.\Delta q.\Delta x$$

The electrical potential difference (p.d.) between A and B, ΔV, is defined by the equation

$$V = \frac{W}{Q}$$

or $$\Delta V = \frac{\Delta W}{\Delta q}$$

for small changes in each quantity.

Thus $$\Delta V = \frac{E\Delta q\Delta x}{\Delta q}$$

$$= E\Delta x$$

$\Rightarrow$ $$E = \frac{\Delta V}{\Delta x}$$

From this result we see that the electric field strength is numerically equal to the potential gradient.

Note that the unit of potential gradient is $1\,\mathrm{V\,m^{-1}}$. This is therefore an alternative to $1\,\mathrm{N\,C^{-1}}$ as a unit of electric field strength.

It is now possible to calculate the strength of an electric field between two parallel plates, if we know the p.d. across them, V, and their separation, d. From our previous discussions we know that the electric field between the plates is uniform and so, therefore, is the potential gradient. Consequently the size of the field between two parallel plates is given by:

$$E = \frac{V}{d}$$

Example A beam of electrons enters a region of uniform electric field between charged parallel plates, travelling at a speed of $3.0 \times 10^7 \, ms^{-1}$, as shown in figure 2. By what distance has the beam been deflected by the time it leaves the field?

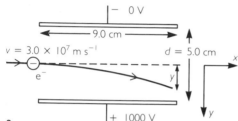

figure 2

The electric field, and thus the force on each electron, act only in the y-direction. Their speed along the x-direction thus remains constant at $3.0 \times 10^7 \, ms^{-1}$. They follow a parabolic path, analogous to that of a projectile near the Earth's surface.

Using the formula to calculate the distance travelled in time t under a constant acceleration a we write:

$$y = \tfrac{1}{2}at^2. \tag{i}$$

t is the time the electrons spend between the plates which is given by:

$$t = \frac{x}{v}$$

$$= \frac{0.090 \, m}{3.0 \times 10^7 \, ms^{-1}}$$

$$= 3.0 \times 10^{-9} \, s. \tag{ii}$$

a is the downwards acceleration of the electron. This can be calculated using $F = ma$:

$$F = ma$$

$$\Rightarrow \qquad a = \frac{F}{m} \tag{iii}$$

But the force F on the particle due to the electric field is:

$$F = Eq = \frac{V}{d} \times q \tag{iv}$$

V is the p.d. between the plates, d is their separation and q the charge of an electron. The formula $E = V/d$

is used for the field strength because we are dealing with a uniform field.

Putting our expression for the force in equation (iv) into equation (iii) gives

$$a = \frac{Vq}{md}$$

We will now calculate the size of the acceleration:

$$a = \frac{1000 \, V \times 1.6 \times 10^{-19} \, C}{9.1 \times 10^{-31} \, kg \times 0.050 \, m}$$

$$= 3.5 \times 10^{15} \, ms^{-2}$$

Finally we may calculate the deflection, y, of the beam:

$$y = \tfrac{1}{2}at^2$$

$$= \tfrac{1}{2} \times 3.5 \times 10^{15} \times (3.0 \times 10^{-9})^2 \, m$$

$$y = 0.016 \, m$$

The beam is deflected by 1.6 cm.

Work done in non-uniform fields

Suppose now a small positive charge q is to be moved from A to B towards a sphere with charge $+Q$ (figure 3). The distance AB is now large and the field cannot be considered to be uniform over that region.

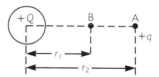

figure 3

To calculate the work done we must calculate the area under a force-against-distance graph. The relevant area is shaded in figure 4. This area can be calculated using calculus, but we have already tackled this sort of problem in the chapter on gravitation (see page 213). Since both the electrostatic and gravitational forces obey an inverse

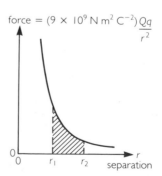

figure 4

square law relationship, we can draw an analogy between the two cases and produce the result:

work done = change in electrical potential energy

$$= (9 \times 10^9 \, \mathrm{N\,m^2\,C^{-2}}) Qq \left[\frac{1}{r_1} - \frac{1}{r_2} \right]$$

and thus the change in potential $\left(\dfrac{\text{work done}}{\text{charge}} \right) = \Delta V$

$$= (9 \times 10^9 \, \mathrm{N\,m^2\,C^{-2}}) Q \left[\frac{1}{r_1} - \frac{1}{r_2} \right] \qquad \text{(v)}$$

Electric potential at a point

The electrical potential V at a point is analogous to the gravitational potential encountered on page 213. It may be defined as the work done per unit charge in moving a positive charge from Earth (or an infinite distance away) to the point. It is thus equal to the potential difference V between the point and the Earth (zero potential).

Potential near an isolated charged sphere

We will now consider the potential near an isolated charged sphere. This sphere is suspended in space, far enough away from any other objects that their effects may be neglected.

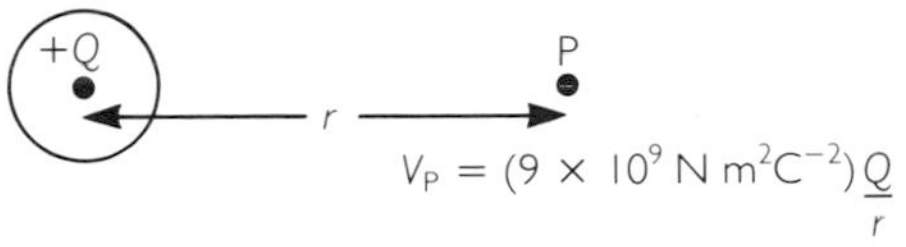

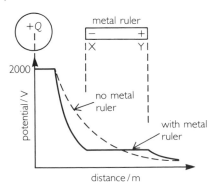

figure 5

We will use equation (v) to calculate the potential difference V_P between a point P and the Earth. The surface of the Earth may be taken to be infinitely distant; so $V_P = \Delta V$ in equation (v) with $r_1 = r$ and $r_2 = \infty$.

Thus $\qquad V_P = (9 \times 10^9 \, \mathrm{N\,m^2\,C^{-2}}) \dfrac{Q}{r}$

It is possible to check that the variation of potential is indeed proportional to $1/r$ by using a flame probe, as in figure 6. The flame probe is a hypodermic needle through which gas can flow; when the gas is lit the probe quickly reaches the potential V at that point. (If the metal needle, with no flame on it, were placed near the positively-charged sphere, a charge would be induced on the tip of the needle. This induced charge would distort the electric field around the needle's tip, and hence, change the potential. When the flame, which is a source of both positive and negative ions, is lit the induced negative

charge on the tip of the needle will be rapidly neutralised. Thus, the electric field of the sphere will no longer be distorted. Hence, the potential at the tip of the needle will be dependent only on the sphere.) This potential is measured on an electroscope, which has been calibrated beforehand to measure potentials. A graph of V against $1/r$ should give a straight line.

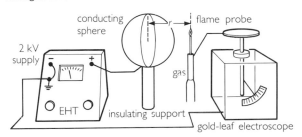

figure 6   Flame probe determination near a charged sphere

In performing this experiment it is vital that the sphere is at a considerable distance from any other objects, so the measurement of r must be made after the value of potential has been recorded. As an example of how errors might arise, suppose a metal ruler was clamped next to the sphere to facilitate the measurement of r (figure 7). The charge on the sphere would induce a negative charge on the end X of the ruler, causing a positive charge to exist at end Y. The negative charge at X would cause the potential there to be lower than expected, whereas the positive charge at Y would lift the potential at Y above its expected value. Since XY is a conductor it must have the same potential along its length (see figure 7).

figure 7

Example 1   What is the potential at a distance of 0.01 m from an isolated small charge of $+10^{-9}$ C?

To calculate the potential near to a point charge we use the formula:

$$V = \frac{(9 \times 10^9 \, \mathrm{N\,m^2\,C^{-2}}) Q}{r}$$

$$\Rightarrow \qquad V = \frac{9 \times 10^9 \times 10^{-9}}{0.01} \text{ V} = 900 \text{ V}$$

Example 2 Figure 8 shows two point charges separated by a distance of 2×10^{-2} m. What is the potential at P?

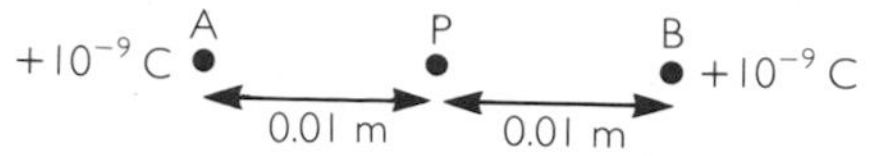

$+10^{-9}$ C   A   P   B $+10^{-9}$ C

0.01 m   0.01 m

figure 8

What would the potential be at P if one of the charges were to become -10^{-9} C?

Remember that the potential at a point is the work done in taking unit positive charge to that point from infinity. In taking a charge up to P, work is done equally against both charges A and B, and so the potential at P is the sum of the potential due to A and the potential due to B. Work is a scalar quantity, and so potentials are added as scalars. Example 1 showed that the potential at a point 0.01 m from a charge of $+10^{-9}$ C is 900 V. The potential at P is thus 1800 V. If one charge was changed to -10^{-9} C, the potential at P would be zero: the work done against the positive charge is cancelled out by the work done by the negative charge.

Lines of equipotential

A line of equipotential, sometimes called an equipotential line, or simply an equipotential, describes the locus of points which are at the same potential, in the same way that isobars on a weather map link together points of equal pressure. No work is done in moving a charged object along an equipotential, so the equipotentials are always at right angles to the field lines. Figures 9 and 10 show the field lines and equipotentials for a parallel plate capacitor and a charged sphere.

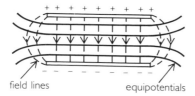

field lines   equipotentials

figure 9

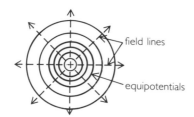

field lines

equipotentials

figure 10

If we draw equipotentials so that the change in potential is always the same between neighbouring lines, then we find that the equipotentials get closer together when the field is stronger; this is because the magnitude of the electric field strength E is equal to the potential gradient dV/dx. In the uniform field produced by the capacitor the equipotentials are evenly spaced; but they get closer together as the surface of the sphere is approached because the field gets stronger.

We may use the idea of equipotentials to explain why a needle attached to a charged object will cause the object to lose its charge rapidly. The equipotentials near to the needle are shown in figure 11.

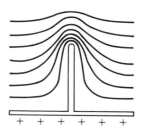

figure 11 Equipotentials near a needle

The needle distorts the field, causing the potential gradient (field) near its point to be very large. When air is subjected to an intense electric field, stray electrons colliding with atoms cause the air to become ionised. Thus charge may be conducted away through the needle.

We shall use the concept of equipotentials to show that when a charge is moved in an electric field between two points, the work done on that charge is independent of the path taken. In figure 12 the work done in taking a small charge q from B to A (directly) is:

$$W = (9 \times 10^9 \text{ N m}^2 \text{C}^{-2}) Qq \left(\frac{1}{r_A} - \frac{1}{r_B} \right)$$

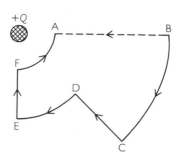

figure 12

The work done in taking the charge along the line BCDEFA is the same. No work is done along BC, DE, or FA, as these are equipotentials. The work done along CD and EF is:

$$W = (9 \times 10^9 \, \text{Nm}^2 \, \text{C}^{-2}) Qq \left(\frac{1}{r_D} - \frac{1}{r_B} \right)$$

$$+ (9 \times 10^9 \, \text{Nm}^2 \text{C}^{-2}) Qq \left(\frac{1}{r_A} - \frac{1}{r_D} \right),$$

(because $r_C = r_B$, $r_F = r_A$, and $r_E = r_D$)

$$\Rightarrow \quad W = (9 \times 10^9 \, \text{Nm}^2 \text{C}^{-2}) Qq \left(\frac{1}{r_A} - \frac{1}{r_B} \right)$$

The work done will always be the same no matter how complicated the path taken. The movement of the charge may always be resolved into movements along or at right angles to equipotentials. Only when the charge is moved at right angles to the equipotentials will work be done and this distance moved will always add up to AB.

An equipotential map provides a means of calculating the electric field strength in complicated situations. Figure 13 shows the equipotentials for an array of four equal positive charges, which were drawn with the help of a computer. It is much easier for the computer to calculate the potential at any point than the electric field, since the potential is the sum of scalar numbers, four in this case, whereas the electric field strength is the sum of four vectors.

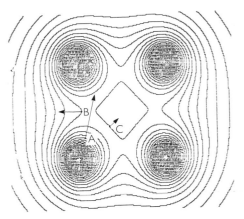

figure 13 Equipotential map, courtesy of A. R. B. Shimmin Esq.

From the equipotential map we can calculate the field. We met earlier the relationship

$$E = \frac{\Delta V}{\Delta x}$$

which tells us that the electric field strength is equal to the potential gradient at any point. So when equipotentials with constant ΔV between them are close together the field is large; when they are far apart the field is weak. In figure 13 the field strength is indicated at three points; notice that the field is always at right angles to the equipotential lines.

Summary

1 Field strength E is related to electrical potential difference ΔV by the equation

$$E = \frac{\Delta V}{\Delta x}$$

2 In a uniform field (as between two flat parallel plates),

$$E = \frac{V}{d}$$

3 Electric potential V at a point is defined as the work done in taking unit charge from a place of zero potential to that point.

$$V = \frac{W}{q}$$

4 The potential at distance r from a point charge q is given by

$$V = (9 \times 10^9 \, \text{NC}^{-2} \text{m}^2) \frac{q}{r}$$

5 Equipotential lines (surfaces in three dimensions) join points of equal potential.

6 Field lines are perpendicular to equipotential surfaces.

Questions

1 A small ball is charged from a high-voltage supply and is suspended between two parallel vertical plates, with a potential difference of 5000 V between them. The ball is moved along the line AB and its deflection from the vertical is the same at all points along that line.

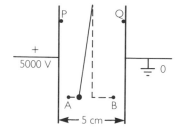

figure 14

a What can you say about the size of the force acting on the ball at any position in the electric field?

b Explain why the electric field is uniform between the plates.

c It is discovered by a separate experiment that the charge on the ball is 10^{-10} C. How much energy will be expended in taking the ball from P to Q?

d Given that the separation of the plates is 5 cm, show that the force acting on the ball is 10^{-5} N.

e The electric field strength is defined as the force per unit positive charge. Use this to calculate the electric field in figure 14.

2 This question is an algebraic version of question 1.

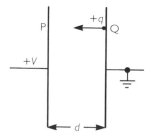

figure 15

a How much energy is expended in taking a charge $+q$ from Q to P, when the potential difference between the plates is V?

b While q goes from Q to P a uniform force F acts on it. Write down an expression for the work done on the charge during this process.

c Equating the work done to the gain in electrical potential energy we get: $Fd = Vq$. Explain how this leads to the relationship $E = V/d$ for a uniform electric field.

d The units of electric field are quoted as either NC^{-1} or Vm^{-1}; show that these two are equivalent.

3 This question is about the deflection of an electron beam in an evacuated tube. In the absence of electric fields the electron beam follows the path OA; when a potential of 3000 V is applied to the top plate the beam deflects upwards along the path OB – figure 16.

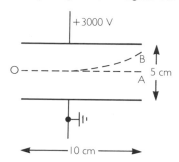

figure 16

a The electrons arrive at O travelling at a speed of $10^8 \, ms^{-1}$. How long do they take to reach A? When the field is switched on, how long do they take to reach B? Justify your answer.

b When the field is switched on what force acts on each electron? Assume that the field is uniform and that the charge on the electron is 1.6×10^{-19} C.

c Given that the mass of an electron is about 10^{-30} kg, calculate its upwards acceleration. Compare your answer with the acceleration due to gravity and hence explain why we have neglected gravitational effects in this question.

d Use your answers to (a) and (c) to show that the distance AB is about 5 mm.

4 A Van de Graaff generator is capable of storing charge on a dome up to a potential of about 500 kV.

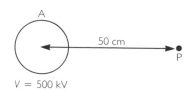

figure 17

a A small particle carrying a positive charge of 10^{-10} C experiences of force towards the generator dome of 4×10^{-5} N. Calculate the strength and direction of the electric field at P.

b Show that the charge on the dome is 1.1×10^{-5} C.

c Calculate the force that would act on a charge of 5×10^{-11} C at a distance of 1 m from the centre of the dome.

d Show that the potential at P is 200 kV.

e Calculate the potential at a point 30 cm from the dome's centre.

f Work out (i) the radius of the dome, (ii) the capacitance of the dome.

g A Van de Graaff generator identical to the first, A, is brought into position B, so that their centres lie 1 m apart and P is equidistant from A and B along a line joining their centres. Calculate the value of (i) electric field strength at P, (ii) potential at P. Explain your reasoning carefully.

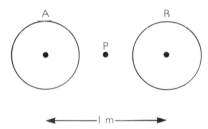

figure 18

5 a A point charge of 10^{-6} C is taken from a place of zero potential to a point P a distance 0.3 m away from the centre of a sphere, radius 0.1 m, which is charged up to 3 kV. How much work is done?

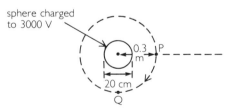

figure 19

b The charge is then moved to point Q. How much work is done this time?

6 From the data for question 5 on page 232, calculate the potential at P.

7 a For the situation described in question 6 on page 232, sketch diagrams to show how the potential varies along the lines (i) RS, (ii) PQ.

b Repeat part (a) for the situation where B is changed to a negative charge of equal magnitude.

8 For the situation in question 7 on page 232, the potential at A is 100 V. Calculate the potentials at B, C and D.

9 Figure 20 shows equipotentials in a region where there is an electric field.

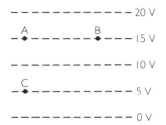

figure 20

a What is an equipotential?

b Copy the diagram and add to it electric field lines.

c A positive charge of 1 nC is placed somewhere in the field. Where is there the greatest force acting on it: at A, B or C?

d Calculate the work done to move the charge from C to B.

e How much work has to be done to move it from B to A?

10 Each diagram in figure 21 represents two conducting wires or surfaces between which a p.d. is maintained by a supply.

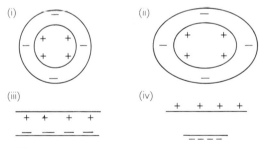

figure 21

a Copy each of the diagrams (i) to (iv), and sketch in field lines and lines of equipotential. Explain why your lines of equipotential must always cut field lines at right angles.

b Use the realtionship $E = \Delta V / \Delta x$ to explain why the electric field is stronger at P than at Q in figure 22 on the next page.

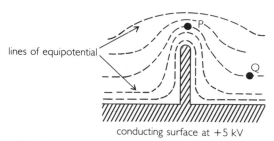

lines of equipotential

conducting surface at +5 kV

figure 22

11 a Figure 23 shows an extremely long row of ions each with a diameter a. They have, alternately, charges $+q$ and $-q$. Show that the energy W required to take one ion at the end of the line away from the others is given by:

$$W = \frac{kq^2 \log_e 2}{a}$$

where $k = 9 \times 10^9 \text{ NC}^{-2}\text{m}^2$

$$\left(\text{Hint: } \log_e(1 + x) = x - \frac{x^2}{2} + \frac{x^3}{3} - \frac{x^4}{4} \cdots\right)$$

a

figure 23

b Calculate the energy required to remove an ion in a solid, using the formula derived above. Assume that $a = 2 \times 10^{-10}$ m, $q = 1.6 \times 10^{-19}$ C, and express your answer in electron volts. Comment on the validity of this formula when applied to a real solid, and suggest how we might set about deriving a more valid formula. (Hint: Real materials are three-dimensional not one-dimensional.)

12 a The Bohr model of the hydrogen atom consists of a stationary proton with an electron in a fixed orbit of radius 0.5×10^{-10} m (figure 24). What is the electric potential energy of the electron–proton pair? (Express your answer in eV.)

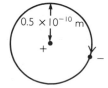

0.5×10^{-10} m

figure 24

b How does your answer to (a) compare with the experimentally measured value of ionisation energy, 13.6 eV?

c The reason for the discrepancy is that the electron has some kinetic energy. Use the fact that the electrostatic force of attraction between the proton and electron must provide the necessary centripetal force to derive an

expression for the kinetic energy of the electron.

d Now use your answers to (a) and (c) to make an estimate for the ionisation energy of the hydrogen atom, using the Bohr theory.

e Given that the mass of an electron is 9.1×10^{-31} kg, calculate the speed of the electron in its orbit and its frequency of rotation.

13 In this question you are asked to write an article on electrostatics for a magazine which is read by sixth form scientists in Central Africa. You need to remember that these students do not have access to the textbooks that you have, so it is essential that you include the following important concepts: Coulomb's law, uniform and radial fields, potential. It would also be a good idea to include an application of electrostatic theory to some problem in atomic physics or crystallography.

14 Justify the following statements:

a The closest distance of approach of a 6 MeV α-particle to a gold nucleus, when making a head-on collision, is 4×10^{-14} m. (The atomic number of gold is 79.)

b The force experienced by the α-particle at its closest distance of approach is 25 N.

15 An extremely keen student, who has enjoyed investigating potentials near to a charged sphere, decides to investigate potentials inside a large cylindrical conductor. She makes an apparatus consisting of a central rod, radius 1 cm, surrounded by a conducting cage of radius 50 cm. The central rod and cage are then connected across a 5 kV supply and a flame probe is used to measure potentials as a function of position from the central rod. Her results are tabulated below.

Potential V/V	5000	4100	2900	2100	1200	600
Distance r from centre/cm	1	2	5	10	20	30

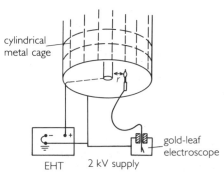

cylindrical metal cage

gold-leaf electroscope

EHT   2 kV supply

figure 25

a Plot a graph of V against r.
b Use the relationship $E = \Delta V/\Delta x$ to tabulate the electric field strength as a function of distance from the centre of the cage.
c Plot a suitable graph to investigate whether the electric field strength E is proportional to $1/r$.

16 The apparatus shown in figure 26 is used to investigate the strength of electric field near to a charged conducting sphere of radius 0.05 m. It is left permanently connected to the positive terminal of an EHT supply. The field strength is measured by recording the force of attraction on a small negatively charged ball, which is attached by a very fine wire to a Van de Graaff generator. The force constant of the thin elastic supporting the ball is known to be 0.2 Nm^{-1}.

In a separate experiment, it was determined by using a coulombmeter that the charge on the small ball was 10^{-7} C. Below are tabulated the results of the investigation, recorded as the test charge approached the small sphere along the line AB.

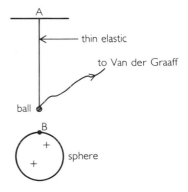

figure 26

Extension x of elastic string/cm	9.8	2.5	1.0	0.6	0.4
Distance r from centre of sphere/cm	5	10	15	20	25

a Plot a graph of the force acting on the ball as a function of distance from the centre of the sphere.
b From the graph, determine the work you would have to do to take the small sphere from a position where it is touching the sphere away to infinity.
c Now determine the potential at the surface of the sphere.

17 A balloon of radius r is covered in a coat of conducting paint. The balloon carries a charge $+Q$, and its surface is at a potential V (figure 27).

figure 27

a What is the electric potential at the centre of the balloon?
b What is the electric field strength at the centre of the balloon?
c The balloon is now blown up to a radius $2r$, with the charge Q remaining constant. What is the new potential of the balloon's surface?
d Write down an expression for the change in electrostatic energy stored on the balloon during the process of blowing up from a radius r to a radius $2r$.
e Has the balloon's electrostatic energy grown or diminished in the blowing-up process?
f When the balloon is charged, is it easier or harder to blow up than when it is not charged?

18 It is often said that there are good analogies between gravitational and electric fields. Think of as many as you can to show the similarities between the two fields, but also point out any differences.

19 *Millikan's oil drop experiment*

This experiment appears here because it is to do with charges and electric fields. However, the conclusion that we can draw from it is very important indeed: by measuring the charge on a large number of oil drops it is possible to show that charge comes in multiples of 1.6×10^{-19} C. We now accept that value as the charge on an electron or proton.

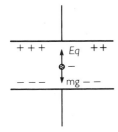

figure 28

The main features of the experiment are shown in figure 28. Oil drops are sprayed into the region between two plates where an electric field acts. These droplets gain charge as a result of friction as the drops pass through a small nozzle. If the drops are negatively charged the field will tend to stop them falling, and if we can adjust the p.d. across the plates carefully it is possible to cause the upwards force due to the electric field to counterbalance exactly the weight of the drop.

We can then write:

$$Eq = mg \qquad \text{(i)}$$

where E is the strength of the field between the plates, q is the charge on the drop and mg is the drop's weight. Since the electric field is uniform we may write:

$$E = \frac{V}{d} \Rightarrow \frac{V}{d} \cdot q = mg$$

and therefore $\quad q = \dfrac{mgd}{V} \qquad \text{(ii)}$

We can now use equation (ii) to calculate the charge on any oil drop when we have made it stationary in an electric field. However, there is a snag and that is the need to calculate the weight of the oil drop.

The way the weight is calculated is by switching the electric field off and allowing the drop to fall through a given distance. The more massive the drop is, the less significant is air resistance, and so the less time it takes to fall that distance. The graph of figure 29 shows the results of a calibration experiment to determine the weight of oil drops which were observed through a microscope as they fell at a constant speed through a distance of 1.0 mm.

(This calibration curve was obtained by applying Stoke's law of terminal velocities for spheres falling through viscous fluids.)

a The points on the graph refer to six different experiments. How fast was the drop travelling in experiment D?
b Why does the drop travel so slowly?
c Why does the drop travel at constant speed?
d Using equation (ii) above, calculate the charge on the oil drop in each of the six cases, using the experimental results tabulated below. You need to know that the separation of the electric field plates was 5.0 mm.

Experiment	A	B	C	D	E	F
p.d. to hold drop still/V	2200	520	280	780	320	170
Time taken to fall 1 mm with field switched off/s	3.65	5.15	6.95	10.05	12.55	15.15

e Express your answers to part (d) as multiples of 1.6×10^{-19} C, and comment on these results.

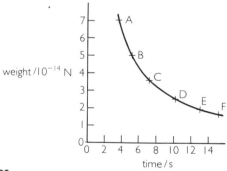

figure 29

CHAPTER 22

The parallel-plate capacitor

Introduction

This chapter develops ideas which we met first in chapter 12, which was about capacitors. Now we explore in detail how the capacitance of a parallel-plate capacitor depends on its geometric dimensions. An important constant, the permittivity of free space, ε_0, appears. This chapter is an integral part of the development of electric field theory; chapter 23 will bring together these ideas with others explored in chapters 20 and 21.

GCSE knowledge

You should remember that:
   only the negatively charged electrons are free to move through metals;
   electric current is the rate of flow of electric charge.
Also you should have a good understanding of chapter 12 (Capacitors).

Measurement of charge on a parallel-plate capacitor

The purpose of studying the parallel-plate capacitor is to learn more about uniform electric fields. In the process we will discover that there is a link between the strength of a uniform electric field and the strength of a radial electric field.

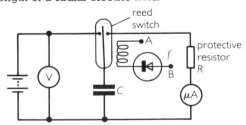

figure 1   Circuit to determine the charge on a capacitor

The charge on a capacitor may be determined using the circuit shown in figure 1, which incorporates a reed switch. The detailed working of this circuit was discussed on page 148. If Q is the charge on the capacitor at each cycle of the switch, which is being driven at frequency f, then the current I in the ammeter is given by

$$I = Qf$$

Thus if we measure I and f we can find Q, using

$$Q = I/f$$

If the charging p.d. V is varied we find that

$$Q \propto V$$

If the separation of the plates d and the p.d. V are kept constant, and the area of overlap A of the plates is varied, we find that

$$Q \propto A$$

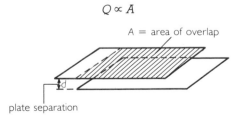

figure 2

Finally, if we fix A and V we find that the charge on the plates varies inversely with the distance of separation d. (The distance between the plates is varied by including small polythene spacers.)

So $$Q \propto \frac{1}{d}$$

However, there is one precaution that we must take to obtain satisfactory results, because even with the plates well separated, current will pass due to stray capacitance associated with the circuit. The simple precaution that we take is to plot graphs; when, for example, we wish to show that $Q \propto 1/d$ we plot a graph of Q against $1/d$, figure 3. The graph is consistent with the capacitance being inversely proportional to the separation, together with the idea that the circuit incorporates a stray capacitance.

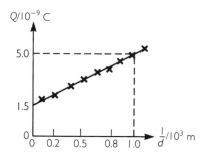

figure 3

We may now summarise the three results above to give:

$$Q \propto \frac{VA}{d}$$

or $$Q = \text{constant} \times \frac{VA}{d}$$

The reed switch experiment may be used to calculate this constant of proportionality. The graph in figure 3 shows a linear variation of Q with $1/d$. The gradient of the graph is:

$$\text{gradient} = \text{constant} \times V \times A$$

Suppose in one particular experiment the area of the plates A was $4.0 \times 10^{-2}\,\text{m}^2$ and the applied voltage V was 10 V. We can now calculate the constant of proportionality.

From the graph the gradient is:

$$\text{gradient} = \frac{(5.0 - 1.5) \times 10^{-9}\,\text{C}}{1.0 \times 10^3\,\text{m}^{-1}}$$
$$= 3.5 \times 10^{-12}\,\text{Cm}$$

since, $\text{gradient} = \text{constant} \times V \times A$

$$\text{constant} = \frac{\text{gradient}}{VA}$$
$$= \frac{3.5 \times 10^{-12}\,\text{Cm}}{10 \times 40 \times 10^{-2}\,\text{Vm}^2}$$
$$= 8.8(5) \times 10^{-12}\,\text{Fm}^{-1}$$

This constant is given the symbol ε_0 ('epsilon nought') and is called the *permittivity of free space*. We will make no distinction between the permittivity of air and that of free space (vacuum), since they are virtually the same. We will say more about permittivity later.

(The above work may have given you the impression that ε_0 is an independent constant, whose value is determined purely by experiment. In fact this is not so. Its value is really decided by the fundamental definitions of the amp, the metre and the second – see page 483.)

So the charge on one plate of our parallel plate capacitor is:

$$Q = \frac{\varepsilon_0 VA}{d}$$

Since the capacitance is given by $C = Q/V$ (see page 147), it follows that

$$C = \frac{Q}{V} = \frac{\varepsilon_0 A}{d}$$

Example   What is the capacitance of a parallel-plate capacitor with circular plates, radius 10 cm, separated by an air gap of 1.0 mm?

$$C = \frac{\varepsilon_0 A}{d} = \frac{\varepsilon_0 \pi r^2}{d}$$
$$= \frac{8.85 \times 10^{-12}\,\text{Fm}^{-1} \times \pi \times (0.10)^2\,\text{m}^2}{1.0 \times 10^{-3}\,\text{m}}$$
$$= 2.8 \times 10^{-10}\,\text{F}$$

Permittivity

If a sheet of insulating material is inserted between the plates of a capacitor of fixed dimensions, then the capacitance is increased. This may be shown by further experiments with the reed switch, using polythene as the insulator.

The reason for this increase in capacitance is as follows. The presence of the electric field will tend to polarise the molecules in the material, so that one end of a molecule gets an induced negative charge and the other end an induced positive charge. As a result the surfaces of the material acquire a charge which is opposite to the charge on the plates next to them.

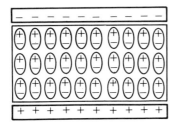

figure 4

These induced charges tend to lower the potential difference between the plates; if the capacitor is connected to a battery of fixed p.d., more charge will have to flow onto the plates to charge the capacitor up to the p.d. of the battery. In this way the capacitor stores more charge for the same applied potential difference, i.e. its capacitance is greater.

When there is only an air gap between the plates of a capacitor the expression for capacitance is

$$C_0 = \frac{\varepsilon_0 A}{d}$$

When there is another material completely filling the space between the plates of the capacitor, we write the following expression for the capacitance:

$$C = \frac{\varepsilon A}{d}$$

ε_0 is the permittivity of air (or free space) and ε is the permittivity of the material. Both ε_0 and ε have the same unit: Fm^{-1}.

The ratio C/C_0 is called the relative permittivity ε_r of the material. ε_r is a pure number, and has no unit. For example, the relative permittivities of polythene and mica are 2.3 and 7.0 respectively. We can deduce that $\varepsilon = \varepsilon_r \varepsilon_0$, and so the capacitance C of any parallel-plate capacitor is given by

$$C = \frac{\varepsilon_r \varepsilon_0 A}{d}$$

Example A parallel-plate capacitor has plates of area 100 cm^2. The separation of the plates is 5.0 mm. Half the capacitor is filled with a substance of relative permittivity 2.0. What is the capacitance of the capacitor and what can you say about the density of charge on each part of the capacitor?

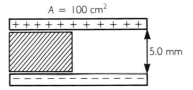

figure 5

The two halves of the capacitor behave like two capacitors C_1 and C_2 in parallel because they have the same p.d. across them. So the total capacitance C is:

$$C = C_1 + C_2$$
$$= \frac{\varepsilon_r \varepsilon_0 A}{d} + \frac{\varepsilon_0 A}{d}$$
$$= \frac{2.0 \times 8.85 \times 10^{-12} \times 50 \times 10^{-4}}{5.0 \times 10^{-3}} \text{ F}$$
$$+ \frac{8.85 \times 10^{-12} \times 50 \times 10^{-4}}{5.0 \times 10^{-3}} \text{ F}$$
$$= 1.8 \times 10^{-11} \text{ F} + 0.9 \times 10^{-11} \text{ F}$$
$$= 2.7 \times 10^{-11} \text{ F}.$$

The capacitance of the part with the material between the plates is twice that of the part with the air gap. Consequently the charge density on the former part must be twice that of the latter; twice as much charge is stored for a given applied potential difference.

Summary

1 The capacitance of a parallel-plate capacitor is given by:

$$C = \frac{\varepsilon_r \varepsilon_0 A}{d}$$

A is the area of overlap of the plates, d is the separation of the plates, ε_0 is the permittivity of

space and ε_r is the relative permittivity of the material between the plates.

2 ε_r is determined by the ease with which a material can be polarised.

Questions

1 a Calculate the capacitance of a parallel-plate capacitor whose plates are separated by 5.0 mm of air and have an area of overlap of 0.20 m².
  b If the p.d. across the plates of this capacitor is 5000 V, how much charge is on each plate of the capacitor?

2 A material with a relative permittivity of 3.0 is now used to fill the gap between the plates of the capacitor described in question 1. Calculate the new capacitance.

3 When a capacitor is charged it stores energy in the electric field between its plates. Use this idea to decide which of the arrangements of parallel-plate capacitors shown in figure 6 stores the most energy. All the capacitors have the same area of plates, but not necessarily the same separation of plates.

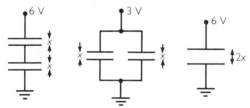

figure 6

4 A parallel-plate capacitor of capacitance 10 pF is charged so that the p.d. across its plates is 10 V. It is now isolated and the separation of its plates is doubled.
  a Calculate the energy stored in its original state.
  b Calculate the new capacitance once the plate separation has been doubled.
  c Work out the p.d. across the plates once their separation has been doubled.
  d How much energy does the capacitor store in its second state? Explain where this extra energy comes from.

5 a When the electric field strength in air exceeds $2 \times 10^6\,\mathrm{NC^{-1}}$ then air will begin to conduct.

Given that the base of a thundercloud measures about 1 km × 1 km, estimate the charge stored on the base of a thundercloud just before a stroke of lightning is emitted from it.
  b If the base of the thundercloud is 300 m above the ground, calculate the potential difference between the cloud and ground.

6 A parallel-plate capacitor has plates of area A separated by a distance x. When the space between the plates is filled with air, the capacitance is C.
  a A piece of material with relative permittivity 5.0 is placed as shown in figure 7. Calculate the new capacitance of the capacitor.

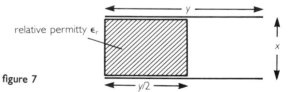

figure 7

  b The resistivity of the material is $10^{14}\,\Omega\,\mathrm{m}$; show that the capacitor will discharge with a time constant of about 1.5 hours.

7 A reed switch circuit, figure 8, is used to calculate the capacitance C of a capacitor. When the reed switch vibrates with a frequency of 100 Hz, the current through the galvanometer is 100 μA.

Use the information given to calculate the capacitance of the capacitor.

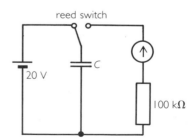

figure 8

Electric flux

Introduction

This chapter ends our section on electrostatics. It is fairly hard, and the questions at the end are quite testing. However, it is worth reading the chapter through once (it is very short) because we show here how the formulae that we met for electric fields near to a sphere and between the plates of a capacitor can be joined together, to form one single rule for calculating field strengths in any situation.

GCSE knowledge

This chapter builds on the previous three chapters, and it is necessary to have a good grasp of those before tackling this work on electric flux.

Electric flux

The idea of electric flux is a very useful and powerful one. We will introduce the concept of flux by first considering light spreading out from a small light source S – see figure 1. A and B are two imaginary spheres surrounding the source. The same amount of light energy (or flux) must pass through each sphere per second, but the energy density (or flux density) is greater through the surface of sphere A since its surface area is smaller.

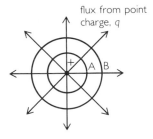

figure 2

figure 1

Similarly, if we consider electric field lines spreading out from a point charge (figure 2) then we see that the number of field lines is the same through each sphere, but the number of field lines per unit area is smaller for the more distant sphere B.

Earlier we discussed how we may represent a stronger electric field by showing more lines per unit area than for a weaker field. Also, a stronger field is produced by a greater charge. From a very bright light we get a high energy density or flux density; in the same way we would expect a high electric flux density from a large point charge. It is therefore reasonable to equate the electric flux through a given surface with the charge enclosed, and this is how electric flux may be defined:

total electric flux through a closed surface = charge enclosed by that surface

The units of flux are therefore coulombs, and the electric flux through the surface A in figure 2 is q. If the sphere A has a radius a, then its surface area is $4\pi a^2$. We can now say that the flux per unit area through the sphere, or *flux density*, is $q/A = q/4\pi a^2$. The units of flux density are Cm^{-2}.

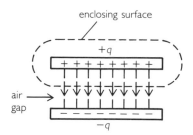

figure 3

We will now apply this idea of flux to a parallel-plate capacitor with an air gap. In figure 3, a dotted line encloses the positively charged plate. To a good approximation, all the electric flux passes straight towards the negatively charged plate as shown, with little flux passing through any other part of the enclosing surface. We can justify the last sentence because we saw in the chopped hair/castor oil demonstration that the field was confined to the region between the electrodes.

The flux density between the plates is q/A. We know, however, that

$$C = \frac{\varepsilon_0 A}{d} \text{ and } C = \frac{q}{V}$$

$$\Rightarrow \quad \frac{q}{V} = \frac{\varepsilon_0 A}{d}$$

$$\Rightarrow \quad \frac{q}{A} = \frac{\varepsilon_0 V}{d}$$

$$\Rightarrow \quad \frac{q}{A} = \varepsilon_0 E \text{ since the field is uniform}$$

This shows us that the flux density $= \varepsilon_0 \times$ electric field.

This idea may be extended for any shape of surface. Thus in figure 4 the total flux passing through the surface A is q.

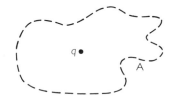

figure 4

So we may write that

$$q = \varepsilon_0 EA$$

or in general, if the charge is surrounded by a medium of relative permittivity ε_r,

$$q = \varepsilon_r \varepsilon_0 EA \tag{i}$$

Equation (i) is known as *Gauss' Theorem*.

Note that the product EA can only be evaluated

easily if E is constant and at right angles to the surface, or else zero. This is so for the three examples discussed in this chapter: a parallel-plate capacitor, a point charge, and a line of charge.

We may use this idea to produce two useful results for a charged sphere of radius b, placed in a vacuum. The flux passing through a spherical surface of radius r surrounding the sphere is q (figure 5).

figure 5

But

$$q = \varepsilon_0 EA$$

$$\Rightarrow \quad E = \frac{q}{4\pi\varepsilon_0 r^2} \tag{ii}$$

This is the same field that would have been produced by a point charge q at the sphere's centre, so we may treat the charged sphere as a point charge.

We may also see immediately that the field inside the sphere is zero, because there is no charge enclosed inside the surface of radius a. A similar result was proved by a slightly different method for the gravitational field inside a hollow shell.

The equation (ii) above is the second one we have met that describes the field near to a point charge. Compare it with the previous one:

$$E = (9 \times 10^9 \, \text{Nm}^2\text{C}^{-2}) \frac{q}{r^2}$$

These two equations can only be consistent if $1/4\pi\varepsilon_0 = 9 \times 10^9 \, \text{Nm}^2\text{C}^{-2}$. The reader should check that this is indeed the case.

We may now write three important equations that we met earlier in another form.

Force F between two charges in air is:

$$F = \frac{q_1 q_2}{4\pi\varepsilon_0 r^2}$$

Electric field E at a distance r from a charge q is:

$$E = \frac{q}{4\pi\varepsilon_0 r^2}$$

Potential V at a distance r from charge q is:

$$V = \frac{q}{4\pi\varepsilon_0 r}$$

If the charges are completely surrounded by a medium of permittivity ε, then ε must replace ε_0 in these formulae.

Example 1 A thundercloud carries a charge of 10 C and has an area of $1.0 \times 10^7 \, \text{m}^2$. What is the electric field intensity underneath the cloud, assuming that the field is uniform in this region?

$$\frac{Q}{A} = \varepsilon_0 E$$

$$\Rightarrow \qquad E = \frac{Q}{\varepsilon_0 A}$$

$$= \frac{10}{8.85 \times 10^{-12} \times 10^7} \text{ Vm}^{-1}$$

$$= 1.1 \times 10^5 \text{ Vm}^{-1}$$

Example 2   A long thin wire carries a charge of 1.0×10^{-9} C per metre. What is the field at a distance of 0.10 m from the wire?

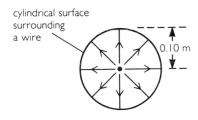

cylindrical surface surrounding a wire

0.10 m

figure 6

The flux from the charge will pass symmetrically outwards from the line of charge as shown in figure 6. We may calculate the field at a distance of 0.10 m from the line if we know the flux density at that point.

$$\frac{Q}{A} = \varepsilon_0 E$$

$$E = \frac{Q}{\varepsilon_0 A}$$

where Q is the charge per metre of the line and A is the area of the cylindrical surface per metre of length.

$$\Rightarrow \quad E = \frac{10^{-9} \text{ (Cm}^{-1})}{8.85 \times 10^{-12} \text{ (Fm}^{-1}) \times 2\pi \times 0.1 \text{ (m)}}$$

$$= 180 \text{ Vm}^{-1}$$

Summary

1   Total electric flux through a closed surface = charge enclosed by that surface.

2   Electric flux can be thought of as 'number of field lines'.

3   Gauss' theorem:

$$q = \varepsilon_0 \varepsilon_r E A$$

4   For a parallel-plate capacitor:

$$\frac{q}{A} = \varepsilon_0 \varepsilon_r E$$

5   The 'k' from Coulomb's law is related to ε_0:

$$k = \frac{1}{4\pi\varepsilon_0} = 9 \times 10^9 \text{ Nm}^2\text{C}^{-2}$$

6   The formulae met previously now become:

$$F = \frac{q_1 q_2}{4\pi\varepsilon_0 r^2}$$

(force between point charges)

$$E = \frac{q}{4\pi\varepsilon_0 r^2}$$

(field due to point charge)

$$V = \frac{q}{4\pi\varepsilon_0 r}$$

(potential near point charge)

Questions

1 a Figure 7 shows two concentric spheres of radii 0.2 m and 0.3 m carrying the charges shown. The value of q is 10^{-9} C. Sketch graphs of (i) the electric field strength, (ii) the potential at any point between O and a point of zero potential a long way from the spheres. Include as much detail as possible on your graph.

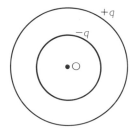

figure 7

b What difference would it make to your graphs if both spheres were to carry a *positive* charge of 10^{-9} C?

2 Figure 8 shows two isolated parallel-plate capacitors, WX and YZ. The area of all the plates is the same. The potential difference between W and X is 100 V.

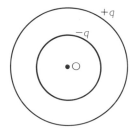

figure 8

a Show that the potential difference between Y and Z is 40 V.
b If Y is connected to W and X to Z, calculate the potential difference between the plates of each capacitor. Compare the energy stored in the capacitors before and after the connection is made.
c The capacitors are now returned to their original state as shown in figure 8. The smaller capacitor YZ is pushed in between the plates WX, so that all the plates lie flush. Draw a diagram to show the electric fields in this case, and show that the p.d. between W and X will be 140 V regardless of how close Y is to W.

3 Two very long, thin threads carry a positive charge ρ per unit length; they are separated by a distance a. Show that the force of repulsion per unit length is

$$F = \frac{\rho^2}{2\pi\varepsilon_0 a}$$

4 A nucleus carries a charge Ze which may be assumed to be uniformly distributed throughout its volume. See figure 9.

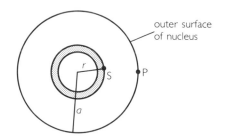

figure 9

a Write down an expression for the potential at P, on the surface of the nucleus.
b Explain why the only charge that makes a contribution to the electric field at S is that charge enclosed within a radius r. Hence show that the electric field E at S is given by:

$$E = \frac{Zer}{4\pi\varepsilon_0 a^3}$$

c Use the relationship $E = \Delta V/\Delta r$ to show that the potential difference between P and S is:

$$\Delta V = \frac{Ze(a^2 - r^2)}{8\pi\varepsilon_0 a^3}$$

d Hence show that the potential at S is

$$\left(3 - \frac{r^2}{a^2}\right)\frac{Ze}{8\pi\varepsilon_0 a}$$

e How does the potential at the centre of the nucleus compare with the potential at its surface?

CHAPTER 24

Magnetic fields near currents

Introduction

In this chapter we meet the ideas of magnetic flux and magnetic flux density. Magnetic flux density is given the symbol B and is often called the B-field. We then look at the factors that affect the magnetic flux density near long solenoids, coils and long straight wires.

GCSE knowledge

We expect that from your GCSE knowledge you will already be able to recall that:

☐ a permanent magnet has both a north-seeking and a south-seeking end (north and south poles);

☐ two like poles repel one another, while two unlike poles attract one another;

☐ compasses and plotting compasses are actually magnets;

☐ the pointing end of a compass is by conventional agreement the north-seeking end.

Magnetic field lines

A magnetic field is a region in space in which a suitable object can experience a magnetic force. We commonly use iron filings and plotting compasses to test for magnetic fields. For example, if you bring a magnet towards an iron filing, the iron filing moves towards the magnet even when they are some distance apart. This illustrates 'action at a distance'. We have already met the idea of 'force field' to explain action at a distance in the chapters on gravitation and electrostatics.

For both gravitational and electric fields, we use 'field lines' to denote certain features of the field. The direction of the line shows the direction of the force (on a point mass and a point positive charge respectively); and the relative density of the lines on a diagram indicates the relative strength of the force. For a magnetic field, the arrow along a field line shows the direction of the force on the north-seeking end of a plotting compass; the south-seeking end experiences a force in the opposite direction. And, as before, the relative density of the lines indicates the relative strength of the force, that is, the field strength.

Water analogy for magnetic field lines

To illustrate the idea of magnetic field lines, let us consider water flowing through a series of interconnected glass tubes of rectangular cross-section, all of the same depth, and all full of water: figure 1.

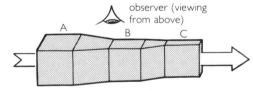

figure 1   Water flowing along a tube

The flow should be slow so that streamline flow occurs, i.e. there is no turbulence. If two very fine glass pipettes release ink into the stream at W and Y, as shown in figure 2, then two ink lines will flow through the tubes as shown.

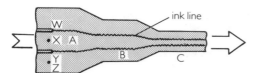

figure 2   This is figure 1 viewed from above

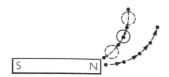

figure 3   Plotting the field round a bar magnet

Consider these questions

☐ Where is the rectangular glass tube narrowest? (At C)

☐ In which section of tube is the water flow fastest, A, B or C? (In C)

☐ What do you notice about the ink lines when the water current is *strongest* (i.e. fastest)? (They are closest together.)

☐ Suppose two more very fine glass pipettes were to start releasing ink very slowly at points X and Z. Would this significantly change the water flowing per second through the large tubular system? (No)

☐ The widest part of the tubular system is A. Would the fact that there are now twice as many ink lines mean that the current is twice as strong as it was before? (No)

☐ If the ink lines in A are three times further apart than the lines in C, does this mean that the water current in C is three times stronger (faster) than that in A? (Yes)

In answering the above questions you should have come to the following conclusions:

☐ When the *water flow* is strongest the *ink* lines are closest together.

☐ The number of lines can be chosen by the experimenter, and does not affect the actual flow.

If the italicised words in the first statement were replaced on both occasions by the words 'magnetic field', both statements would apply to magnetic field lines.

Permanent magnetic fields

The magnetic field around a bar magnet can be plotted by placing the bar magnet on a piece of paper and using a plotting compass to find the direction of the magnetic field; a dot at either end of the compass needle records the direction at that position, as shown in figure 3. By connecting up all the corresponding dots the magnetic field pattern of the bar magnet can be obtained.

Figure 4 shows the results obtained by two different students who used the *same magnet* for their experiments. The diagram made by student B has many more field lines than that of student A. This was not because the magnet had become stronger, but

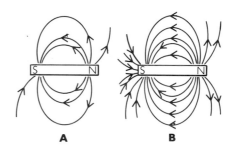

figure 4

because student B chose to draw more lines. Similarly with the water flow analogy, introducing four ink lines instead of two does not make the water current twice as strong.

If a student does not introduce ink jets into the water in the water flow experiment, then there will be no ink lines, but there will still be a flow of water along the rectangular tubes. If a student does not draw magnetic field lines there will be no magnetic field lines, but there will be a flux, or 'flow', of magnetism.

Magnetic flux density, *B*

Consider figure 1 again. The amounts of water flowing per second through A and C are equal. To compare the strengths of the two currents at A and C, we could compare the *rates of water flow through unit areas* in A and C. We would choose unit areas perpendicular to the flow lines for this purpose.

In magnetism the corresponding quantity is the *magnetic flux density B*. The total magnetic flux Φ through any area is measured in webers, Wb. Hence the magnetic flux through unit area, i.e. the magnetic flux density, is measured in $Wb\,m^{-2}$. The SI unit for B is the tesla, T, defined so that $1\,T = 1\,Wb\,m^{-2}$. Precise formal definitions of B and the tesla are given later on page 256.

Thus $$\Phi = BA$$

If the area you are considering is not perpendicular to the flow lines, then the total flux through the area is given by

$$\Phi = BA\cos\theta$$

where θ is the angle indicated in figure 5(b).

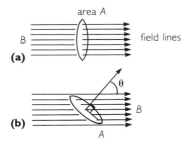

figure 5

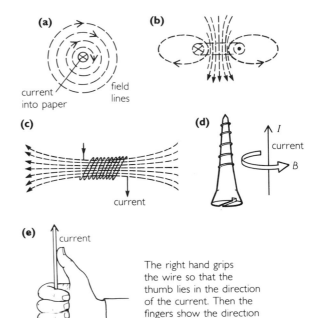

figure 8

Magnetic flux near current-carrying wires

Consider a current in an infinitely long straight wire. We would expect the magnetic field strength round the wire to be symmetrical, i.e. the magnetic flux densities at the points A, B, C and D in figure 6 should all have the same magnitude. The magnetic field patterns round the wire can be found by using plotting compasses, as in the case of the bar magnet discussed earlier (figure 7). Alternatively, iron filings can be used.

The magnetic field patterns obtained around (a) long straight wires, (b) plane coils, and (c) long solenoids are illustrated in figure 8.

The direction of the magnetic flux lines around a current-carrying wire can be found by *Maxwell's right hand screw or grip rule*, which is illustrated in figure 8(d) and (e).

The right hand grips the wire so that the thumb lies in the direction of the current. Then the fingers show the direction of the magnetic field lines.

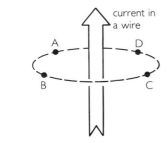

figure 6

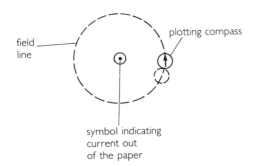

figure 7   Plotting the magnetic field round a current-carrying wire

Magnetic flux density *B* inside solenoids

The magnetic flux density inside a solenoid can be measured by a device called a 'Hall probe'. The way in which this operates will be explained in the following chapter, but for the moment it is sufficient to say that the p.d. generated across the Hall probe is proportional to the magnetic flux density B.

If a series of different solenoids is set up as shown in figure 9, then the Hall probe can be inserted into each one in turn. The Hall probe element must be perpendicular to the axis of the solenoid for a maximum reading to be obtained. By measuring the variation of the Hall p.d., V_H, it is possible to see how the magnetic flux density B varies with current I and position.

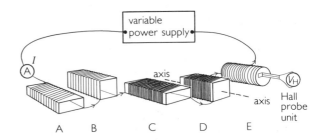

figure 9   Measuring B inside solenoids with a Hall probe

The following results are obtained from this experiment:

1 | Variation of *B* along the axis of one of the solenoids

The features to notice are:
a) *B* is almost constant well within the solenoid;
b) *B* drops to 50% at the ends.

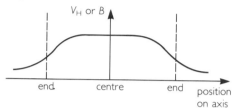

figure 10 Variation of *B* along axis of solenoid

2 | Variation of *B* perpendicular to solenoid's axis

The feature to notice here is that *B* is constant across the width of the solenoid.

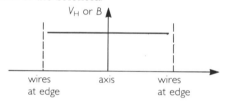

figure 11 Variation of *B* across solenoid

3 | Variation of *B* with current *I*

Doubling and trebling the current doubles and trebles V_H.

$$\Rightarrow \qquad B \text{ is proportional to } I$$

4 | Variation of *B* with the length ℓ of the solenoid and the number of turns *N* on the solenoid

It turns out that V_H does not depend separately on these two quantities. If the values of V_H measured at the *centres* of each of the coils A, B, C, D and E (figure 9) are compared, it is found that

$$V_H \propto \frac{N}{\ell}$$

If we let n = number of turns per metre = N/ℓ, then

$$B \propto \text{number of turns per metre, } n$$

The above results apply to all long solenoids regardless of their shape and cross-sectional area. Therefore, at the centre of a long solenoid:

$$B = \text{constant} \times n \times I$$

More often we meet this equation in the form:

$$B = \mu_0 nI$$

where our constant of proportionality μ_0 ('mu nought') is called the *permeability of free space*. μ_0 has the same significance for magnetic fields as ε_0 has for

electric fields. ε_0 is a constant whose value determines the strength of an electric field in a vacuum near to a charged object; μ_0 is a constant whose value determines the magnetic flux density, or *B*-field, in a vacuum close to a wire carrying a current.

Once we have a way of measuring *B* – developed in the next few pages – then we can obtain an experimental value for μ_0. However, this 'experimental' value is somewhat misleading, as it was for ε_0 – see page 244. In fact the value of μ_0 is decided by the way we define the amp: this is discussed more fully on page 258. The definition of the amp means that $\mu_0 = 4\pi \times 10^{-7}\,\mathrm{Tm\,A^{-1}}$.

The same value of constant will do to calculate the magnetic flux density in either air or a vacuum, since the magnetic properties of air are virtually indistinguishable from those of a vacuum. However, a different constant has to be used to calculate the magnetic flux density inside a solenoid that is filled with iron, or other magnetisable materials. The flux density in an iron-filled solenoid is much greater than that of an air-filled solenoid, when they carry the same current. We describe this difference by introducing the idea of *relative permeability*, μ_r.

Thus $$B = \mu_r \mu_0 nI$$

Inserting an iron core into a solenoid can increase *B* by a factor of about 1000. Thus for iron, $\mu_r \approx 1000$.

As with ε_r (page 245), μ_r is a pure number, with no units. The solenoid must be completely filled with the material for this equation to apply.

Unfortunately, *B* inside an iron-filled solenoid is not completely predictable using the above equation, because two other factors are involved. First, iron becomes magnetically 'saturated', so that *B* is not proportional to *I* at higher *I* values; second, *B* depends to some extent on the recent history of the iron. For these two reasons, μ_r is not a very reliable 'constant' for any material.

Interacting magnetic fields

Magnetic fields interact and distort one another. This results in forces on the sources of the magnetic fields. For example, the case illustrated in figure 12 shows how a strip of aluminium foil carrying a current of about 2 A is forced upwards by the permanent magnetic field between the poles of the horseshoe magnet.

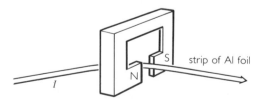

figure 12 Effect of a magnet on a current-carrying strip of foil

This can be pictured in terms of the distorted magnetic field between the poles of the magnet and around the current-carrying foil. Figure 13 illustrates the distorted magnetic field. The magnetic field is distorted to look like the rubber of a catapult pulled downwards. The current-carrying wire is 'catapulted' upwards.

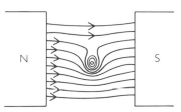

figure 13   'Catapult' left hand (motor) rule

Fleming's left hand (motor) rule is used to predict the behaviour of current-carrying wires in such cases:
    First finger = **F**lux (magnetic)
    Centre finger = **C**urrent
    Thumb = **T**hrust (force)
This is illustrated in figure 14.

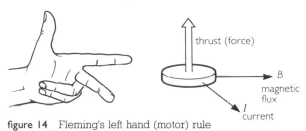

figure 14   Fleming's left hand (motor) rule

The equation for the magnitude of the force on the current-carrying wire can be deduced experimentally using the current balance shown in figure 15.

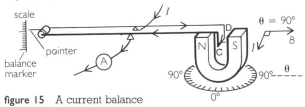

figure 15   A current balance

The apparatus 'weighs' the upward magnetic force on the horizontal wire CD when a current passes from D to C. This is done by putting a small piece of fine wire, as a rider of known weight W, on CD, then increasing the current through CD until the downward gravitational force W is balanced by the upward magnetic force F.

Experiment 1: *To show $F \propto I$*   Put riders of different weights W on CD, and adjust the current I for balance each time. Plot I against W ($W = F$). Typical results are illustrated in figure 16.

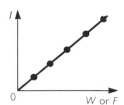

figure 16

Experiment 2: *To show $F \propto \sin\theta$*   Keep the current I constant. Put riders of different weights W on CD, and obtain balance by rotating the magnet about its vertical axis. Note the value of θ for which balance is obtained in each case. (θ is the angle between the magnetic field and the wire.) Plot $\sin\theta$ against W. Typical results are illustrated in figure 17.

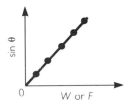

figure 17

Experiment 3: *To show $F \propto \ell$*   In the apparatus illustrated in figure 15, ℓ is the length of wire CD. To see how the force F depends on the length ℓ we must keep all other variables constant. Therefore the magnetic flux density B must be kept constant along the length of CD, no matter how long we make it. This is not possible with the apparatus illustrated in figure 15, but it is possible inside a rectangular solenoid. Thus we use a current balance of the type shown in figure 18.

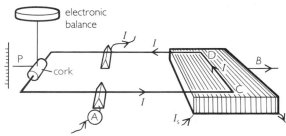

figure 18   A fixed current is passed through the solenoid to give a constant magnetic flux density

Current passes along the wire CD, of length ℓ, and produces a downward force F on CD. The other end of the balance, P, is attached to a frame resting on an electronic balance. When P tries to move upwards the reading on the electronic balance will reduce by the value of F.

Balances with different lengths CD are used, but all other variables are kept constant. If ℓ is plotted against W then a graph like the one shown in figure 19 will be obtained.

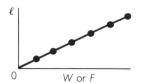
figure 19

Field	Definition	Unit
Gravitational field	$g = F/m$	Nkg^{-1}
Electric field	$E = F/q$	NC^{-1}
B-field	$B = F/I\ell \sin\theta$	$NA^{-1}m^{-1}$

Summary of fields

Experiment 4: To show $F \propto B$ This can be done using the apparatus illustrated in figure 18. It was shown earlier on page 254 that the magnetic flux density B along the axis of a solenoid is proportional to the current in the solenoid. Therefore this experiment is carried out by keeping all other variables constant except the current I_s in the solenoid. A set of readings of force F against solenoid current I_s should be obtained and plotted as displayed in figure 20.

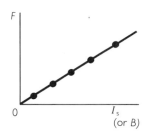
figure 20

Summary

Experiment 1:   force ∝ current
(i.e. $F \propto I$)
Experiment 2:   force ∝ sin θ
(i.e. $F \propto \sin\theta$)
Experiment 3:   force ∝ length
(i.e. $F \propto \ell$)
Experiment 4:   force ∝ magnetic flux density
(i.e. $F \propto B$)

Therefore $F = \text{constant} \times BI\ell \sin\theta$

The thrust F is measured in N;
the current I is measured in A;
the length ℓ is measured in m.
The unit chosen for the magnetic flux density B makes the constant in the above equation equal to unity; this unit is called the tesla (T), and thus $1\,T = 1\,NA^{-1}m^{-1}$.

   Thus $F = BI\ell \sin\theta$

This equation defines B. You can see why this quantity is often called the B-field: it is defined in terms of a force per unit current-length, and it is therefore a similar concept to the gravitational and electric fields that we met earlier. These were defined in terms of force per unit mass and force per unit charge respectively. This is illustrated by the table below.

Example 1   Consider the apparatus shown in figure 21. $\theta = 90°$. The length of the wire CD is 20 cm and the current along CD is 10 A. When a current of 10 A is passed through the solenoid the electronic balance indicates that the magnetic force is 10 mN. Calculate the value of B.

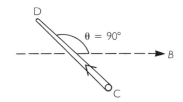
figure 21

$$F = BI\ell \sin\theta$$
$$10 \times 10^{-3}\,N = B \times 10\,A \times 0.20\,m \times \sin 90°$$
$$\Rightarrow \qquad B = 5.0 \times 10^{-3}\,T$$

Example 2   If the solenoid in example 1 has 400 turns m^{-1}, calculate a value for μ_0.

The flux density in the solenoid is given by:

$$B = \mu_0 \times n \times I$$
$$\Rightarrow \quad 5.0 \times 10^{-3}\,T = \mu_0 \times 400\,m^{-1} \times 10\,A$$
$$\Rightarrow \qquad \mu_0 = 1.3 \times 10^{-6}\,TmA^{-1}$$

Example 3   Consider a solenoid of circular cross-section, diameter 10 cm. What will be the magnetic flux density B at the centre of the solenoid, which is 50 cm long and has a total of 100 turns, when a current of 1.0 A passes round the solenoid?

$$B = \mu_0 \times n \times I$$
$$= (1.3 \times 10^{-6}\,TmA^{-1}) \times \frac{100}{0.50\,m} \times 1.0\,A$$
$$\Rightarrow \quad B = 2.6 \times 10^{-4}\,T$$

Equations for magnetic flux densities B near wires carrying currents

Now that we have an equation for the magnitude of the magnetic flux density B at the centre of a long solenoid ($B = \mu_0 nI$), we can calibrate a Hall probe by finding the voltage V_H obtained in a magnetic flux density of known magnitude. V_H is proportional to B, therefore the variation of B in other regions can be investigated.

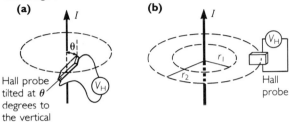

figure 22 Investigating the B-field near a wire

If we place a Hall probe a few centimetres from a long straight wire carrying a large current, say 10 A, and then rotate it as shown in figure 22(a), we find that V_H is a max when $\theta = 0$, and $V_H = 0$ when $\theta = 90°$. This shows that the magnetic field lines must be flat circles in planes perpendicular to the current-carrying wire (cf. Maxwell's right hand screw rule).

If we set up the Hall probe perpendicular to these magnetic flux lines, as shown in figure 22(b), we can obtain a set of readings of V_H against r, where r is the distance from the wire to the centre of the probe. A plot of V_H against r will then show how B varies with r.

Example 1 *Hall probe calibration*
A Hall probe is placed in the centre of a solenoid, which has 100 turns m^{-1}. With a current of 1.0 A in the solenoid, the Hall probe is adjusted to give $V_H = 6.0\,mV$, then no further adjustments are made. What is the magnetic flux density corresponding to 6.0 mV? Take $\mu_0 = 4\pi \times 10^{-7}\,TmA^{-1}$.

$$B = \mu_0 n\,I$$
$$= (4\pi \times 10^{-7}\,TmA^{-1})nI$$
$$= 1.3 \times 10^{-6}\,TmA^{-1} \times 100\,m^{-1} \times 1.0\,A$$
$$\Rightarrow \quad B = 1.3 \times 10^{-4}\,T$$
$$\text{corresponds to } V_H = 6.0\,mV$$

Example 2 When a current of 20 A is passed along a thin metal rod, as in figure 22(b), the following values of V_H are obtained at different distances r from the rod.

r/cm	3.0	5.0	8.0	10.0	12.0
V_H/mV	6.2	3.7	2.3	1.8	1.5

Work out the corresponding value of B for each value of r, then plot a suitable graph to show how B varies with r.

$$V_H \propto B$$
$$\Rightarrow \quad \frac{V_H}{6.0\,mV} = \frac{B}{1.3 \times 10^{-4}\,T}$$
$$\Rightarrow \quad B = \left(\frac{V_H}{6.0\,mV}\right) \times 1.3 \times 10^{-4}\,T$$

By inspection V_H appears to halve as r is doubled, therefore one should suspect that $B \propto 1/r$.

The above table gives the folowing data for a graph.

r/cm	3.0	5.0	8.0	10.0	12.0
$B/10^{-5}T$	13	8.0	5.0	3.9	3.3
$\dfrac{1}{r}$/m^{-1}	33	20	13	10	8.3

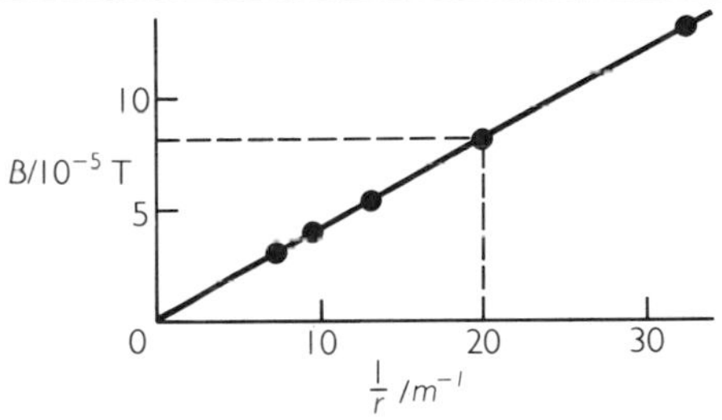

figure 23 Variation of B with $1/r$

In figure 23,

$$\text{gradient} = \frac{8.0 \times 10^{-5}\,T}{20\,m^{-1}}$$
$$= 4.0 \times 10^{-6}\,Tm$$
$$\Rightarrow \quad B = \frac{4.0 \times 10^{-6}\,Tm}{r}$$

Experimentally, it is found that the magnetic flux density B at a point near a wire is proportional to the current in the wire. If this observation is taken along with the equation obtained at the end of the previous example, then

$$B \propto \frac{I}{r}$$

For $I = 20$ A, figure 23 gave
$$B = \frac{4.0 \times 10^{-6}\,Tm}{r}$$

for $I = 1$ A,
$$\Rightarrow \quad B = \frac{4.0 \times 10^{-6}\,Tm}{r} \div 20$$
$$= \frac{2.0 \times 10^{-7}\,Tm}{r}$$

Generally, for current I through a long straight wire:

$$B = \frac{(2.0 \times 10^{-7}\,\mathrm{Tm\,A^{-1}})I}{r}$$

For a plane circular coil of N turns with radius r a similar experimental procedure will reveal:

$$B = \frac{(6.3 \times 10^{-7}\,\mathrm{Tm\,A^{-1}})NI}{r}$$

at the centre of the coil.

Summary of experimental results

The results of our experimental observations for the solenoid, circular coil and long straight wire are given below. The equations have also been written in terms of μ_0 in order to simplify and unite them.

At the centre of a long solenoid of n turns per unit length,

$$B = (1.3 \times 10^{-6}\,\mathrm{Tm\,A^{-1}})nI$$

$$= \mu_0 nI \qquad \text{(i)}$$

At the centre of a flat, plane, circular coil of radius r with N turns:

$$B = \frac{(6.3 \times 10^{-7}\,\mathrm{Tm\,A^{-1}})NI}{r}$$

$$= \frac{\mu_0 NI}{2r} \qquad \text{(ii)}$$

At a distance r from a very long, straight wire:

$$B = \frac{(2.0 \times 10^{-7}\,\mathrm{Tm\,A^{-1}})I}{r}$$

$$= \frac{\mu_0 I}{2\pi r} \qquad \text{(iii)}$$

Note that each of these results for B, though apparently obtainable by experiment, is in reality related to μ_0 via the definition of the ampere.

If you inspect the above summary, and also that on page 248 for electric field equations, you will see an interesting consequence of the SI unit system. For both electric and magnetic fields:

☐ equations relating to situations with *spherical* symmetry contain 4π (e.g. equation (ii) on page 248);

☐ equations relating to situations with *cylindrical* symmetry contain 2π (e.g. equation (iii) above);

☐ uniform or non-symmetry situations do not contain π (e.g. equations (i) and (ii) above).

The definition of the ampere (amp)

The unit of electric current is considered to be one of the seven basic units of physics. Many early experiments on the forces of attraction and repulsion between two current-carrying conductors were carried out by Ampère. Each wire was acted on by the magnetic field produced by the other, as illustrated in figure 24.

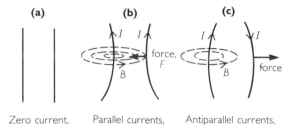

(a)	(b)	(c)
Zero current, zero force	Parallel currents, attractive forces	Antiparallel currents, repulsive forces

figure 24 Forces between current-carrying conductors (try the left hand rule to confirm the directions of the forces)

Currents in the same direction attract each other, while currents in opposite directions repel one another. To honour Ampère's work, the basic unit of electricity is called the ampere or amp, A.

The ampere is defined by the statement:
'One ampere is that constant current which, when passing through two infinitely long, parallel, straight wires of negligible circular cross-section, separated by a distance of one metre in vacuum, produces a force between the wires of 2×10^{-7} newtons/metre'. In practice, of course, it is not possible to measure the ampere using infinitely long wires and a current balance of finite dimensions is used instead.

Once the ampere has been defined in this way, it is possible to deduce theoretically a value for μ_0. Figure 25 shows two long parallel wires a distance r apart carrying currents of I_1 and I_2 respectively.

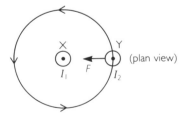

figure 25

Using equation (iii) from the summary of experimental results, the flux density at a distance r from wire X is given by:

$$B = \frac{\mu_0 I_1}{2\pi r} \qquad \text{(iii)}$$

We may now write an expression for the size of the force F on wire Y. We start with the equation:

$$F = BI_2\ell \qquad \text{(ii)}$$

ℓ is the length of wire Y, so it follows that the force F_m acting per unit length on wire Y is:

$$F_m = BI_2$$

However, using equation (iii) we see that:

$$F_m = \frac{\mu_0 I_1 I_2}{2\pi r}$$

If we use the definition of the amp we know that this force is $2 \times 10^{-7}\,\text{Nm}^{-1}$ if $I_1 = I_2 = 1\,\text{A}$ and r is a distance of 1 m. Under these circumstances:

$$F_m = 2 \times 10^{-7}\,\text{Nm}^{-1}$$

$$= \frac{\mu_0 \times 1\,\text{A} \times 1\,\text{A}}{2\pi \times 1\,\text{m}}$$

$$\Rightarrow \mu_0 = 4\pi \times 10^{-7}\,\text{NA}^{-2}$$
$$(= 1.3 \times 10^{-6}\,\text{NA}^{-2})$$

This is the value that we measured experimentally before.

Notice that we have used here a new unit for μ_0: NA^{-2}. Can you show that this is the same as the unit we used before, TmA^{-1}?

Ampère's law

Earlier in the chapter we compared stream lines in flowing water with magnetic field lines. This was quite a useful idea to help us understand the concept of magnetic flux density. It must be emphasised, however, that magnetic flux does not really flow; but it provides us with a good working model if we treat flux as if it did flow. We now take this model a little further.

Consider a very long solenoid of cross-sectional area A, length ℓ and having N turns: figure 26.

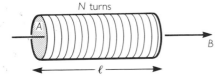

N turns

figure 26

If it carries a current I, the flux density B in its centre is given by:

$$B = \mu_0 \left(\frac{N}{\ell}\right) I$$

so the total magnetic flux 'flowing' through the solenoid is:

$$\Phi = BA = \frac{\mu_0 A N I}{\ell}$$

We may rearrange this formula in the form:

$$NI = \Phi \times \frac{\ell}{\mu_0 A}$$

It is most instructive if we compare this equation with the very familiar equation:

$$V = I \times R = I \times \frac{\rho \ell}{A} = I \times \frac{\ell}{\sigma A}$$

This equation describes the 'flow' of current I when a potential difference V is applied across a resistance R. The resistance R can be expressed in terms of length ℓ, cross-sectional area A, and resistivity ρ or conductivity σ. Electrical conductivity is the reciprocal of resistivity.

Comparison of the equations describing the 'flow' of current and the 'flow' of flux shows us that:

☐ V and NI are analogous;

☐ I and Φ are analogous;

☐ $\ell/\sigma A$ and $\ell/\mu_0 A$ are analogous.

We call NI *current turns*, or *magneto-motive-force* (m.m.f.). This emphasises that it is the current that is responsible for creating the magnetic flux. (Compare this idea with the electromotive force or e.m.f. of a battery.) The quantity $\ell/\mu_0 A$ is called *reluctance*; it is of course a similar idea to that of resistance. If the area of the solenoid is large, then the reluctance is small and more flux can flow; μ_0 could be called the magnetic conductivity. We can summarise this in the word equation:

$$\text{m.m.f.} = \text{flux} \times \text{reluctance}$$

This is a simple statement of Ampère's law.

Summary

1 There is a useful analogy between magnetic field lines and lines of water flow.

2 $\Phi = BA \cos\theta$ (see figure 5(b)).

3 Maxwell's right-hand grip rule. See figure 8(e).

4 B inside a long solenoid:

B constant everywhere except near ends;
$$\begin{matrix} B \propto I \\ B \propto n \end{matrix} \Rightarrow B = \mu_0 nI$$

5 With magnetic material in the solenoid:

$$B = \mu_r \mu_0 nI$$

μ_r is roughly constant for the material.

6 Fleming's left hand (motor) rule:

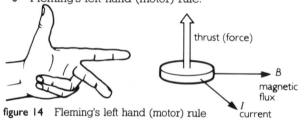

thrust (force)

B magnetic flux

I current

figure 14 Fleming's left hand (motor) rule

7 Force F on a wire carrying current I in a magnetic field B:

$$\begin{matrix} F \propto I \\ F \propto \sin\theta \\ F \propto \ell \\ F \propto B \end{matrix} \Rightarrow F = BI\ell \sin\theta$$

This defines B.

8 B near a long straight wire:

$$B = \frac{\mu_0 I}{2\pi r}$$

9 The definition of the ampere fixes a value for μ_0:

$$\mu_0 = 4\pi \times 10^{-7}\,\text{N}\,\text{A}^{-2}$$

10 One ampere is that constant current which, when passing through two infinitely long, parallel, straight wires of negligible circular cross-section, separated by a distance of one metre in vacuum, produces a force between the wires of 2×10^{-7} newtons/metre.

11 Ampère's law:

$$NI = \frac{\ell}{\mu_0 A} \cdot \Phi$$

Questions

1 A long solenoid has a circular cross-section of diameter 6.0 cm. It is 30 cm long and has a single layer of 750 turns of copper wire. For the case when it carries a current of 2.0 A calculate:
a the magnetic flux density at the centre of the solenoid;
b the magnetic flux through the solenoid.

2 How does the light intensity W fall off with distance r from the centre of a *long*, fluorescent, striplight?
   Explain how this can be used as an analogy to show how magnetic field strength B varies with distance from a long, straight, current-carrying wire.

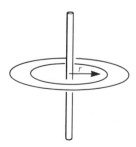

figure 27

3 A long, straight wire carries a current of 4.0 A. Calculate:
a the magnetic flux density at 15 cm from the wire;
b the force on a short wire (30 cm long) which is 15 cm from the first wire and parallel to it, if the second wire carries a current of 2.0 A.

4 Mains-operated electric lamps (240 V, 50 Hz in UK) are connected by lengths of twin-flex cable to the wall sockets. There will be a force between the live wire and the neutral return wire, due to the magnetic fields they produce. Which of the following best illustrates how this force varies with time? (Take positive force to be an attractive force.)

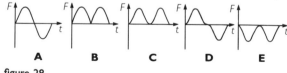

figure 28

5 What current must be passing through a flat, circular coil consisting of 30 turns, of average diameter 15 cm, if the magnetic flux density at the centre of the coil is 3.0×10^{-3} T?

6 Figure 29 shows a form of current balance. The copper wire frame is situated centrally between the two magnadur magnets, which form a horseshoe magnet. The larger faces of the magnadur magnets are the poles. The wire frame is connected from A to B by an insulating rod which rests on a sensitive electronic balance.

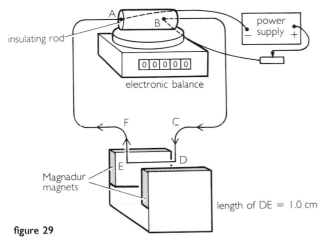

length of DE = 1.0 cm

figure 29

Current passes through the frame from a 4 V d.c. supply. A variable resistor is used to vary the current. An ammeter records the magnitude of the current. When a current passes through the frame from B to D to E to A, a downward magnetic force causes the reading on the electronic balance to increase.

A typical set of values is given in the following table:

Current/A	0	1.0	2.0	3.0	4.0
Extra force/10^{-5}N	0	21	39	60	79

a The pieces of wire labelled CD and EF are in the field of the horseshoe magnet. Explain why the forces acting on them produce no noticeable effect on the electronic balance.

b Draw a graph of the above data and use it to prove that the magnetic force on the length of wire labelled DE is proportional to the current through it.

c The length of DE used in obtaining this data was 1.0 cm. Use this information and the data given to determine the magnetic flux density B in the vicinity of the wire.

d A student who has read in a textbook that the magnetic force is proportional to the length of wire DE decides to test the truth of what he has read. He makes another frame from copper wire, but this time DE is 4.0 cm long. He obtains an extra force reading of 2.81×10^{-3} N when a current of 4.0 A passes through the wire. Discuss the test and explain his result.

7 Figure 30 illustrates a form of current balance that can be used to investigate the field inside a solenoid. The distance PX equals PC.

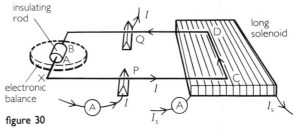

figure 30

The following results are for a frame which had CD equal to 20 cm. The current through the wire CD was kept constant at 5.0 A, whilst the current through the solenoid was varied from zero to 8.0 A. CD was positioned at the centre of the solenoid.

I_s/A	0	2.0	4.0	6.0	8.0
Extra force/10^{-4} N	0	2.51	4.99	7.48	9.98

a Draw a graph of the above data and explain how it proves that the magnetic flux density B inside the solenoid is proportional to the current through each of its turns.

b Explain why we can ignore the parts of PC and DQ that are within the solenoid when we calculate the magnetic flux density inside the solenoid. That is, explain why we can assume that all the extra force detected by the electronic balance was due to CD only.

c The extra force was 4.99×10^{-4} N when a current of 4.0 A passed through the solenoid. Use this information to determine the magnetic flux density B at the middle of the solenoid.

d What would the magnetic flux density be at the middle of the solenoid when the current I_s through the solenoid is 1.0 A?

This apparatus can be used to determine how the magnetic flux density varies along the axis of the solenoid. In order to investigate this variation, the solenoid should be moved small distances to the right when both currents are kept constant.

e A student finds that the extra force is 2.50×10^{-4} N when the solenoid has been moved so far to the right that the wire CD is exactly in line with the left hand edges of the solenoid. This value occurred for a current of 5.0 A through CD and 4.0 A through the solenoid. Calculate the magnetic flux density of the solenoid in this position.

f As the solenoid was long, explain why you would expect the result that the student obtained in part (e).

8 *Moving-coil loudspeaker*

The essential features of a moving-coil loudspeaker are illustrated in figure 31. A circular coil of fine copper wire is wound round a plastic or cardboard tube. The tube is glued to the narrow end of a flexible cone of stiff paper. The tube is free to move in the space between two pieces of iron. There is a *radial* magnetic field in the space.

Consider a particular speaker which has a magnetic flux density B of 0.50 T in the space, and a coil of 100 turns. Radius of coil is 1 cm.

a What is the magnetic force on the coil when a current of 20 mA passes through it?

b If the mass of the cone and coil together is 15 g, what will be the acceleration of the coil?

c Describe the motion of the cone when the current through the coil varies sinusoidally.

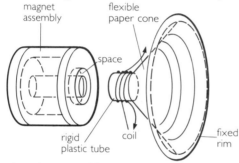

(a) Loudspeaker assembly

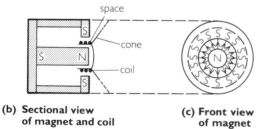

(b) Sectional view of magnet and coil   **(c) Front view of magnet**

figure 31   The moving-coil loudspeaker

The natural frequency at which something oscillates depends on a restoring force and an inertial factor. This can generally be expressed by

$$\text{natural frequency } f_0 \propto \sqrt{\frac{\text{restoring force}}{\text{inertia}}}$$

(see page 83)

d What will be the inertial factor in the case of the above loudspeaker?

e What will be the restoring factor in the case of the above loudspeaker?

f Clearly there must be some frequency at which this loudspeaker will resonate, which will distort its performance. Discuss in some detail how this problem is overcome.

g The fine copper wire used for making the coil is covered in a thin film of insulating varnish. Discuss and explain what happens to a loudspeaker's performance if it is driven too hard.

9 a Consider two vertical wires just a few centimetres apart. Draw two diagrams to show the magnetic field patterns in a horizontal plane when (i) the currents both pass down the wires (parallel currents), (ii) one current passes up and the other down (antiparallel currents).
Use arrows on each diagram to indicate clearly the direction of the magnetic fields.

b What is the magnetic flux density at a distance of 5.0 cm from a very long, straight wire carrying a current of 10 A? Assume the wire is in a vacuum.

c Calculate the force per unit length on a similar wire carrying a current of 10 A, when it is placed 5.0 cm from the first wire.

d Explain how this idea of a force between two conductors is used to define the ampere. How is this definition realised as a practical measurement?

10 Consider a rectangular coil of N turns suspended with its axis horizontal in a uniform horizontal magnetic field B. The axis of the coil is perpendicular to the magnetic field (figure 32).

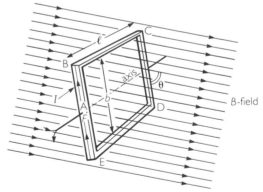

figure 32   Rectangular current-carrying coil in a uniform magnetic field

a Write an expression for the magnetic force on each of the following sections of the coil, and in each case indicate the direction of the force: (i) AB, (ii) BC, (iii) CD, (iv) DE, (v) EF.

b Calculate the net magnetic torque on this coil due to these forces.

c For what value of θ will this torque be greatest?

d Explain why a circular coil which has an area equal to that of the above rectangular coil will experience an identical magnetic torque when placed at the same angle θ.

11   *The moving-coil galvanometer*
The essential features of a moving-coil
galvanometer are illustrated in figure 33. The coil
is rectangular and consists of many turns of fine
copper wire. It is suspended in the radial
magnetic field that exists in the cylindrical gap
between the iron cylinder and the curved pole
pieces of the horseshoe magnet.

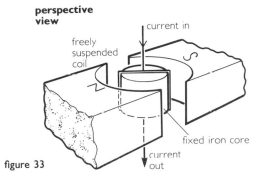

figure 33

a Figure 34 is a plan view of the galvanometer.
Copy the part representing the coil and mark
on it the direction of the forces on each side
wire. The current flows down on the right-hand
side and up on the left.

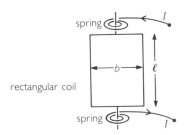

figure 34

b The coil has breadth b and length ℓ. If there
are N turns of fine copper wire on the coil,
what is the magnetic torque that it experiences
when a current I passes through it? Take the
magnetic field in the gap to be B.

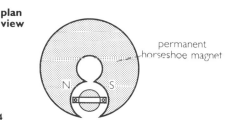

figure 35

The coil will turn under the influence of this
torque. The idea behind a moving-coil meter is
that the angle through which it rotates should be
proportional to the current in its coil. A large,
loosely wound, weak spring is normally used to
resist the rotation of the coil.

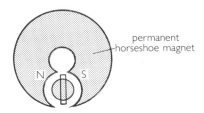

figure 36

c Explain why the coil should never be allowed
to reach the position shown in figure 36.
d Look back at figure 34. What approximately is
the largest angle through which the coil should
be allowed to rotate?
e Explain why the pole pieces are shaped the
way they are, and why a solid cylinder of iron
is used as a core.
f What causes the coil to return to its zero
position when no current is passing?
g How does the electric current get into and out
of the coil?

Your answer to part (b) should indicate that the
magnetic torque on the coil is proportional to the
current in the coil. The springs should have a
torsion constant c which is constant over the
angle through which the coil rotates. Then the
mechanical torque of the spring will be equal to
$c\theta$, where θ is the angle through which the coil
has turned.

h Use your answer to part (b) to show that a
steady deflection of the coil through an angle θ
means that the current I is given by

$$I = \left(\frac{c}{BAN}\right) \cdot \theta$$

where A is the area of the coil ($A = b.\ell$).
i Obtain an expression for the *current sensitivity*,
S_I, which is defined as $S_I = \theta/I$, and discuss the
conditions for high current sensitivity. (That is,
discuss how varying the values of B, c, A and
N affects the value of S_I.)
j The *voltage sensitivity*, S_V, is defined by
$S_V = \theta/V$. For a coil of resistance R, $V = IR$.
Obtain an expression for the voltage sensitivity.
k The expression you obtained in part (j) should
show that $S_V \propto N$, and $S_V \propto 1/R$. These two
proportionality statements seem to be in
conflict. The first statement, $S_V \propto N$, indicates
that a large voltage sensitivity will be achieved
if as many turns as possible are fitted onto the
coil. This would mean using a very long length
of very fine wire. The second statement,
$S_V \propto 1/R$, indicates that a large voltage
sensitivity will be achieved if the resistance of
the coil is as small as possible. This would
mean using a very short, thick piece of wire
for the coil. Discuss the problem and decide
on the best compromise.

CHAPTER 25

Magnetic force on charged particles

Introduction

We start with the equation $F = BI\ell \sin\theta$ which we considered in chapter 24. We then combine this idea with the model of a wire as a tube containing many small, 'freely' moving electrons. This leads to the equation for the magnetic force on one moving electron. The idea is then extended to electron guns, particle accelerators and the Hall effect.

GCSE knowledge

☐ A plotting compass points away from the north pole of a bar magnet.

☐ The force on a current-carrying conductor placed in a magnetic field is perpendicular to the plane containing the current and the field.

Electrons in magnetic fields

We now use the idea of electron flow through a wire to look at the force that an individual particle experiences when it moves through a magnetic field. Figure 1 shows electrons moving from left to right through a magnetic field whose direction is into the plane of the paper. We may use the left hand rule to determine the direction of the force on the electrons, bearing in mind that the conventional current flows from right to left.

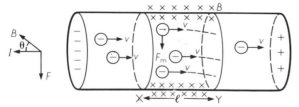

The magnetic field acts over the region between X and Y.
In this region each electron experiences a sideways thrust F_m.

figure 1   Electrons moving in a magnetic field

In chapter 9 on electrical circuits, we met the equation $I = nAve$, where

I = current,
n = number of 'free' electrons per unit volume,
A = cross-sectional area of the wire,
v = drift velocity of 'free' electrons,
e = charge on each electron.

The electrons 'drift' along the wire at speed v under the influence of the external cell. We imagine that the sideways force experienced by the wire in a magnetic field is due to the sum of the individual forces experienced by the electrons.

i.e.   force on wire = number of 'free' electrons $\times$ force on one electron

$$\Rightarrow \qquad BI\ell \sin\theta = N \times F_m$$

where θ is the angle between the wire and the B-field.

But $\qquad\qquad I = nAve$ and $N = n \times A\ell$

$$\Rightarrow \qquad B(nAve)\ell \sin\theta = nA\ell \times F_m$$

Therefore,

force on one electron $F_m = Bev \sin\theta$ \qquad (i)

It has been assumed in figure 1 that the force F_m acts perpendicular to the electron's velocity v. If this is always the case then the *speed* of an electron will not change, but its direction will. Equation (i) shows that the force F_m is constant if velocity v is constant. This is exactly the condition for centripetal motion. Therefore, we can test Fleming's left hand (motor) rule for individual electrons by using a 'fine beam tube', because a circular trajectory will indicate that the force is always perpendicular to B and v.

The fine beam tube

The fine beam tube is a piece of apparatus used as a teaching aid, so that students can see the trajectories followed by electrons when they move through magnetic fields. It is a transparent glass tube containing an electron gun and is filled with argon gas at low pressure. The purpose of the argon gas is to make the trajectory of the electrons visible. The construction of such a tube is illustrated in figure 2.

When the magnetic field is not switched on, the electrons hit the glass tube at P. Every time an electron collides with an argon atom a photon of light is emitted, hence the trajectory of the electrons shows up as a coloured line.

The magnetic field is produced by a pair of coils designed to produce a uniform field, and when this is switched on the electrons follow a trajectory similar to trace Q in figure 2(c), i.e. a circular trajectory of radius r.

In figure 2 the electron gun is shown firing its electrons perpendicular to the magnetic field. The

fine beam tube can be rotated about the axis shown in figure 2(b). When this is done the illuminated trajectory changes from a flat circle into a helix. This is illustrated in figure 3.

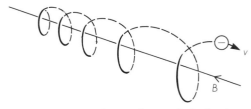

figure 3 Twisting the tube produces a helical trajectory

If the tube is rotated by 90° the electron beam then lies parallel to the magnetic field. Under these circumstances the electron beam is not deflected at all because there is only a force on a moving particle in a magnetic field if it is cutting across lines of flux. The spiral may thus be explained: a component of the electron's motion lies parallel to the field and is undeflected, another component lies perpendicular to the field. A spiral is a combination of travelling in a circle and in a straight line.

Figure 4 summarises the motor rule for charged particles.

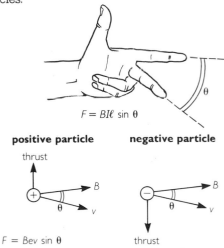

figure 4 Left hand (motor) rule for charged particles

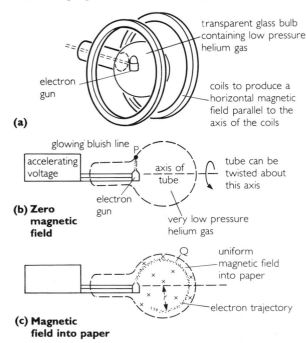

(a)

transparent glass bulb containing low pressure helium gas

electron gun

coils to produce a horizontal magnetic field parallel to the axis of the coils

(b) Zero magnetic field

glowing bluish line

accelerating voltage

axis of tube

tube can be twisted about this axis

electron gun

very low pressure helium gas

(c) Magnetic field into paper

uniform magnetic field into paper

electron trajectory

figure 2 Fine beam tube

Example Figure 5 illustrates the equatorial circular trajectory followed by a proton trapped by the Earth's magnetic field, 3600 km above the Earth, i.e. in an orbit of radius 10 000 km. Calculate the speed of the proton, and say whether it is rotating east to west or west to east.

Would it be possible for protons to be trapped in such orbits at only 600 km above the Earth's surface?

mass of proton $\quad m_p = 1.7 \times 10^{-27}\,\text{kg}$
charge on proton $\quad e = +1.6 \times 10^{-19}\,\text{C}$
radius of Earth $\quad R = 6.4 \times 10^6\,\text{m}$
magnetic flux density at orbit $\quad B = 1.0 \times 10^{-8}\,\text{T}$

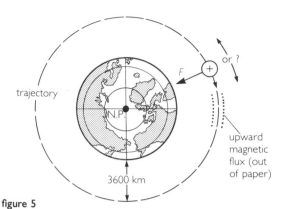

figure 5

The proton is undergoing circular motion, and the force due to the magnetic field, Bev, provides the necessary centripetal force.

i.e.
$$Bev = m\left(\frac{v^2}{r}\right)$$

$$v = \frac{Ber}{m}$$

$$= \frac{(1.0 \times 10^{-8}\,\text{T}) \times (1.6 \times 10^{-19}\,\text{C}) \times (10 \times 10^6\,\text{m})}{(1.7 \times 10^{-27}\,\text{kg})}$$

$$= 9.4 \times 10^6\,\text{ms}^{-1}$$

Fleming's left hand (motor) rule indicates that the proton must travel clockwise in figure 5, i.e. east to west.

At 600 km above the Earth the magnetic flux density would be much greater. For example, in London at sea-level it is about 6×10^{-5} T.

$$\Rightarrow \qquad v = \frac{Ber}{m}$$

$$\approx \frac{(6 \times 10^{-5}\,\text{T}) \times (1.6 \times 10^{-19}\,\text{C}) \times (7 \times 10^6\,\text{m})}{(1.7 \times 10^{-27}\,\text{kg})}$$

$$= 4 \times 10^{10}\,\text{ms}^{-1}$$

This speed is greater than the speed of light. Therefore it is not possible.

The Hall effect

Figure 6 illustrates a thin sheet of conductor lying in a magnetic field. An electric current passes through the sheet from X to Y.

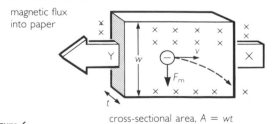

cross-sectional area, $A = wt$

figure 6

The 'free' electrons drifting with velocity v will experience a force F_m due to the magnetic field. The effect of this force will be to deflect the drifting electrons downwards as shown in figure 6. This will result in a build up of negative charge on the lower edge of the conductor as shown in figure 7. A voltmeter connected between the top and lower edges will register a p.d., V_H, known as the Hall potential. The negative charge on this lower edge will produce an electrostratic field which will cause an upward force, F_e, on the drifting electrons. The build up of negative charge will cease when the upward electrostatic force F_e is equal and opposite to the downward magnetic force F_m.

i.e.
$$F_e = F_m$$

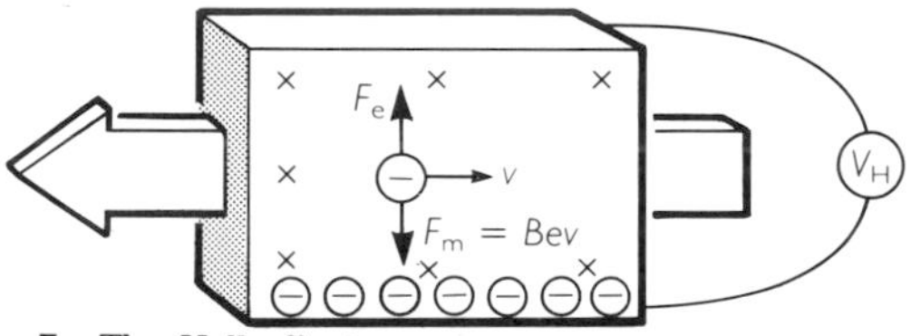

figure 7   The Hall effect

The magnetic force $F_m = Bev$, and the force on the electron which is in a uniform electric field E is:

$$F_e = eE = \frac{eV_H}{w}$$

$$\Rightarrow \qquad e\frac{V_H}{w} = Bev$$

Hall potential $V_H = Bvw$ \qquad (ii)

This is not a particularly useful equation, because the drift velocity v is not easily known. The current I, however, is easily measured; and current and drift velocity are linked by the transport equation:

$$I = nAve$$

$$\Rightarrow \qquad v = \frac{I}{nAe}$$

$$\Rightarrow \qquad V_H = B\left(\frac{I}{nAe}\right)w$$

where $A = t.w$ from figure 6.

$$V_H = \frac{BIw}{n(tw)e}$$

$$= \left(\frac{1}{ne}\right)\left(\frac{BI}{t}\right)$$

i.e. \qquad Hall potential $V_H = \left(\frac{1}{ne}\right) \times \left(\frac{BI}{t}\right)$ \qquad (iii)

The factor $R_H = (1/ne)$ is a property of the material from which the sheet of conductor is made. The other terms in the equation above are controllable variables. The constant R_H is called the *Hall coefficient* of the conducting material.

The magnitude of Hall potentials

Equation (iii) gives $V_H \propto 1/n$.

Therefore, if the number of 'free' electrons per unit volume, n, is large, the Hall potential will be small. This is the case in metals.

In semiconductors n is very much smaller than it is in metals. Therefore the Hall potential will be larger in semiconductors. Semiconductor Hall probes are used for comparing the strengths of magnetic fields.

Example   A thin slab of semiconducting material is to be used as a Hall probe. Its thickness is 1.0 mm and it is in a B-field of 1.0×10^{-2} T; the current through it is 0.10 A. We are required to calculate the Hall p.d. across it, given that there are 1.0×10^{22} free electrons per m³ in the material.

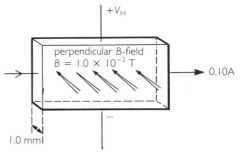

figure 8   A thin slab of semiconducting material

$$V_H = \frac{BI}{net}$$

$$= \frac{(1.0 \times 10^{-2}\,\text{T}) \times (0.10\,\text{A})}{(1.0 \times 10^{22}\,\text{m}^{-3}) \times (1.6 \times 10^{-19}\,\text{C}) \times (1.0 \times 10^{-3}\,\text{m})}$$

$$\Rightarrow \qquad V_H = 6.3 \times 10^{-4}\,\text{V}$$

This is a small but measurable p.d.

Summary

1   Force F on a particle with charge e and velocity v in a magnetic field B is given by:

$$F = Bev \sin \theta$$

figure 9

2   Charged particles moving perpendicular to a uniform field move in a circle:

$$Bev = \frac{mv^2}{r}$$

$$r = \frac{mv}{Be}$$

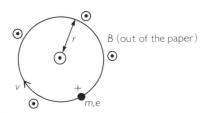

figure 10

3   Hall voltage $V_H = Bvw$

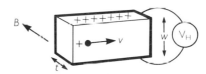

figure 11

Also $V_H = \left(\dfrac{1}{ne}\right) \times \left(\dfrac{BI}{t}\right)$.

4   Hall coefficient,

$$R_H = \left(\frac{1}{ne}\right).$$

Questions

1 Consider a cathode ray oscilloscope which has an electron gun operated at 3000 V. Electrons have a mass of 9.1×10^{-31} kg and a charge $e = -1.6 \times 10^{-19}$ C.
   a Calculate the speed of the electrons emitted from the electron gun.
   b Calculate the maximum magnetic force on the emitted electrons due to the Earth's magnetic field. [$B = 6.0 \times 10^{-5}$ T in the south of England.]
   c Explain why we say that this magnetic force on the electrons is a centripetal force which does not alter the speed of the electrons.
   d Calculate the radius of curvature of the trajectory followed by the electrons as they pass through the Earth's magnetic field, assuming they are moving at right angles to it.
   e Will the magnetic force due to the Earth's field produce a noticeable deflection on a typical school CRO operating at 3000 V? Justify your answer.
   f What will happen to the curvature of the trajectory if the accelerating voltage in the electron gun is increased?

2 From earlier work (see page 110) we have this equation for the current I through a conductor:

$$I = nAvq$$

where n = the number of 'free' charge carriers per unit volume,
   A = the cross-sectional area of the conductor,
   v = the drift velocity of the 'free' charge carriers,
   q = charge on each carrier.

The magnetic force on a piece of wire of length ℓ carrying an electric current I is given by: $F = BI\ell \sin\theta$, where θ is the angle between the field and the wire. Consider the electric current I passing through the length of copper wire illustrated in figure 12 to be a flow of drifting electrons.

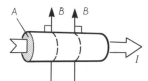

figure 12

   a Show that the force F_e on an individual drifting 'free' electron is

$$F_e = Bqv.$$

The following questions are about electrons moving through a vacuum. Assume that the same equation $F_e = Bqv$ applies in this case.
   b Calculate the force on a free electron travelling at 2.0×10^6 ms^{-1} through a magnetic field of 0.50 T, when the electron's path is at right angles to the magnetic field lines.
   c Explain why this magnetic force does not increase the speed of the electron.
   d If the magnetic field is uniform and extensive then the electron follows a circular trajectory. Explain why this is so, and determine the radius of this circular path.
   e What would the value of the magnetic force on the electron be if its velocity made an angle of 30° with the magnetic field lines?
   f Describe the subsequent trajectory of the electron through an extensive uniform magnetic field. Give as much detail as possible.

You may find the following data useful: the charge on an electron, $e = -1.6 \times 10^{-19}$ C; the mass of an electron, $m_e = 9.1 \times 10^{-31}$ kg.

3 A synchrotron is a large machine used to accelerate protons (and other small charged particles). In its simplest form the protons travel in a circle and are given a small 'push' once every rotation. Strong magnets are used to deflect the high speed protons so that they travel in circular paths. See figure 13.

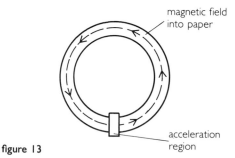

figure 13

   Consider a proton travelling round a synchrotron of diameter 1.0 km, at 3.0×10^8 ms^{-1}.
   a Calculate the magnetic force experienced by the protons, each of charge $e = +1.6 \times 10^{-19}$ C, when the magnetic flux density is 5.0×10^{-1} T.
   b The magnetic force on the protons is a centripetal force. Use your answer to part (a) to determine the mass of each proton as it circles this synchrotron.
   c If the above synchrotron were used to accelerate electrons, what alterations would have to be made to it?

4 The fine beam tube can be used in a school laboratory to determine the charge-to-mass ratio (e/m) of electrons. Alternatively, if known values of the charge ($e = -1.6 \times 10^{-19}$ C) and the mass ($m_e = 9.1 \times 10^{-31}$ kg) are used, it can be used to determine the strength of a magnetic field.

Read the description of the fine beam tube on page 265.

a When the experiment is carried out the trajectories followed by the electrons are visible as coloured lines. Explain how these visible traces are produced.

b The kinetic energy E_k of the electrons leaving the gun results from the accelerating voltage.
(i) Calculate E_k when the accelerating voltage is +400 V.
(ii) Calculate the speed of these electrons.

c Use your answer to part (b) to determine the magnetic flux density B which causes these electrons to travel in a circle of diameter 10 cm.

5 Figure 14 illustrates the workings of a mass spectrometer. The whole of the apparatus beyond the accelerating tube is subject to a strong magnetic field out of the paper. Consider the situation when the apparatus is used to study positively charged magnesium ions.

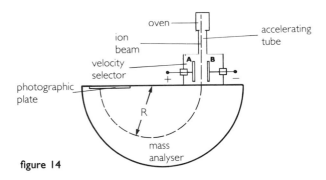

figure 14

a The ion beam arriving at the velocity selector will contain ions which have a range of speeds. Explain how the crossed magnetic and electric fields in the space between plates A and B act as a velocity selector.
Draw a diagram to show what happens to slower and faster ions.

b Calculate the velocity selected if the magnesium ions are singly charged, with a charge of 1.6×10^{-19} C, when the electric field intensity is 2.0×10^4 V m^{-1}, and the magnetic flux density is 0.30 T.

c Explain why the magnesium ions travel along circular trajectories when they are inside the mass analyser.

d Naturally-occurring magnesium has three isotopes, containing 24, 25 and 26 nucleons (protons and neutrons) respectively. What is the separation of adjacent isotope lines on the photographic plate?
(The mass of a nucleon may be taken as 1.67×10^{-27} kg.)

6 This question is about a magnetic lens which is used to focus electrons. The lens consists of an aluminium tube on which ten layers of thick copper wire are wound.
The first layer is hollow copper tubing, along which cold water is pumped. The dimensions of the lens are shown in figure 15.

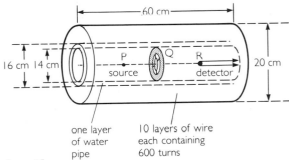

figure 15

The space inside the aluminium tube is evacuated. In the diagram, P represents a radioactive source which is emitting electrons with a kinetic energy of 2.0×10^{-15} J, Q represents a baffle in which is cut a narrow annular aperture, and R is a small detector onto which the electrons are focused. The baffle is mid-way between P and R, which are 30 cm apart.

a Estimate the magnetic flux density at the centre of the lens, when a current of 1 A is passing in the coils. State what assumptions you have made.

b What would occur if cold water were not pumped through the copper tubing?

c Consider an electron which leaves the source P at right-angles to the axis of the lens as illustrated in figure 16. Show that the electron travels in a circle, the plane of which is perpendicular to the axis of the solenoid.

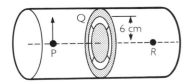

figure 16

(i) Calculate the radius of the circle.
(ii) Obtain an expression for the time taken by the electron to complete the circle. Hence, show that the time-period is independent of the velocity v.

(iii) Obtain a numerical value for the time taken by the electron to complete the circle, if $B = 6.4 \times 10^{-3}\,\text{T}$. (Take the mass of an electron as $9.1 \times 10^{-31}\,\text{kg}$.)

d Draw a diagram, where the plane of the paper represents the plane of the circle in which the above electron rotates. Start this diagram with the point source P, and draw in the circular trajectory followed by the electron which left P in a vertical direction, as seen by an observer at Q.

On the same diagram draw in the trajectories of electrons which left the source (i) vertically downwards, (ii) horizontally to the left, and (iii) horizontally to the right, as seen from Q.

e Describe the trajectory of an electron which leaves the source P at an angle α to the axis of the solenoid. Choose an electron which, at the moment it leaves source P, is travelling in a vertical plane containing the axis of the solenoid, and which leaves at the correct angle α so that it passes through the annular aperture in the baffle Q, after it has rotated through half a rotation as seen from R (figure 17).

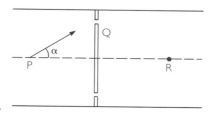

figure 17

f Copy figure 17 and draw on it the trajectory of the electron discussed in part (e). Explain why it is possible to say that the solenoid focuses the electrons on R, i.e. that R is an image of P.

7 Figure 18 represents a Hall probe made from n-type germanium. X and Y represent leads attached to the exact mid-points of the two edges. When the probe passes a current of $100\,\mu\text{A}$, and is placed in a magnetic flux density B of $2.0 \times 10^{-3}\,\text{T}$, it produces a maximum Hall p.d. of $200\,\text{mV}$.

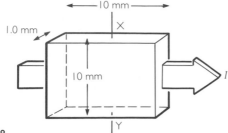

figure 18

a Explain, with the aid of a diagram, how the probe will be orientated with respect to the magnetic field when the Hall p.d. is a maximum.

b Indicate on the diagram used to answer part (a) which side of the piece of germanium is positive.

c Determine the Hall coefficient, R_H, for the material of this probe.

d Given that the charge on an electron is $1.6 \times 10^{-19}\,\text{C}$, determine n, the number of free charge carriers per unit volume for the material of this probe. (Hint: Use your answer to part (c).)

8 Figure 19 represents a strip of copper. The magnetic flux density B is $5.00\,\text{T}$ and the current through the copper is $100\,\text{A}$. With the arrangement shown in figure 19(a), the Hall potential is found to be $2.28 \times 10^{-6}\,\text{V}$.

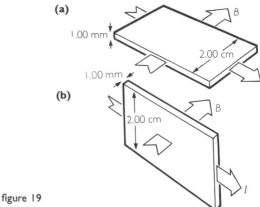

figure 19

a What will be the Hall potential in the arrangement illustrated in figure 19(b)?

b Use the above data to deduce how many 'free' electrons there are per unit volume of copper.

c When the Hall effect is investigated in semiconductors it is discovered that the p.d.s generated are much larger than those found across metals. Why is this so?

d Suppose that figure 19(b) represented an n-type semiconductor. What are the main carriers of current in an n-type semiconductor? In which direction is the magnetic force on these charge carriers? In which direction do these charge carriers move due to this magnetic force? Does a positive or a negative charge build up on the upper edge of figure 19(b)?

e What are considered to be the main carriers of current when talking about a p-type semiconductor? In which direction is the magnetic force on these charge carriers? In which direction do these charge carriers move due to this magnetic force? Does a positive or a negative charge build up on the upper edge of figure 19(b)? In actual fact, only electrons move within the semiconductor; have they moved to the top edge or towards the bottom edge? What does this tell you about the mass of valence bond electrons?

CHAPTER 26

Electromagnetic induction

Introduction

Whenever a conductor cuts across a magnetic flux, or a magnetic flux moves so that it cuts across a conductor, an electromotive force (e.m.f.) is induced in the conductor. We start by developing this idea from an experimental point of view. Later we consider the model of 'free' electrons moving within the wire, and relate the induced e.m.fs to the forces on the individual electrons.

GCSE knowledge

We expect that from your GCSE work you will already know that:

☐ a plotting compass points away from the north pole of a magnet;

☐ the magnetic force on a current-carrying conductor placed in a magnetic field is perpendicular to the plane containing the current and the magnetic field;

☐ whenever a conductor cuts across a magnetic flux, or a magnetic flux moves so that it cuts across a conductor, an e.m.f. is induced in the conductor (Faraday's first law);

☐ the magnitude of the induced e.m.f. is proportional to the rate at which the magnetic flux is cut (Faraday's second law);

☐ if the induced e.m.f. can make a current flow, then there will be a force on the induced current which will oppose the cutting of the flux, or the changing of the flux (Lenz's law).

The experimental laws of electromagnetic induction

The apparatus illustrated in figure 1 is suitable for deducing the laws of electromagnetic induction. If the wire AB is moved in the direction of the y-axis, then a current flows in the sensitive galvanometer. When the wire AB is moved in the direction of the z-axis or the x-axis then the galvanometer registers no current. Thus,

'whenever the wire AB moves so that it cuts across the magnetic flux an induced e.m.f. is produced in the wire'.

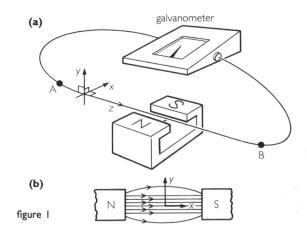

(a)

(b)

figure 1

This last sentence is known as *Faraday's first law of electromagnetic induction*.

Faraday's second law of electromagnetic induction states that

'the magnitude of the induced e.m.f. in the wire AB is proportional to the rate at which AB cuts across the magnetic flux'.

As a simple experimental justification for this law, the wire AB in figure 1 can be moved more and more rapidly across the magnetic flux; this results in larger and larger induced currents. A more quantitative justification of this law can be carried out using the apparatus depicted in figure 2. The magnitude of the induced e.m.f. is proportional to the amplitude of the trace shown on the CRO. Doubling the speed of rotation of the wire coil causes the amplitude of the trace to double, which confirms Faraday's second law.

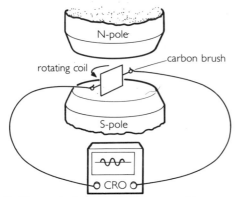

figure 2   Demonstration of Faraday's second law

The final law of electromagnetic induction is known as *Lenz's law*, which relates the direction of the induced e.m.f. to the relative motion of the conductor (wire AB) and the magnetic flux. The apparatus in figure 1 can be used to discover this relationship. If the wire AB is lifted rapidly upwards across the magnetic flux then it is found that the induced e.m.f. causes a current to pass from B to A. Figure 3 illustrates this situation.

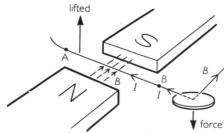

figure 3   Applying Fleming's left hand rule

If Fleming's left hand (motor) rule is applied, it can be seen that the induced current results in a downward force opposing the upward lifting of the wire.

Thus Lenz's Law states that

'the direction of the induced e.m.f. is such that any current caused by it results in opposition to the motion causing the induced e.m.f.'.

Calculation of the magnitude of the induced e.m.f.

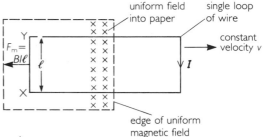

figure 4

Figure 4 shows a rectangular coil being pulled at a constant velocity v through a magnetic flux of density B. Only in the length XY will an e.m.f. be induced. Suppose that the velocity v induces an e.m.f. E which causes a current I to pass. Then, by Fleming's left hand (motor) rule there will be a magnetic force F_m opposing the motion, where

$$F_m = BI\ell$$

The external agent pulling the wire loop at constant velocity must be doing work against this force.
The rate of working (power) of the agent $= F_m . v$
The rate of working (power) of the e.m.f. $= E . I$

Thus: $\quad\quad E . I = F_m . v = (BI\ell)v$

$\Rightarrow\quad$ The induced e.m.f. $E = B\ell v$ (i)

In time Δt the length XY sweeps out an area $\ell \Delta x$, where $\Delta x = v\Delta t$ (figure 5).

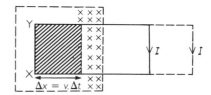

figure 5

Therefore the flux, $\Delta\Phi$, cut by the conductor in time Δt is given by

$$\Delta\Phi = \text{area} \times B$$

$$= (\ell . v . \Delta t) . . B$$

$$\Rightarrow\quad \frac{\Delta\Phi}{\Delta t} = B\ell v$$

But we know from equation (i) that $B\ell v = E$.

$$\Rightarrow \quad \text{induced e.m.f. } E = \frac{d\Phi}{dt} \quad \text{(ii)}$$

or in words: induced e.m.f. = rate of cutting of magnetic flux.

Example A spaceship passes over the magnetic north pole where the B-field has a strength of 6.0×10^{-5} T; the field is vertical. The spaceship travels with a velocity of 8000 ms^{-1}. Calculate the magnitude of the induced e.m.f. between opposite sides of the spaceship, which are 4.0 m apart. Explain why a cosmonaut could not measure this e.m.f.

figure 6   A spaceship passing over the magnetic north pole

$$E = \frac{d\Phi}{dt} = B\ell v$$

$$E = (6.0 \times 10^{-5}\,\text{T}) \times (4.0\,\text{m}) \times (8000\,\text{ms}^{-1})$$

$$= 1.9\,\text{V}$$

Any voltmeter would have wire leads which would also cut across the magnetic flux. The induced e.m.fs oppose one another, so there is no current.

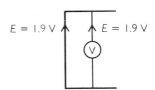

figure 7   The e.m.f. in the voltmeter opposes the e.m.f. across the spaceship

The wire frame illustrated in figure 5 consisted of only *one* loop of wire, but suppose the frame were like the one shown in figure 8.

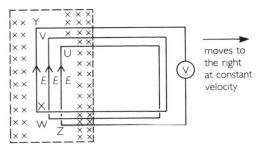

figure 8   A coil with three loops

Then the induced e.m.f. in XY would be E, and the induced e.m.fs in WV and ZU would also be E in each case. All the e.m.fs would be in series therefore the total induced e.m.f. would be $3E$. More generally, for N turns the induced e.m.f. would be N times greater than for 1 turn:

$$\text{induced e.m.f. } E = N\frac{d\Phi}{dt} \quad \text{(iii)}$$

This is sometimes known as *Neumann's equation*.

Sometimes the equation is written as $E = -N(d\Phi/dt)$ to remind you that the induced e.m.f. causes opposition (Lenz's law).

Example 1 Consider a rotating disc placed in a uniform magnetic field as illustrated in figure 9. If the magnetic flux density B is 2.0 T, and the disc, of radius 20 cm, is rotating at 360 revmin^{-1}, what will be the voltmeter reading?

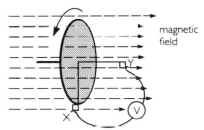

figure 9   A rotating disc in a magnetic field

$$\text{Induced e.m.f. } E = N\frac{d\Phi}{dt}$$

$$= \frac{d\Phi}{dt}$$

as $\qquad N = 1$

The radius R cuts across the magnetic flux, and sweeps out an area ΔA in time Δt (figure 10).

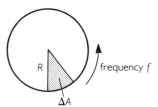

figure 10

$$\text{Rate of cutting of flux} = \frac{\Delta\Phi}{\Delta t}$$

$$= B\frac{\Delta A}{\Delta t}$$

$$\Rightarrow \quad \text{Induced e.m.f. } E = B\frac{dA}{dt}$$

But $\qquad \dfrac{\mathrm{d}A}{\mathrm{d}t}$ = rate of sweeping of area

$$= (\pi R^2) \times f$$

where f is the frequency of rotation.

$$f = 6.0 \text{ revs}^{-1}$$

$\Rightarrow \quad E = B\pi R^2 f$

$$= (2.0 \text{ T}) \times \pi \times (0.20 \text{ m})^2 \times (6.0 \text{ s}^{-1})$$

$$= 1.5 \text{ V}$$

Example 2   A bar magnet falls through a large coil as shown in Figure 11(a). Sketch graphs to show (i) the galvanometer deflection, (ii) the velocity of the magnet, (iii) the acceleration of the magnet as it falls through the coil.

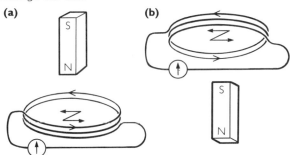

figure 11   A bar magnet falling through a coil of wire

As the magnet enters the coil an e.m.f. will be induced. The current will be in such a direction as to oppose the motion of the magnet. So the top side of the coil will behave like a north pole. When the magnet is in the middle of the coil, the effects of the two poles are to produce equal e.m.f.s in opposite directions, and there is no current. As the magnet leaves, current again passes, and sets up a field which opposes the change of flux. Thus the underside of the coil behaves like a north pole to attract the south pole of the magnet. So the top of the coil as we see it in figure 11(b) will look like a south pole. The galvanometer reading will vary as in figure 12(a).

When currents pass in the coil there is a small upwards force on the magnet, so its acceleration will be slightly less than the normal gravitational acceleration (figure 12(b)).

The magnet's speed increases as it falls, so the induced e.m.f. is slightly bigger as it leaves the coil. The rate of increase of speed is less when the retarding forces (due to the current) act on the magnet (figure 12(c)).

Generators

Consider a conducting coil of N turns rotating with angular velocity ω in a uniform magnetic field of magnetic flux density B (figure 13).

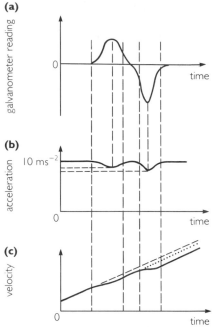

figure 12   Variation in (a) galvanometer reading, (b) acceleration and (c) velocity as the bar magnet falls through the coil

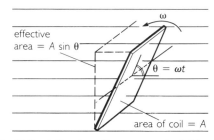

figure 13   Coil rotating in uniform magnetic field

The magnetic flux Φ through each turn is given by

$$\Phi = BA \sin\theta = BA \sin\omega t$$

$\Rightarrow$ induced e.m.f. in each turn $= \dfrac{\mathrm{d}\Phi}{\mathrm{d}t} = BA\omega \cos\omega t$

$\Rightarrow$ for N turns:

induced e.m.f. $E = N\dfrac{\mathrm{d}\Phi}{\mathrm{d}t} = NAB\omega \cos\omega t$

Figure 14 illustrates how the magnetic flux through the coil of the dynamo varies as the coil rotates. When the ends of the coil are cutting across the magnetic flux most rapidly then the induced e.m.f. is a maximum.

When the coil rotates from the position shown in figure 14(a) to that in figure 14(b), the magnetic flux through the coil decreases. Therefore, according to Lenz's law the current must pass around the coil in such a direction as to attempt to maintain the magnetic flux.

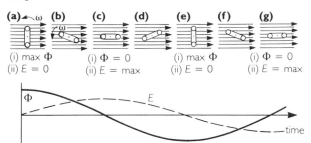

(a) ω (b) (c) (d) (e) (f) (g)

(i) max Φ (i) $\Phi = 0$ (i) max Φ (i) $\Phi = 0$
(ii) $E = 0$ (ii) E = max   (ii) $E = 0$ (ii) E = max

figure 14   Variation of magnetic flux through the coil of a dynamo

The alternating e.m.f. generated in the rotating coil can be led to an external circuit in two different ways:

☐   by having two slip rings as shown in figure 15;

☐   by having a single split-ring commutator as shown in figure 16.

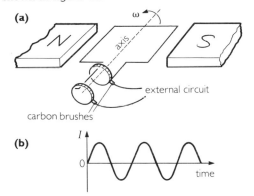

figure 15   An a.c. generator

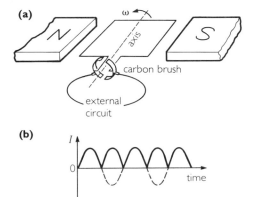

figure 16   A d.c. generator

If the arrangement shown in figure 15 is used, then the sinusoidally varying e.m.f. generated in the coil is transferred directly to the external circuit (figure 15(b)). Thus figure 15 is an a.c. generator (dynamo). Whereas, if the arrangement shown in figure 16 is

used, the sinusoidally varying e.m.f. generated in the coil is transformed into a pulsating direct current as shown in figure 16(b).

Electromagnetic induction in terms of electrons

Imagine a wire full of 'free' electrons being pulled upwards through a magnetic field. This is illustrated in Figure 17 where v represents the upwards velocity of the wire.

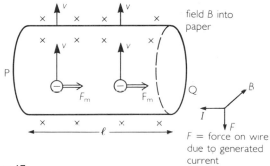

figure 17

Every electron in the wire will be moving upwards with the same velocity v. Therefore each 'free' electron will experience a sideways force $F_m = Bev$. This means that electrons are being 'pumped' along the wire in the direction P to Q. Therefore, conventional current is said to be passing from Q to P; application of Fleming's left hand rule again shows that there will be a force on the wire acting downwards, in agreement with Lenz's law.

We can also use this idea of electron movement to deduce the size of the induced e.m.f. Suppose now we imagine that the wire is not part of a circuit, so that a current cannot pass. Electrons will still move towards Q as a result of the force $F_m = Bev$, but Q will become negatively charged and therefore the electrons will experience a force due to an electric field in the opposite direction. As in the Hall effect, the electrons will cease to move when the force due to the electric field, eE, is equal and opposite to the force from the magnetic field.

Thus $$Bev = eE = \frac{eV}{\ell}$$

$$\Rightarrow \qquad V = B\ell v$$

This result is in agreement with our earlier one.

Summary

1   Faraday's first law: 'Whenever a magnetic flux cuts across a conductor an e.m.f. is induced in the conductor.'

2   Faraday's second law: 'The magnitude of the induced e.m.f. is proportional to the rate at which the magnetic flux is cut.'

3   Lenz's law: 'The direction of the induced e.m.f. is such that any current caused by it would result in opposition to the motion.'

4   Neumann's equation:

$$\text{induced e.m.f. } E = N\frac{d\Phi}{dt}$$

5   *Example 1:* (metal bar)
    induced e.m.f. $E = B\ell v$
    Example 2: (rotating disc)
    induced e.m.f. $E = B\pi R^2 f$
    Example 3: (rotating coil)
    induced e.m.f. $E = NAB\omega \cos\omega t$

Questions

1   Consider a metal girder, of length 15 m, falling from a crane. The girder falls with its length pointing east–west, in a region where the Earth's magnetic field is horizontal ($B = 4.0 \times 10^{-5}$ T). Calculate:
    a   The speed of the girder 2.0 s after the rope snapped;
    b   the magnitude of the e.m.f. induced between the ends of the girder 2.0 s after its release.

2   Consider a railway train travelling at 200 km h^{-1} along the line which runs east to west from London to Bristol. Magnetic field lines in the London area make an angle of 70° to the horizontal and the magnetic flux density $B = 6.0 \times 10^{-5}$ T.
    a   Use the data above to calculate the magnitude of the e.m.f. induced between opposite ends of each axle of the train, if the length of an axle is 1.5 m.
    b   How would you set about measuring this induced e.m.f.?

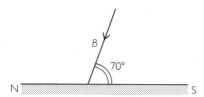

figure 18

3   A small metal-skinned aeroplane flying horizontally over the Earth's magnetic north pole has an e.m.f. induced between the outer ends of its wings. The aeroplane has a wing span of 7 m.
    a   Calculate the magnitude of the induced e.m.f. when the plane is travelling at 600 km h^{-1} at a height where the Earth's magnetic flux density is 8×10^{-5} T.

The pilot decides to try and measure this e.m.f. by using a sensitive voltmeter and two long pieces of copper wire connected to the wing tips. To his surprise he obtains a zero indication on his meter.
    b   Explain why the voltmeter reading is zero.

A friend of the pilot tells him that it is a mistake to connect the copper leads to the aeroplane. The pilot is then advised to attach the *insulated* copper leads along the wings of the aeroplane and allow the free ends to stick out from the wing-tips. The friend tells the pilot that he will obtain a voltmeter reading if he paints the two free ends of the leads with radioactive paint, because the radioactive emissions will ionise the air around the wing-tips.
    c   Say why this advice is sensible, and explain how it results in a voltmeter reading.

4   Consider a large school electromagnet with pole-pieces 15 cm in diameter, and flux density 1.2 T. A student places a square aluminium frame between the poles and tries to move it across the magnetic field. As predicted by Lenz's law he experiences opposition to his action.

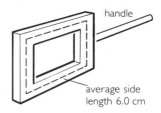

figure 19

    a   Calculate the approximate flux through the frame (average length of side = 6 cm).
    b   The student succeeds in moving the frame steadily at 10 cm s^{-1} across the magnetic field.

Calculate the rate at which the magnetic flux is being cut, by the side still in the field.
c What is the magnitude of the induced e.m.f. in the frame?
d The aluminium from which the square was cut was 5.0 mm thick. The sides of the frame are 2.0 cm wide. Show that the effective electrical resistance of this frame is approximately $7 \times 10^{-4}\,\Omega$. (For Al, $\rho \approx 3 \times 10^{-8}\,\Omega m$.)
e Use your answer to part (c) to determine the current flowing round the frame.
f Determine the magnitude of the magnetic force resisting the student's action.
g Comment on the magnitude of this force.

5 Consider a small wheel, diameter 6.0 cm, spinning between the poles of a large electromagnet.

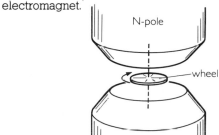

figure 20

The strength of the magnetic flux density is 0.80 T. Calculate the magnitude of the e.m.f. induced between its rim and its axle when it is spinning at 300 rev min^{-1}.

6 A cyclist travels at 45 km h^{-1} along a horizontal road from east to west, in the London area.
a Explain why an e.m.f. will be induced between the axle and rim of his rear wheel.
b The wheels of his bicycle have a diameter of 66 cm (26 inches), and the magnetic field lines make an angle of 70° to the horizontal (see figure 18). Calculate the magnitude of this induced e.m.f. if the magnetic flux density $B = 6.0 \times 10^{-5}\,T$.
c The cyclist decides to check the result of your calculation in part (b). To do this he attaches a voltmeter between the axle and rim of his rear wheel; the contact for the rim is a metal brush which allows the wheel to rotate but still gives electrical contact with the voltmeter. With the voltmeter clipped to his handlebars he again cycles east to west at 45 km h^{-1}.
   Comment on the result he obtains.
d Which is positive, the axle or the rim?
e At what speed would he have to cycle in order to operate his 6 V bicycle lamp from this induced e.m.f.?

7 a State Faraday's laws of electromagnetic induction.
   b State Lenz's law.

c Consider a 26 inch bicycle wheel (diameter 66 cm) rotating at 15 revs^{-1} in a uniform magnetic field of flux density 0.30 T. Calculate the magnitude of the e.m.f. induced between the axle and rim of the wheel when (i) the axle is parallel to the magnetic field, (ii) the axle makes an angle of 25° with the magnetic field.
d Estimate the e.m.f. induced between the extended arms of a sky diver falling at a terminal velocity of 190 km h^{-1} above London, where the magnetic flux density B is $6.0 \times 10^{-5}\,T$, and the angle between the field and the horizontal is 70°.

8 a State Lenz's law of electromagnetic induction.

Consider the coil of wire shown in figure 21. The sense of the windings is indicated in the diagram and the positive and negative terminals of the centre-zero ammeter are coloured red and black respectively.

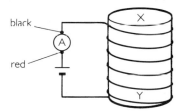

figure 21

b Will the ammeter indicate a positive or a negative reading with the circuit shown?
c The coil is effectively an electromagnet. Does the top of the coil labelled X behave as a north or a south pole?

Now consider what happens when the cell and ammeter are replaced by a centre-zero microammeter, and a bar magnet, north pole facing downwards, is plunged rapidly into the coil.
d Will a north or a south pole be induced in the top of the coil at X as the magnet plunges downwards towards it?
e Will the microammeter indicate a positive or a negative reading as the magnet is plunged into the coil?

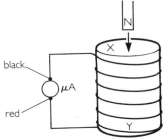

figure 22

f What reading does the microammeter register when the bar magnet is stationary within the coil?

g If a bar magnet, with its south pole uppermost, is rapidly withdrawn from the bottom of the solenoid, will the bottom of the coil, labelled Y, have a north or a south pole induced in it (figure 23)?

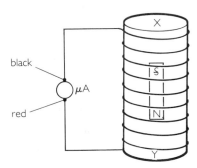

figure 23

h Will the induced current in part (g) cause the microammeter to give a positive or a negative reading?

Imagine a bar magnet falling at a constant (terminal) velocity through a long solenoid which is attached to the microammeter (figure 24).

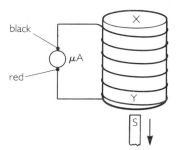

figure 24

i The north pole of the bar magnet is falling towards the bottom end of the coil at Y. Will a north or a south pole be induced at Y?

j The south pole of the magnet is falling away from the top end of the coil at X. What pole will the falling magnet induce in the solenoid at X?

k All the time the magnet is falling through the solenoid, it will be inducing magnetic poles in the coil ahead of itself and behind itself. What can you say about the magnitude of the current registered on the microammeter while the magnet is falling *through* the solenoid, i.e. while the whole of the magnet is between X and Y?

Figure 25 illustrates the bar magnet at five positions as it falls.

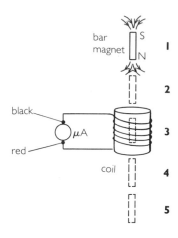

figure 25

l Sketch a graph of induced current against time for the case when the magnet falls at a constant (terminal) velocity. Mark the times corresponding to the positions 1, 2, 3, 4 and 5 on the time axis.

m Sketch a graph of induced current against time for the case when the magnet is released from position 1. Again mark the times corresponding to the positions 1, 2, 3, 4 and 5 on the time axis.

9 Consider a conducting coil of N turns rotating with angular velocity ω in a uniform magnetic field of magnetic flux density B. The area of the coil is A.

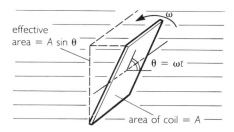

figure 26

a Give an expression for the magnitude of the magnetic flux, Φ, through each turn when the coil is in the position shown in figure 26.

b Write an expression for the induced e.m.f. in this position in terms of N, ω, t, B and A.

Figure 27 illustrates different positions of the coil.

c What will be the induced e.m.f. at each position?

The diagram at the beginning of this question did not indicate how the electric current was obtained from the coil. Figure 28 shows two ways of doing this.

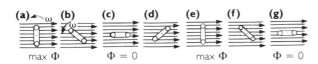

figure 27

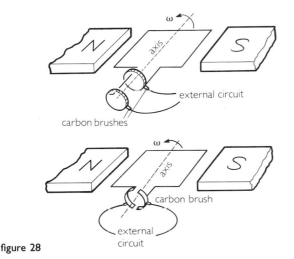

figure 28

d Sketch a graph to show how the e.m.f. across the external circuit varies with time when method (i) is employed.
e Now sketch external e.m.f. against time for method (ii).
f Imagine that a fellow student of yours attended the lesson when method (i) was discussed, but was absent from the lesson on method (ii). Write, for him, a short piece on its operation. This should include a section when you explain the form of the external e.m.f.-against-time graph.

10 *Magnetic monopoles*
Figure 29 shows a bar magnet falling at speed through a coil of wire. The sense of the windings is indicated in the diagram.
a Sketch the trace produced on the screen of the CRO when the time base is switched on and is set so that the spot crosses the screen in the same time that it takes the bar magnet

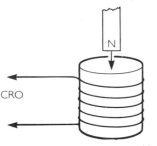

figure 29

to fall from position 1 to position 2. Explain your answer.
b What would be observed on the screen if the magnet fell sideways through the coil? Explain your answer.

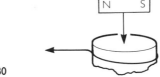

figure 30

One of the pursuits of theoretical physicists is to try and unify theories. For example, there used to be a theory of magnetism and a separate theory of electricity. The theory of magnetism said there were such things as north magnetic poles and south magnetic poles. James Clerk Maxwell unified these two theories and showed that magnetism could be attributed to moving electric charges. He produced a unified theory called the *theory of electromagnetism*, which did away with the idea of separate north and south poles. In the middle of the 1970s, Alexander Polyakov in the USSR showed that electromagnetism, the strong nuclear force that holds protons and neutrons together, and the weak nuclear force that controls radioactive decay could be unified into one theory. This theory is known as the *grand unified theory* (GUT). His theory predicted that there should be such things as isolated north and south poles, i.e. monopoles. If you placed a plotting compass near a monopole that was a north pole the compass would point away from the monopole no matter which side of the monopole you placed it. The theory predicts that there will be very few of these monopoles in the universe. In 1982 Blas Cabrera, at Stanford University, detected what looked like a magnetic monopole.

figure 31 A magnetic monopole

c Consider such a magnetic monopole travelling at speed through a solenoid (figure 32). Sketch the trace produced on the screen of the CRO by the passage of the monopole.

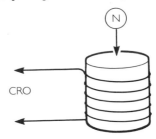

figure 32

CHAPTER 27

The applications of electromagnetic induction

Introduction

In this chapter we put the ideas of the last chapter into action. We look at the important phenomena of self and mutual induction, which lead us on to transformer theory. Finally, we look briefly at d.c. motors.

GCSE knowledge

In addition to the laws of electromagnetic induction summarised at the beginning of the last chapter, it will help in understanding this chapter if you are familiar with the transformer and the d.c. electric motor. You should know that a transformer can step up or step down a.c. voltages, and that it has important applications in the distribution of electrical power throughout the country.

Self inductance

In this section we are concerned with the behaviour of inductors. Figure 1 shows two similar inductors. They are simply coils of wire; one is air-cored and the other has a core of iron through its centre. We use the letter L for the self inductance of an inductor. The symbols used to represent an inductor in a circuit are shown in figure 2.

air-cored inductor   inductor with iron core

figure 2   B.S. symbols for inductors

We can illustrate the behaviour of an inductor with a simple experiment. Figure 3 shows two light bulbs connected in parallel to a battery. The inductor and resistor are chosen carefully so that when the light bulbs light they have the same brightness. However, immediately after the switch is closed something rather unexpected occurs. The bulb that is in series with the resistor lights immediately but the one that is in series with the inductor lights a second or so after the other. So the inductor delays the growth of the current.

We can examine the growth of the current by putting an oscilloscope across AB in figure 3. The oscilloscope will measure the p.d. across the light bulb which is roughly proportional to the current.

figure I   An air-cored inductor and an iron-cored inductor

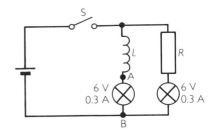

figure 3

Figure 4 shows graphically the way in which the current is found to change. Graph X shows the rise of current when the iron core of the inductor is fully in position, and graph Y shows the rise of current when the iron core is partially removed. The current rises faster in the second case.

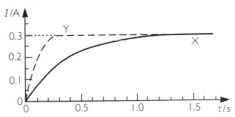

figure 4

We can explain the behaviour of an inductor using the laws of electromagnetic induction. Consider the circuit of figure 5.

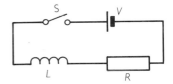

figure 5

As soon as the switch is closed there is a changing current through the inductor. Associated with this changing current is a changing magnetic flux. By Lenz's law there will be an induced e.m.f., E, across the coil which will be in such a direction as to oppose that change of flux. We can write this equation:

$$E = N\frac{d\Phi}{dt}$$

where N is the number of turns in the coil. Clearly, the size of the induced e.m.f. is affected by the physical size and properties of the coil. (The important factors are the number of turns, the cross-sectional area and length of the coil, and the magnetic properties of the coil's core.) However, once we choose a particular inductor all these factors are fixed, and the only factor that can affect the induced e.m.f. is the rate of change of current. This enables us to define the *self inductance* for the coil, L, as follows:

$$E = N\frac{d\Phi}{dt} = L\frac{dI}{dt} \qquad (i)$$

The unit of self inductance is called the henry, H. An inductor has a self inductance of 1 henry if an e.m.f. of 1 V is induced across it when the current changes through it at a rate of $1\ \text{As}^{-1}$.

We may now return to the circuit shown in figure 5. When the switch is closed there is an e.m.f. induced across the inductor in such a direction as to oppose the growth of current. We can therefore write:

$$V - L\frac{dI}{dt} = IR \qquad (ii)$$

The e.m.f. $L(dI/dt)$ is called a 'back' e.m.f., and it is often given a minus sign to remind us that it opposes the change in flux. Equation (ii) shows us that the net p.d. across the resistor is less than the applied e.m.f. V. This was also the situation with the light bulb in figure 3. Initially there is a large back e.m.f. as the rate of change of current is large and so the p.d. across the light bulb is small. After a while, however, the rate of change of current is smaller and so the back e.m.f. is small, and therefore the p.d. across the light bulb is large and it lights normally.

The purpose of this section on self inductance has been to explain the action of an inductor using the laws of electromagnetic induction. The uses of inductors are explored more fully in chapters 13 and 14.

Example 1   Calculate the self inductance of an air-filled solenoid with 10 000 turns. The length of the solenoid is 0.30 m and its cross-sectional area is $0.010\ \text{m}^2$.

From equation (i) we may write:

$$N\frac{d\Phi}{dt} = L\frac{dI}{dt}$$

$$\Rightarrow \qquad L = \frac{N\Phi}{I} \qquad (iii)$$

This is a useful expression which shows us that inductance can be thought of as 'flux-linked with the whole coil per unit current'.

We know from our work on magnetic fields that the flux density in a solenoid is:

$$B = \mu_0\frac{N}{\ell}I$$

where N is the number of turns and ℓ is the length. Thus the total flux through the solenoid is:

$$\Phi = BA = \mu_0\frac{N}{\ell}AI$$

Using this equation in equation (iii) gives:

$$L = \frac{N}{I}\frac{(\mu_0NAI)}{\ell}$$

$$\Rightarrow \qquad L = \frac{\mu_0N^2A}{\ell} \qquad (iv)$$

Thus $L = \dfrac{(4\pi \times 10^{-7}\,\text{NA}^{-2}) \times (1.0 \times 10^4)^2 \times (0.010\,\text{m}^2)}{0.30\,\text{m}}$

$= 4.2\,\text{H}$

From equation (iv) we can see that:

$$\mu_0 = \dfrac{L\ell}{N^2 A}$$

Since N is just a number, it follows that it is also possible to express μ_0 in units of Hm^{-1}.

Example 2 An iron-cored inductor has an inductance of 15 H. It is placed in series with a resistance of 30 Ω in the circuit shown in figure 6. Calculate the initial rate of growth of current as soon as the switch is closed and calculate the final current. Explain why the current will grow more rapidly if the iron core is removed from the inductor.

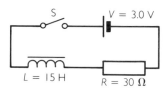

figure 6

We can solve this problem using the equation:

$$V - L\dfrac{dI}{dt} = IR$$

Initially the current is zero
⇒ the initial rate of change of current is:

$$\dfrac{dI}{dt} = \dfrac{V}{L}$$

$$= \dfrac{3.0\,\text{V}}{15\,\text{H}}$$

$$= 0.2\,\text{As}^{-1}$$

When a steady current passes, $(dI/dt) = 0$
⇒ the final current is given by:

$$I = \dfrac{V}{R}$$

$$= \dfrac{3.0\,\text{V}}{30\,\Omega}$$

$$= 0.10\,\text{A}$$

When the iron core is removed the self inductance decreases because the flux linked per unit current is less. If the self inductance is small the initial rate of rise of current is increased:

$$\dfrac{dI}{dt} = \dfrac{V}{L}$$

Mutual inductance

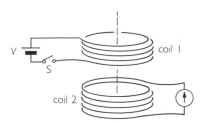

figure 7 Co-axial coils

Figure 7 shows an experiment to demonstrate mutual inductance. Coil 1 can be connected to a cell by closing a switch. Coil 2 is not connected to a cell, but its ends are connected to a sensitive galvanometer. Experiment shows that the galvanometer registers nothing when a *steady* current passes through coil 1. However, the galvanometer 'kicks' when S is opened and closed. The explanation is that opening and closing the switch causes a changing current I_1 in coil 1. A changing current sets up a changing magnetic field; this changing magnetic field reaches as far as coil 2 and so an e.m.f. E_2 is induced in coil 2 as a result of the changing flux through it.

We can define the mutual inductance M of the two coils by the equation:

$$E_2 = M\dfrac{dI_1}{dt} \qquad\qquad \text{(v)}$$

The unit of mutual inductance is the same as that of self inductance: the henry, H.

Transformers

Consider a coil of wire carrying an alternating current, as shown in figure 8. The alternating current produces a varying magnetic flux density, B_1.

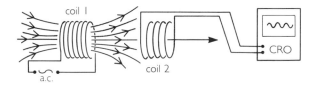

figure 8

If this varying magnetic flux passes through a second coil, then an e.m.f. will be induced in this second coil. In figure 8 the CRO is being used to measure the magnitude and frequency of the e.m.f. induced in coil 2.

Coils 1 and 2 together constitute a transformer. An ideal transformer would transfer 100% of the power in coil 1 across to coil 2 via the varying magnetic flux. There are various experiments that you can carry out to find how the greatest efficiency can be achieved.

Clearly some of the magnetic flux emitted by coil 1 never goes through coil 2; this represents a loss of energy in the transfer process. Two modifications would improve this:

☐ put coils 1 and 2 closer together;

☐ thread an iron bar through them to carry a high fraction of the flux from coil 1 to coil 2.

This is illustrated in figure 9.

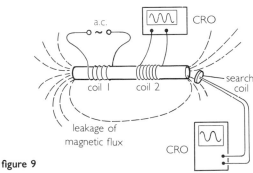

figure 9

If a small search coil is placed at the end of the iron bar then a signal is picked up by it. This means that not all the energy is being transferred from coil 1 to coil 2; some of it is being emitted into the surrounding space.

If the two coils are wound on a complete C-core as shown in figure 10, then the search coil detects very little flux leakage. Magnetic flux flows much more readily through iron than it does through air, therefore most of the magnetic flux is trapped to circulate in the iron ring. The output of coil 2, shown on the CRO, is greatest when the iron ring is as short as possible, and when the ring is laminated. (The point about laminations will be discussed on page 284.)

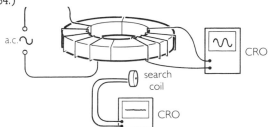

figure 10   Coils wound on a C-core

Experimental relationship between output and input voltages

To carry out this investigation, two coils should be wound on C-cores as illustrated in figure 10. To start with, 60 turns on each coil is a suitable number. Voltmeters should be attached to measure the input voltage to coil 1, and the output voltage from coil 2. Figure 11 shows the circuit; the line in the

transformer symbol represents the laminated iron core. If some of the turns are unwound from coil 2, it is possible to construct a table of data showing how the secondary voltage, V_2, depends on the number of turns in the primary (coil 1); such results are illustrated in the table below.

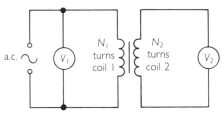

figure 11

N_1	N_2	V_1/V	V_2/V
60	60	10	10
60	120	10	20
60	30	10	5.0
30	120	10	40
120	24	10	2.0

These experimental results are summarised by the equation:

$$\frac{V_1}{V_2} = \frac{N_1}{N_2}$$

Theoretical relationship between output and input voltages

In this section we wish to apply the laws of electromagnetic induction to derive theoretically the result obtained at the end of the last section. However, it must be emphasised that the theory only applies to *good* transformers. A 'good' transformer is one which has large numbers of turns on primary and secondary. This ensures that the self inductance of each coil is large, so that for each coil reactance $(2\pi fL) \gg$ resistance (R) (see page 170).

For the primary circuit (figure 12), we can write:

$$V_1 - L\frac{dI}{dt} = IR$$

where V_1 is the applied e.m.f. and L and R are the self inductance and resistance of the primary coil. If L is large then the back e.m.f. will also be large, and virtually no current will flow when the secondary coil is on open circuit as shown in figure 12. (Alternatively, a large reactance means that the current will be very small – see page 172.)

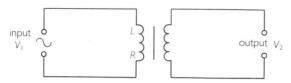

figure 12

Under these circumstances we may write:

$$V_1 \simeq L\frac{dI}{dt} = N_1\frac{d\Phi}{dt} \qquad \text{(i)}$$

Provided that the transformer is well made so that no flux escapes, we know that all the flux produced by the primary coil is linked to the secondary coil.

The secondary voltage is therefore given by:

$$V_2 = N_2\frac{d\Phi}{dt} \qquad \text{(ii)}$$

Dividing equation (i) by equation (ii) gives:

$$\frac{V_1}{V_2} = \frac{N_1}{N_2}$$

A transformer delivering power

If a small search coil is wrapped around the core of a transformer, as in figure 13, it can be used to measure the changing flux through the transformer. The e.m.f. across the coil can be displayed on an oscilloscope and we know that the e.m.f. is proportional to the rate of change of flux.

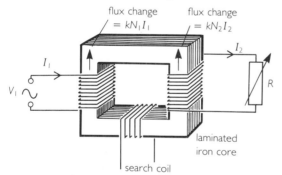

figure 13

Experiment shows that the e.m.f. induced in the search coil remains constant as the current drawn from the secondary is steadily increased. This means that the total flux through the core is unaffected by the secondary current. The explanation is that the changing flux due to the secondary current induces an e.m.f. in the primary coil so that the primary circuit draws more current. By Lenz's law the flux change from the primary circuit will be such that it tends to cancel that of the secondary. The flux change produced by the secondary is proportional to $N_2 I_2$, and the flux change produced by the primary is

proportional to $N_1 I_1$. (The flux through a coil is proportional to the number of turns and the current.)

Thus we can write:

$$\Delta\Phi_1 = kN_1 I_1$$

and

$$\Delta\Phi_2 = kN_2 I_2$$

$\Delta\Phi_1$ and $\Delta\Phi_2$ are the flux changes due to the currents I_1 and I_2. k is a constant that depends on the size and magnetic properties of the core.

In an efficient transformer these flux changes are equal and opposite, and so it follows that

$$N_1 I_1 = N_2 I_2$$

or

$$\frac{N_1}{N_2} = \frac{I_2}{I_1} \qquad \text{(i)}$$

From our previous results we know that:

$$\frac{N_1}{N_2} = \frac{V_1}{V_2}$$

Combining this with equation (i) gives:

$$V_1 I_1 = V_2 I_2$$

So in an *efficient* transformer the power supplied to the primary is equal to that supplied by the secondary.

The result could have been deduced using the principle of conservation of energy had we started by assuming that our transformer was 100% efficient.

Causes of inefficiency in a transformer

Most practical transformers are highly efficient (95% or more). The small losses that do occur are due to three main causes.

1   *Heating of the windings due to the currents in them.* The resistances of primary and secondary cannot be made zero (except perhaps in superconducting transformers of the future). Hence the currents in them cause some heating, according to the usual formula RI^2.

2   *Eddy currents in the iron core.* If the core was solid iron, the changing flux through it would induce e.m.f.s and hence 'eddy' currents in loops within the iron itself. See figure 14(a). These eddy currents would heat the iron.

To counteract this possibility, iron cores are always laminated – that is, sliced up, as shown in figure 14(b). Interleaved with the slices of iron are very thin sheets of non-conducting material. The laminations do not affect the flux at all, but they intersect the paths which the current would otherwise take. By making the slices thin enough, the designer can reduce the heating caused by eddy currents to negligible proportions.

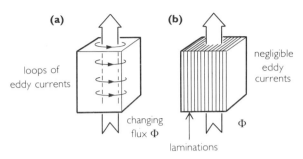

figure 14 Eddy currents in (a) solid metal; (b) laminated metal

figure 16

3 *Hysteresis.* Some energy is required to make the core go through repeating cycles of magnetisation and demagnetisation. This effect is called hysteresis. The energy needed appears as heat within the core.

A fourth possible cause of energy loss is incomplete flux linkage between primary and secondary already referred to on page 283. This possibility can be almost totally eliminated by winding the two coils on top of each other, on a core shaped like the one in figure 15.

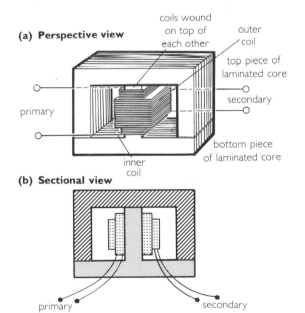

figure 15 An a.c. transformer

The photograph (figure 16) shows a small transformer of this type cut in half. You can clearly see the laminations of the core; the small number of thick copper windings for the low p.d./high current side; and the larger number of thinner copper windings for the high p.d./low current side.

Simple d.c. motors

A coil carrying a current in a magnetic field experiences a torque. The torque turns the coil until the plane of the coil is at right angles to the magnetic field lines as shown in figure 17.

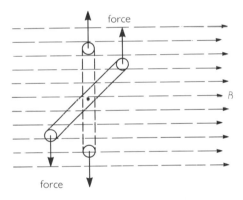

figure 17 Torque on a current-carrying coil in a magnetic field

If the d.c. dynamo with the split-ring commutator, illustrated in the last chapter, figure 16, is attached to a battery, then it becomes a d.c. motor (figure 18).

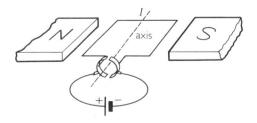

figure 18 A d.c. motor

The action of a commutator in producing continuous rotation is shown in figure 19.

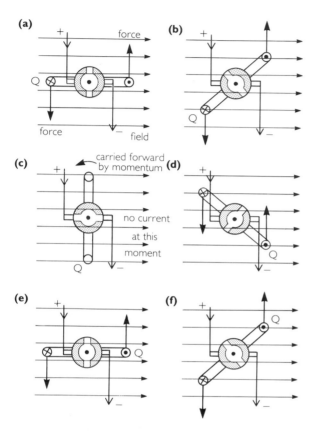

figure 19   Action of a split-ring commutator

Note in figure 19 that the current in the side of the coil labelled Q is reversed every half rotation, so that the direction of the torque producing the rotation is maintained. The coil's momentum carries it through the position shown in figure 19(c), when there is, for a short time, no current.

The back e.m.f. induced in the motor

An e.m.f., E_B, is induced in the rotating coil of a motor, which, according to Lenz's law, will oppose the rotation of the coil, i.e. E_B will oppose the passage of the current in the coil.

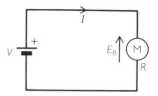

figure 20

Figure 20 shows a power supply of negligible internal resistance producing an e.m.f. V.

$$\text{net e.m.f.} = \text{current} \times \text{resistance of motor}$$

$$\Rightarrow \qquad V - E_B = IR$$

This equation leads to the relationship:

$$VI = E_B I + I^2 R$$

VI is the total power taken from the supply. I^2R is the power dissipated in the motor; this is wasted as heat. $E_B I$ is the useful power which is transferred to the load. As the load on the motor increases, its rate of rotation slows, and so the back e.m.f. diminishes. This causes the current in the motor to increase. An increasing current allows the motor to do more work on the load unless it rotates too slowly.

Although the motor can do more work as it draws more current, it becomes less efficient.

$$\text{efficiency} = \frac{\text{power out}}{\text{power in}}$$

$$\Rightarrow \qquad \text{efficiency} = \frac{E_B I}{VI} = \frac{E_B}{V} = \frac{E_B}{E_B + IR}$$

This equation shows that as I increases, the efficiency decreases, so a greater fraction of the power will be wasted as heat in the resistance of the motor.

Summary

1 Self inductance L is defined by the equation

$$E = L\frac{dI}{dt}$$

2 For a circuit containing inductance and resistance,

net e.m.f. $= IR$

$$V - L\frac{dI}{dt} = IR$$

3 Mutual inductance M between two coils is defined by:

$$E_2 = M\frac{dI_1}{dt}$$

4 The unit of both self inductance and mutual inductance is the henry (H).

5 For transformers which are 100% efficient:

$$\frac{V_1}{V_2} = \frac{N_1}{N_2}$$

$$V_1 I_1 = V_2 I_2$$

6 Energy loss in a transformer may be due to
   a heating of the windings by currents;
   b eddy currents heating the core;
   c hysteresis in the core;
   d incomplete flux linkage.

7 Power relationship in a motor:

$$\begin{matrix} VI & = & E_\text{B}I & + & RI^2 \\ \text{power in} & & \text{useful} & & \text{heat loss} \\ & & \text{power} & & \end{matrix}$$

Questions

1 Consider a toroidal coil of average radius 10 cm and cross-sectional area 4 cm². There are 2000 turns of wire in the toroid. It is wound on a hollow plastic former.

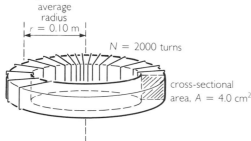

figure 21   Toroidal coil

a Write down an expression for the magnetic flux density B inside a long solenoid of length ℓ with N turns, when a current I passes through it.

b Substitute the numerical data for the above toroid into your answer to part (a), and obtain an expression for the magnetic flux density B in terms of the current I.

c Obtain an expression for the magnetic flux Φ through one turn of the above toroid when a current I is passing.

d If the magnetic flux changes by a small amount $\delta\Phi$ in a small time δt, then the back e.m.f. E_B induced in any solenoid is given by

$$E_\text{B} = N\frac{\delta\Phi}{\delta t}$$

Also the induced back e.m.f. E_B is given by

$$E_\text{B} = L\frac{\delta I}{\delta t}$$

where δI is the small change in current that produced the small change in flux $\delta\Phi$ in time δt.

   Use these two expressions to calculate a value for the self inductance L of this toroidal solenoid.

e Iron has a relative permeability $\mu_\text{r} \approx 1000$. What would the self inductance have been had the toroid been wound on a laminated iron ring of the same dimensions?

Now consider a circuit containing the toroidal inductor we have just been discussing. The circuit diagram (figure 22) shows a cathode ray oscilloscope (CRO) attached across the resistor.

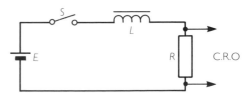

figure 22

Because $V = IR$, the variation in the potential difference registered on the CRO will indicate how the current through the inductor changes.

The general form of the variation is illustrated in figure 23.

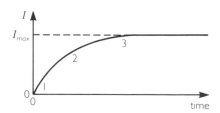

figure 23

f Assume that in the circuit of figure 22 the cell and inductor have zero resistance. Apply Kirchoff's second circuit law to this circuit, and write down an expression for the back e.m.f. E_B in terms of E, I and R.

g Earlier we said that the self-inductance L is defined by $E_B = L(dI/dt)$. Use this and your answer to part (f) to show that the rate of growth of current dI/dt, in the circuit is given by

$$\frac{dI}{dt} = \frac{(E - IR)}{L}$$

The graph in figure 23 showing the growth of current with respect to time has three regions, labelled 1, 2 and 3. The next few questions refer to these labelled regions.

h What is the value of the current *immediately* after the circuit has been switched on (region 1)?

i Use the answer to part (g) to obtain an expression for the rate of growth of current, dI/dt, at the moment the switch closes. How does the growth of the current depend on the self inductance?

j After some time the current's growth ceases (region 3). Obtain an expression for the steady value of the current, I_{max}, when this occurs.

k The equation for the growth of current against time is

$$I = I_{max}[1 - e^{-(R/L)t}]$$

Obtain an expression for the current when $t = L/R$. (This is known as the time constant for the circuit and is given the symbol $\tau_L = L/R$.) Work out the ratio I/I_{max} when $t = 5(L/R)$, i.e. $t = 5\tau_L$.
(5(L/R) is normally taken as the time for the current in the circuit to stop changing.)

l What effect will increasing the resistance R in the circuit have on (i) the initial rate of growth of the current, (ii) the time taken for the current to become nearly constant?

m What effect will increasing the inductance L in the circuit have on (i) the initial rate of growth

of the current, (ii) the time taken for the current to become nearly constant?

2 a Write down an expression for the magnetic flux density B at the centre of a long air-filled solenoid. The solenoid has a length ℓ, carries a current I, has N turns and a cross-sectional area A.

b Express the magnetic flux Φ through the centre of the solenoid as a function of B, N, ℓ, I and A.

c Show that the self inductance L for this air-filled solenoid is

$$L = \frac{\mu_0 N^2 A}{\ell}.$$

d Calculate the self inductance L for such an air-filled solenoid, when $N = 23000$ turns, $\ell = 0.10$ m and $A = 3.0 \times 10^{-4}$ m^2.

e The relative permeability μ_r of iron is about 1000. Discuss how you think the self inductance would increase if the *inside* of the solenoid were filled with laminated iron.
($\mu_0 = 4\pi \times 10^{-7}$ Hm^{-1})

3 Consider a stone released at the surface into a deep tank of water. Initially the stone will not be moving (velocity $v = 0$), but it will accelerate (acceleration $a = dv/dt$). As its speed increases so will the water resistance. It is reasonable to assume that the resistive force of the water is proportional to the velocity. That is, resistive force = kv, when k is a constant of proportionality.

a Write an equation relating the net force F acting on the falling stone, its mass M, and its acceleration a.

b Sketch a graph to show how the velocity of the stone varies with time.

c What will the resistive force be equal to when the stone reaches its terminal velocity?

d Express the terminal velocity v_t in terms of the mass M, the constant k, and the gravitational field strength g.

e What was the initial rate of change of velocity?

Now consider the circuit illustrated in figure 24. Assume that the cell and inductor have zero resistance. A changing current through the coil produces a back e.m.f.: $E_B = L (dI/dt)$. Note that L is analogous to the constant k above.

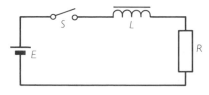

figure 24

f   Write an equation relating the net e.m.f. E_{net} acting in the circuit to the current I and the resistance R, once the switch is closed.

g   Sketch a graph to show how the current varies with time.

h   Express the final steady current I_{max} in terms of the constants of the circuit.

i   What is the initial rate of change of current?

In answering questions (a) to (e) you were dealing with the rate of change of velocity (the acceleration), and how it depends on the inertia (the mass) of the stone and the resistive forces (viscous drag) on it. The idea behind (f) to (i) was to draw your attention to the similarities between the two systems.

j   Discuss carefully, by considering the effect of increases and decreases of the quantities, the analogy between these two systems.

4   a   Calculate the self inductance of an iron-cored inductor of length 0.50 m, diameter 5.0 cm, and 20000 turns. Take the relative permeability of iron to be $\mu_r = 1000$.

b   Calculate the initial rate of growth of current on closing the switch when the above inductor is connected in the circuit shown in figure 25.

figure 25

c   Estimate how long it will take the current to reach 99% of its final value. That is, calculate how long it is before the current is effectively constant.

d   What will the final steady value of the current be?

Now consider the inductor connected as shown in figure 26. The switch is in position 1 and the current has reached its steady value, therefore there is a steady magnetic flux through the coil.

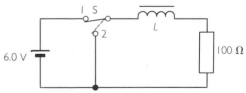

figure 26

e   When the switch is instantaneously moved from position 1 to position 2 can the magnetic flux instantaneously disappear?

f   Calculate the initial rate of fall of current when the switch contacts position 2.

g   How long will it be before the current has dropped to I_{max}/e where $e = 2.7$ and I_{max} is the initial current when the switch contacts position 2.

Imagine that the switch is repeatedly moved between positions 1 and 2. Every time it is in position 1, a p.d. of 6.0 V is applied to the LR circuit; every time it is in position 2, zero volts are applied. This can be achieved by attaching a square wave generator as shown in figure 27.

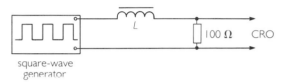

square-wave generator

figure 27

h   For the inductor we have been considering, sketch what you will see on the CRO if the square wave generator is set on 10 Hz.

i   Sketch what you will see on the CRO if the frequency is changed to 50 Hz.

5   Consider the circuit shown in figure 28. There is no connection between X and Y.

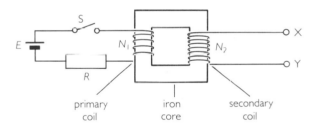

figure 28

a   Will a current pass in the secondary coil when the switch is closed?

b   The primary coil has a self inductance L_1. Apply Kirchoff's second law to the primary circuit and obtain an expression for the current I_1 in the primary coil, in terms of E, R, L_1 and dI_1/dt.

The magnetic flux produced by the current in the primary coil flows along the iron core and through the secondary coil. Therefore, the magnetic flux through the secondary coil will change when the current in the primary coil changes. Hence, an e.m.f. will be induced across the secondary coil.

c   Explain why the induced e.m.f. E_2 across the secondary coil will be proportional to the rate of change of current, dI_1/dt, in the primary coil. That is, explain why $E_2 = M(dI_1/dt)$, where M is known as the mutual inductance between the coils.

d Write down an expression for the magnetic flux density B inside the primary coil in terms of the number of turns N, the length of the coil ℓ, and the primary current I_1. Assume it is effectively a long solenoid.

e Assume that all the magnetic flux produced by the primary coil flows along the iron core and through the secondary coil. If the core has the same cross-sectional area A all along its length, write an expression for the magnetic flux through the secondary coil due to the primary current.

f The magnitude of the e.m.f. E_2 induced in the secondary coil is given by the equation $E_2 = N_2(d\Phi/dt)$, where N_2 is the number of turns in the secondary coil. Use your answer to part (e) to obtain an expression for the e.m.f. E_2 induced across the secondary coil, in terms of N_2, N_1, A, ℓ and dI/dt.

g In part (b) you obtained an expression for the primary current I_1. At the instant the switch in the primary circuit is closed the primary current will be zero. Write an expression for the initial back e.m.f. E_1 across the primary coil, in terms of N_1, ℓ and A.

h Divide your answer to part (f) by your answer to part (g) to show that

$$\frac{E_2}{E_1} = \frac{N_2}{N_1}$$

6 The ignition system of a car is illustrated in figure 29.

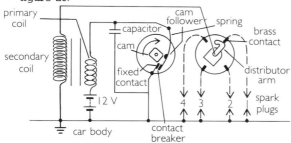

figure 29   Ignition system of a car

   A p.d. of several thousand volts is needed across the terminals of the spark plugs. The cam is a bar of rectangular cross-section which has rounded corners. The distributor is mounted on top of the cam and rotates with the cam in time with the engine. The cam-follower is attached to the spring of the contact breaker and slides over the surface of the rotating cam.

a Explain how the system generates a high voltage across a spark plug.

b Explain how the system produces the high p.d. at each spark plug in turn.

c What do you think is the purpose of the capacitor across the contact breaker?

d Explain what might happen if the capacitor was removed?

7 This question is about the principles behind a particle accelerator known as the Betatron. The principle is explained to you by a series of questions. Figure 30 illustrates the basic structure, which is a strong magnetic field between two large electromagnets. Electrons travel in circular paths through the vacuum between the magnets.

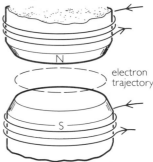

figure 30

a Work out an expression for the momentum $p(= mv)$ for an electron of mass m travelling in a circular orbit of radius r, when the magnetic flux density is B (electronic charge $= e$).

b If the strength of the magnetic field is slowly increased, what happens to the radius r of the electron's trajectory?

c When the magnetic flux density is increased, what happens to the magnitude of the magnetic flux Φ through the original circular orbit of the electron? Will this change in the magnetic flux through the orbit induce an e.m.f. which accelerates or decelerates the electron?

d The e.m.f. E induced is the work done in taking unit charge once round the circuit. Therefore, the work done on the electron by this induced e.m.f. is eE for one complete orbit. The equation for this induced e.m.f. is $E = d\Phi/dt$, and 'work done' is equal to 'force × distance moved'. Use the above information to find an expression for the tangential force F acting on the electron.

e Now reconsider your answers to parts (a) and (b). These need to be thought about more carefully.

   Write down again your expression for the momentum, $p = mv$. The tangential force acting on the electron can be obtained by considering the rate of change of momentum, $(d/dt)(mv)$. The Betatron is designed so that the tangential force $F = er(dB/dt)$. What does this tell you about the radius of the orbiting electron as the magnetic flux density is increased?

f You now have two different expressions for the tangential force acting on the electron:

   from (e) $F = er\dfrac{dB}{dt}$

and from (d) F equals an expression in terms of dΦ/dt. Look at these two answers and say how it is possible for the magnetic field to increase with time in such a way as to keep r constant and accelerate the electrons.
g What can you say about the magnetic flux density at the centre compared with the magnetic flux density in the region of the electron's trajectory?

8 Say what effect the fatness of a transformer core has on the e.m.f. induced in the secondary winding, and use the following experimental set-ups to explain the reasoning behind your answer.

(a)

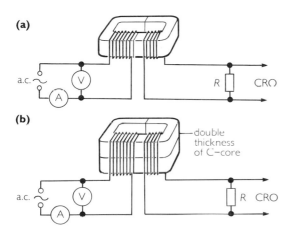

(b)

figure 31

a Consider what will happen to the output e.m.f. in each of the circuits in figure 31 if the *current* in the primary is made the same for each circuit.
b What will happen to the output e.m.f. in each circuit if the p.d. across the primary is made the same in each circuit?

9 The principles of a simple motor are outlined in figure 32. The rotating part of an electric motor is called the armature.

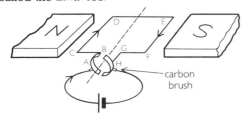

figure 32

a When the coil is in the position illustrated in the diagram, in what direction will the magnetic forces on the coil act?
b The magnetic forces will produce a magnetic torque which will cause the motor to rotate.

Explain why a split-ring commutator is necessary in the case of a d.c. motor.

In practice the armatures of commercial motors consist of several angled, independent coils wound on a single iron core (figure 33). Each coil has its own commutator, and the carbon brushes make contact with each pair of contacts in turn as the armature rotates.

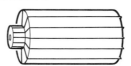

figure 33

c What are the advantages of having an iron core?
d What are the advantages of using an armature with several, angled, independent coils, rather than just one coil?
e Commercial motors also have curved pole pieces on their permanent magnets (figure 34). Explain why this is so.

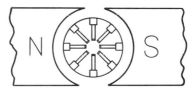

figure 34

As the armature rotates in the magnetic field the coils of wire cut across the magnetic flux lines. This will induce an e.m.f. in the rotating coils which is known as a back e.m.f., E_B, because, according to Lenz's law of electromagnetic induction, it opposes the current producing the rotation.
f Explain why you would expect the magnitude of the induced back e.m.f. to be proportional to the speed at which the coil is rotating.
g Use Kirchhoff's second law for electric circuits to obtain an expression for the current through the motor in terms of the e.m.f. E of the cell, the back e.m.f. E_B and the electrical resistance R of the motor.
h Discuss how the current through a d.c. motor varies as the speed of the motor increases from zero.

Kirchoff's second law for this circuit gives

$$E - E_B = IR$$

If we multiply this equation by I we obtain

$$EI = I^2R + E_BI$$

The term EI represents the total power supplied by the cell to the motor.
i What does the term I^2R represent?
j If energy is being supplied at the rate EI, and some of it is being wasted in the form I^2R,

what must the rest of it be doing? What does the term $E_B I$ represent?

k The efficiency of a motor is defined as

$$\text{efficiency} = \frac{\text{mechanical power obtained}}{\text{total power supplied}} \times 100\%$$

Express the efficiency in terms of E, I and R.

The magnetic fields of large electric motors are normally produced by electromagnets. In a *series-wound motor* the magnetic field coils are in series with the armature, therefore the same current passes through each. *Shunt-wound motors* have their field coils in parallel with the armature.

l Explain why a series-wound motor will produce a large torque at low speeds which diminishes considerably as its speed increases, whereas the torque from a shunt-wound motor is far less variable.

10 Following the nuclear power station disaster at Chernobyl in 1986, there has been a great increase in the building of magnetohydrodynamic (MHD) generators, in both the USSR and the USA. The essential feature of these MHD generators is that a conducting fluid flows through a magnetic field. Figure 35 outlines the process. It shows a square-section tube of width w, carrying liquid mercury at a speed v through a magnetic flux density B.

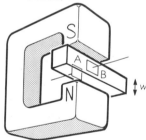

figure 35

a Explain how this device generates an e.m.f.
b Obtain an expression for the e.m.f. induced between the electrodes A and B.

There is a tale that Lord Kelvin attempted to measure the e.m.f. induced across the Thames between the Houses of Parliament on the north bank and St Thomas' hospital on the south bank. The conducting fluid in this case is the water of the Thames.

c In the London area the Earth's magnetic flux density is 6.0×10^{-5} T, and the angle between the field lines and the horizontal is 70°. Estimate the magnitude of the e.m.f. that Kelvin was measuring and say how it varies during the course of the day. (The Thames is tidal at this point.)

Power stations which burn fossil fuel drive steam turbines. The efficiency of steam turbines in modern power stations is between 35% and 40%. This low efficiency is due to the limitation put on the operating temperature and pressure (560° C and 165 atmospheres) by the materials available.

d What are the chief pollutants emitted by power stations burning fossil fuels?

Commercial magnetohydrodynamic (MHD) generators are a spin-off from the space industry. In building the rocket motors for the space shuttle, and rockets in the USSR, the engineers have developed ceramic conductors that can operate at 3500 K. They have also learned how to control flames at such high temperatures.

e Imagine a fixed rocket motor burning kerosene and oxygen, and firing the exhaust gases along a pipe which passes between the poles of a large electromagnet. Explain how this generates d.c. power.

f How will the d.c. power produced by an MHD generator depend on the speed of the hot gas?

g The MHD generators so far built convert energy with an efficiency of about 15%. The problem is that reasonable quantities of electrical energy cannot be extracted from the gas once its temperature drops below about 2000 K. Explain why you think this is the case.

h A gas stream at 2000 K is a potent source of heat, and can be used to boil water for a conventional steam turbine. What will be the efficiency of a power station which combines MHD with a conventional steam turbine?

11 Electrical devices such as d.c. electric motors cause problems to TV and radio receivers because sparks are produced at the commutators. Every time contact is broken, a spark occurs between the brush and the split ring. The spark lasts for a very short time and causes high-frequency oscillations which feed back along the leads. In your own house these high-frequency oscillations feed along the mains leads and into the radio or TV, where they produce interference. In motor cars the high-frequency oscillations in the wires radiate a high-frequency electromagnetic wave which is picked up by the radio aerials of nearby cars.

A typical type of suppressor circuit is shown in the figure 36. Explain why the slow variations in the mains supply pass through the inductors to the motor, but the high-frequency oscillations from the sparks at the motor do not pass back into the mains.

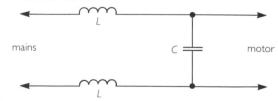

figure 36 Suppressor circuit

SECTION D

WAVES

This is the nice side of waves...

. . . and the nasty side of waves

28   Wave speeds   295
29   Electromagnetic waves   304
30   Diffraction   316
31   Interference   326
32   Reflection and refraction   345

CHAPTER 28

Wave Speeds

Introduction

We start by reviewing the basic vocabulary of wave physics. Next we derive an equation for the speed of a transverse mechanical wave travelling along a wire. We follow that by modelling a longitudinal wave passing through a solid in terms of trolleys interlinked with springs, and arrive at an equation for the speed of longitudinal waves. The longitudinal wave concept is then applied to sound waves in gases. We finish the chapter by studying the Doppler effect.

GCSE knowledge

It will be assumed that you can recall that:

☐ a transverse wave has oscillations perpendicular to the direction in which the wave travels;

☐ a longitudinal wave has oscillations parallel to the direction in which the wave travels;

☐ the wavelength λ is the distance between successive crests;

☐ the speed of waves v is given by speed = frequency × wavelength; $v = f\lambda$.

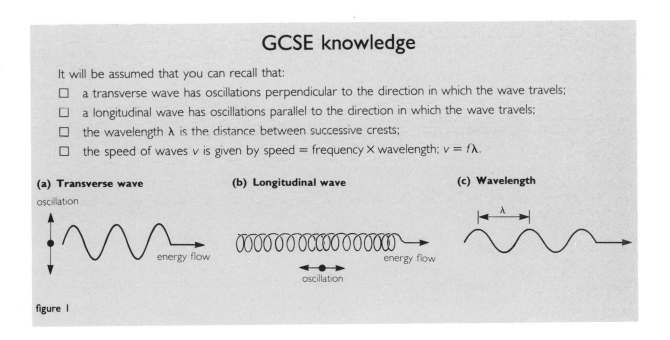

(a) Transverse wave

oscillation

energy flow

(b) Longitudinal wave

energy flow

oscillation

(c) Wavelength

λ

figure 1

The basic vocabulary of waves

A *progressive transverse* wave is shown in figure 2 travelling along a long, narrow spring.

Progressive means that it is a travelling wave, travelling along the spring away from the hand in this case. *Transverse* means that the oscillations of the spring are perpendicular to the direction in which the

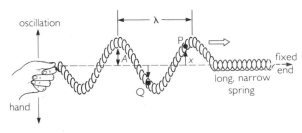

oscillation

λ

P

A

x

Q

hand

fixed end

long, narrow spring

figure 2   Progressive transverse wave

wave is travelling. The system carrying the wave is known as the *medium*, which in our diagram is the spring.

The *displacement*, x, at any point on the spring refers to the transverse displacement of that point on the spring from its undisturbed, or equilibrium, position. The displacement of P is labelled x in figure 2. Displacement is a vector. Hence, if x is positive for P it must be negative for Q. The greatest value of the displacement is known as the *amplitude*, A, and is indicated under the first crest in figure 2. The distance along the wave between successive points which are moving in phase with one another is called the *wavelength*, λ. This is illustrated as the distance between successive crests in figure 2.

A progressive *longitudinal* wave is illustrated in figure 3. *Longitudinal* means that the oscillations of the spring are parallel to the direction in which the wave is travelling.

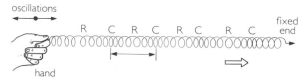

figure 3   Progressive longitudinal wave. C is a compression, R is a rarefaction

The *frequency*, f, is the number of crests or compressions passing a fixed point per unit time; and the *time-period*, T, is the time taken by any point on the spring to complete one oscillation; $T = 1/f$.

Note that x, A, f and T have similar meanings when relating to oscillations – see chapter 7.

Transverse waves

The experimental observations that we have to take into account when finding an expression for the speed of transverse waves are the following:

1   The speed of the wave along a spring, a string, or a rubber tube is independent of the shape of the wave. This is illustrated in figure 4.

2   If the tension T in the spring, string or rubber tube is increased, then the speed of the waves increases.

3   If the rubber tube is filled with sand, so becoming more massive, then the speed of the waves decreases.

We may investigate quantitatively how factors 2 and 3 above affect the speed of the transverse waves along a string by using the standing wave experiment from chapter 8. See figure 5.

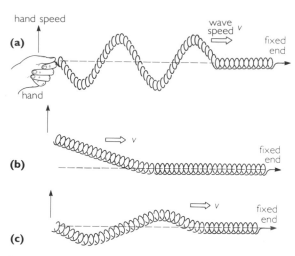

figure 4   Various transverse waves travelling along a spring

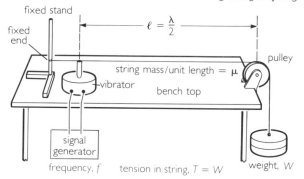

figure 5   Standing wave experiment

In general $\qquad v = f\lambda = 2f\ell$

for the fundamental standing wave.

1   We will keep ℓ constant, in which case $v \propto f$.

2   We vary T by varying the weight, W, attached to the end of the string. We find that when we adjust f so that the fundamental standing wave appears, then

$$f \propto \sqrt{W}$$

We conclude that $v \propto \sqrt{T}$.

3   We can vary the mass/unit length, μ, by selecting different strings, keeping ℓ and W constant. We find in this case:

$$f \propto \frac{1}{\sqrt{\mu}}$$

$$\Rightarrow \qquad v \propto \frac{1}{\sqrt{\mu}}$$

Combining these results leads to

$$v \propto \sqrt{\frac{T}{\mu}}$$

Measuring each of the quantities in this equation leads to the conclusion:

$$v = \sqrt{\frac{T}{\mu}}$$

$$v = \sqrt{\frac{\text{tension}}{\text{mass/unit length}}}$$

This expression is true for transverse waves along taut springs, strings or wires.

Example A guitar string has a mass of 25 g and a total length of 75 cm. Calculate the frequency of the note produced when the length is limited to 50 cm, tension = 6000 N, and it is plucked.

$$\mu = 0.025 \text{ kg}/0.75 \text{ m} = 0.03 \text{ kg m}^{-1}$$

$$\Rightarrow \qquad v = \sqrt{\frac{T}{\mu}}$$

$$= \sqrt{\frac{6000 \text{ N}}{0.03 \text{ kg m}^{-1}}}$$

$$= 450 \text{ ms}^{-1}$$

standing wave, $\ell = \lambda/2$

$\Rightarrow$ wavelength $\lambda = 1$ m

$\Rightarrow$ frequency $f = v/\lambda$

$$= 450 \text{ Hz}.$$

Longitudinal waves

Consider the row of trolleys in figure 6. They are all interlinked with pairs of spiral compression springs.

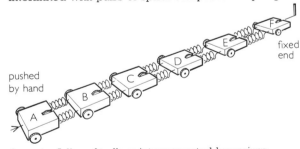

figure 6 A line of trolleys interconnected by springs

When trolley A is suddenly displaced a short distance towards B and then held stationary in its new position, you will notice that successive trolleys B, C, D, etc. move on with a constant time-lag between their motions. That is, a wave moves along the system at constant speed. To analyse the progress of the wave we should consider the system as a line of coupled oscillators. The motion of any trolley, e.g. C, will be determined by the tensions in the springs immediately before and after it.

We shall treat two springs joining trolleys as one spring of spring constant k. The motion of the trolleys will also depend on the mass M of each trolley. Let

us think about what happens to C as the wave arrives. B will approach C, thus compressing the spring between B and C. Hence, the force to the right on C will increase, and C will start accelerating to the right. The movement of C compresses spring CD, which produces a force on C to the left. Therefore, the acceleration of C is reduced (see figure 7). A full mathematical analysis leads to the expression

$$\text{velocity of waves } v = x\sqrt{k/M} \qquad \text{(i)}$$

where x is the distance between the centres of successive trolleys in the undisturbed row; see figure 7.

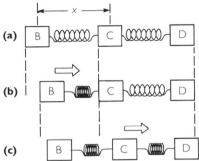

figure 7 A longitudinal wave passing along a row of trolleys

The speed of sound in a solid

Picture a solid as consisting of atoms, into which the mass is concentrated, interlinked by springy electronic bonds (figure 8(a)). We wish to convert equation (i) into a more useful form in terms of the Young modulus and the density.

$$\text{The Young modulus } E = \frac{\text{longitudinal stress}}{\text{longitudinal strain}}$$

$$= \frac{\text{force/area}}{\text{extension/original length}}$$

The above equation involves stress and strain. Therefore, it is valid for a solid of any dimensions. We shall simplify the discussion by considering the solid to be made up of a single line of atoms as shown in figure 8(b)).

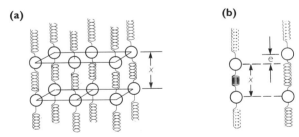

figure 8 A solid considered as an array of atoms linked by springy electronic bonds

Strain is independent of length, therefore we limit the discussion to two neighbouring atoms.

i.e.
$$E = \frac{F/A}{e/x} = \frac{Fx}{Ae}$$

where
- e = the extension of a bond,
- F = the tension in the stretched bond,
- x = the equilibrium separation of atoms.

We shall assume that the springy electronic bonds obey Hooke's law: $F = ke$; and that the area available for each atom is $A = x^2$.

Thus
$$E = \frac{Fx}{Ae}$$
$$= \frac{(ke)x}{x^2 e}$$
$$= \frac{k}{x}$$

which gives $k = Ex$.

If M is the mass of an atom then the density $\rho = M/x^3$, which gives $M = \rho x^3$.

Substituting the above expressions for k and M in the earlier equation (i) gives

$$v = x \sqrt{\frac{Ex}{\rho x^3}}$$

i.e.
$$v = \sqrt{\frac{E}{\rho}} \qquad (ii)$$

Example For steel: the Young modulus $E = 2.0 \times 10^{11}\,\mathrm{Nm^{-2}}$
The density $\rho = 7.8 \times 10^3\,\mathrm{kgm^{-3}}$
Hence the speed of sound in steel, c, is

$$c = \sqrt{\frac{E}{\rho}}$$
$$= \frac{2.0 \times 10^{11}\,\mathrm{Nm^{-2}}}{7.8 \times 10^3\,\mathrm{kgm^{-3}}}$$
$$= 5.1 \times 10^3\,\mathrm{ms^{-1}}$$

Sound waves in gases

Sound exhibits all the properties of waves except polarisation. For example, it is diffracted through open doorways, exhibits interference when two loudspeakers are fed from one oscillator, and passes through a closed window without producing a net displacement of the window. This implies that sound is a form of wave motion. Since sound cannot travel through a vacuum, but it can travel through rigid solids, it must be a mechanical wave involving the molecules or atoms of the medium through which it is travelling.

In order to arrive at a mental image of a sound wave, let us consider the effect of the loudspeaker in figure 9. When the air is pushed to the right by the cone of the loudspeaker it is compressed a little. The molecules in the immediate vicinity of the loudspeaker-cone will gain momentum to the right.

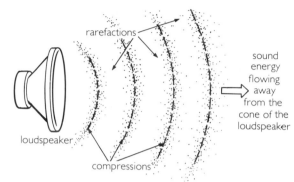

figure 9 A loudspeaker emitting sound

At standard atmospheric temperature and pressure there are about 2×10^{25} gas molecules per cubic metre and they are undergoing about 10^{10} collisions per second with other gas molecules. Thus this extra momentum will be passed on from molecule to molecule by the intermolecular collisions. The speed at which this momentum is passed on must depend on the speed at which the molecules are moving. The mean speed for air molecules at s.t.p. is $4.5 \times 10^2\,\mathrm{ms^{-1}}$.

Now consider a molecule at A in figure 10. It picks up the extra momentum at A, then travels at the average speed of $4.5 \times 10^2\,\mathrm{ms^{-1}}$ in any direction such as 1, 2, 3 or 4.

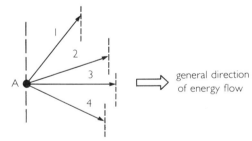

figure 10 Free paths of molecules

If the air molecule follows path 3, then the extra momentum to the right travels to the right at $4.5 \times 10^2\,\mathrm{ms^{-1}}$; but if it follows any of the other paths then the extra momentum (the sound) will progress to the right at less than $4.5 \times 10^2\,\mathrm{ms^{-1}}$. Hence, the average rate of progress of the extra momentum is less than $4.5 \times 10^2\,\mathrm{ms^{-1}}$. Therefore, we should expect the speed of sound to be less than $4.5 \times 10^2\,\mathrm{ms^{-1}}$. Experimentally it is found to be about $3.3 \times 10^2\,\mathrm{ms^{-1}}$, which is in good agreement with our model.

Measurement of the speed of sound in air

We can measure the speed of sound using the apparatus illustrated in figure 11.

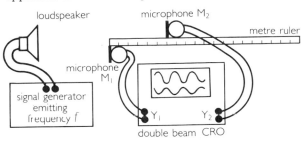

figure 11   Measurement of the speed of sound

We set the signal generator to emit a steady note, say 3000 Hz. We set the two microphones side by side initially, with the CRO triggering from trace Y_1 (to which M_1 is connected). The two traces will appear exactly in phase – above one another – since the waves take the same time to reach M_1 and M_2. We now move M_2 slowly away from the speaker. Sound now takes longer to reach M_2, and the Y_2 trace moves to the right. By the time M_2 has moved exactly one wavelength, the traces will again line up above each other.

In the experiment, we move M_2 through several wavelengths, counting each time the traces line up. Thus we can get an accurate measurement of the wavelength λ.

For example, suppose that in moving through eight wavelengths M_2 moves a distance of 89 cm.

Then $\lambda = \dfrac{89\ \text{cm}}{8} = 11.1\ \text{cm}$

and $v = f\lambda \quad = 3000\ \text{Hz} \times 11.1\ \text{cm}$

$= 333\ \text{ms}^{-1}$

The Doppler effect

When a train travelling at speed passes through a station, a stationary observer on the platform will notice that the note emitted by the train's whistle appears to drop suddenly as it passes. This is an example of the Doppler effect. When dealing with sound waves the relative motion of source and observer can be separated into two parts, that due to the motion of the source, and that due to the motion of the observer.

Moving source of sound   Figure 12
illustrates successive compression waves emitted by source S as it moves towards observer O at a speed

of v_s. The outermost wave was emitted by S when it was at position 1. The waves will pass the observer at the normal speed of sound, v, but their wavelength will be reduced due to the source's motion towards O.

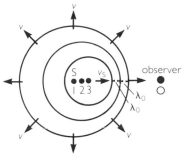

figure 12   Moving source

Since the wavelength of the waves is reduced, the observer will hear a higher frequency f_o than the actual frequency f_s being emitted by the source. The observed wavelength, λ_0, will be less than the wavelength, λ_s, which is produced when the source is stationary.

$$\lambda_0 = \lambda_s - v_s T$$

Whether the source is moving or stationary, the waves pass the observer with speed v, therefore, $v = f_o \lambda_0$ and $v = f_s \lambda_s$. Substituting these in the above equation gives

$$\frac{v}{f_o} = \frac{v}{f_s} - v_s \cdot \frac{1}{f_s}$$

Hence f_o and f_s are related by this equation:

$$f_o = f_s\left\{\frac{v}{v - v_s}\right\}$$

In this equation, $v_s > 0$ when the source approaches the observer, and $v_s < 0$ when the source recedes from the observer.

Moving observer   Figure 13 illustrates
successive compression waves moving out from a stationary source. Because the observer is moving with speed v_o towards the source, he will cross more compression waves in unit time than if he were stationary. Each time he moves one more wavelength towards the source he will cut an extra compression.

A stationary observer would intercept f_s waves per second, but an observer moving towards the source would intercept an extra v_o/λ_s per second. Therefore the observed frequency, f_o, is given by

$$f_o = f_s + \frac{v_o}{\lambda_s}$$

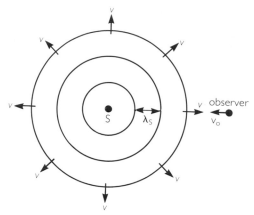

figure 13 Moving observer

Hence, because $v = f_s\lambda_s$, f_o and f_s are linked by the equation:

$$f_o = f_s\left\{\frac{v + v_o}{v}\right\}$$

In this equation $v_o > 0$ for an observer approaching the source, and $v_o < 0$ for an observer receding from the source.

Example A sports car is approaching a stationary observer. If the car is travelling at 70 mph ($31\,\mathrm{ms}^{-1}$), what frequency will an 'A' note (440 Hz) produced by the car's horn appear to have to the observer? Take the speed of sound in air to be $330\,\mathrm{ms}^{-1}$.

$$f_o = f_s[v/(v - v_s)]$$

i.e.

$$f_o = \frac{(440\,\mathrm{Hz}) \times (330\,\mathrm{ms}^{-1})}{(330\,\mathrm{ms}^{-1} - 31\,\mathrm{ms}^{-1})}$$

$$= \frac{(440\,\mathrm{Hz}) \times (330\,\mathrm{ms}^{-1})}{(299\,\mathrm{ms}^{-1})}$$

$$\approx 486\,\mathrm{Hz}$$

This is almost a whole note higher. (A note 'B' has a frequency of 494 Hz.)

Doppler effect in light

The theory of special relativity leads to the following equation for the observed frequency f_o, when a source and observer approach each other with a relative velocity v:

$$f_o = f_s\left\{\frac{\sqrt{(1 - v^2/c^2)}}{1 - v/c}\right\}$$

where v is the speed of light.

If $v \ll c$ then $(1 - v^2/c^2) \simeq 1$ and the above equation simplifies to

$$f_o = f_s[c/(c - v)]$$

Police radar traps for speeding motorists use this last equation, but astonomers use the full equation when looking at the spectra from distant galaxies, which are receding from us at more than 90% of the speed of light.

Summary

1   The basic vocabulary of waves:
   a *Progressive* means that it is a travelling wave.
   b *Transverse* means the oscillations are perpendicular to the direction in which the wave is travelling.
   c *Longitudinal* means the oscillations are parallel to the direction in which the wave is travelling.
   d The *medium* is the system carrying the waves.
   e The *displacement* of any point in a wave is the distance of that point from its undisturbed position. Displacement is a *vector*.
   f *Amplitude A* is the greatest displacement.
   g *Wavelength* λ is the distance along the wave between successive points which move in phase with one another.
   h *Frequency f* is the number of complete waves passing a fixed point per unit time.
   i *Time-period T* is the time taken by any point in the wave to complete one oscillation; $T = 1/f$.

2   Speed of transverse waves along a wire:
   a The speed is independent of the shape of the wave.
   b Speed $v = \sqrt{T/\mu} = \sqrt{\text{tension}/(\text{mass/length})}$.

3   Speed of longitudinal waves:
   a Along a line of interlinked trolleys the speed v is
   $$v = x\sqrt{k/M}$$
   b Speed of sound through a solid $v = \sqrt{E/\rho}$.

4   Sound waves in gases:
   a The speed of sound is slightly less than the mean speed of the gas molecules: for example, $330\,\text{ms}^{-1}$ compared with $450\,\text{ms}^{-1}$ for air at s.t.p.
   b The speed of sound can be measured using a double-beam CRO.

5   Doppler effect:
   a Moving source, stationary observer (sound):
   $$f_o = f_s\left\{\frac{v}{v - v_s}\right\}$$
   b Moving observer, stationary source (sound)
   $$f_o = f_s\left\{\frac{v + v_o}{v}\right\}$$
   c Doppler shift of light:
   $$f_o = f_s\left\{\sqrt{\frac{(1 - v^2/c^2)}{1 - v/c}}\right\}$$
   d Doppler shift of light: $(v \ll c)$
   $$f_o = f_s[c/(c - v)]$$

Questions

1   Consider this simplified model of a caterpillar. It is 4 cm long, and consists of 12 loosely-connected sections. When it is moving forward, one compression pulse starts at its tail every second, and moves forward towards its head. There is one pulse travelling at any one time; as one ends, the next one starts. During a pulse, the forward movement of a section is 5 mm; at any one moment during a pulse there are two adjacent sections in motion. See figure 14.

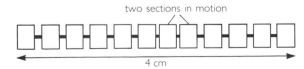

two sections in motion

4 cm

figure 14

   a For what fraction of a second is any one section in motion?

   b Assuming the section instantaneously acquires a steady velocity, which it keeps until it stops again, what is that steady velocity?
   c Draw a graph of speed against time for an individual section, for a period of 3 s. Mark on your graph any numerical information you can.
   d Draw a similar graph of displacement against time for a section.
   e At what speed does the caterpillar advance?

2   Figure 15 shows a hand sending a wave along a length of sand-filled rubber tubing.

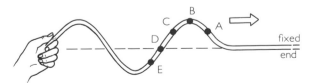

figure 15

a Describe the subsequent motion of the point labelled A.

b If *up* the page is taken as *positive* and *down* the page is taken as *negative*, state whether the velocities and accelerations of each of the labelled points are positive, zero or negative at the moment illustrated.

c This experiment is often used to demonstrate waves in classrooms. Why do teachers often fill the rubber tubing with sand?

d Explain *qualitatively* how increasing the tension in the tubing results in an increased wave-speed.

3 During the World Cup football matches in Mexico in 1986, the crowd in a stadium became the medium through which transverse waves passed. Consider a simplified model of this wave. Assume that it takes a spectator 2 s to get up from a seat and then sit down again. As each person will not want to move out of turn, he will not contemplate moving until he sees his neighbour move. A reasonable assumption for the reaction time of a spectator is 0.25 s. Take the distance from the centre of one seat to the centre of the next seat as 0.6 m.

a Draw a sketch, using squares to represent people, of an instantaneous situation as the wave passes through a row of 20 people.

b Draw a graph of displacement against time for one person. Put appropriate numerical values on the axes.

c Determine the speed of the wave.

d Although this wave running round the stadium is an isolated pulse, it can be thought of as part of the continuous wave which would result from the people repeatedly rising and sitting. Determine its effective wavelength.

4 a Distinguish between transverse and longitudinal waves.

b Explain why transverse waves cannot pass through a gas.

c A violin string has a mass of 12 g and a length of 36 cm. Calculate the frequency of the note produced when the length is limited to 15 cm and the tension is set at 580 N.

d A neighbouring string on the violin has a mass of 15 g for a length of 36 cm. Calculate the tension needed if the note produced is 334 Hz when the length is limited to 15 cm.

5 The speed at which waves travel through a medium depends on two factors, a *force factor* and an *inertial factor*. The force factor represents the restoring forces which are brought into being when a wave displaces parts of the medium from their equilibrium positions. The inertial factor represents the inertia of the system to respond to these forces. The general expression is of the form

$$\text{wave speed} = \sqrt{\frac{\text{force factor}}{\text{inertia factor}}}$$

a Explain by a *qualitative* argument why this ratio gives the speed of the waves.

b What are the force factor and inertial factor for transverse waves travelling along a stretched wire?

c Consider the steel wires of different diameters, but each stretched by a load of 500 N. If wire A has twice the diameter of wire B, what will be the ratio of the frequencies emitted by each when they are plucked?

d What are the force factor and inertial factor for longitudinal waves passing along a metal rod?

e Consider two parallel lines of cars in a traffic jam. One line consists of high performance sports cars, each containing only the driver, the other line consists of large family saloons, each containing a large family. When the leading vehicle in each line moves forward, the second, third, and so on, will follow. This will result in a 'decompression' wave moving along each line. Discuss what factors determine the speeds of these pulses, and say which one will be the quicker.

6 A person standing on a platform in a railway station hears the whistle of an approaching express train. She notices that there is a sudden drop in frequency as the train passes her.

a Explain why she hears an increased frequency when the train approaches her, and a decrease in frequency when it is going away from her.

b Calculate the frequencies she observes when the train's whistle sounds at 800 Hz and it is travelling at 200 km h^{-1}. (The speed of sound in air is 330 m s^{-1}.)

c A person standing 400 m from the railway track hears the same train approach and pass. What frequencies will he observe when the train's whistle is 566 m from him?

7 a Explain what is meant by the *Doppler effect*.

b Explain why a cyclist hears a note of lower frequency when he cycles away from a siren.

c Consider a cyclist who approaches a factory which is sounding a siren at a frequency of 200 Hz. Calculate the frequency observed by the cyclist if he is travelling at 50 km h^{-1}.

d What frequency would the cyclist observe if he were capable of cycling at 330 m s^{-1} (i) towards the source, (ii) away from the source? (The speed of sound is 330 m s^{-1} through air.)

Consider a police car following an ambulance. Both have sirens which operate at 300 Hz and both are travelling at 100 km h^{-1}.

e If the ambulance is sounding its siren, what frequency will the policeman hear, assuming that his own siren is switched off?

f If the police car is sounding its siren, what frequency will the ambulance driver hear when his own siren is off?

8 The spectrum of the Sun is a continuous spectrum on which certain dark lines appear. These lines are known as Fraunhofer lines after the man who first discovered them. They are due to atomic gases in the upper atmosphere of the Sun. The atoms *absorb* certain wavelengths from the light escaping from the Sun's surface, and the lines are characteristic of the elements (e.g. sodium) in the upper atmosphere of the Sun. All stars will have similar elements in their upper atmospheres. Therefore, they will produce continuous spectra crossed by dark lines in the same positions. When a telescope collects light from a distant galaxy and passes it through a diffraction grating, the dark lines appear to have a similar pattern to those obtained from the Sun, but they are all displaced towards the red end of the spectrum.

a Assume that this shift in the dark lines is due to the Doppler effect. Explain what you can deduce from it.

b What would a shift towards the blue end of the spectrum indicate?

Figure 16 represents two spectra. The upper one was obtained from the Sun, the lower one was obtained from a not-too-distant galaxy. Note that the line labelled C appears in the same place in each spectrum. It is due to oxygen in the Earth's atmosphere.

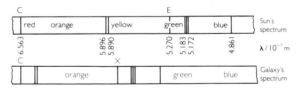

figure 16

c Explain why the C-line appears in the same place in each spectrum.

d In the sun's spectrum the line labelled E has a measured wavelength of 5.270×10^{-7} m and appears in the green part of the spectrum. This line is due to iron atoms. In the galaxy's spectrum it is labelled X, has a measured wavelength of 5.809×10^{-7} m, and appears in the yellow part of the spectrum. Calculate the relative velocity of the Earth and the galaxy.

e The absorption lines on the galaxy's spectrum are all slightly broader than those on the Sun's spectrum. Why might this be so?

9 A piano tuner uses the principle of beats to bring the strings of an instrument to the correct frequency. He has a tuning fork which gives an accurate frequency of 440 Hz (an 'A'). Imagine the retaining screw has slipped slightly and the string no longer gives 440 Hz.

a Will the frequency of the note be more or less than 440 Hz? Explain your answer. Figure 17 represents the two notes produced when the piano note and the tuning fork are sounded simultaneously.

b Copy figure 17 and, by applying the principle of superposition, add the two waves together to show what the resultant oscillations entering the ear of the piano tuner will look like.

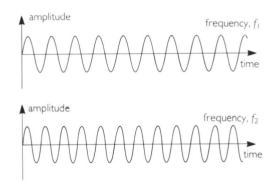

figure 17

c If the piano note has slipped to 436 Hz how many times per second will the combined tuning fork and piano notes be in phase? (That is, how many times will they 'beat' loudly per second.) This is known as the *beat-frequency*.

d By what percentage should the piano tuner change the tension in the wire when he tunes the piano to the correct frequency?

CHAPTER 29

Electromagnetic waves

Introduction

We start by analysing in detail what we mean by an electromagnetic wave. Then we introduce plane-polarisation and how it can be used to test the transverse nature of electromagnetic waves. This leads into the rotation of the plane of polarisation and Malus's law. We follow with a section on the production and uses of polarisation which includes Brewster's law. We follow this with the measurement of the speed of light and finish with the tests used to identify electromagnetic waves.

GCSE knowledge

It will be assumed that you can recall:

☐ the magnetic field pattern around a wire carrying a current;

☐ that when a magnetic field moves relative to a wire an e.m.f. is induced in the wire;

☐ that a transverse wave is one in which the oscillations are perpendicular to the direction in which the wave is travelling;

☐ that a longitudinal wave is one in which the oscillations are parallel to the direction in which the wave is travelling.

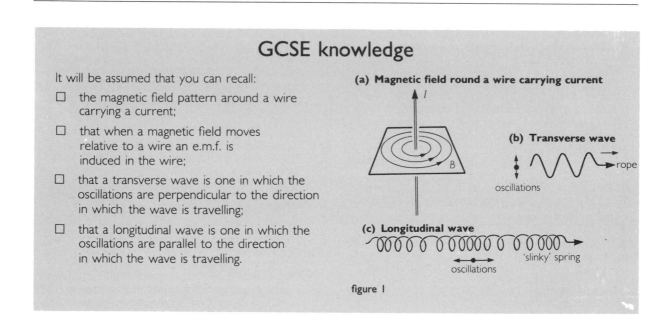

(a) Magnetic field round a wire carrying current

(b) Transverse wave

(c) Longitudinal wave

figure I

Electromagnetic wave

What do we mean by an electromagnetic wave? In order to make this clear, let us consider the signal emitted from a transmitting aerial at a radio station. We shall first of all consider the transmission of the signal, and then the reception of the signal.

The emission from a transmitting aerial In figure 2(a) and (b) we illustrate the magnetic fields which exist around long, straight wires carrying electric currents. The rule to remember for predicting the direction of these magnetic fields is Maxwell's right hand screw (or grip) rule which is illustrated in figure 2(c). We met it earlier in chapter 24.

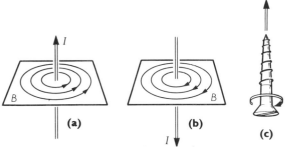

figure 2 Maxwell's right hand screw rule

Consider the long straight piece of wire in figure 3. When the switch S is closed, a large, steady d.c. current passes vertically upwards through the wire. Usually the region around the wire is only affected by the weak magnetic field of the Earth, which causes all the small plotting compasses A, B, C etc., to line up in a long vertical straight line with all their needles pointing north, as shown.

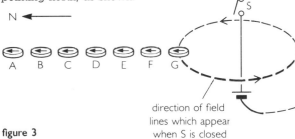

figure 3

When the switch is closed, the needles will all point out of the paper towards you. That is, they will line up with the magnetic field around the wire as indicated in figure 2(a).

Next, imagine that the long, straight, vertical wire is fed from a square-pulse generator as depicted in figure 4(a). The current in the wire keeps on reversing, therefore the magnetic field around the wire keeps on reversing. Hence, there must be rings of magnetic field around the wire, each successive ring pointing in the opposite direction as shown in figure 4(b).

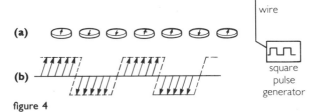

figure 4

The magnetic field will move away from the aerial in rings, just like waves on the surface of a pond when a rod is repeatedly pushed into and then pulled out of the surface of the pond.

Now let us consider what is produced when the square-pulse generator feeding the wire is replaced with a sine-wave generator. The strength of the

magnetic field emitted from the wire is proportional to the magnitude of the electric current in the wire. In figure 5 the instantaneous strength of the magnetic field at each point is represented by the length of the arrow. The fact that the magnetic B-field pattern is moving away from the wire is indicated by the heavy arrow in the diagram. Thus the wire is really acting as an aerial sending out waves of B-field.

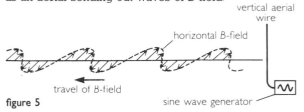

figure 5

The reception at a receiving aerial

Let us start by considering what happens inside a wire when a strong horseshoe magnet is pulled rapidly away from it. This is shown in figure 6. The 'free' charge carriers inside the wire, that is, the electrons, will experience a force which drives them along the wire. We say that an e.m.f. is induced in wire. This was discussed earlier in chapter 26.

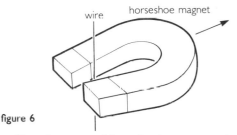

figure 6

Next, let us consider what happens to the e.m.f. induced in the wire when the array of north and south poles shown in figure 7 is pulled rapidly past the wire. The direction of the e.m.f. induced in the wire will reverse as each successive pair of north and south poles passes the wire.

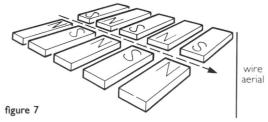

figure 7

Now let us consider what happens to the e.m.f. induced in the vertical wire by the passage of the varying magnetic field shown in figure 5: that is, by the passage of the varying magnetic field travelling from the transmitting aerial to the wire. This sinusoidally varying magnetic field will induce a sinusoidally varying e.m.f. in the wire. Hence, we call the wire a receiving aerial. The process is illustrated in figure 8.

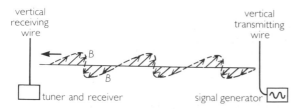

figure 8

Now here is rather a subtle idea. We needed to imagine putting a receiving wire in the path of the B-field waves, in order to provide some free electrons which could be made to move along it by the induced e.m.f. What causes the electrons to move vertically? Answer – an induced electric field. Now imagine that the wire isn't there – is there still an induced electric field? Answer – yes; provided there is a changing B-field at a particular point in space, then there must always be an electric E-field associated with it.

Hence, we call the wave travelling from transmitter to receiver in figure 8 an *electromagnetic wave*. In diagrams we usually represent it as shown in figure 9.

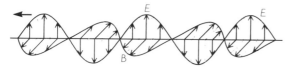

figure 9   An electromagnetic wave

Polarisation

In the previous section we have built up the idea that electromagnetic waves behave as transverse waves. In order to confirm this idea we need a test to distinguish between transverse and longitudinal waves.

A mechanical analogue of what we shall discuss for electromagnetic waves is depicted in figure 10, where a long spring connects points A and B.

Waves are sent along the spring by moving the end A. The spring passes between two closely spaced wooden blocks, C and D. When the wooden blocks are oriented as in figure 10(b), the slit between the blocks is perpendicular to the plane of the transverse wave, which will stop energy flowing through to B. However, with longitudinal waves, there is no orientation of the blocks which has this effect. So, if we can devise a similar test for electromagnetic waves we can check that they are transverse waves.

The test we have just discussed illustrates the idea of polarisation. We say that if *all* the oscillations in a *transverse* wave are in *one plane*, then the wave is a *plane-polarised wave* .

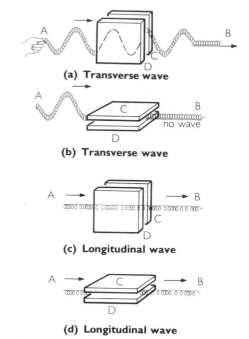

(a) Transverse wave

(b) Transverse wave

(c) Longitudinal wave

(d) Longitudinal wave

figure 10

Plane-polarisation of electromagnetic waves

We have just been saying that, when the varying horizontal magnetic field passes the vertical receiving wire in figure 8, then an e.m.f. is induced in the wire. But will the vertical wire have an e.m.f. induced between its ends when a *vertical* magnetic wave passes by? From the laws of electromagnetic induction (chapter 26) we know that the e.m.f. induced in this case will be zero.

Now, returning to the horizontal B-field and vertical aerial, let us consider what occurs when an e.m.f. is induced in the wire. An electric current oscillates up and down in the wire, which means that energy has been absorbed from the wave into the wire. The resistance of the wire will turn some of the energy into heat, but the oscillating current will radiate a circular magnetic wave pattern. This idea is illustrated in figure 11.

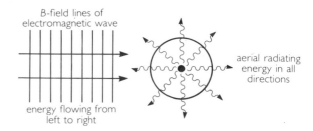

figure 11

Thus, energy that was originally flowing from left to right will, after absorption, be re-radiated equally in all directions. So, the amount of energy radiated by the aerial in a direction exactly parallel to the original direction of energy flow will be very small indeed. Hence, we should expect waves with a horizontal B-field to be absorbed by and then re-radiated from a grill of vertical wires, whereas waves with a vertical B-field should just pass straight through the grill. These ideas are illustrated in figures 12(a) and (b). A practical demonstration, using microwaves, is depicted in figures 12(c) and (d).

The fact that the waves fail to pass through the grills when they are orientated vertically as in (a) and (c) of figure 12, yet do pass through in the other two situations, indicates that the radiowaves in (a) and (b) and the microwaves in (c) and (d) are transverse waves.

Rotation of plane-polarised waves

We have seen that a horizontal B-field wave is absorbed by the vertical grill of metal rods, whereas a vertical B-field wave passes through the grill. If the horizontal and vertical waves, of equal amplitude and frequencies, are transmitted simultaneously, with zero phase difference, the net effect will be a plane wave at 45° to the vertical. The amplitude of this wave will be $\sqrt{2}$ times the amplitude of the vertical (or horizontal) wave, as shown in figure 13(a).

When this wave (at 45° to the vertical) tries to pass through a vertical grill of wires, as shown in figure 13(b), what form does the wave have as it passes out on the right hand side of the grill? The horizontal component of the magnetic wave will be absorbed and the vertical component will pass through.

Now let us ask the question 'What effect has the vertical polarising grill on the incident, plane-polarised wave?' We could answer this question by saying that the grill allows only the vertical component of the magnetic field to pass through it. Alternatively, we might say that the vertical grill does two things to the electromagnetic wave:

☐   it rotates the plane of the magnetic vector to be vertical;
☐   it reduces the initial amplitude A_0 of the wave by a factor of $\sqrt{2}$; or multiplies it by cos 45°, to A_0 cos 45°:

$$A_1 \text{ (amplitude leaving grill)} = A_0 \cos 45°$$

In figure 13(b) we were dealing with a wave whose magnetic vector was at 45° to the vertical of the grill.

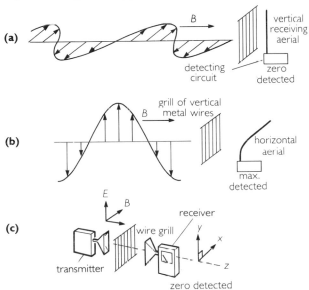

figure 12

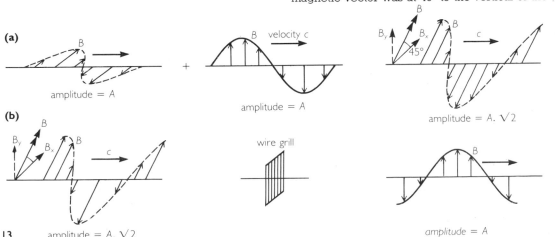

figure 13

If the angle between the magnetic vector and the grill is θ then the second statement above will become

☐ 'it will multiply the amplitude of the wave by a factor of cos θ'.

Since the electromagnetic waves oscillate with a simple harmonic motion, the energy in the oscillations is proportional to the (amplitude)^2. Hence, in general,

$$\text{intensity } I \propto A^2$$

(Intensity I is defined as 'energy flow through unit area per unit time'. It thus has units of Wm^{-2}.)

Now
$$A_1 = A_0 \cos\theta$$
$$\Rightarrow \qquad A_1{}^2 = A_0{}^2 \cos^2\theta$$
$$\Rightarrow \qquad I_1 = I_0 \cos^2\theta \qquad (i)$$

where I_0 is the intensity of the plane-polarised wave incident upon the grill, and I_1 is the intensity of the wave leaving the grill. Equation (i) is known as Malus's Law. Sheets of Polaroid film can be used to demonstrate this effect with light, as shown in figure 14.

The fact that the Polaroid rotates the plane of the electromagnetic wave is proved by the result in figure 14(c). In experiments such as that shown in figure 14(c) it is usual to refer to the first sheet of Polaroid, A, as the *polariser* and the final sheet, C, as the *analyser*. The nature of the light emerging from C gives information about the *sample* placed at B.

Production of plane-polarised light

1 *By Polaroid sheets* We mentioned this at the end of the last section. Polaroid sheets are sheets of a particular plastic in which tiny crystals of iodine-quinine sulphate are all aligned in the same direction. The effect of the aligned crystals on light waves is similar to the effect of the wire grills on microwaves, which we have just been discussing.

2 *By reflection* Consider unpolarised light incident on the glass surface in figure 15. There will always be a reflected ray and a refracted ray. If the reflected ray passes through a sheet of Polaroid, its polarisation can be checked by rotating the Polaroid sheet as indicated in the diagram.

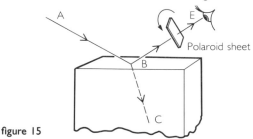

figure 15

If you carry out this experiment you will discover that the intensity of the light reaching E will vary as the Polaroid sheet is rotated. At one particular angle of incidence, about 57° for ordinary glass, we find that it is possible by rotating the Polaroid sheet to prevent any of the reflected light from reaching E. This must mean that the reflected light is plane-polarised. This angle is known as the *polarising angle* or the *Brewster angle* θ_B; Brewster discovered that when the reflected beam is plane-polarised the angle between the reflected and refracted rays is 90°.

If you consider the top molecules of the surface of the glass, they must have vibrations in them which are parallel to those in the refracted ray, i.e. both in the plane of the paper *and* perpendicular to it, as shown in figure 16. The component of the oscillations inside the glass surface which is *perpendicular* to the plane of the paper in figure 16, can be transmitted along the reflected ray as *transverse* oscillations. Therefore, this component is reflected. But the component of the oscillations which is in the glass surface and *in* the plane of the paper would have to be transmitted along the reflected ray as longitudinal oscillations. However, we have already shown that electromagnetic waves are transverse oscillations. Therefore the reflected ray can only have oscillations perpendicular to the plane of the paper. Thus whenever the angle between the reflected and refracted rays is 90° we should expect the reflected ray to be completely plane-polarised, as Brewster discovered.

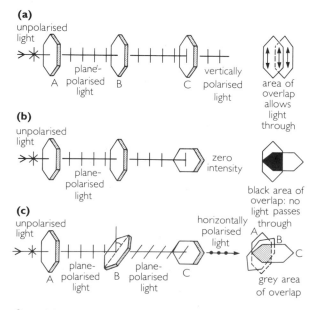

figure 14

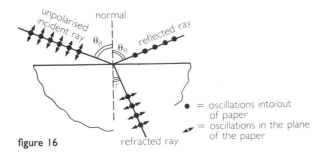

figure 16

= oscillations into/out of paper
= oscillations in the plane of the paper

Let us now apply Snell's law of refraction to figure 16.

refractive index $n = \dfrac{\sin \theta_B}{\sin r}$

Also $\theta_B + 90° + r = 180°$

$\Rightarrow$ $r = 90° - \theta_B$

$\Rightarrow$ $\sin r = \sin(90° - \theta_B) = \cos \theta_B$

Hence refractive index

$$n = \dfrac{\sin \theta_B}{\cos \theta_B}$$

$\Rightarrow$ $n = \tan \theta_B$

This is known as *Brewster's law*.

Uses of polarisation

1 *Sun glasses* Light reflected by many surfaces is either partially or completely plane-polarised. Hence, the dazzling brightness of reflections can be reduced by Polaroid sunglasses if the orientation of the Polaroid crystals is correct.

2 *Optical activity* Sugar and other solutions rotate the plane of vibration of polarised light as the light passes through them. Chemists use an instrument which depends on this effect, called a *polarimeter*, to measure concentrations of solutions.

3 *Stress analysis* If a piece of glass, or of certain plastics such as Perspex, is stressed, it will cause the plane of plane-polarised light to rotate. Different colours are rotated by different amounts. If the stressed specimen is viewed in white light between two crossed Polaroids, coloured fringes are seen. This is known as *photo-elasticity*, and is used to analyse stresses in plastic models of real objects.

Measurement of the speed of electromagnetic waves

From interference experiments we believe that light, radio waves, microwaves and X-rays are all electromagnetic waves of different wavelengths. If they are magnetic waves (i.e. electromagnetic waves)

travelling through space they should all have the same speed. Therefore we shall now consider methods of measuring their speeds. Two methods of measuring the speed of light are given here; further experiments are described in questions 9, 10 and 12.

Light waves (rotating mirror method)

Figure 17(a) illustrates one apparatus, developed by the Nuffield Advanced Physics Project from the original experiment carried out by Foucault in 1850.

Light travels from the source slit S, through the semi-silvered glass plate G, and after reflection from the rotating mirror R it traverses the 2 metres to the stationary concave mirror, M. If R is also stationary, the light retraces its path and an image of S is formed at E_0. Consider what happens if we now rotate the small mirror R at high speed. It will have turned through a small angle α by the time the light returns from M. This is shown in figure 17(b). Since the normal to R will have rotated through α, the light beam reflected back from R will be turned through 2α. Hence, the image of the slit S seen through the eyepiece will move from E_0 to E_1. If we measure $E_0 E_1$, and we know $(RG + GE_0)$, then we can calculate the angle α in radians.

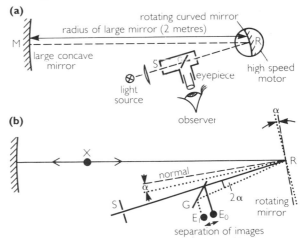

figure 17 Foucault's method for the velocity of light

The speed of rotation of the small mirror can be measured by placing a photo-transistor in the light-beam at X, in figure 17. If the output from the transistor is fed into a scaler then the number of rotations per second is obtained. Hence we can calculate the time taken by R to rotate through angle α; and we can now calculate the speed of the light travelling from R to M and back to R.

Light waves (pulses along an optical fibre)

In this method, we use a pulse generator to send short bursts of current (at 1 MHz or higher) through a light emitting diode (LED). The light passes through a long optical fibre, and is detected at the other end by a photodiode. We display the transmitted and received pulse on the two traces of a double-beam CRO. See figure 18.

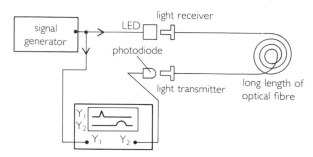

figure 18   Measuring the velocity of light with pulses along an optical fibre

The displacement of the second trace relative to the first enables us to work out the time delay, using the CRO's time base. In practice, the electronic circuitry also contributes to the delay; so we use a very short length of optical fibre to do a control measurement, which allows us to determine the delay due to the fibre.

Example   From the following data, calculate the speed of light in the optical fibre:

time base setting $0.1\,\mu\mathrm{s\,div}^{-1}$
short length of fibre $0.20\,\mathrm{m}$
displacement of trace using short length $2.1\,\mathrm{div}$
long length of fibre $20\,\mathrm{m}$
displacement of trace using long length $3.1\,\mathrm{div}$.

Distance travelled by light $= 20\,\mathrm{m} - 0.20\,\mathrm{m}$
$$\approx 20\,\mathrm{m}$$
Time $= (3.1 - 2.1)\,\mathrm{div} \times 0.1\,\mu\mathrm{s\,div}^{-1}$
$$= 0.1\,\mu\mathrm{s}$$
$$\mathrm{speed} = \frac{\mathrm{distance}}{\mathrm{time}} = \frac{20\,\mathrm{m}}{0.1\,\mu\mathrm{s}}$$
$$= 2 \times 10^{8}\,\mathrm{ms}^{-1}$$

Of course, this is the speed of light in the material of the fibre. To obtain the speed of light in air, we would need to know the refractive index n of the material, since:

$$\frac{\mathrm{speed\ of\ light\ in\ air}}{\mathrm{speed\ of\ light\ in\ material}} = n$$

Identification of electromagnetic waves

We attribute the following properties to electromagnetic waves.

☐ They are transverse waves. The waves can be plane-polarised, therefore this property is proved.

☐ The waves can travel through a vacuum.

☐ The speed of the waves through a vacuum is $3.00 \times 10^{8}\,\mathrm{ms}^{-1}$.

Summary

1 The magnetic field pattern transmitted by a long straight wire aerial which carried a sinusoidal electric current looks like figure 5.

2 E.m.f.s are induced in long straight wires by the passage of plane magnetic waves:

e.m.f. induced

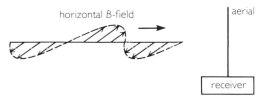

e.m.f. not induced

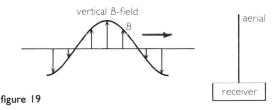

figure 19

3 Plane polarisation occurs in transverse waves but not in longitudinal waves.

4 A plane-polarised wave is one in which the B-field oscillates in only one plane, the E-field oscillates in a plane at right angles to this. See figure 9.

5 The plane of polarisation can be rotated by a metal grill. See figure 13(b).

6 The intensity of the electromagnetic wave leaving the grill is given by

$$I_1 = I_0 \cos^2\theta \quad \text{[Malus's Law]}$$

7 Light can be plane-polarised by reflection from a surface at the Brewster angle, θ_B.
Brewster's law: $\tan\theta_B = n$
See figure 16.

8 Uses of polarisation:
  a sun glasses;
  b optical activity and polarimeters;
  c stress analysis.

9 *Measurement of the speed of light*
  a Foucault's method, see figure 17(a).
     speed $= 2r/t$, where $(t = \alpha/2\pi) \times$ time for one rotation of small mirror
     Time for one rotation is determined by a phototransistor and scaler.
     Angle α, in radians, is determined from displacement observed through the eyepiece.
  b Timing pulses travelling along an optical fibre.

10 The essential properties of electromagnetic waves are
  a they are transverse waves (*test:* polarisation);
  b they can pass through a vacuum;
  c speed in a vacuum equals $3.00 \times 10^8\,\text{ms}^{-1}$.

Questions

1 a Explain what is meant by 'electromagnetic wave'.
   b If you discovered a new type of wave, what tests would you carry out in order to decide if it were an electromagnetic wave?
   c Explain what is meant by plane polarisation.
   d Explain why longitudinal waves cannot be plane-polarised.

2 a Explain as fully as you can the nature of plane-polarised radio waves.
   b Explain how a vertical transmitting aerial generates radio waves, and how they pass from the transmitting station to a distant radio receiver.
   c When people wish to pick up FM signals (Frequency Modulated signals) on their transistor radios they pull out a telescopic aerial and make sure that it is vertical. Explain what happens to the strength of the signal

detected by the radio receiver as its aerial is slowly tilted over to the horizontal.

3 Figure 20 shows a horizontal magnetic wave moving through space towards a vertical wire.

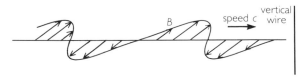

figure 20

  a Will the vertical wire have an e.m.f. induced between its ends when the varying magnetic field cuts it horizontally?
  b Will an e.m.f. be induced between the ends of the vertical wire when a varying vertical magnetic field cuts it (figure 21)?

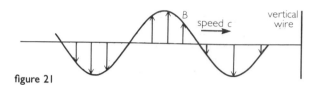

figure 21

c If an e.m.f. is induced between the ends of a vertical wire, then energy has been absorbed from the wave into the wire; in which of the cases shown in figure 22 might the detected signal be zero?

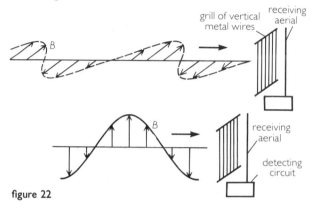

figure 22

If the horizontal and vertical waves shown in figure 22 are transmitted simultaneously, with zero phase difference, the net effect will be a plane wave at 45° to the vertical, of amplitude $= \sqrt{2} \times$ (amplitude) of the vertical (or horizontal) wave.

d When this wave (at 45° to the vertical) tries to pass through a vertical grill of wires, as shown in figure 23, what form does the wave have as it passes out of the right hand side of the grill?

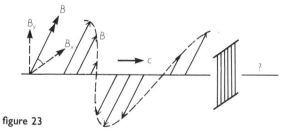

figure 23

e How is the amplitude of the wave on the right hand side of the grill related to the amplitude of the wave on the left hand side of the grill?

4 The diagrams in figure 24 represent microwave transmitters and receivers. Deduce what you can from each pair of diagrams. The meter on the receiver indicates the strength of the detected signal in each case; full-scale deflection to the right signifies maximum intensity. You should assume that the separation of transmitter and receiver is the same in each case.

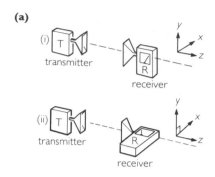

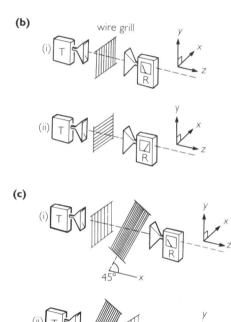

figure 24

5 Figure 25 illustrates the types of microwave transmitter and receiver that are used as teaching aids in schools.

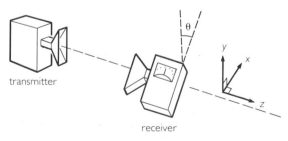

figure 25

The current shown on the microammeter varies as the receiver is rotated about the z-axis. A student obtained the following results from such an experiment:

Angle $\theta/°$		0	40	70	90
Current detected/μA		82	48	10	0

a Check these results to see if they agree with Malus's law.
b The student then sets θ to zero and places a horizontal grill of aluminium rods between the transmitter and receiver. He finds that the receiver detects 77 μA when the horizontal grill is in place, whereas without the grill it detected 82 μA. When the grill is vertical the receiver detects zero current. Explain these observations in some detail.
Next, the student removes the grill and rotates the receiver so that θ = 90°. The receiver then detects zero current.
c With two grills between the detectors as shown in figure 26, the receiver detects a current of 56 μA. When grill 2 is removed the current drops to 16 μA. Explain these observations in some detail and check them numerically with the predictions of Malus's law.

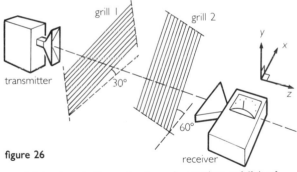

figure 26

6 Dichroism is the selective absorption exhibited by certain crystals. Iodine-quinine sulphate, used in sheets of Polaroid, is such a crystal. It has the property of selectively absorbing one of the components of ordinary light, and thus transmitting a plane-polarised beam. The crystals illustrated in figure 27 are all 'perfect polarisers', i.e. dichroic crystals which transmit a plane-polarised beam whose intensity is 50% of the incident, unpolarised intensity (actual Polaroid sheet only transmits 35%).
a When an unpolarised beam of intensity I_0 is incident on a pair of crossed polarisers (i.e. mounted with their transmission axes at 90° to one another) the intensity transmitted is zero. Explain this fact.
b When a third 'perfect polarizer' is placed between the crossed polarisers as shown in figure 27, then some light passes through the system. Explain this fact, calculate the intensity

of the transmitted beam, and say what the orientation of the B-field oscillations must be in the transmitted beam.

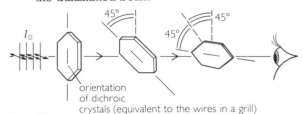

figure 27

c What happens to the intensity of the transmitted beam if the middle polariser is gradually rotated so that the angle θ between the plane of B-field vibration in the incident wave and the transmission axis of the middle polariser varies from 0° to 180°?
d A student notes that when viewing a white-hot tungsten filament through a pair of crossed polarisers she sees a blue filament. Explain what can be deduced from this.

7 The crystals in the following diagrams are dichroic crystals. That is, they selectively absorb one of the two components of ordinary light. Hence, the beam of light that they transmit is plane-polarised. Assume that these crystals are perfect polarisers, i.e. dichroic crystals which transmit 50% of the incident, unpolarised, light intensity.
a A beam of plane polarised light is shown in figure 28.
(i) If light of intensity I_0 is incident at the left, what will be the intensity detected by the observer?

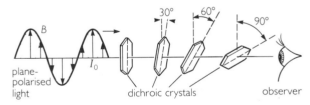

figure 28

(ii) If the incident wave is plane-polarised in a vertical plane, then what will be the polarisation of the wave detected by the observer?
b If the light incident at the left is unpolarised but of intensity I_0, what is the intensity detected by the observer?
c Explain what is meant by 'dextro-rotatory' sugar.
d Explain why with certain clear adhesive tapes (old style Sellotape) a sequence of colours is observed when the right-hand Polaroid sheet is rotated in the experimental set up illustrated in figure 29.

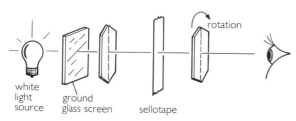

figure 29

8 Brewster's law tells us that unpolarised light incident at angle i on a plane block of transparent material is completely plane-polarised on reflection if $\tan i$ equals the refractive index n of the material.
   a Describe how you would verify this law experimentally.
   b What is the angular relationship between the reflected and refracted beams when the reflected beam is plane-polarised?
   c Explain why the reflected light *cannot* contain any oscillations perpendicular to the ones actually present in its plane-polarised beam.
   d Explain why it is not possible to produce interference effects with two beams of plane-polarised light if their planes of polarisation are at right angles to one another, even though the beams overlap.

9 The first successful measurement of the speed of light was carried out by Fizeau in Paris in 1849. The essential features of his apparatus are illustrated in figure 30.

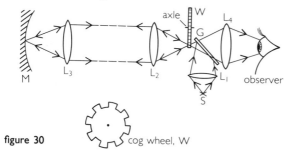

figure 30

   Light from the source S was converged by lens L_1. Some of the light was reflected by the glass block G, and then passed between the cogs of the rotating wheel W. Lens L_2 made the light parallel and after travelling about 9 km it was reflected back by mirror M. Some of the light passed through G to the observer.
   a Explain what was seen by Fizeau as he slowly increased the speed of rotation of the cog wheel.
   b Fizeau was able to determine the speed of rotation of the cog wheel accurately. Explain how he determined the time taken for light to travel from W to M and back to W.
   c Use the following data of Fizeau to calculate

the speed of light in air:
   distance W to M = 8633 m
   number of teeth on cog wheel = 720
   minimum speed of rotation to block out light = 12.6 revs^{-1}

10 Figure 31 shows a powerful positron source at the centre of a large cylindrical block of lead. The positrons, when produced, immediately attach themselves to electrons and annihilate one another, simultaneously emitting identical photons in opposite directions (by the conservation of linear momentum). The γ-ray beams are collimated by the 2 mm-diameter hole running along the axis of the lead cylinder. The ray travelling towards detector A will be detected first. The output from detector A triggers a transistor switch and starts the capacitor-charging process. The output from detector B triggers a second transistor switch and stops the capacitor-charging process.

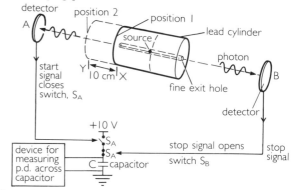

figure 31

   The potential difference across the capacitor will rise in an approximately linear manner provided that the charging time is small compared with the CR-value of the charging path (figure 32). When this is the case, halving the charging time will halve the potential difference produced across the capacitor.

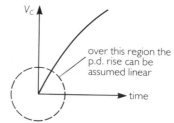

figure 32

   Suppose that when the source is in position 1 the p.d. measured across the capacitor is 1.000 V.
   a When the source is moved to position 2, what happens to (i) the distance that the photon travelling to detector A has to travel, (ii) the distance that the photon travelling to detector B has to travel?

b Will the p.d. generated across the capacitor, when the source is in position 2, be larger than, smaller than, or equal to 1.000 V?

c With the source still in position 1, a piece of cable which delays the signal by 1.000 ns is inserted between detector B and switch B. The p.d. across the capacitor is now found to reach 2.000 V. Explain why the p.d. reached has increased.

d When the source is moved from position 1 to position 2, the magnitude of the final p.d. across the capacitor increases from 1.000 V to 1.667 V. Calculate the speed of γ-rays.

e The measured distance between positions 1 and 2 was 10.00 cm. Estimate the uncertainty in the value that you obtained for the speed of γ-rays.

11 Figure 33 shows two crossed polarisers A and B. If a liquid which does not polarise light is placed between A and B, then no light will pass through to the observer.

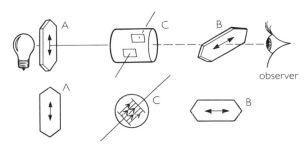

figure 33

For certain liquids, e.g. nitrobenzene, an electric field applied between the metal plates in cell C will cause the liquid to polarise the light. Such a cell is known as a Kerr Cell. The individual molecules of nitrobenzene are anisotropic in their interactions with light; also they are polar molecules.

a Explain what is meant by a polar molecule.

b Explain how it is that light passes through the system when an electric field is applied between the metal electrodes in cell C, yet no light passes through when the field is removed.

12 The problem with many early methods of measuring the speed of light was that the light had to pass through many kilometres of varying atmospheric conditions. These large distances had to be used because of the slow rate at which their mechanical systems chopped the light. Because the speed of light varies with atmospheric conditions these measurements did

not achieve reproducible results. This problem has been overcome by using a Kerr Cell containing nitrobenzene, across which a high-frequency square-wave voltage is applied. The effect of this is to gate the light, allowing it to pass only for short periods when the voltage is on. The essential features of the apparatus are shown in figure 34.

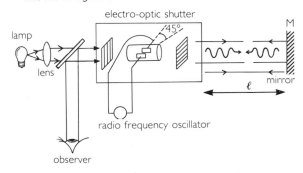

figure 34

Pulses of light travel a measured distance ℓ to a mirror, which sends the light back along its original path. If the p.d. on the Kerr Cell is zero when the pulse of light returns, the cell will be opaque and the observer will not see the light.

a Measurements are made by varying the frequency of the applied p.d. until the observer sees darkness. Explain whether the observer should record the *lowest* or the *highest* frequency at which he is able to achieve darkness, in order to calculate the time taken by the light to travel to mirror M and back.

b An experimenter obtains darkness when the radio-frequency oscillator is producing 5.98 MHz, and decides that this is the correct frequency for the calculation. If the distance from the electro-optic shutter to the mirror M is 25.00 m, what value did she obtain for the speed of light?

c The experimenter then introduced a pipe full of water into the path of the light pulses (figure 35). She found that the appropriate frequency for extinguishing the light was 5.624 MHz. Use this and your answer to part (b) to determine a value for the speed of light through water.

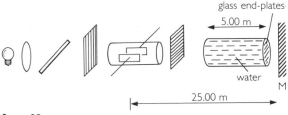

figure 35

CHAPTER 30

Diffraction

Introduction

In this chapter we pick up some ideas you may have met qualitatively before in your GCSE course about diffraction or spreading out of waves after they pass through a gap. We develop a model to explain the patterns found, and consider measurements which support the model. We see that one consequence is that there is a limit on the fine detail that your eye, or any instrument depending on information from waves, can resolve. Finally we consider briefly how diffraction affects the design of aerials.

GCSE knowledge

To understand this chapter, it will help if you recall any qualitative ideas about diffraction which you met at GCSE level. You probably did experiments with water ripples passing through gaps, and may have discussed other examples, particularly light and sound. You will also need to understand the terms 'wavelength' and 'constructive superposition'. If the latter term is unfamiliar, you might do well to read the section about superposition in chapter 31 before you tackle this chapter.

Diffraction of waves

Diffraction, as you may recall from your GCSE course, describes how waves behave when they pass through gaps or round obstacles. We all take examples of diffraction for granted in everyday life. Consider the situation which occurs when you are standing outside the door of a room: see figure 1.

You are perfectly able to hear the conversation going on between the two people present, yet you cannot see them at all. You hear them because the sounds they make diffract through the gap of the door. Perhaps the real question is, why can't you see them? After all, light is a wave and should diffract as well.

The answer, of course, lies in the wavelengths of sound and light in relation to the width of the door.

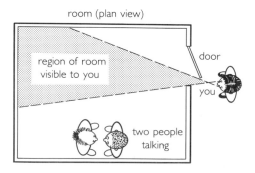

figure 1   To illustrate diffraction of sound through a doorway

The sounds have wavelengths comparable with the door gap, whereas light has very short wavelengths and hardly diffracts at all through this gap. Yet light can diffract in a very spectacular way – see the photograph, figure 2.

figure 2   Light diffracted past the edge of a razor blade

Furthermore, as we shall see later, diffraction of light is actually very important to all of us. For it determines how well our eyes can resolve detail at a distance. Why, for example, can you not read the print of this book from across a room? The answer lies partly in diffraction of the light as it passes through the gap of your eye's pupil; and partly in the structure of the eye itself, which has evolved over the centuries in response to the phenomenon of diffraction.

It is worth just reminding ourselves how ripples on a tank diffract – see figures 3 and 4.

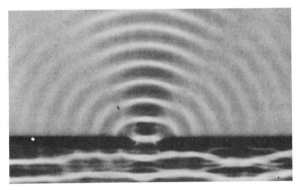

figure 3   Diffraction of waves in a ripple tank

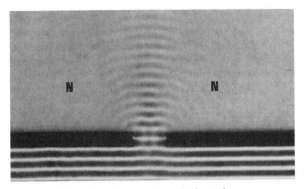

figure 4   Diffraction of waves in a ripple tank

Notice that when the gap is very small – one wavelength or less (figure 3) – the waves spread out in semicircles, with almost uniform amplitude in all directions. This is an important fact, and we will make use of it several times when we analyse different types of wave behaviour: diffraction, refraction and reflection.

Notice also that with a gap somewhat wider than one wavelength – figure 4 – the pattern as the waves emerge is more complex. There are directions on either side of the straight-on direction where the water is quite still – marked N on the photo. There are other directions where the wave amplitude is quite significant. We will see a similar type of behaviour when we examine the diffraction of light.

Diffraction of light

It is worth recalling, or doing now, some simple experiments on diffraction. A possible arrangement is shown in figure 5.

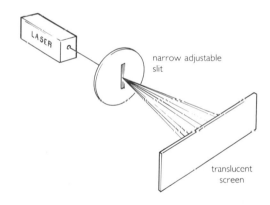

figure 5

If you look through the adjustable slit at the lamp, and gradually reduce the width of the slit, you see bright beams of light that spread out on both sides. If you look carefully you can see some dark bands crossing the beams near the middle. As you narrow the slit these bands spread wider; and finally, just before the slit closes altogether, these bands spread sideways out of sight, and you see only a faint, fairly uniform horizontal beam. Compare this with the ripple experiments. The final phase corresponds to figure 3, whereas the banded structure corresponds to figure 4.

We can do more quantitative work using a modified version of the above experiment – see figure 6. The pattern you see on the screen appears as in figure 7; we will use the measurements in the next section.

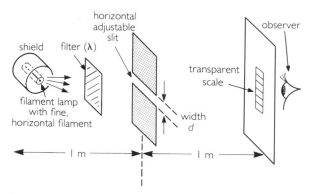

figure 6 Arrangement for making measurements about diffraction

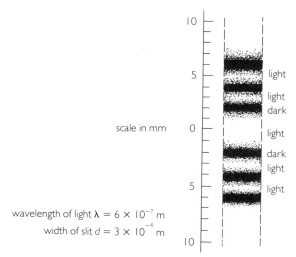

figure 7 The appearance of the diffraction pattern on the screen

Notice in particular:

☐ the central band of light is twice as wide as the side bands, and much brighter;

☐ the brightness and clarity of the side bands decreases rapidly away from the middle of the pattern.

Theory of diffraction by a single slit

This piece of theory will explain why we obtain a pattern like figure 7 when light diffracts through a slit. It will allow us to predict quantitatively where the light and dark bands appear, and it will be useful when we consider instruments like your eyes and their success in resolving detail.

Consider light waves arriving at a single slit of

width d. Now imagine that the slit is divided into 20 subslits; each of these is narrow enough for the light to diffract in a semicircular manner, like the ripples in figure 3. (This idea is formally called *Huygens' principle*. The number 20 is an arbitrary but convenient choice.) See figure 8.

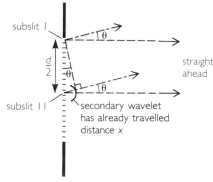

figure 8 Using Huygens' principle to analyse diffraction by a single slit

Consider the waves travelling from subslits 1 and 11 in a direction at angle θ away from the initial straight-on direction. The extra path x which waves from slit 11 must travel in order to reach a distant receiver is given by

$$x = \frac{d}{2}\sin\theta$$

If $x = \lambda/2$, then waves from subslits 1 and 11 will superpose destructively, when they meet at a distant point.

In the same direction, waves from all similar pairs of subslits (2 and 12, 3 and 13, etc.) will superpose destructively. Thus no wave energy will travel in direction θ, where

$$\frac{d}{2}\sin\theta = \frac{\lambda}{2}$$

or

$$\sin\theta = \frac{\lambda}{d}$$

In much work with light and single slits, $\lambda \ll d$, so $\sin\theta$ is small, and thus $\sin\theta \approx \theta$ (in radians); therefore no light goes in direction θ, when

$$\theta = \frac{\lambda}{d}$$

If we choose pairs of subslits carefully, we find that no light goes in any direction θ given by $\theta = n\lambda/d$, where n is any whole number except zero.

For example, if we consider slits 1 and 6 (figure 9), then they would have an extra path x equal to $\lambda/2$ when θ is described by

$$\frac{d}{4}\sin\theta = \frac{\lambda}{2}$$

$$\Rightarrow \qquad \sin\theta = \frac{2\lambda}{d}$$

$$\Rightarrow \quad \theta \approx \frac{2\lambda}{d} \text{ (provided } \theta \text{ remains small)}$$

(a)

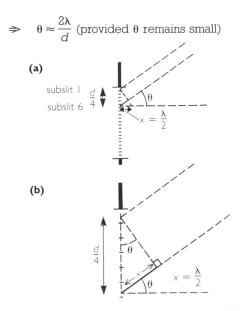

(b)

figure 9 To illustrate the direction of the second diffraction zero: (a) general; (b) detail of the important part of (a)

Thus subslits 1 and 6 superpose destructively and cancel in that direction; so do 2 and 7, 3 and 8, and so on. So too do 11 and 16, 12 and 17, etc. In fact, all the subslits cancel in that direction, so in that direction also there is no light.

The complete result is that the intensity (energy arriving) is related to angle from straight-on as shown in figure 10.

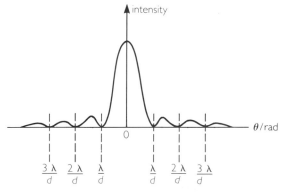

figure 10 Graph of intensity against angle for diffraction by a single slit

Compare this with the pattern illustrated in figure 7. The centres of the first dark bands come 2 mm on either side of the middle. The screen is 1 m from the slit, so the angle θ for these bands is given by

$$\theta = \frac{2\,\text{mm}}{1\,\text{m}} = 2 \times 10^{-3}\,\text{rad}$$

Figure 11 illustrates the relationship between angle, band size and distance between slit and screen.

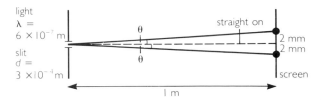

figure 11 How θ relates to the measurements in the experiment

The theory predicts the first dark band at θ given by

$$\theta = \frac{\lambda}{d}$$
$$= \frac{6 \times 10^{-7}\,\text{m}}{3 \times 10^{-4}\,\text{m}}$$
$$= 2 \times 10^{-3}\,\text{rad}$$

So the theory accords with the experiment.

Note that the above theory only works if the slit obeys two conditions. It must be:

☐ long ($\gg d$) in the direction at right angles to the paper;

☐ a long distance ($\gg d$) from the screen or detector.

Diffraction by a circular hole

We are probably more accustomed to waves passing through circular holes than through slit-shaped gaps – for example, our eyes, lenses, telescopes. In this case, we obtain a ring pattern; see figure 12.

figure 12 Laser diffraction pattern from a circular hole

This pattern has zeros of intensity at small angles θ from the centre given by

$$\theta = \frac{1.22n\lambda}{d}$$

where d is now the diameter of the hole.

In other words, the pattern is a modified version of

figure 7, with rotational symmetry about the central point, and all the angles multiplied by this factor 1.22 (which arises from the more complex geometry of the situation).

Resolution of two sources

If a receiving system can *resolve* two sources of waves, then it can see them as two distinct separate sources as opposed to a single source. If the sources are close together, this may become difficult because the diffraction patterns of the waves from the two sources may overlap. Diffraction occurs because the first stage of any receiving system (such as your eye, or a camera, or a telescope) is an aperture through which the waves are collected.

Figure 13 shows how the diffraction patterns created by waves from two sources would overlap and gradually merge if the sources moved closer together.

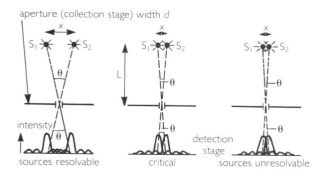

figure 13 To illustrate that the overlap of their diffraction patterns increases as the angle separating two sources becomes smaller

At some critical angular separation θ of the sources, the detector in the system (for example, the retina in your eye) cannot distinguish between the two different diffraction patterns because they overlap too much.

Figure 14 shows the critical picture in more detail, from a computer simulation. The critical stage is reached approximately when the central peak of S_1's pattern is over the first minimum of S_2's pattern. This criterion for resolution is called *Rayleigh's criterion*.

From figure 13 you can see that the angular separation of the two peaks on the detector is the same as the angular separation of the sources. Thus a system with a circular aperture, diameter d, will be able to resolve two sources at a distance ℓ and separated by a distance x if

$$\theta \geqslant \frac{1.22\,\lambda}{d}$$

But
$$\theta = \frac{x}{\ell}$$

$$\Rightarrow \qquad \text{for resolution } \frac{x}{\ell} \geqslant \frac{1.22\lambda}{d}$$

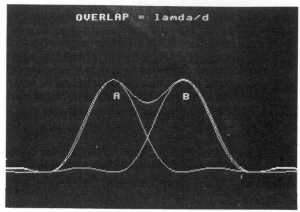

figure 14 A computer simulation of Rayleigh's criterion

Notice three points about this result.

☐ An instrument will in general have better resolution if its aperture d is large, and the wavelength λ which it detects is small. (The better the instrument, the *smaller* the angular detail it can resolve.)

☐ Two sources, or fine details, are more likely to be resolvable if their separation x is large, and they are a small distance ℓ away. Thus it helps you to read small print if it is close to your eye.

☐ The criterion $\theta \geqslant 1.22\lambda/d$ for resolution sets an upper limit on the ability of the instrument to resolve. Any imperfections in the system, such as scattering between the sources and the instrument, poor optics within the instrument, or noise in amplifiers, for example radar, will make the resolution actually achieved worse.

Let us consider two examples, both of them simple experiments you may have done.

Your eye has been mentioned several times as an example already. Now let us do a little calculation. The diameter of your pupil is about 3 mm, and the average wavelength of light can be taken as about 5×10^{-7} m. So you should be able to resolve angles θ given by

$$\theta \geqslant \frac{1.22\,\lambda}{d}$$

$$\Rightarrow \qquad \theta \geqslant \frac{1.22 \times 5 \times 10^{-7}\,\text{m}}{3 \times 10^{-3}\,\text{m}}$$

$$\Rightarrow \qquad \theta \geqslant 2 \times 10^{-4}\,\text{rad}$$

If you rule two very fine black lines 1 mm apart (x) on a sheet of white paper, you can just about tell them apart at 4 m distance (ℓ). Their angular separation θ would be given by

$$\theta = \frac{x}{\ell}$$

$$= \frac{1 \times 10^{-3}\,\text{m}}{4\,\text{m}}$$

$$= 2.5 \times 10^{-4}\,\text{rad}$$

This angle is close to, though larger than, that calculated above for the resolution your eye ought to have. But remember that this criterion sets the *best* possible resolution: in practice your eye won't do quite as well, so the result for the black lines is about right.

2 In this experiment, you look at a row of little bulbs through a filter and an adjustable slit – see figure 15.

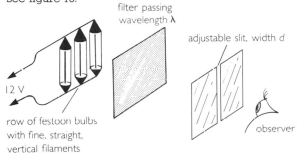

figure 15 A simple experiment showing how resolution depends on λ

With a green filter in position, you reduce d until the bulbs appear just to merge into one blob of light (the critical situation). If you replace the green filter by a red one, the bulbs appear to merge more completely (because red light has a longer wavelength than green light, so resolution is worse). If, on the other hand, you use a blue filter, your eye can resolve the bulbs into separate sources, because the blue wavelength is less than that of green light.

Transmitting aerials

Transmission of waves also involves diffraction, since the waves must emerge from their source via some kind of aperture. Most of the energy of waves being diffracted goes into the central bright section; so most of the energy transmitted by a circular dish aerial goes into a cone with its apex at the dish and half-angle equal to 1.22 λ/d. See figure 16.

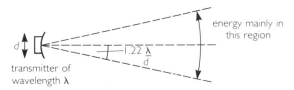

figure 16 Diffraction by a transmitting aerial

For example, consider a satellite 3.6×10^7 m above the Earth's surface. If it transmits waves of wavelength 1.0 mm down to the Earth from a dish with diameter 50 cm, over what region will they spread?

The cone into which the waves spread has half-angle θ given by

$$\theta = \frac{1.22\,\lambda}{d}$$

$$= \frac{1.22 \times 1.0 \times 10^{-3}\,\text{m}}{50 \times 10^{-2}\,\text{m}}$$

$$= 2.4 \times 10^{-3}\,\text{rad}$$

The waves spread to cover a circular path on the Earth's surface (see figure 17), with radius r given by

$$r = \theta \times \text{height of satellite}$$

$$- 2.4 \times 10^{-3}\,\text{rad} \times 3.6 \times 10^7\,\text{m}$$

$$= 8.6 \times 10^4\,\text{m}$$

$$= 86\,\text{km}$$

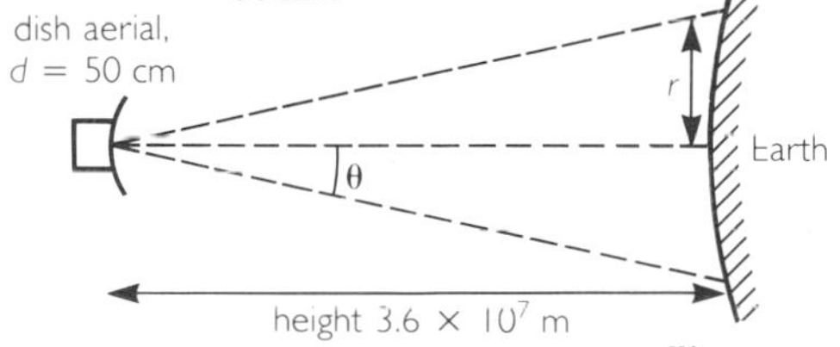

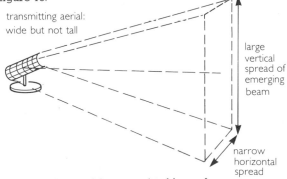

figure 17 Transmission to Earth from a satellite

Another example of diffraction by a transmitting aerial occurs with radar antennae. An antenna which is wide horizontally but not very tall will send a horizontally narrow but vertically spreading beam – good for locating the direction of an aircraft, but not its height. See figure 18.

figure 18 Shape of the transmitted beam from a rectangular aerial

The photograph of figure 19 shows such an aerial, part of an aircraft-tracking system called 'Martello', made by Marconi p.l.c. The aerial is 7.1 m high and 12.2 m wide. The radar system uses a wavelength of 23 cm. The makers claim that the horizontal beam width is 1.4°: can you check that this is about what you would expect?

To create a beam with a narrow vertical spread, clearly you require a tall aerial.

figure 19   The aerial of a Martello aircraft-tracking system made by Marconi

Summary

1 Diffraction effects occur when waves pass through an aperture or round an obstacle.

2 Waves passing through a gap comparable in size to their wavelength spread out nearly uniformly in all directions.

3 Diffraction by a single slit: zeros of intensity occur at:

$$\sin \theta = \frac{n\lambda}{d} \quad (\theta \simeq \frac{n\lambda}{d} \text{ if } \theta \text{ is small})$$

4 Rayleigh's criterion for resolution:

$$\theta \geqslant \frac{1.22\,\lambda}{d} \text{ (circular aperture)}$$

5 Most of the energy of a wave passes into the central diffraction region after diffraction (important for aerials).

Questions

1 Figure 20 shows two different experimental arrangements for taking photographs of a narrow source slit, A.

(i)
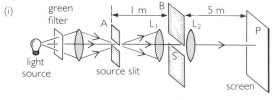

(ii)
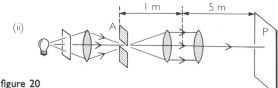

figure 20

a Which of the following patterns, X and Y, is obtained on the screen P in each case? Explain your choices.

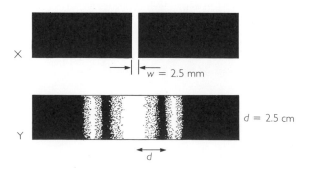

figure 21

b Estimate the width of the slit source A.
c Estimate the width of the slit S placed between the lenses in arrangement (i).
d Say what differences you observe if the slit S is replaced by a slit of double its width.
e Say what differences you observe if the slit A is replaced by a slit of four times its width.
f Say what differences you observe if the 5 m distance is increased to 10 m.

2 A laser emits an intense narrow beam of wavelength 6×10^{-7} m. It is mounted on a turntable which revolves once every hour. In front of the laser beam is placed a slit of width 0.1 mm. See figure 22.

figure 22

Describe what light a distant observer will see if he looks towards the laser. Plot an intensity-against-time graph for the period of time when the laser is pointing towards the observer.

3 A laser, emitting light of wavelength 600 nm, is mounted behind an adjustable slit. Two metres away a light sensor is set 5 mm to one side of where the laser beam arrives with the slit wide open. See figure 23.

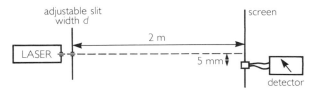

figure 23

The slit is first closed completely. It is then slowly opened. Sketch on axes like those shown in figure 24 the intensity the sensor will receive, plotted against the width d of the slit.

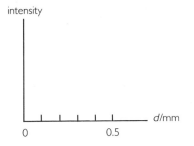

figure 24

4 A radar set is placed 10 m away from a straight road, with the centre of its beam at an angle of 25° to the road – see figure 25. If the radar aerial is circular, with diameter 20 cm, and the wavelength of the waves is 4 cm, estimate the length of road along which cars can be detected.

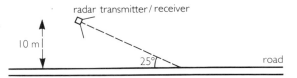

figure 25

5 Imagine using a pinhole camera to take a photograph of a distant star, see figure 26.

figure 26

a The light from the star will make a small spot on the film. What will be the diameter of the spot, if d is quite large (say 1 cm)?
b The smaller you make d, the better is the image of the star. But if you make d too small, the spot of light from the star starts to become larger and fuzzier. Explain why.

The optimum size for the pinhole is when the geometric size of the spot is equal to the size of the central diffraction blob.
c Explain this statement more fully.
d Find the optimum value of d, if $\lambda = 5 \times 10^{-7}$ m, and $\ell = 20$ cm.

6 This question is about the resolution obtainable by your eye, and the retina which detects the light arriving. The following data may be relevant:
average wavelength of light $\quad 5 \times 10^{-7}$ m
approximate diameter of pupil $\quad$ 4 mm
distance from pupil to retina $\quad$ 17 mm
diameter of nerve receptor cells (cones) at yellow spot (fovea) $\quad$ 1.5 μm
a Using Rayleigh's criterion, what angle of separation of sources should your eye in theory be able to resolve?
b What separation would this give between the adjacent points of brightness on the retina?
c In order for the retina to distinguish these bright points, there must be at least one cone between them which will signal the relative darkness which it receives. In view of this, comment on the size of the cones in your retina.

7

figure 27

The world's largest fully-steerable radio telescope dish is at Effelsburg, in West Germany. It is 100 m in diameter, and is considered to have reached the practical limit in size which is mechanically feasible at present. See figure 27.

a Estimate the wavelength at which this telescope would have to operate in order to resolve as small an angle as the unaided human eye can.

b Working at $\lambda = 1.0$ cm, can this telescope show that Venus is a disc and not a point? (Diameter of Venus $= 1.2 \times 10^7$ m; distance to Venus varies between 4.1×10^{10} m and 2.6×10^{11} m.)

8 Your two ears, acting together, and the associated analysis which goes on in your brain, are pretty good at locating the direction from which a sound is coming. This is thought to be due in part to the difference in intensity of the sounds reaching your two ears, because of your head shielding the ear on the opposite side to the sound. Suggest why high frequencies may be easier to locate than low frequencies.

9 For amateur astronomers, a 4 inch (10 cm) diameter aperture reflecting telescope is relatively inexpensive and easy to operate. (Average wavelength of light $= 5 \times 10^{-7}$ m; ignore atmospheric blurring.)

a What size of crater would the astronomer be able to distinguish on the Moon's surface (distance 3.8×10^8 m)?

b Would he be able to distinguish Neptune (distance 4.5×10^{12} m, diameter 4.4×10^7 m) as a disc rather than a point of light?

c At about what distance should he be able to read a car number plate?

10 Spy satellites used by both the USA and the USSR operate at about 1.6×10^5 m above the Earth's surface. Assuming they use light with wavelength about 5×10^{-7} m, what aperture would a camera need to be able to distinguish objects about 1 m in size? (Ignore atmospheric blurring.)

11 The Mount Palomar optical reflecting telescope has a 5 m aperture. But its ability to resolve detail is reduced by a factor of about 10 because of atmospheric blurring. Assume that the visible light range extends between 400 and 700 nm.

a Estimate the size of the smallest detail the telescope can detect on the surface of the Moon. (Distance to Moon $= 3.8 \times 10^8$ m.)

b Can this telescope detect that Pluto is a disc and not a point?
(Distance to Pluto $= 5.9 \times 10^{12}$ m; diameter of Pluto $= 6 \times 10^6$ m.)

c If this telescope examines the Andromeda Galaxy, 2.2 million light years away, how far apart must two stars in it be to be distinguishable?

The Space Telescope, due to be launched into orbit in the early 1990s, will have a 2.4 m aperture. It will not suffer at all from atmospheric blurring, and should be able to resolve detail right to its theoretical limit. Answer parts (a) and (c) above for this new telescope.

12 Part of a plan to set up a 5 GW solar-energy station in space involves beaming the energy back to Earth using microwaves with $\lambda = 10$ cm. The transmitting dish aerial would have a diameter of 1000 m. The station itself would be in synchronous orbit at a height of 3.6×10^7 m.

a What diameter on the Earth would the central diffraction spot of this wave energy have?

b What would be the average intensity (power/area) of the arriving energy?

13 Light from a school laser effectively emerges from a hole about 2 mm wide, with wavelength 6×10^{-7} m and power 2 mW.

a If such a laser is pointed at the Moon, what would be the diameter of the spot of light produced? (Distance to Moon $= 3.8 \times 10^8$ m.)

b What intensity (power/area) would be received back on the Earth? Assume that the light is scattered from the Moon's surface into a hemisphere, without loss.

c If one photon has energy 2.8×10^{-19} J, estimate how many photons would enter your eye each hour from this light, if you looked towards the Moon.

14 Read the article below, from *Waves* by David Chaundy (Longman, 1972); then answer the questions which follow.

St. Paul's Cathedral, showing three of the present column loudspeakers

St. Paul's Cathedral, built in the 17th century, is probably one of the most difficult buildings to speak in and it is impossible for an unaided speaker to be clearly audible in all parts of the cathedral. When it is empty the reverberation time is 11 seconds and when full it is still as long as 6 seconds. This is because the volume is large and the hard walls are efficient reflectors of sound. In addition the concave dome produces strong echoes. But in 1951 a carefully calculated system of amplification was installed in the cathedral with eight column loudspeakers, one of which can be seen in the photographs.

In the vertical direction this line loudspeaker sends out a very narrow beam so that very little sound is sent up towards the roof to produce reverberation. But in horizontal directions the sound spreads out to right and left so that it reaches all the congregation.

You will notice that the column loudspeaker contains two lines of loudspeakers: the large ones deal with the lower frequencies from 250 to 1 000 Hz, while the shorter line of smaller loudspeakers deals with the higher frequencies from 1 000 to 4 000 Hz. When, as an experiment, all the frequencies from 250 to 4 000 Hz were fed only through the large loudspeakers, the beam of high frequencies was so narrow that when people stood up their heads were above the high notes and speech became unintelligible. But with the higher frequencies coming from a short column and the lower frequencies coming from a long column the two beams have the same height.

One of the original column loudspeakers for St. Paul's Cathedral

a What do you think is meant by 'reverberation time'?

b Explain why 'in horizontal directions the sound spreads out to right and left'.

c The short column of small loudspeakers is actually 0.85 m long. Make suitable measurements to find out how long the long column is (on the photograph).

d What range of wavelengths does each column of speakers emit? (Speed of sound $= 340\,\mathrm{m\,s^{-1}}$.)

e What angular spread would the 4000 Hz sounds have when emitted from the long column? Is it possible that when people stood up 'their heads were above the high notes'?

f Assume that all the sound energy goes into the central diffraction maximum at each wavelength. Show by making suitable calculations that the higher frequency range from the short column and the lower frequency range from the long column both spread out to reach the same parts of the cathedral.

CHAPTER 31

Interference

Introduction

This chapter deals with a most important property of waves – superposition – and some of its consequences. First the idea itself is discussed at length, referring to one area of physics, standing waves, where you may already have used the idea. Then we discuss superposition effects and experiments for a variety of wave types – microwaves, sound, light. The rest of the chapter considers light in more detail: first the famous Young's slits interference experiment, then what happens when you have a larger and larger number of slits, finishing with a diffraction grating. We look at applications of gratings in spectroscopy and finish by considering very briefly what makes laser light so special.

GCSE knowledge

You may or may not have explored briefly the idea of superposition during your GCSE course; though you ought to be familiar with the fact that waves pass through each other without distortion, as on a slinky or in a ripple tank. You may also have looked at interference, on a ripple tank, or with sound, or light. The chapter is written assuming that you know nothing about these things; but the ideas are quite hard, and anything you know already will come in very useful.

Superposition of waves

If you sit by a pond on a still summer day, and watch ripples spreading out on its otherwise calm surface, you will probably notice that they pass through other ripples and continue outwards, unaffected by the ripples coming in from another direction. See figure 1.

This ability to pass through each other unaffected is a property of all waves, and distinguishes them from particles, which bounce off each other exchanging momentum. You have probably seen demonstrations of this happening on a slinky also. If it did not happen, light would not reach our eyes from distant objects clearly, but in a hopeless jumble.

This property of waves depends on the principle of superposition. The idea is that when waves from two or more sources meet, the total displacement of any point at any instant is the sum of the displacements

figure 1   Ripples on a pond spreading out through each other

each individual wave would cause. You can observe this if you look very carefully at two slinky waves crossing each other. Figure 2 is a series of photographs taken from a computer simulation of this happening.

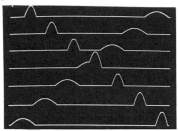

figure 2   Computer simulation of two 'slinky' waves meeting

Let us now consider in some detail the principle of superposition of waves, using as an example waves which are generated in phase from each end of a slinky as suggested in figure 3. Suppose the wavelength of the waves is 2 m, their amplitude is 15 cm, and the distance between S_1 and S_2 is 10 m.

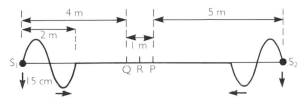

figure 3   To illustrate the formation of standing waves

At point P, the waves arrive from S_1 and S_2 having travelled the same distance. They therefore arrive in phase. That means that two peaks, each with displacement $+15$ cm, arrive at P simultaneously – followed half a cycle later by two troughs, each with displacement -15 cm. P therefore oscillates with amplitude 30 cm. We say there is a *maximum* at P, or, in standing wave terms, an *antinode*. This is called *constructive* superposition.

To reach Q, waves from S_1 have to travel 4 m, while those from S_2 travel 6 m. Since this difference in distance is exactly one wavelength, the waves will again arrive in phase: Q will also oscillate with large amplitude, being another maximum or antinode.

To reach R, on the other hand, waves have to travel $4\frac{1}{2}$ m from S_1 and $5\frac{1}{2}$ m from S_2. The difference in path is thus 1 m, or half a wavelength. This means that at the very moment that a $+15$ cm displacement arrives from S_1, the displacement due to S_2 is -15 cm. These two add together to give zero displacement. Half a cycle later the displacement from S_1 is -15 cm, while that from S_2 is $+15$ cm: but they still add to give zero. In fact, point R always has zero displacement. It is at a *minimum*, or *node*, and the situation here is called *destructive* superposition.

When you have to decide whether a point is at a maximum, a minimum or something in between, the

crucial factor to work out is the path difference. Waves always reach the point by two routes, and you must find the difference in distance along the two routes. For example, for a general point X in figure 3, the path difference x would be given by

$$x = S_1X - S_2X \quad \text{(see figure 4)}$$

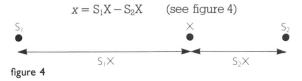

figure 4

If the sources are oscillating in phase, then the condition for a maximum is given by

$$x = n\lambda$$

where n is any whole number

$$\Rightarrow \qquad S_1X - S_2X = n\lambda$$

For a minimum,

$$S_1X - S_2X = (n+\tfrac{1}{2})\lambda$$

Some of these ideas are also discussed in chapter 8, in the section on standing waves. Figure 12 on page 98 illustrates successive positions of the slinky.

Superposition – some laboratory examples

This section gives a selection of examples of superposition which you may have met during your laboratory work.

I   *Microwaves* ($\lambda = 3$ cm)

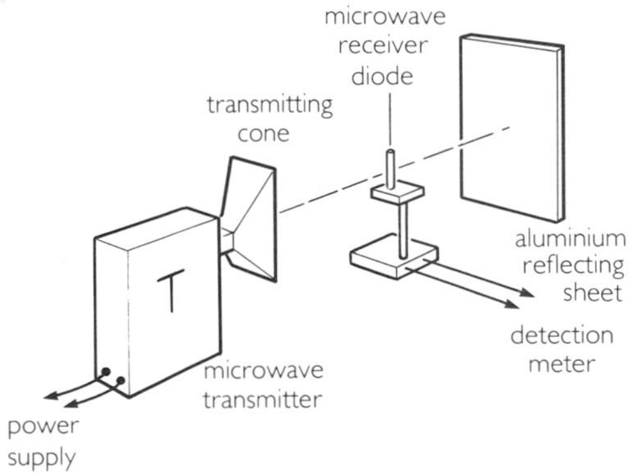

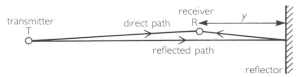

figure 5   Microwave superposition experiment (3-D view of experiment)

figure 6   Superposition experiment for 3 cm microwaves (plan view of experiment)

In the situation illustrated in figures 5 and 6, there are two routes for the waves to follow from the transmitter to the receiver – one direct, one by reflection. The path difference is $2y$. Unfortunately there is one slight complication. If you push the reflector right up close to the receiver R, so that the path difference is zero, you get almost zero on the receiver, instead of the maximum you probably expected. This is because the reflection causes the wave being reflected to change phase by half a cycle. This phase change occurs when waves are reflected from a medium denser than the one in which they are travelling. So we have to allow for this when writing down the condition from maxima and minima. In fact, the effect of this reflection exactly reverses the conditions. For a maximum we need the path difference $2y$ such that

$$2y = (n + \tfrac{1}{2})\lambda$$

For a minimum we need it such that

$$2y = n\lambda$$

If we have the reflector set for a maximum, and we then move it to the position giving the next maximum, we will have to change the path difference by one wavelength. This means changing $2y$ by one wavelength, or y by half a wavelength. In other words, half a wavelength movement of the reflector changes the received signal from one maximum, through a minimum to the next maximum.

If the reflector is moving quickly over a distance of many wavelengths, a series of maxima is detected by the receiver. The frequency of these maxima depends on the speed of the reflector. This is the principle of measuring speed by radar.

For example, suppose $\lambda = 0.03$ m, and the frequency of received maxima is 600 Hz. How fast is the reflector moving?

To change the signal from one maximum to the next, the reflector must move $\tfrac{1}{2}\lambda$, or 0.015 m. So it moves 0.015 m 600 times each second; so its speed is 0.015 m $\times 600$ s^{-1}, or 9 ms^{-1}.

2 Sound

The arrangement shown in figure 7 is precisely that discussed for microwaves. The microphone will encounter a maximum every time it moves half a wavelength. Hence we can measure the wavelength of the sound.

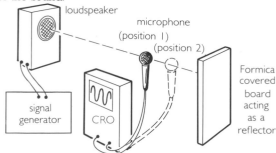

figure 7 Laboratory sound experiment

3 Light

Superposition of light passing through two or more slits is discussed at length in the following pages. In this section we will consider a different light situation: why colours appear on soap bubbles and oil films.

Consider light arriving at the surface of an oil film, and being reflected. Some will reflect from the top surface; some will pass down into the film, be reflected from the bottom surface, then pass back up and out. See figure 8.

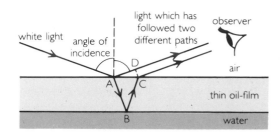

figure 8 Superposition of light reflected by a film of oil

Thus, by the time the waves reach the eye they have travelled different distances, and therefore may superpose constructively or destructively. Since white light contains light of many different wavelengths, one particular wavelength is likely to be constructively superposing at one particular angle of incidence, and so that colour will be seen at that angle. Nearby, at very slightly different angles, other colours will be seen.

In a laboratory, you can set up a different arrangement, which also uses a thin film to create a path difference. In this case the 'film' is actually air, sandwiched between two glass slides. See figure 9.

The enlarged view of figure 10 shows how superposition occurs.

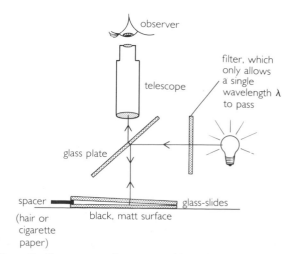

figure 9 Arrangement for superposition using an air-wedge

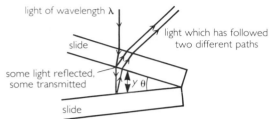

figure 10   Enlarged (and exaggerated) view of figure 9 to show how superposition occurs (N.B. In practice θ is very small and the incident and reflected paths are very nearly perpendicular to the surfaces)

The path difference at the place illustrated is $2y$. But to the right of this position the path difference is less than $2y$; and it decreases steadily down to zero. Using light of a single wavelength, therefore, lines of constructive and destructive superposition occur parallel to the touching edges of the slides, i.e. lines for which y is constant. See figure 11.

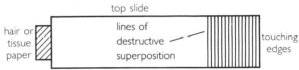

figure 11   Superposition effects of an air wedge – viewed from above

Intensity and amplitude

We need to look at a small but important point about waves at this stage. Consider the sound experiment shown in figure 12. The microphone, midway between S_1 and S_2, receives an amplitude A which is twice the amplitude it would receive with one speaker alone – this is the principle of superposition. The p.d. V it creates will therefore be double – and you can see this is so when you look at the CRO trace.

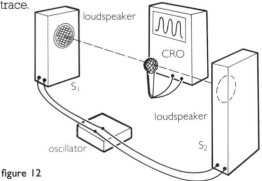

figure 12

However, if you were to connect the microphone to a resistor, R, then what power would be converted to heat in R? The answer must be V^2/R; therefore if V is double when both speakers are on, the power being converted must be four times as great.

The *intensity* I of a wave is a measure of the rate at which the wave transfers energy: that is, the power of the wave. So with light, for example, intensity and

brightness are closely related. So we conclude from the above argument that

$$I \propto A^2$$

You may wonder how putting on a second identical speaker can increase the energy flow, or power, fourfold. The answer is that in some places – the minima – the power is zero. Fourfold at the maxima, zero at the minima: on average, therefore, the power could be twice as great, as we would expect.

Two-slit interference with light

The word interference is used to refer to the observable effects of superposition. We discussed above two examples of superposition with light. Much of the rest of the chapter is devoted to other cases of superposition with light, in which the light passes through slits. This work is important for two main reasons.

☐   It is of major historical significance, since Young's two-slit experiment in 1801 strongly suggested that light behaves like a wave. The experiment enabled Young to measure its wavelength.

☐   It leads to the idea of a diffraction grating, which is a very important tool for physicists who want to measure the wavelength of waves of all sorts.

You can observe two-slit interference (Young's experiment) using the apparatus of figure 13.

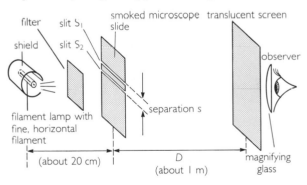

figure 13   Experimental arrangement to observe two-slit interference

We can derive a theory to predict the results using figure 14.

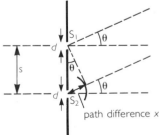

figure 14   To derive the theory for two-slit interference

At a particular angle θ, the path difference x between light from S_1 and S_2 to a distant observer is given by

$$x = s \sin \theta$$

For constructive superposition, $x = n\lambda$,

$$\Rightarrow \qquad n\lambda = s \sin \theta$$

or

$$\sin \theta = \frac{n\lambda}{s}$$

For most double-slit experiments, $\lambda \ll s$, so for constructive superposition

$$\theta \simeq \frac{n\lambda}{s}$$

There is a snag, however. Energy does not spread evenly in all directions from S_1 and S_2. Diffraction occurs at both S_1 and S_2.

The distribution of diffracted energy with angle for each slit is given by figure 10 on page 319. This distribution therefore forms an 'envelope' for the interference pattern. A graph of intensity against angle for a full double-slit pattern is given in figure 15.

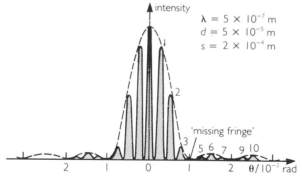

figure 15 Graph of intensity against angle for two-slit interference

A photograph of a pattern, produced using a laser as the light source, is shown in figure 16; you can clearly see the places where the diffraction envelope prevents the appearance of interference maxima – marked X.

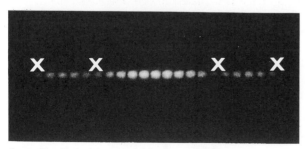

figure 16 A double-slit interference pattern

Let us consider how this pattern arises in more detail. The diffraction minima occur at angles θ given by

$$\theta = \frac{n\lambda}{d} \qquad \text{(where } n = 1, 2, 3, \text{ etc.)}$$

Therefore, using the values in figure 15, we will have *diffraction minima* at

$$\theta = \frac{1 \times 5 \times 10^{-7}\,\mathrm{m}}{5 \times 10^{-5}\,\mathrm{m}}, \frac{2 \times 5 \times 10^{-7}\,\mathrm{m}}{5 \times 10^{-5}\,\mathrm{m}}, \text{ etc.}$$

$$= 1 \times 10^{-2}\,\mathrm{rad}, 2 \times 10^{-2}\,\mathrm{rad}, \text{ etc.}$$

See the dotted line on figure 15, marking the diffraction envelope of the pattern.

The *interference maxima* occur at angles θ given by

$$\theta = \frac{n\lambda}{s} \quad (n = 0, 1, 2, \text{ etc.})$$

$$= 0, \frac{1 \times 5 \times 10^{-7}\,\mathrm{m}}{2 \times 10^{-4}\,\mathrm{m}}, \frac{2 \times 5 \times 10^{-7}\,\mathrm{m}}{2 \times 10^{-4}\,\mathrm{m}}, \text{ etc.}$$

$$= 0, 2.5 \times 10^{-3}\,\mathrm{rad}, 5 \times 10^{-3}\,\mathrm{rad}, \text{ etc.}$$

These maxima are shown by the solid line in figure 15, labelled 1, 2, 3, 5, 6, 7 etc.

The interference maxima are produced by interference from the two slits; the spacing of these fringes is controlled by the slit separation s. Consideration of the two-slit interference would lead us to expect a maximum at an angle of 10^{-2} rad; however we do not see one at this angle because it is the angle at which each slit gives out no light (the diffraction minimum). So a fringe is missed out from the interference pattern.

If you carry out the Young's experiment, you will probably measure the fringe spacing, W, produced on a screen rather than their angular spacing θ. In that case,

$$\theta = \frac{W}{D},$$

and thus

$$\frac{W}{D} = \frac{\lambda}{s}$$

Figure 17 illustrates the geometry of the double slit pattern.

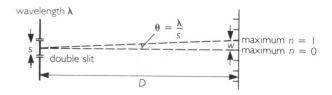

figure 17 The relationship between angular separation of fringes and their separation on a screen

Multiple-slit interference

With three or more equally-spaced slits, there is the same path difference x between each pair of adjacent slits – figure 18

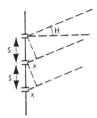

figure 18   Path differences for three slits

Thus you get superposition maxima in the same directions as with two slits, at angles θ given by

$$\sin\theta = \frac{n\lambda}{s}$$

or $\qquad\qquad \theta = \dfrac{n\lambda}{s}$ if θ is small

Consider figure 18 again. At angles which make x equal to $\frac{1}{2}\lambda$, $1\frac{1}{2}\lambda$, $2\frac{1}{2}\lambda$, etc., light from the top two slits would be exactly out of phase, and cancelling. That would leave some light from the third slit to go on in that direction. These angles are exactly half-way between the main maxima given by $\theta = n\lambda/s$. Thus between each pair of main maxima there is a secondary maximum. Figure 19 illustrates how the intensity varies with angle for three slits, and figure 20 is a photograph of the pattern achieved.

In general, if there are N equally-spaced slits, there will be $(N-2)$ secondary maxima between each pair of main maxima. As N increases, the secondary maxima decrease in width and brightness compared with the main maxima. See figure 21. This leads to a narrowing of the main maxima.

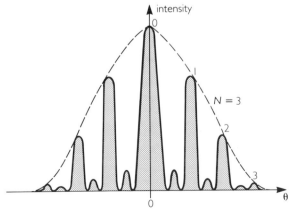

figure 19   Graph of intensity against angle for three-slit interference

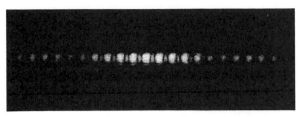

figure 20   Three-slit interference pattern obtained using a laser

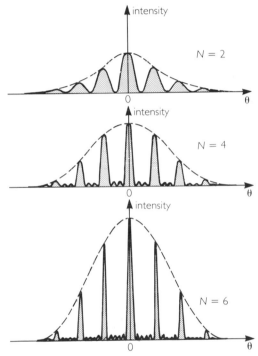

figure 21   Graphs of intensity against angle for different numbers, N, of slits (s remaining constant). The intensity scale is different in each graph

Diffraction gratings

When the number of evenly-spaced slits becomes very large, we have a diffraction grating. Following on from the multiple-slit interference described above:

☐ the secondary maxima are so numerous and faint that they do not appear at all;

☐ the main maxima are very narrow;

☐ with a finely-ruled grating, the slit separation s is so small that the assumption $\lambda \ll s$ may no longer be true. Therefore, θ is no longer small and $\sin\theta \neq \theta$ is no longer valid. Hence we must use:

$$\sin\theta = \frac{n\lambda}{s}$$

or
$$s \sin\theta = n\lambda$$

With a very finely-ruled grating and a particular value of λ, there may only be a few main maxima before θ becomes 90°. (We can now forget about the secondary maxima, as they are very faint.)

As an example, consider a grating with 400 lines mm^{-1}. At what angles will light with $\lambda = 7.0 \times 10^{-7}$ m appear?

$$s = \frac{1 \times 10^{-3}\,\text{m}}{400} = 2.5 \times 10^{-6}\,\text{m}$$

$$\sin\theta = \frac{n\lambda}{s}$$

$$\Rightarrow \sin\theta = \frac{\lambda}{s} = \frac{7 \times 10^{-7}\,\text{m}}{2.5 \times 10^{-6}\,\text{m}}$$

$$= 0.28$$

Thus the following angles will be possible:

$n = 1$	$\sin\theta = 0.28$	$\theta = 16°$
$n = 2$	$\sin\theta = 0.56$	$\theta = 34°$
$n = 3$	$\sin\theta = 0.84$	$\theta = 57°$
$n = 4$	$(\sin\theta = 1.12)$	no θ is possible.

Note that n is called the 'order' of the maximum being discussed.

Resolution of a grating

The usual purpose of a diffraction grating is to separate, and measure, all the different wavelengths of waves arriving at it. The resolution of a diffraction grating is its ability to create, from two slightly different wavelengths, two maxima which are separate enough to be distinguished. (See page 320, Rayleigh's criterion for resolution.)

Diffraction gratings are good at resolving two narrowly separated maxima, because the interference maxima formed by the grating are themselves so narrow. Figure 22 helps to show why the maxima are narrow.

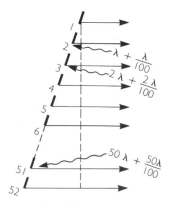

figure 22 A grating with 100 lines: path differences between slit 1 and other slits

Figure 22 shows part of a grating with 100 lines, We know that when the path difference between two adjacent slits is λ then the grating will produce an interference maximum. But what happens when the path difference is $\lambda + \lambda/100$ between adjacent slits? For only two slits we would expect an interference maximum under these circumstances, but with 100 slits we actually get an interference minimum. Cancellation occurs between slit 1 and slit 51 because the path difference between them is $50\lambda + \lambda/2$; similar cancellations occur between slits 2 and 52, etc. Figure 23 contrasts the interference fringes produced by two slits with those produced by 100 slits.

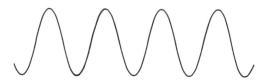

(a) Graph of intensity against angle for 2 slits

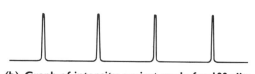

(b) Graph of intensity against angle for 100 slits

figure 23 N.B. The two graphs do not have the same intensity scale and the 'envelope' is ignored

We can use Rayleigh's criterion to tell us how good a grating will be at resolving two maxima of wavelengths λ and $\lambda + \delta\lambda$. They will just be resolved when the maximum of one source coincides with the first minimum of the other as shown in figure 24.

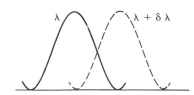

figure 24 Chromatic resolution

Suppose the principal maximum of order n, for light of wavelength $\lambda + \delta\lambda$, is formed at angle θ. We can then write

$$n(\lambda + \delta\lambda) = s \sin\theta$$

If the two wavelengths are barely resolved in the spectrum of order n, the first minimum for wavelength λ, in that order, must occur at the same angle, θ (Rayleigh's criterion). In the section above we explained that when this occurs the path difference between rays from neighbouring slits is $n\lambda + \lambda/N$, where N is the number of slits in the grating. Thus for the path difference we may also write

$$n\lambda + \frac{\lambda}{N} = s\sin\theta$$

From these two equations it follows that

$$n\lambda + \frac{\lambda}{N} = n(\lambda + \delta\lambda)$$

$$\Rightarrow \qquad \frac{\delta\lambda}{\lambda} = \frac{1}{nN}$$

So the resolution of a grating is high if both n, the order of the interference, and N, the number of slits, are large.

Spectroscopy

Spectroscopy is the study of electromagnetic radiation emitted or absorbed by material. Scientists can gain much information about the material concerned from such study.

The general arrangements for emission or absorption spectroscopy are shown in figure 25.

For an *emission* spectrum, you give the material excess energy in some way, e.g. by heating it. It then emits radiation, at wavelengths which may be characteristic of the material.

For an *absorption* spectrum, you direct radiation over a broad band of relevant wavelengths at the sample. The sample may absorb certain characteristic wavelengths, and the radiation which passes into the analyser can be seen to be missing these wavelengths.

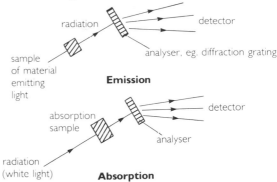

figure 25   To illustrate the principles of emission and absorption spectroscopy

A *line* spectrum (emission or absorption) is one which contains certain precise wavelengths. If the material is gaseous and in the form of single atoms it produces a line spectrum. A *continuous* spectrum contains all wavelengths in a broad range, and is more likely to be produced by a sample of solid material.

The table below gives a summary of the principle types of spectroscopy, across the range of the electromagnetic spectrum.

In your own laboratory, you are most likely to do experiments with optical (visible light) spectroscopy.

Type	Detector/analyser	Material	Comments
radio	aerial + tuned circuit	gas in the universe	clouds of hydrogen can be located because of a characteristic wavelength
infrared	reflection grating + photographic paper	organic molecules	frequencies absorbed relate to springiness of bonds; certain chemical groups identifiable
visible light	grating + spectrometer	hot solid / hot gas	continuous spectrum (as from a filament light bulb) / line spectrum characteristic of the elements in the gas
X-ray	crystal + ionisation chamber or photographic film	atoms bombarded with electrons	continuous emission spectrum with additional lines characteristic of the element
γ-ray	scintillation counter + energy analyser	nuclei undergoing change	line spectrum characteristic of the nucleus

The experimental arrangement, called a spectrometer, is shown in figure 26.

When we use a spectrometer, there is an elaborate setting-up procedure detailed in the manufacturer's handbook. The main idea, though, is simple. The telescope pivots around an axis through the centre of the grating; so you simply line up through the telescope the spectral line you wish to study, and read θ from the scale provided.

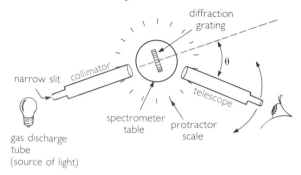

figure 26 The important features of a spectrometer (plan view)

Lasers

Several of the interference photographs in this chapter have been taken using a laser. Why is it easier to do interference experiments with a laser than with normal light from a bulb?

The answer is that the laser light is very pure and very coherent. 'Pure' means that it contains one single wavelength only. You don't need a filter at all; and anyway, most filters don't really give a single wavelength. 'Coherent' means that the waves are in phase right across the whole beam; so when the beam passes through the Young's double slits, waves come from both slits precisely in phase – as the theoretical analysis assumes.

These properties of a laser's light are what gives it such practical importance: for example, for reading bar-codes in a supermarket, for creating holograms, and for measuring very small distances very accurately.

Beats

The principle of superposition leads to one other interesting effect we should consider briefly – beats. Beats occur when waves arrive at one point P from two sources which have slightly different frequencies, f_1 and f_2. By the principle of superposition, the displacements you would get at P due to each source alone add together at every moment. The result is an oscillation at P with a basic frequency at the mean of f_1 and f_2, i.e. $(f_1 + f_2)/2$. The amplitude of this oscillation rises and falls at a frequency equal to $(f_1 - f_2)$. This is called the 'beat frequency' of the oscillation. See page 303.

As an example, suppose two loudspeakers are giving out pure notes at 220 Hz and 224 Hz, with equal amplitude. See figure 27.

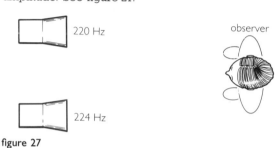

figure 27

The observer hears a note at 222 Hz, which rises and falls away four times each second. We say the sound has a 'beat frequency' of 4 Hz.

A graph of displacement against time for the observer looks like that in figure 28.

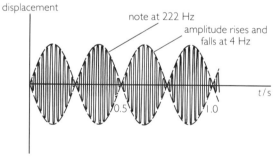

figure 28 Beats

Summary

1 Waves pass through each other and emerge without distortion, due to superposition.

2 The principle of superposition is that the total displacement at a point where waves meet is at any moment the sum of the displacements that the individual waves could cause.

3 Constructive superposition gives rise to maxima or antinodes.

4 Destructive superposition gives rise to minima or nodes.

5 a For waves following two routes to a point, their path difference is crucial in determining the outcome there.

 b In general,
 path difference $= n\lambda$ for a maximum;
 $= (n + \frac{1}{2})\lambda$ for a minimum.

6 At some reflections the waves are changed in phase by half a cycle; this phase change occurs when the waves are reflected by a *denser* medium.

7 Intensity $I \propto A^2$.

8 For two-slit interference:

$$\theta = \frac{n\lambda}{s} \text{ for maximum } (s \text{ is slit-separation})$$

or $\dfrac{W}{D} = \dfrac{\lambda}{s}$ ($W =$ fringe spacing and $D =$ distance from slits to screen)

9 The interference pattern is enveloped by the diffraction pattern of each individual slit.

10 With N slits, there are $(N-2)$ secondary maxima between each pair of main maxima.

11 For a diffraction grating,

$$s \sin\theta = n\lambda$$

12 The resolution of a grating is given by

$$\frac{\delta\lambda}{\lambda} = \frac{1}{nN}$$

13 Emission and absorption spectra are widely used in various branches of science.

14 Light from a laser is monochromatic and coherent.

15 Two sources with frequencies f_1 and f_2 cause beats at frequency $(f_1 - f_2)$.

Questions

1 Draw a series of displacement-against-position diagrams to describe the next 5 seconds of the hypothetical wave situation shown in figure 29.

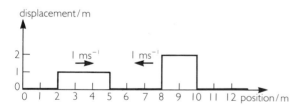

figure 29

2 The graphs in figure 30 show hypothetical wave profiles, drawn at time $t = 0$. Draw displacement-against-time graphs for the points A, B and C, showing as much numerical information as you can.

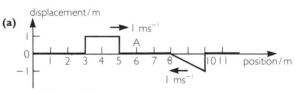

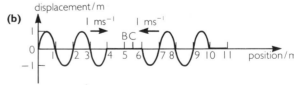

figure 30

3 This question is about a simplified version of an 'interferometer' built by Michelson, the same scientist who measured the speed of light really accurately for the first time (in 1926). He also gave his name to the Michelson–Morley experiment (1881) which provided the basis for Einstein's theory of special relativity.

The principle is that light is partially reflected, and after that follows two different paths to the detector. See figure 31.

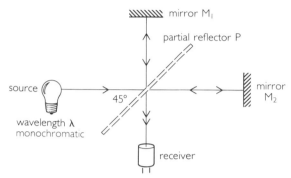

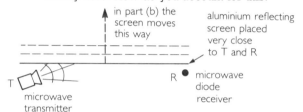

figure 31

a What is the condition for constructive superposition at the receiver?

If M_2 is moved slowly away from the partial reflector, the intensity at the receiver will fluctuate between maximum and minimum.

b How far would M_2 have to move for one complete fluctuation (maximum–minimum–maximum)?

c The relationship in (b) can be the basis for very accurate measurements *either* of wavelength *or*, if a known wavelength is used, of distance. In one experiment, using a laser with a wavelength of 632.8 nm, M_2 is moved so that 200 cycles of intensity are observed. How far is M_2 moved?

4 Two loudspeakers A and B are positioned in a field as shown in figure 32, and are connected to the same oscillator. On a windless day in October a man walks from O, where the sound is a maximum, along a line parallel to AB until he finds a point P where the sound first becomes a maximum again. He later finds that AP = 67 m and BP = 64 m.

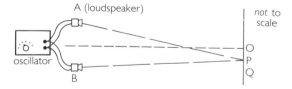

figure 32

a If the frequency of the oscillator is 110 Hz, deduce a value for the speed of sound in the air.

b The man now walks to a new point Q so that OP = PQ. Will Q be exactly at a maximum? Explain your answer.

c If the connections to one of the speakers were reversed, would there be any difference in the sound heard by the man (i) at O, (ii) at other places along the line OQ? (Reversing the connections alters the phase of the sound wave from the speaker by half a cycle.)

In the next two questions, T is a transmitter transmitting 3 cm radio waves (microwaves), and R is a receiver.

5 a With T and R as shown in figure 33, R receives a normal signal with *no* screen present. But when the screen is positioned as shown, very close to T and R, the signal received falls to nearly zero. How do you account for this?

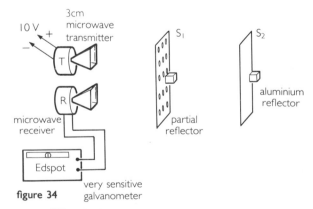

figure 33

b If the screen is moved away from T and R (see figure 33), the signal first rises to a maximum, then alters successively through further minima and maxima. Explain this.

c If the distance between T and R is 12 cm, calculate how far the screen has been moved when the received signal first becomes a minimum again.

d Explain why the maxima and minima become less distinguishable as the screen moves further away from T and R.

6 a In the situation shown in figure 34, it is found that when S_1 is moved towards or away from R and T, the signal at R fluctuates through maxima and minima. Explain this.

figure 34

b What is the shortest distance S_1 needs to be moved to change the signal from a maximum to a minimum?

c What will happen if S_1 is kept stationary and S_2 is moved instead?

d The same effect as in (a) can often be found if S_2 is removed altogether. Suggest a reason for this.

e The experiment in (a) is repeated, with S_1 immersed in a tank of paraffin. To alter the signal from one minimum to the next minimum, S_1 needs to be moved 0.2 cm through the paraffin. Calculate (i) the wavelength of the waves in the paraffin, (ii) the refractive index n of the paraffin for these waves ($n = c_{air}/c_{paraffin}$)

7 The audience at an outside concert sits on a hill opposite the stage as shown in figure 35.

figure 35

a Describe two possible routes by which the sound could travel to a girl seated on the hill at the same vertical height as the group's loudspeaker.

b The girl finds that notes of about 170 Hz sound noticeably quiet. Explain quantitatively why this is so. (Assume that the land between the stage and the hill is flat and perfectly reflecting. The reflection alters the phase of the waves: it is equivalent to adding an extra half-wavelength to the distance travelled by the reflected wave. Speed of sound = 340 m s^{-1}.)

8 Imagine a transmitter on one side of a tidal river broadcasting across the river to a receiver on the other side. The system is illustrated in figure 36. The transmitter and receiver are both at the high-water-mark heights. The wavelength of the waves is 30 cm.

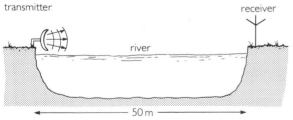

figure 36

a Explain why the intensity of the signal received may vary with the height of the tide.

b Explain why the signal is a minimum when the tide is high. Assume water does not enter the transmitter.

c If the water level drops 5 m from high to low tide, find how many maxima the received signal will have passed through.

9 This question is about the 'bloom' on the front of good-quality camera lenses. A thin film is deposited on the surface of the lens. The film is designed to be one quarter of a wavelength thick, for light in the middle of the visible range. In addition, the film is made of material having 'optical density' midway between air and glass. This means that light reflected back upwards at either surface has its phase changed by half a cycle, whereas light reflected back downwards does not. See figure 37.

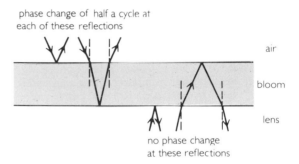

figure 37

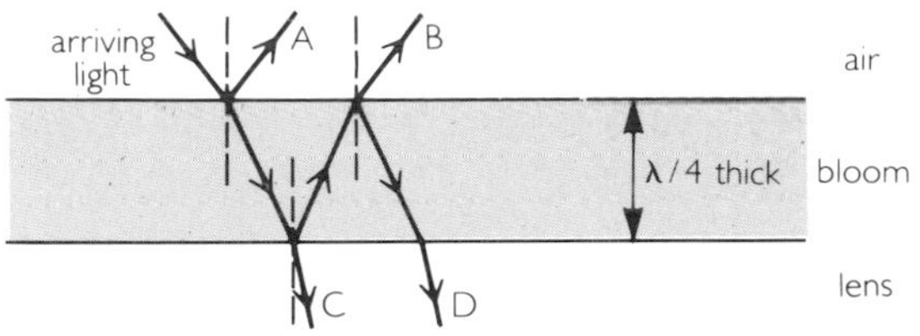

figure 38

a Explain why waves following paths A and B in figure 38 are out of phase.

b Explain why waves following paths C and D in figure 38 are effectively in phase.

c Hence explain why the bloom reduces the amount of light being reflected, and increases the amount being passed into the lens.

d The bloom can only be $\lambda/4$ thick for one wavelength, which is chosen to be in the middle of the visible range of wavelengths. Explain why good-quality lenses usually look purple.

10 In some circumstances, radio waves travel from a transmitter to a receiver by two different paths: directly along the ground (the 'ground wave') and by reflection off a layer of the ionosphere (the 'sky wave'). Consider the particular situation in which $\lambda = 100$ m, the direct distance between transmitter and receiver is 500 km, and the ionosphere is effectively at a height of 200 km. See figure 39.

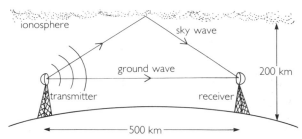

figure 39

a Work out the path difference between the sky wave and the ground wave.
b The ionosphere is often, in fact, moving in a vertical direction. This leads to a rising and falling in the amplitude of the received wave. Explain why the amplitude rises and falls.
c Suppose that, from the situation given, the ionosphere moves upwards far enough for one complete cycle of rising and falling amplitude to occur. What would be the new path difference at the end of this process?
d Calculate how far the ionosphere would have to rise for the situation in (c) to occur.

11 An interesting example of a 'stellar interferometer' is located on a cliff 100 m high, near Sydney harbour. The 1.5 m waves from the Sun are being studied. They reach the receiver by two paths, illustrated in figure 40.

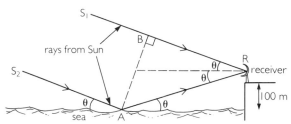

figure 40

a In terms of the letters on the diagram, what is the path difference which may give rise to interference?
b Can you show that this path difference is equal to $200\,\text{m} \times \sin\theta$?
c Write down a condition for the receiver to detect an intensity maximum. (Remember there is a phase change on reflection of these waves by the sea.)
d Explain why the intensity received goes through a series of maxima and minima as the Sun sets.
e Will the changes between maximum and minimum be most rapid when the Sun is directly overhead, or when it is nearly set? Explain.

12 A householder is watching television in his house, which is set back 100 m from a road running north–south. See figure 41.

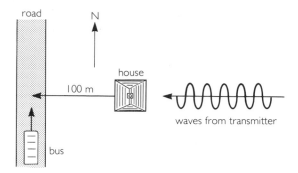

figure 41

The frequency of the signal is 60 MHz, and the transmitting station is due east of his house. He notices that when a bus drives at steady speed northwards, along the road, the signal strength fluctuates. The fluctuations slow down then stop briefly as the bus passes closest to the house, then continue.
a Explain why the signal strength fluctuates as the bus moves.
b Calculate the wavelength of the waves.
c Is the signal strength as the bus passes closest to the house a maximum or a minimum?
d Why do the fluctuations stop as the bus passes closest to the house?
e How many cycles of maximum and minimum will the signal pass through as the bus moves from being south-west of the house to being due west of the house?

13 A two-slit interference arrangement is constructed so that one slit is four times the width of the other. See figure 42.

figure 42

At the screen where the light is received, what will be the ratio between maxima and minima of (i) the amplitude; (ii) the intensities?

14 Michael and Sita are having a discussion about two-source interference. It is about the intensity of the interference bands near the centre of the pattern, and its relationship to the width of the slits. They start from the situation in figure 43.

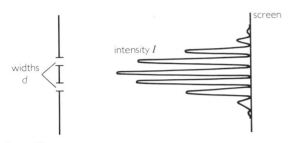

figure 43

Michael (1): If we make each slit have width $2d$, then we are letting twice as much light energy through; so the intensity of maxima must become $2I$.

Sita (2): But if each slit width doubles, the amplitude from each slit doubles; so the amplitude at a maximum doubles, and we get intensity $4I$.

Michael (3): That must mean we are getting four times as much energy arriving, for twice as much light. That doesn't make sense.

Sita (4): But making each slit $2d$ wide reduces the number of interference maxima, because they are restricted by the diffraction pattern of either slit. So there will be roughly half as many interference maxima, each with intensity $4I$.

Michael (5): Does that mean that the total energy arriving will be twice as much, as we would expect?

a The discussion hinges on the difference between *amplitude* and *intensity*. Explain each term, and say how they are related.

b Explain how Sita is using the superposition principle in statement 2.

c Explain the point Sita is making in statement 4 more fully.

d Sketch two graphs, on comparable axes, of intensity against position along the screen for the two situations. How might Sita use the comparative areas under each graph to answer Michael's final question?

15 Examine the photograph in figure 44, which shows interference from two dippers making circular ripples in phase on a tank.

a In terms of wavelength, what must be the distance $S_2P - S_1P$?

b Measure that distance on the photo, and also the size of one wavelength. Show that they relate as you stated in (a).

c Now verify, by making other measurements, that the formula derived for the Young's slits experiment on page 329 works.
(The formula is $\theta = n\lambda/s$ or $W/D = \lambda/s$.)

16 White light containing all wavelengths in the range 400–800 nm strikes two slits 1 mm apart. Which of the wavelengths in this range will be (i) constructively interfering, (ii) destructively

interfering at each of these angles of observation:

a 4×10^{-4} rad

b 6×10^{-4} rad

c 8×10^{-4} rad

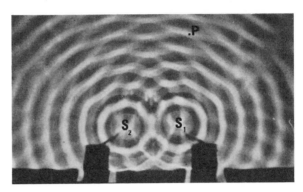

figure 44

17 Light of wavelength 5×10^{-7} m falls on the double-slit arrangement shown in figure 45.

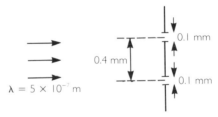

figure 45

a On a copy of the axes illustrated in figure 46, draw the graph of intensity against angle which would result if one slit was covered up.

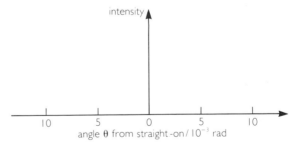

figure 46

b Now draw, in a different colour, the graph of intensity against angle which will result if both the slits are open. For the sake of clarity, scale this second diagram so that the intensities of the two patterns appear at the same height at $\theta = 0$.

c In fact the intensities at $\theta = 0$ would not be equal. By what factor would the intensity be greater at θ = 0 when both slits are open?

d Now suppose the central 0.3 mm of opaque material separating the slits is removed. By what factor will the intensity at θ = 0 now be greater than when there was only one narrow slit, as in (a)?

18 Two transmitters 6 m apart emit radio waves of frequency 10^{10} Hz.

a Describe the pattern which you would receive along a line 100 m away from the transmitters, if they emit in phase. See figure 47.

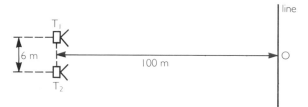

figure 47

b What change will occur to the pattern in (a) if the phase of T_2 is advanced to be one quarter of a cycle ahead of T_1?

c An observer is monitoring the signal at O. What will she notice if the transmitter T_2 is now adjusted to transmit at a frequency of 1 Hz higher than T_1?

d What happens to the pattern viewed as a whole if the frequency of T_2 is 1 Hz higher than that of T_1?

19 Figure 48 shows how an interference pattern produced by the two slits at B may be observed on a screen, P.

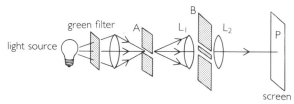

figure 48

The two diagrams X and Y in figure 49 show photographic results obtained on the screen, P. For each of the two patterns, three different exposures are shown, to bring out the details of the faint and the strong parts of the pattern. All six photographs are to the same scale.

a Deduce from the photographs in figure 49 the following ratios:

(i) $\dfrac{\text{Slit separation in case X}}{\text{Slit separation in case Y}}$

(ii) $\dfrac{\text{Slit width in case X}}{\text{Slit width in case Y}}$

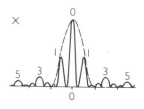

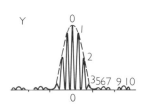

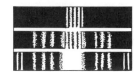

figure 49

b In what way would the experimental arrangement have to be altered in order to obtain pattern (i) U (ii) V (iii) W?

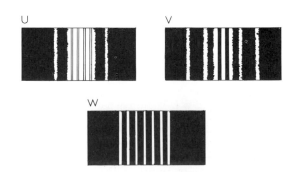

figure 50

20 Three radio transmitting aerials X, Y and Z are placed in a north–south line, with spaces of 1 m between them. They emit with wavelength 0.1 m, with equal amplitudes and in phase. An observer with a receiver walks along a north–south line 50 m to the east. See figure 51.

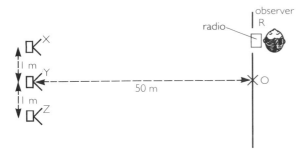

figure 51

a Describe how the intensity of signal received by the observer varies with his distance from O. Sketch a graph, with values, of intensity against distance.

b Transmitter Y is put out of phase with the other two. Describe on another graph how this changes the observed signal pattern.

c With Y still out of phase with X and Z, the amplitude of Z is doubled. Show on a third graph the resulting intensity-against-position pattern.

21 Two aircraft are flying along, one behind the other, and approach a radar station so as to pass it at a distance of 8000 m. The aircraft are 20 m apart. See figure 52. The radar has wavelength 1.0 m.

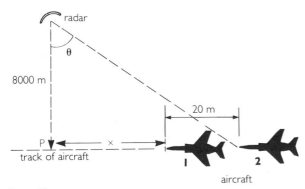

figure 52

The radar waves travel out to each aircraft, are reflected, and some return to the receiver. Consider the situation when the line joining the radar to the aircraft makes an angle θ as shown.

a What, approximately, is the total difference in path of the waves reflected back by aircraft 1 and by aircraft 2?

b What condition is required for the radar operator to receive maximum reflected intensity?

c How does the path difference change as the aircraft travel along their track at a steady speed?

d Describe how the intensity of reflected waves varies as the aircraft fly along. Sketch a graph of intensity against distance away from P, x, in the range 0 < x < 1000 m. Mark on carefully any distances you can calculate.

22 a What is Rayleigh's criterion for resolution?

b If two vertical-filament lamps are observed from a distance of 5 m, two close but distinct images are obtained on the eye's retina. Estimate how far from the lamps the observer will be when he can just no longer distinguish two separate filaments (assuming Rayleigh's criterion applies).

c Calculate the minimum angle of resolution for a radio-telescope which has a dish of diameter 30 m, when it is being used to observe emissions of wavelength 21 cm from distant galaxies.

d Draw a diagram to show how the intensity detected by the radio-telescope in part (c) varies with angle as the telescope's dish scans across the region of the sky containing a point radio source.

e Say how the observations obtained in part (d) would alter if another identical telescope were to be lined up with the first one to form a two-aerial interferometer telescope. See figure 53.

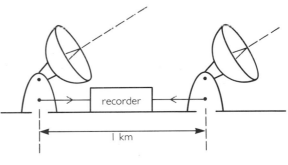

figure 53

f Say how the observations would alter if two more identical radio-telescopes were to be linked up with the first two, as shown in figure 54, to form a four-aerial interferometer telescope.

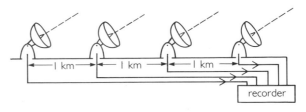

figure 54

23 Read the article below, and then answer the questions which follow it.

Radio telescopes

If you could look at the night sky with radio eyes, Cygnus A would be the second brightest compact object you could see. Yet it is not a nearby star, but a galaxy so distant that its visible emission is too faint to detect except in a large telescope. Its radio emission is enormous, of the order 10^{37} W (1960 estimate), as large as the whole emission from a normal galaxy. Because it is so distant, the power falling on the whole Earth is only about 10 W, and the power collected by a telescope as large as that at Jodrell Bank is only about 10^{-10} W. This figure includes the whole range of radio frequencies. The power within a band of, say, 1 MHz close to a typical frequency of reception of about 80 MHz, is as small as about 10^{-10} W. Remembering that Cygnus A is a 'bright' object, as astronomical radio objects go, it is clear that not only must a radio telescope have a large area, but also an extremely sensitive amplifier system.

Not only does the radio astronomer have to detect very weak sources, but he has also to cope with the problem of obtaining good resolution despite the long wavelength of the radiation involved. For example, the pair of sources which constitute the source Cygnus A subtend an angle of less than one-thousandth of a radian at the Earth.

The Jodrell Bank telescope is not unlike a scaled-up version of an optical reflecting telescope, but its performance in seeing the detail of compact, distant objects is much the worse of the two. The shortest wavelength at which it will work efficiently is 0.3 m, so that its 80 m diameter is less than 300 wavelengths across, and objects within an angle of 1/300 radian will not be distinguished. This corresponds to the width of a car at a distance of 600 m, or to a fly seen across a

room. The naked eye can do much better than that (with visible light, naturally). Figure 1 shows the variation of the response of such a telescope, as it sweeps past a single, small source. It is, of course, just the usual single-aperture, diffraction pattern.

The pattern can be made narrower by making the telescope wider, but to achieve a resolution of, say, 10^{-4} radian, would require a Jodrell Bank type of dish 10 000 wavelengths across, which is a diameter of 3 km at a wavelength of 0.3 m. This is scarcely practicable, but fortunately it is not necessary.

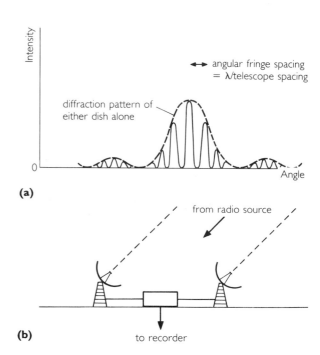

(a)

(b)

Figure 2
a Intensity variation from a two-aerial interferometer (single, point source).
b A pair of dishes connected together.

A pair of dish aerials placed a large distance apart will do instead. Figure 2 shows how the intensity from such a pair of aerials varies as a single small source crosses the line along which they point.

If the signals from the two telescopes are combined, the record shows closely spaced rises and falls of intensity, exactly analogous to the interference fringes that would be produced if the aerials were transmitting a signal. Because the fringes are much narrower than the diffraction pattern, the central fringe identifies the direction of a single source much more accurately than either dish can do on its own.

(Adapted from *Nuffield Advanced Science, Physics Students' Book Unit 8* (Longman, 1972).)

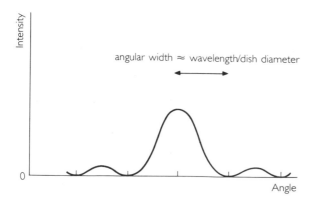

Figure 1
Intensity variation from a single-aperture telescope (single, point source).

The Cambridge five-kilometre telescope (see figure 3) is of this type. Another variation is the MERLIN system. (Multi-Element Radio-Linked Interferometer Network). MERLIN uses the signals collected by six different radio-telescopes, of which Jodrell Bank is one; the combined result is equivalent to a single dish 133 km across.

figure 3

In the future, radio astronomers would like to extend their baseline still further by launching a receiving dish into space, to orbit the Earth. The project is called QUASAT. Working in conjunction with ground-based dishes, this would give a baseline of about 20 000 km. Using radio waves of wavelength 1.35 cm, QUASAT should make it possible to resolve structures nearly as small as the distance between the Earth and the Sun (eight light-minutes) as far away as the centre of our Galaxy (30 000 light years).

a Assume that the radio emission from Cygnus A is the same in all directions, and none is absorbed on its way to the Earth. Estimate the distance to Cygnus A, using the information in paragraph one. Give your answer in (i) metres, (ii) light years.

b The diameter of the Earth is about 1.3×10^7 m, and that of the Jodrell Bank dish aerial is 80 m. Use this information to explain why the power the aerial collects from Cygnus A is only about 10^{-13} W.

c (i) Explain, using the figures in paragraph three, why the Jodrell Bank telescope cannot distinguish between objects within an angle of 1/300 rad.

(ii) Making suitable estimates, show why this corresponds to the car and the fly as stated.
(iii) Within what approximate angle can the naked eye resolve two objects?

d Using a wavelength of 0.3 m, what angular resolution is possible with (i) the Cambridge five-kilometre telescope, (ii) MERLIN, (iii) a baseline equal to the Earth's diameter, (iv) QUASAT?
  Compare each of these angular resolutions with that possible using the 5 metre optical telescope at Mount Palomar.

e Show by doing suitable calculations that the statement about QUASAT's resolving ability is approximately correct.

24 A grating has 300 lines mm^{-1}, and is being used with a laser emitting light with $\lambda = 7 \times 10^{-7}$ m.
a What is the slit spacing s for this grating?
b If the laser light is incident on the grating at right angles, how many orders of diffraction are visible, and in what directions do they occur?

25 A long line of radio transmitters, each 1 m apart, is broadcasting on a wavelength of 0.25 m.
a Each transmitter is in phase with all the others. In what directions relative to the line of transmitters are maximum intensities received?
b Alternate transmitters are switched off. In what directions are maximum intensities now received?
c The transmitters switched off in (b) are now made to transmit exactly out of phase with the others. What effect does this have on the signals received in various directions?

26 A grating with slit-separation 2.0×10^{-6} m is being used to analyse a helium spectrum. The operator observes that at 42° there are a blue and a red line very close together. What order is each of these lines?
  (range of red light   645–770 nm
  range of blue light   440–490 nm)

27 A student is attempting to resolve the two D-lines in the sodium spectrum, which have wavelengths 589.0 nm and 589.6 nm. She has two diffraction gratings available:
  grating A has 2000 lines in total, spread across 20 mm;
  grating B has 2000 lines in total, spread across 5 mm.
Using a spectrometer, parallel light is passed through the grating position across a width of about 30 mm, and thus will pass through all the slits of either grating.
a At what angle do the first-order images of the D-lines appear, using (i) grating A, (ii) grating B?
b Will these first-order lines be resolved using (i) grating A, (ii) grating B?

Before using the spectrometer, the student holds each grating up to her eye in turn, and looks at the sodium light as it emerges from a narrow slit. She observes that with grating B the D-lines are distinct in the first order, whereas with grating A the D-lines only become distinct at about the fourth-order spectrum.

c Explain these observations as fully as you can. Assume that her eye has a pupil diameter of about 3 mm.

28 If microwaves are passed through water, their intensity decreases with distance, because they gradually lose energy to the water. They are said to be 'attenuated'. In one experiment, waves of frequency 1 GHz are passed along a length of coax-type cable full of water. See figure 55.

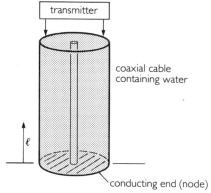

figure 55

The end of the coax is short-circuited, making a wave node. The experimenters can measure the wave intensity at distances ℓ from the end of the coax. When they do so, a graph of intensity I against ℓ looks like figure 56.

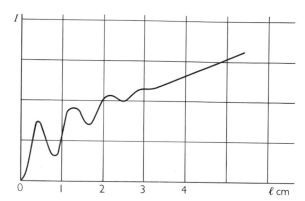

figure 56

a Explain why the intensity fluctuates at low values of ℓ.

b Why does the intensity not fluctuate with higher values of ℓ?

c Why does the intensity graph tend to increase as ℓ increases?

d Use the graph to find the wavelength of the waves in water.

e Find the speed of these waves in the water.

CHAPTER 32

Reflection and refraction

Introduction

In this rather short chapter, we look at the effects of reflection and refraction – which you will have met and studied before. We examine how a wave model can explain both effects, and predict correctly that light is refracted because of a change in its speed of travel.

GCSE knowledge

To follow this chapter, you should recall from GCSE what you learnt about reflection: how straight wavefronts on ripple tanks are reflected from a straight barrier, and how rays of light are also reflected, so that their angle of reflection is equal to their angle of incidence.

You may also have done work on refraction: probably observing the effect on a ripple tank of waves changing speed and bending when the depth of water changes; and possibly observing rays of light bending as they enter and leave a glass block. Anything you can remember about this will help in this chapter.

Huygens' principle

We recalled in chapter 30, page 317, that if waves pass through a small gap (of similar size to their wavelength), they spread out nearly into semicircles.

This idea was used by Huygens (a contemporary of Newton) to create a model to explain certain features of wave behaviour. We have already used it once, to explain the single-slit diffraction pattern – page 318.

Huygens' principle can be stated like this.

'Every point on a wavefront may be treated as a source of secondary spherical wavelets spreading out from the point. The new wavefront is the envelope of these secondary wavelets.'

It is a valuable model because it also helps to explain reflection and refraction effects.

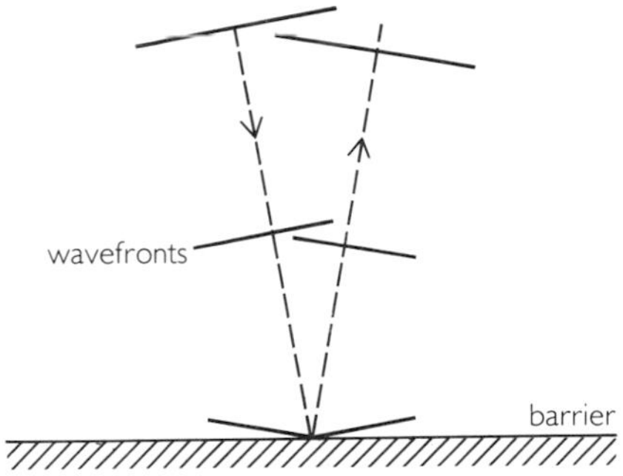

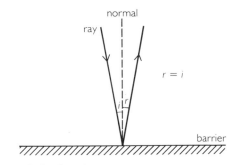

figure 1   Reflection of a wavefront

Reflection

Reflection occurs when a wave arrives at a barrier and rebounds. You probably recall the details of what happens on a ripple tank, and with light.

With ripples we are looking at the wavefront itself: we can see the displacements of the water surface making up the wave. A ray of light, on the other hand, simply tells us the direction of movement of the wavefront – which is always at right angles to the wavefront itself. Figures 1 and 2 show reflection of waves (perhaps ripples) and the corresponding ray picture.

figure 2   Reflection of a ray

Using Huygens' principle, we can explain reflection like this. See figure 3.

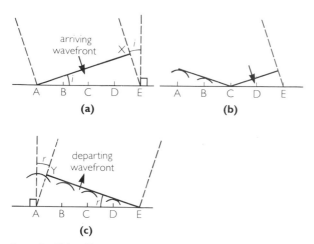

figure 3   Using Huygens' principle as a model for reflection

In (a) the wavefront is just arriving. Since it reaches A first, secondary wavelets start first from A; (b) and (c) show how points B, C and D successively act as sources of secondary wavelets, so that the complete wavefront is eventually reflected.

If the time interval between figures 3(a) and (c) is t, then

$$\text{distance XE} = c \times t$$

where c is the speed of the wavefront,

$$\text{and distance AY} = c \times t$$

This means that triangles AXE and EYA are congruent; thus $r = i$, so the model successfully predicts the law of reflection.

Refraction

Refraction occurs when waves alter speed as they pass from one medium into another. As a result, the direction of wave travel alters. Figure 4 shows a photograph of refraction occurring on a ripple tank.

figure 4   Refraction in a ripple tank

Figure 5 shows the corresponding ray diagram, such as you may have obtained from experiments with rays passing through a glass block.

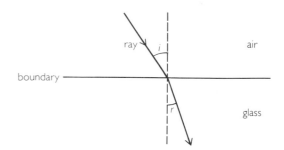

figure 5   Refraction of a ray of light

If you obtain quantitative results for the glass block experiment, you can use them to verify Snell's law of refraction. It is this:

$$\frac{\sin i}{\sin r} = \text{constant}$$

The constant is called the *refractive index* (symbol n) for glass. See page 361 for a detailed description of this experiment.

This law, when discovered, was an empirical law. That is, it was derived from the experimental data alone, without any theoretical basis. With the ripple tank experiment, you can see clearly that the waves are travelling more slowly in the lower part of the tank. (This occurs because the water is shallower there.)

If we apply Huygens' model to the refraction of the wavefronts, we obtain an interesting result. See figure 6.

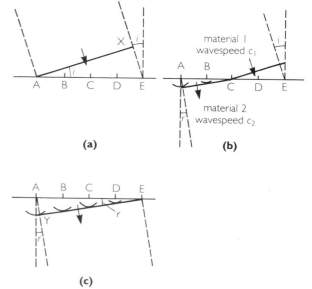

figure 6   Using Huygens' principle as a model for refraction

If the time interval between figures 6(a) and (c) is t, then

$$XE = c_1 t$$

$$\Rightarrow \qquad \sin i = \frac{XE}{AE} = \frac{c_1 t}{AE}$$

and

$$AY = c_2 t$$

$$\Rightarrow \qquad \sin r = \frac{AY}{AE} = \frac{c_2 t}{AE}$$

Thus

$$\frac{\sin i}{\sin r} = \frac{c_1 t}{AE} \div \frac{c_2 t}{AE}$$

$$= \frac{c_1 t}{c_2 t}$$

i.e.

$$\frac{\sin i}{\sin r} = \frac{c_1}{c_2}$$

But c_1 and c_2 are the wavespeeds in the respective materials, and are therefore both constant. So the fraction $\sin i / \sin r$ should also be constant. So now this piece of theory provides a theoretical basis for Snell's empirical law.

Furthermore, we now have the prediction that, since $i > r$, c_1 is greater than c_2, and therefore light should travel faster in air than in glass (or water). When this was first discovered to be the case, by Foucault in 1850, it added strong support to the opinion that light was a wave-like phenomenon.

Dispersion

In vacuum, all electromagnetic waves travel at the same speed, whatever their frequency. In a transparent medium this is not always so. For example, in normal (crown) glass the visible colours travel at a range of speeds, the slowest being blue and the fastest red. As a consequence, the value of n is different for each different frequency. This phenomenon is called dispersion.

One useful effect of dispersion is that it allows us to create a spectrum using a prism: see figure 7.

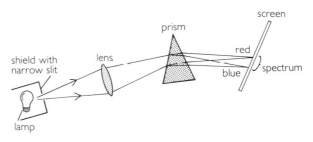

figure 7 Dispersion of light by a prism

Another effect of dispersion is to cause the colours in rainbows: see question 5 of chapter 34.

Dispersion can also be a nuisance. For example, a simple glass lens will have a slightly different focal length for each colour; this is known as *chromatic aberration*. In the design of optical instruments it is a problem. In high-quality cameras, for example, the lens is made up of a number of different elements. Some of the elements are made from crown glass, and some from flint glass, which has slightly different dispersive characteristics. By careful design of the elements of the lens, it is possible to compensate for chromatic aberration.

Summary

1 Huygens' principle: every point on a wavefront acts as a source of secondary wavelets.

2 a Experimental reflection law: $i = r$
   b Huygens' wave principle predicts $i = r$

3 a Experimental refraction law (Snell):

$$\frac{\sin i}{\sin r} = n \text{ (constant)}$$

b Huygens' wave principle predicts:

$$\frac{\sin i}{\sin r} = \frac{c_1}{c_2}$$

4 In a medium, the speed of waves and hence the refractive index of the medium, may be dependent on frequency. This is called dispersion.

Questions

1 A person stands beside a fence which has vertical slats 0.2 m apart. He makes a sharp tap with his walking stick on the ground.
   a Explain why he hears an echo which gradually dies away, and why the echo has a single frequency.
   b Calculate the frequency of the echo. (The speed of sound in air is 330 ms^{-1}.)

2 We can use an approach similar to Huygens' principle to explain the formation of a shock wave by an aircraft travelling faster than sound; and also to explain the wake caused by boats. Suppose an aircraft flies at 550 ms^{-1}; the speed of sound in air is 330 ms^{-1}.
   The aircraft has been flying in a straight line for 5 s, passing successively points P, Q, R, S and T, to arrive at the present moment at point U. See figure 8, which is drawn to scale, 1 mm representing 50 m.

figure 8

When the aircraft was at P, its engines emitted sound.
   a How far has that sound travelled in the 5 s since it was emitted?
   b Draw the diagram to the same scale. Add a line (a circle with centre at P) to show the present position of the sound the aircraft emitted when it was at P.
   c Now draw a similar circle for the sound it emitted when it was at each point Q, R, S and T.

It should be clear from your diagram that the envelope of these 'wavefronts' of sound is a cone (in three dimensions) with the present position of the aircraft at the apex. See figure 9.

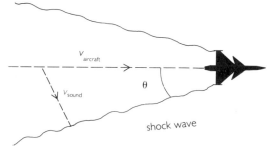

figure 9

d Measure the apex angle θ of the cone (from *your* diagram).
e Explain why the angle θ should be given by

$$\sin \theta = \frac{v_{\text{sound}}}{v_{\text{aircraft}}}$$

f Check whether this is true in your drawing.
g Explain why the sound of the 'boom' is so much more intense than the noise of the aircraft's engines when it is flying just sub-sonically.

A similar explanation can be used to account for the V-shaped waves which spread out from a boat.
h What quantities would be analogous to (i) the speed of the aircraft, (ii) the speed of sound through air?

3 This question refers to the photograph of ripple refraction, figure 4 on page 346.
   a Measure carefully the wavelengths of ripples (i) in the deep water, (ii) in the shallow water.
   b Explain why the ratio of the wavelengths must be equal to the ratio of the wave speeds.
   c Calculate the effective refractive index n for this boundary. (n = speed in deep water/speed in shallow water.)
   d Measure the appropriate angles to obtain i and r.
   e Show that the ratio $\sin i / \sin r$ is equal to n as found above.

4 The speed of light in a vacuum is $3.00 \times 10^8\,\mathrm{ms}^{-1}$; the wavelength in air of a certain colour of yellow/orange is $6.0 \times 10^{-7}\,\mathrm{m}$. For these waves: n for glass $= 1.50$, n for water $= 1.33$.
  a What is the speed of light in glass?
  b What is the wavelength of this colour in glass?
  c What is the speed of light in water?
  d What is the wavelength of this colour in water?
  e We tend to associate colour with wavelength. In view of this question, is that correct? Would you see a different colour under water, because the wavelength is different? Or what do you think colour does depend on?

5 Figure 10 shows a ray of light striking a water surface, then passing through glass below the water, and finally coming out into air again (travelling at the same angle as it did originally). In this question, $_a n_g$ means the refractive index for light travelling between air and glass; $_a n_w$ means refractive index between air and water.

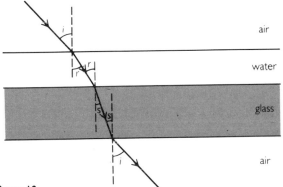

figure 10

  Can you show that the refractive index for light crossing from water into glass, $_w n_g$, is given by

$$_w n_g = \frac{_a n_g}{_a n_w}?$$

6 When white light passes through, or is reflected from, a soap or oil film, the film often appears brightly coloured. This is because light can reach your eye by two different routes – and so, as with other superposition situations, there is a path difference. Figure 11 shows how this can occur for light passing through a film.

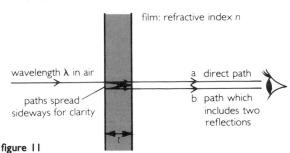

figure 11

  a What is the path difference for routes a and b in figure 11?
  b If the waves have wavelength λ in air, what is their wavelength in the film?
  c How many wavelengths fit into the path difference between routes a and b?

If this number of wavelengths is an integer, then constructive superposition occurs, for this wavelength. It may well be that there is only one wavelength in the visible spectrum for which constructive superposition occurs; and in this case the intensity of that wavelength is enhanced.
  d Suppose $t = 1.15 \times 10^{-6}\,\mathrm{m}$ and $n = 1.3$. In this case only one colour within the visible range can have constructive superposition in the manner of figure 11. Can you find its wavelength (in air)? What visible colour is it? (The visible range in air is 400–700 nm.)
  e Here is another problem on this theme. Suppose you find that constructive superposition occurs for wavelengths (in air) 500 nm and 600 nm, but no wavelengths in between. If the refractive index of the liquid is 1.33, can you find the film thickness?
  f Explain why the enhanced colour predominates in reflected light but not in transmitted light.

7 One way of measuring the refractive index of a gas is to use a variation of the Young's slits experiment. A tube with thin transparent covers at each end is placed in front of one of the two slits, so that the light from this slit has to pass through the tube. See figure 12. Suppose the gas in this tube has a lower refractive index than that in front of the other slit; then the part of a wavefront passing through the tube speeds up and emerges ahead of its corresponding part from the other slit. This causes the interference pattern to be displaced along the screen.

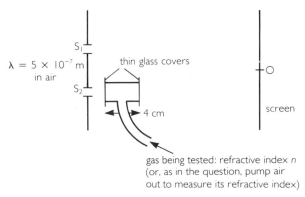

figure 12

Refractive index of air $= \dfrac{\text{speed of light in vacuum}}{\text{speed of light in air}}$

$= \dfrac{\text{wavelength in vacuum}}{\text{wavelength in air}}$

$= n$

Suppose that as the air is pumped out of the tube, an observer watches carefully the point O (where initially the central bright fringe of the pattern appears). By the time there is a complete vacuum in the tube, 24 bright fringes have passed O.

a Explain why the bright fringes move past O. Do they move upwards or downwards?

b How many complete wavelengths fit into 4 cm of air?

c When the tube is fully evacuated, how many complete wavelengths must fit into the 4 cm of the tube?

d What is the wavelength of the waves in the vacuum?

e What is the refractive index of air from this experiment?

SECTION E

GEOMETRICAL OPTICS

*The 100 inch Hooker telescope at Mount Wilson,
California, is one of the world's best designed
optical instruments.*

The design of this eye is every bit as good as the telescope. But who or what designed it?

33   Rectilinear propagation of
     light   353
34   Lenses and prisms   361
35   Optical instruments   373

CHAPTER 33

Rectilinear propagation

Introduction

Anyone who has seen light streaming into a room through a window will have the idea that light travels in straight lines or 'rays'. This chapter explains what is meant by the phrase 'ray of light'. The concept of ray of light is used to explain the behaviour of a pinhole camera, and then the discussion is extended to cover reflection at plane mirrors, cylindrical mirrors and spherical mirrors.

GCSE knowledge

We will assume that you can recall:

☐ that light can be thought of as rays which travel in straight lines;

☐ what happens to light rays when they are reflected by plane mirrors.

Light rays

Often when studying the behaviour of light on a large scale, we choose to think of light travelling in straight lines, i.e. rays of light. What do we understand by the term 'ray of light'? Experimentally it can be defined by the following apparatus, shown in figures 1 and 2.

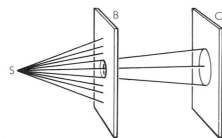

figure 1

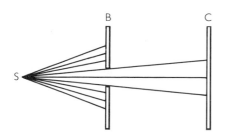

figure 2

A very small, very bright source of light at S gives out light in all directions. An opaque screen at B only allows some of the light to reach the screen at C. Only the part of the screen lying between the straight lines drawn in the figure will be illuminated. We explain this by saying that only the rays not intercepted by screen B reach the screen at C. We might hope to isolate a single ray by making the hole in screen B vanishingly small. The experiment shows, however, that when the hole in B is a few tenths of a millimetre in diameter, the bright spot on C starts to widen. The failure of this attempt to isolate a single ray is due to diffraction. It places a limit on the applicability of the ray model. A light ray represents the direction in which optical energy is flowing.

Rectilinear propagation of light

In teaching laboratories, this light-ray approach is illustrated by using a narrow laser beam, or by using a cylindrical lens and a comb as illustrated in figure 3.

The fact that light travels in a straight line through air or water can be illustrated by using a laser. Dust or smoke blown into the path of the laser beam in a darkened room will scatter light in all directions, thus allowing us to see the straight path along which the light travels. In a similar way the straight path through water can be illustrated by using a weak solution of Dettol in a tank.

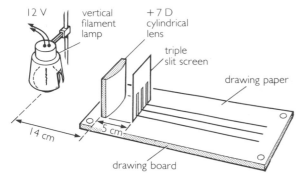

figure 3 Apparatus for producing 'rays' of light

Rectilinear propagation is a technical term which means the same as straight-line travel.

The pinhole camera

The pinhole camera relies on the rectilinear propagation of light. It is not much used today because modern lenses are well designed. But even today it has several advantages over modern cameras. In particular, it does not suffer from

☐ chromatic aberration (see page 365);

☐ spherical aberration (see page 365).

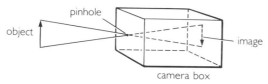

figure 4 A pinhole camera

All modern camera lenses have these aberrations to some extent, though usually they are not noticeable.

The pinhole in a pinhole camera should be about 0.5 mm–1 mm in diameter. If the hole is made smaller, then diffraction becomes a problem. If the hole is made larger, then point images overlap and blurring occurs. With small pinholes and exposure times of 2 hours, people have taken excellent, distortion-free photographs of large buildings.

Laws of reflection

These laws can be illustrated by using the apparatus of figure 5(a). The plane mirror should be a highly polished sheet of metal, so that reflection occurs from the front surface.

The angle between the normal and incident ray is called the *angle of incidence*, and is labelled i in figure 5(b). The angle of reflection is labelled r. The experimental results are:

LAW 1: The incident ray, the normal and the reflected ray are all in the same plane.

LAW 2: $r = i$.

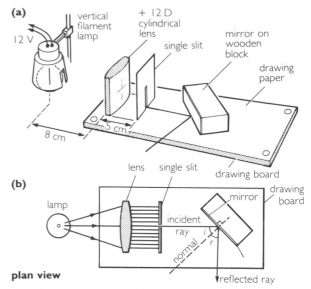

plan view

figure 5 Demonstration of the laws of reflection

The laws of reflection are obeyed in all cases, although they appear not to be obeyed in the case of diffuse reflection. Diffuse reflection occurs from 'rough' surfaces and the comparison with regular reflection is illustrated in figure 6.

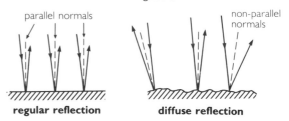

regular reflection   diffuse reflection

figure 6

Reflection at a plane mirror

Figure 7 illustrates rays of light from an object, O, being reflected from a plane mirror (the shading denotes the silvering on the other side of the mirror). The light rays OA and OB undergo regular reflection at the mirror. The light rays AE_1 and BE_2 are diverging when they enter the eye.

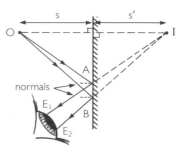

figure 7 Formation of a virtual image by a plane mirror

The brain, because of previous experience, concludes that the rays are coming from the point I. Therefore, the brain perceives an image at the point I. The rays of light do not come from I, so the image at I is purely imaginary. Physicists call such an image a *virtual image*.

We can see from figure 7 that the image point I of object O lies on the normal from the object point to the mirror. Also the image appears to be as far behind the mirror as the object is in front: $s' = s$.

If you look into a plane mirror and raise your right hand, it will appear to be the left hand of your image that is raised. This is known as *lateral inversion*; the image is said to be *laterally inverted*.

Reflection at a cylindrical mirror

This can be studied by using the apparatus illustrated in figure 8. The lower diagram depicts a concave mirror and the paths taken by incident and reflected rays. P is the *pole* of the mirror, F is the *principal focus*, and XFP is the *principal axis*. The distance FP is called the *focal length*, *f*, of the mirror.

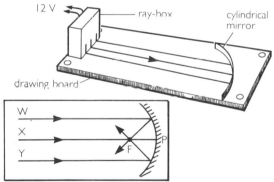

figure 8   A cylindrical mirror

Figure 9 shows a circular mirror of radius *R*, with its concave side towards the light source. The centre of curvature of the mirror's surface is at C. Consider an object represented by the arrow LM in figure 9. Two light rays have been drawn to illustrate the formation of the image at the point L'. The ray LP is reflected according to the laws of reflection, therefore angle *r* = angle *i*. The ray LCQ goes through the centre of curvature, C.

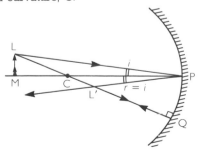

figure 9   Formation of a real image by a cylindrical mirror

Therefore, it will strike the mirror normally at Q, and is reflected directly back along its path. The reflected rays *really do meet* at L'. Therefore, the point L' is a *real* image of the object point L.

Spherical aberration with mirrors

Consider what would be observed if you used the apparatus illustrated in figure 8, but this time you used a metal comb with five slits in it. Figure 10 shows what would happen. Note that the outer rays, 1 and 5, meet at a point nearer to the mirror than do the inner rays, 2 and 4. This is undesirable because all the parallel rays do not pass through the same point after reflection. Not only does this defect occur in cylindrical mirrors, but it also occurs in spherical mirrors; it is known as *spherical aberration*.

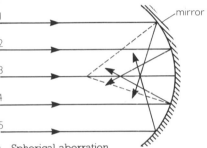

figure 10   Spherical aberration

This can be very easily observed on a sunny day by taking a cup of tea into the sunlight. Figure 11 depicts what you will observe. The light rays do not all come to a single focus, they cross in different places forming a curve. The curve seen on the top of your tea is called the 'caustic curve'.

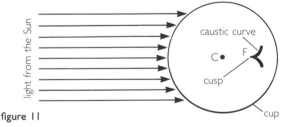

figure 11

Spherical aberration can be minimised by using only the part of the mirror close to the principal axis. Alternatively, if a parabolic mirror is used then there is no caustic curve, and all the light rays come to a focus at F. This is illustrated in figure 12.

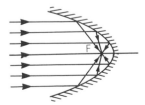

figure 12   A parabolic mirror

Image formation by a converging (concave) spherical mirror

Experiments with the apparatus of figure 8 show us that rays which spread out from a point on the axis (M in figure 13) are in general converged again to another point on the axis (M′ in figure 13). This occurs provided:

☐ the mirror is small compared to its radius of curvature, so that only rays near the axis are used – otherwise spherical aberration occurs;

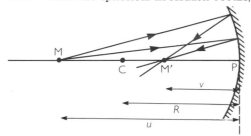

figure 13 Image formation by a concave spherical mirror

☐ M is such that M′P $> \frac{1}{2}$CP. The reason for this requirement will appear below.

By applying the law of reflection, $r = i$, to the geometry of the situation, it is possible to predict the relationship between R, u and v which is,

$$\frac{2}{R} = \frac{1}{u} + \frac{1}{v} \qquad \text{(i)}$$

We can verify this equation experimentally using the apparatus of figure 8.

Points to note

1 *Sign convention* The sign convention for this equation is that R is positive for a concave mirror and v is positive for a real image but negative for a virtual image. u is positive.

2 *Focal length* When the point object M is at a very large distance from the mirror, all the rays from M which strike the mirror will be parallel to one another, and to the axis. Hence the image will be formed at the focal point F. That is, v = focal length f when $u = +\infty$.

From equation (i)

$$\frac{2}{R} = \frac{1}{\infty} + \frac{1}{v} = \frac{1}{f}$$

Therefore the magnitude of the focal length of a concave mirror equals one-half of its radius of curvature:

$$f = \frac{R}{2}$$

The formation of an image by a concave mirror for extended objects at various distances from the mirror is illustrated in figure 14.

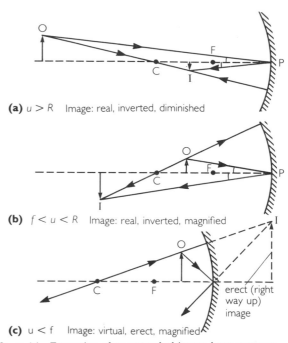

(a) $u > R$ Image: real, inverted, diminished

(b) $f < u < R$ Image: real, inverted, magnified

(c) $u < f$ Image: virtual, erect, magnified

figure 14 Formation of an extended image by a concave mirror: R=radius of curvature, f=focal length

Reflection by a diverging (convex) spherical mirror

The ray diagram for the general situation is shown in figure 15.

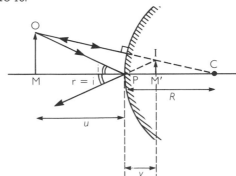

figure 15 Formation of an extended image by a convex spherical mirror

Equation (i) also applies in this case:

$$\frac{2}{R} = \frac{1}{u} + \frac{1}{v}$$

However, to make it work properly we must extend the sign convention to this situation: for a diverging mirror, R is negative.

Thus suppose $u = \infty$. Then

$$\frac{2}{R} = \frac{1}{\infty} + \frac{1}{v}$$

But by definition $v = f$ in this case; so

$$\frac{2}{R} = \frac{1}{f}$$

$$\Rightarrow \qquad f = \frac{R}{2} \text{ as before}$$

However, since R is negative, f also has a negative value. i.e. a virtual focal point behind the mirror.

Example   An object is placed 40 cm in front of a convex mirror with radius of curvature 20 cm. Where is the image? Is it real or virtual?

To solve this problem, put

$$R = -20 \text{ cm}$$

$$u = 40 \text{ cm}$$

$$v = ?$$

$$\frac{2}{R} = \frac{1}{u} + \frac{1}{v}$$

$$\Rightarrow \qquad \frac{2}{-20 \text{ cm}} = \frac{1}{40 \text{ cm}} + \frac{1}{v}$$

$$\Rightarrow \qquad \frac{1}{v} = -\frac{2}{20 \text{ cm}} - \frac{1}{40 \text{ cm}}$$

$$= -\frac{5}{40 \text{ cm}}$$

$$\Rightarrow \qquad v = -8 \text{ cm}$$

Thus the image is 8 cm behind the mirror, and is virtual.

Summary

1   We can think of light as travelling in straight lines or 'rays'.

2   'Rectilinear propagation of light' means light travels in straight lines.

3   The pinhole camera. See figure 4.

4   The laws of reflection:
   a   The incident ray, the normal and the reflected ray are all in the same plane.
   b $r = i$

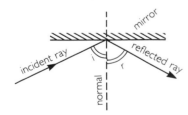

figure 16

5   Reflection at a plane mirror:
   a   The image is virtual.
   b $s' = s$

   See figure 7.

6   Reflection at a concave mirror:
   a   The ray diagram. See figure 9.

   b   The equation:

   $$\frac{2}{R} = \frac{1}{u} + \frac{1}{v} = \frac{1}{f}$$

7   Spherical aberration in curved mirrors. See figure 10.

   a   This is illustrated by the cylindrical mirror of a tea-cup.
   b   Parabolic mirrors give spherical-aberration-free reflections.

   See figure 12.

8   Reflection at a convex mirror:
   a   The ray diagram. See figure 15.
   b   The equation:

   $$\frac{2}{R} = \frac{1}{u} + \frac{1}{v} = \frac{1}{f}$$

9   Sign convention:
   For a concave mirror f and R are positive.
   For a convex mirror f and R are negative.
   For a real image v is positive.
   For a virtual image v is negative.
   For a real object u is positive.

Questions

1 A tailor wants to buy a new mirror for his fitting room. He wishes to spend as little money as possible. Therefore he should buy the shortest mirror that will fulfil his needs. His tallest customer, 1.97 m tall, and his shortest customer, 1.46 m tall, must be able to see all of themselves in the mirror.
   a Draw a ray diagram to show the mirror required just to enable the tallest customer to see all of himself. Assume that his eyes are 15 cm below the top of his head. Calculate the length of the mirror and determine how far above the floor it must be mounted.
   b Calculate the length of the mirror required just to enable the smallest customer to see all of himself, and determine how far above the floor it must be mounted.
   c How big is the mirror that the tailor buys, and how far above the floor must its bottom edge be?

2 Figure 17 illustrates a person looking into two plane mirrors which are mounted, touching each other, with an angle of 90° between them.
   a What will the person notice about the image of himself when he looks into the mirror as illustrated here?
   b Draw a ray diagram to explain your answer.

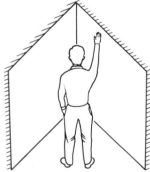

figure 17

3 Figure 18 illustrates the essential features of a simple periscope. It consists of two 45° glass prisms mounted at either end of a tube. The long face of each prism acts as a mirror, because of total internal reflection, see page 362.
   a Draw a ray diagram to show how rays from both the top and bottom of the object being viewed reach the observer's eye. Indicate on your drawing where the final image is and make it clear whether it is an erect or inverted image.

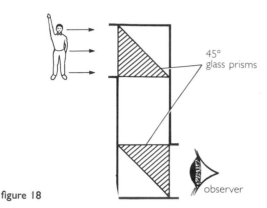

figure 18

   b Will the image be laterally inverted, i.e. will the right hand of a man being viewed appear to be his left hand? Use a diagram to explain your answer.
   c What are the advantages of using prisms instead of ordinary plane mirrors?

4 Car driving mirrors can be of two types: plane mirrors or convex (diverging) mirrors.
   a Draw ray diagrams for both types to explain the differences observed by a driver looking into each one.
   b What are the advantages and disadvantages to the driver when he uses a convex mirror rather than a plane one?

   Interior drivers' mirrors are of the anti-dazzle type. They consist of a mirror mounted behind a sheet of glass. A plane mirror type is illustrated in the diagram. The mirror swings backwards on its hinge when in the anti-dazzle mode.

normal use   observer   antidazzle use   observer

figure 19

   c Explain how it works.

5 In amusement arcades at fairgrounds, there is often a hall of mirrors. People stand about 2 m in front of the mirrors and see distorted images of themselves.
   a Say what a person would see in each of the following cases (explain your answer in each case).
      When the mirror is:
      (i) the convex surface of a vertical cylinder of radius 5 m;
      (ii) the convex surface of a horizontal cylinder of radius 5 m;
      (iii) a convex spherical surface of radius 5 m;

(iv)   the concave surface of a vertical cylinder of radius 5 m;

(v)   a concave spherical surface of radius 5 m.

b Describe the sort of mirror that would make you appear to have a short fat body with long thin legs.

6 a Explain what is meant by a virtual image.

A student wishes to produce an image of a distant object on a screen. He decides to use a converging (concave) mirror. He chooses a spherical mirror with radius of curvature 1000 mm. The object for which he wishes to obtain an image is 100 m away, and has a height of 2 m.

b How large will the image be? Will it be erect or inverted?

c What is meant by 'spherical aberration' of a concave mirror?

d What defects will the student *notice* in the image due to the spherical aberration produced by his spherical mirror?

The student decides to make a camera 'lens' out of his mirror by using a plane mirror in addition, and mounting them both in a tube as illustrated in figure 20. Also he has to drill a hole, 2 cm in diameter, through the centre of the large mirror, to allow light to reach the film.

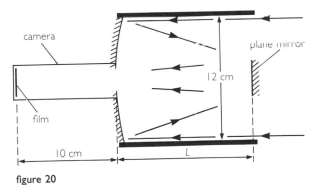

figure 20

e How far apart, L, must the mirrors be if a sharp image is to be focused on the film?

f What is the smallest size of plane mirror that can be used for this purpose, if all the light hitting the mirror from a distant point on the principal axis is to be focused on the film? (The aperture of the concave mirror is 12 cm.)

g Explain all the differences in the final image that will result from placing the small mirror in the position shown in figure 20 compared with the image obtained as in part (b) of the question.

h Discuss how using (i) a convex mirror, (ii) a concave mirror, instead of the plane mirror will affect the size of the final image, and the length L of the lens system.

7 An object 5 cm tall (your chin) is placed 20 cm from a concave mirror with radius of curvature 120 cm.

a Draw a ray diagram to illustrate the formation of the image produced.

b Calculate the size and position of the image, and say whether the image is real or virtual, erect or inverted, magnified or diminished.

8 An object 10 cm tall is placed 1 m from a convex mirror with radius of curvature 50 cm. Draw a ray diagram to illustrate the formation of the image produced. Calculate the position and size of the image, and say whether the image is real or virtual, erect or inverted, magnified or diminished.

9 Figure 21 shows an accurately drawn parabola. It represents a parabolic mirror. The radius of curvature of the mirror surface, at the point where the principal optic axis cuts it, is 3 cm.

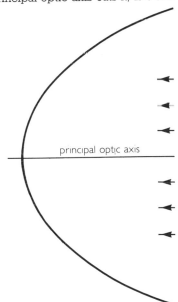

principal optic axis

figure 21

a Copy the parabola accurately onto a piece of paper (tracing paper would be easiest). Then construct accurately the paths taken by the six rays shown in the diagram. They are 0.5 cm, 1.5 cm and 2.5 cm from the principal axis.

b Next, draw accurately a circular mirror of radius 3 cm, and then construct accurately the paths taken by the same six rays that were incident on the parabolic mirror.

c Discuss the differences and similarities in performance of the two mirrors.

d Again copy the parabola accurately onto a piece of paper. This time, draw accurately the paths taken by six parallel rays which are *not* parallel to the principal axis, but make an angle of 15° with it, each ray being spaced 1 cm from its neighbour (see figure 22).

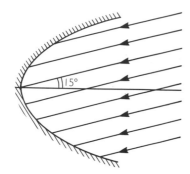

figure 22

e Compare the performances of the parabolic mirror in the two cases, i.e.
(i) when the light is parallel to the principal axis,
(ii) when the light is a parallel beam making an angle to the principal axis.

f You have probably observed on a sunny day that a pattern appears on top of the tea in a tea-cup. The form of the pattern observed, and its position on the surface of the tea are shown in the diagram. Explain its formation and say what the cusp of the curve represents.

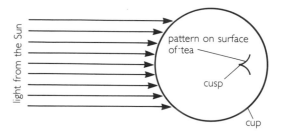

figure 23

CHAPTER 34

Lenses and prisms

Introduction

This chapter covers refraction at glass–air interfaces. The simple ray-optics approach is used. This consists of thinking of rays of light changing direction as they change media. The discussion covers rectangular glass blocks, semicircular glass blocks, prisms, converging lenses and diverging lenses. We make no attempt to explain the refraction of light in terms of waves, as this has already been done in chapter 32.

GCSE knowledge

We will assume that you can recall:

☐ that light travels more slowly through glass than through air;

☐ that this difference in speed explains the change in direction which light rays undergo at air–glass interfaces;

☐ that light rays deviate towards the normal when they pass from air to glass;

☐ the 'principle of reversibility' which applies to the paths followed by light rays. This means that if a light ray passes through a lens or prism, starting at A and finishing at B, then it is also possible for a ray to start at B and finish at A.

Laws of refraction (Snell's laws)

In figure 1 we see a light ray going from air to glass; its direction is changed, and this phenomenon is what physicists call *refraction*.

figure 1   Refraction

The 'normal' is an imaginary line which is perpendicular to the surface at the point where the light ray strikes. The ray is bent towards the normal, so that angle r is smaller than angle i. The effect can be investigated experimentally by using the apparatus illustrated in figure 2. The cylindrical lens is used to produce a parallel beam.

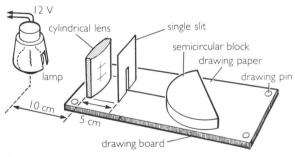

figure 2   Investigating refraction

The semicircular glass block should have a roughened lower surface, but all other surfaces should be polished. The block should be positioned so that a ray emerges from the *centre* of the flat face, as illustrated in figure 3. The roughened surface allows the path of the ray within the block, i.e. from X to Y, to be seen.

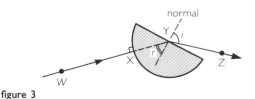

figure 3

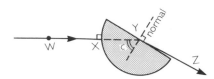

figure 5   The critical angle

The path followed by a light ray, initially travelling along WX, is the same as the one taken by a ray initially travelling along ZY. Because of this reversibility we have labelled the angles in figure 3 with 'i' in the less dense medium, so that the convention illustrated in figure 1 is maintained.

From such an experiment, Snell, a Dutch professor, discovered in 1620 the following laws of refraction. (A theoretical justification is given in chapter 32.)

LAW 1: The incident ray, the normal and the refracted ray all lie in the same plane.

LAW 2: For two given media, and a particular colour of light, the ratio $(\sin i)/(\sin r)$ is a constant. This constant is known as the *refractive index*, n, for the two given media and that frequency.

Hence

$$n = \frac{\sin i}{\sin r}$$

Total internal reflection

Consider again the apparatus represented in figure 2. If the glass block is rotated further than illustrated in figure 3, i.e. angle r is increased, then the ray YZ does not emerge from the flat face; it is reflected back as shown in figure 4. This is called *total internal reflection*.

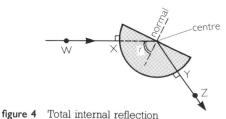

figure 4   Total internal reflection

The smallest value of r for which total internal reflection occurs is called the *critical angle*. If angle r is decreased very slightly below the critical angle, then the ray CD emerges almost parallel to the flat surface as shown in figure 5. We then have $r \approx c$ and $i \approx 90°$.

This gives us a relationship between c and n:

$$n = \frac{\sin i}{\sin r} = \frac{\sin 90°}{\sin c} = \frac{1}{\sin c}$$

Example   For a block of glass the critical angle is found to be 43°. What is the refractive index of the glass?

$$n = \frac{1}{\sin c}$$

$$\Rightarrow \qquad n = \frac{1}{\sin 43°}$$

$$\Rightarrow \qquad n = 1.47$$

The prism (triangular)

The effect of a triangular prism on the path of a ray of light can easily be illustrated by using either apparatus similar to that of figure 2, or a laser. Smoke or chalk dust blown into the laser beam makes it visible; the result is illustrated in figure 6.

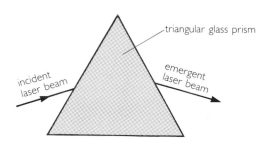

figure 6   A triangular prism

Minimum deviation   The deviation in the direction of the light beam is illustrated in figure 7.

The light ray starts off in direction LMNO, but is refracted through an angle $(i_1 - r_1)$ at the air–glass

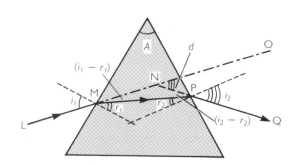

figure 7   Deviation of light ray by a triangular prism

interface at M. At the glass-air interface at P, it is refracted through a further angle $(i_2 - r_2)$. Therefore, the total deviation d is given by

$$d = (i_1 - r_1) + (i_2 - r_2)$$

If the prism is rotated backwards and forwards it is found that the angle of deviation, d, varies as illustrated in figure 8. This shows that there is a minimum deviation which the prism can produce. By the principle of reversibility of light rays, this minimum deviation D must occur when the light ray passes symmetrically through the prism. In other words, if the minimum deviation occurs when the light goes along L to Q, this deviation must also occur when the light goes from Q to L. We will only get the same deviation when $i_1 = i_2$.

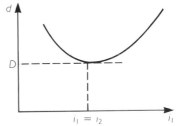

figure 8   Minimum deviation

If we pursue the geometry further, and apply Snell's law at an appropriate point, we reach this equation:

$$n = \frac{\sin[(A + D)/2]}{\sin(A/2)}$$

In this equation, D is the minimum deviation, and A is the apex angle of the prism – see figure 7.

Refraction through thin lenses

The basic behaviour of lenses can be investigated experimentally by using the apparatus illustrated in figures 9 and 10.
   Figure 10 illustrates the converging nature of a cylindrical lens which is thicker at its centre. Figure 11 shows how three parallel rays are brought to a focus. The distance from the lens to the point at which parallel rays are brought to a focus is called the *focal length, f.*

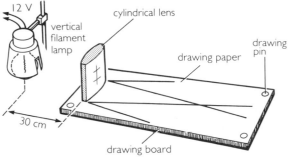

figure 9   A cylindrical converging lens

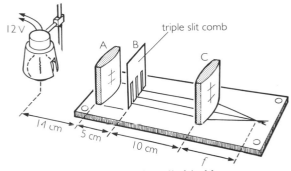

figure 10   The focal length of a cylindrical lens

We may think of thin lenses as being made up of a parallel-sided glass block and glass prisms as illustrated in figure 11.

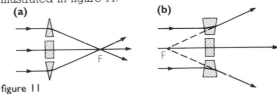

figure 11

Image formation by converging lenses

When working out the position and size of the image formed by a converging lens, the paths followed by just three rays need be considered. The three rays usually chosen are shown in figure 12.

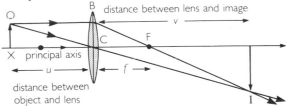

figure 12   Image formation by a converging lens

☐ *Ray XC:* a ray along the principal axis, i.e. through the centre of the lens and perpendicular to the lens, goes straight on.

☐ *Ray OB:* a ray parallel to the principal axis goes through the focal point, after it has passed through the lens.

☐ *Ray OC:* a ray through the centre of the lens goes straight on.

   When we look at figure 13, we can understand why rays that pass through the centre of the lens may be considered to be undeviated. The centre of the lens is like a thin parallel-sided prism. Figure 13 shows that the emerging ray CD has been displaced a little by its passage through the block, although it is still parallel to the incident ray AB. If the lens is thin, this displacement is small and we may neglect it to a good approximation.

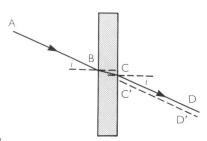

figure 13

The positions and sizes of the images formed by a converging lens for different object-to-lens distances u are illustrated in figure 14.

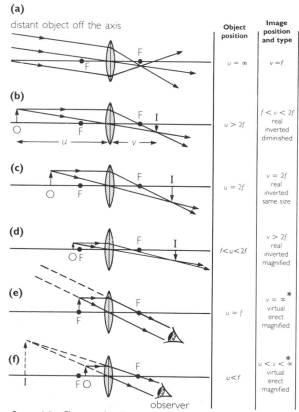

figure 14 Converging lens
* This statement describes the magnitude of v. However, by the sign convention for virtual images, v is regarded as having a negative value

Equation linking u, v and f

If we do some geometry based on figure 12, we can arrive at an algebraic relationship between u, v and f. It is this:

$$\frac{1}{u} + \frac{1}{v} = \frac{1}{f}$$

Notice that all the results of figure 14 can be obtained using either the drawing method of figure 12 or this algebraic relationship. For example, suppose $u \rightarrow \infty$; then $1/u \rightarrow 0$, and so $v \rightarrow f$ as it must do (since this is the way f is defined).

Sign convention The sign convention for this equation is that f is positive for a converging lens; u is positive for a real object; and v is positive for a real image, but negative for a virtual image.

This equation also applies for a diverging lens, if we use the convention that in this case f is negative.

Example Find the position and size of the image formed by a thin converging lens of focal length 10 cm if an object 1 cm tall is placed 14 cm from the lens.

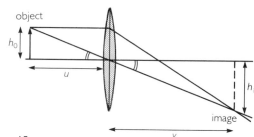

figure 15

We shall use the equation

$$\frac{1}{f} = \frac{1}{u} + \frac{1}{v}$$

where focal length, $f = +10$ cm (converging lens) and the object distance $u = +14$ cm. Substituting gives

$$\frac{1}{+10\,\text{cm}} = \frac{1}{+14\,\text{cm}} + \frac{1}{v}$$

$$\Rightarrow \quad \frac{1}{v} = \frac{1}{10\,\text{cm}} - \frac{1}{14\,\text{cm}}$$

$$= \frac{7-5}{70\,\text{cm}}$$

$$= \frac{2}{70\,\text{cm}}$$

$$= \frac{1}{35\,\text{cm}}$$

$$\Rightarrow \quad v = +35\,\text{cm}$$

Magnification $\qquad m = \dfrac{\text{image size}}{\text{object size}} = \dfrac{h_1}{h_0}$

$$= \frac{v}{u} \text{ (by similar triangles)}$$

$$m = \frac{35\ cm}{14\ cm}$$

$$= 2.5$$

Thus, the image is real, inverted, 2.5 cm tall, and 35 cm from the lens, on the opposite side to the object.

Diverging lenses

figure 16 Diverging lenses

A beam of parallel rays of light incident on any of the lenses illustrated in figure 16 becomes a diverging beam after refraction. Such lenses are called diverging; all of them are thinner in the middle than they are at the edges.

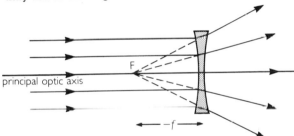

figure 17 A diverging lens has a virtual focal point

The diverging action of such a lens is depicted in figure 17. The focal point F in figure 17 is an imaginary (virtual) focal point; the rays of light which started off parallel to the principal axis only *appear* to come from F, they do not actually go through F.

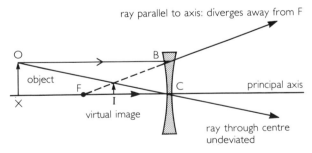

figure 18 Image formation by a diverging lens

We can construct a diagram to show how a diverging lens forms an image, just as we did for a converging lens (figure 12). See figure 18.

In every case a virtual upright image is obtained. This image lies somewhere between F and C, and is always smaller than the object.

Defects (aberrations) of lenses

1   *Chromatic aberration* is caused by light of different colours travelling at slightly different speeds through glass, so that a lens will tend to split up white light into its component colours in the same way that a prism does. This results in an individual lens having slightly different focal lengths for different colours.

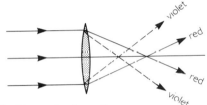

figure 19 Chromatic aberration

2   As it does with mirrors, *spherical aberration* tends also to occur in lenses of large diameter. Unless the lens is very carefully and precisely constructed, rays entering the lens near its edge are deviated more than those passing through the centre of the lens. The result of spherical aberrations is blurring of the image.

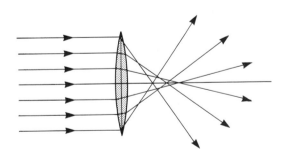

figure 20 Spherical aberration

Summary

1 Laws of refraction (Snell's laws):
a The incident ray, the normal and the refracted ray all lie in the same plane.
b For two given media and a given colour, the ratio $(\sin i)/(\sin r)$ is a constant. This constant is known as the refractive index, n, for the two given media for that colour, (frequency).

$$n = \frac{\sin i}{\sin r} = \text{constant}$$

2 The smallest value of r for which total internal reflection occurs is called the *critical angle c.*

$$n = \frac{1}{\sin c}$$

3 The prism (triangular):
Total deviation $d = (i_1 - r_1) + (i_2 - r_2)$

$$n = \frac{\sin[(A+D)/2]}{\sin(A/2)}$$

where D = minimum deviation.

4 The converging lens:
a Ray diagrams (see figure 14).
b Magnification $m = v/u$.
c $(1/u) + (1/v) = (1/f)$
f is positive
u is positive
v is positive for real images
v is negative for virtual images.

5 The diverging lens:
a Ray diagram (see figure 18).
b $(1/u) + (1/v) = (1/f)$
f is negative
u is positive
v is negative for virtual images.

6 Lens aberrations:
a chromatic aberration;
b spherical aberration.

Questions

1 Consider a ray of light incident on the flat surface of a glass block, such that the angle between the ray and the glass surface is 60°. (Refractive index of glass with respect to air = 1.5.)
a (i) Calculate the angle of refraction.
(ii) Calculate the angle through which the ray is deviated on entering the glass.
(iii) If the glass block has parallel sides, at what angle will the ray emerge from its base?

Now consider the same ray incident at the same angle, 60°, on a flat surface of water.
(Refractive index of water with respect to air = 1.33.)

(a)

(b)

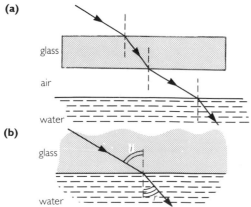

figure 21

b (i) Calculate the angle of refraction.
(ii) Calculate the angle through which the ray is deviated on entering the water.
c Next consider the ray of light passing first through the glass block and then into the water as shown in figure 21(a).
(i) What happens to the angles of incidence and refraction if the air gap between the glass block and the water is made smaller? Only consider the case when the glass block remains horizontal.
(ii) Imagine the glass block just touching the water's surface (figure 21(b)). Use your answers to parts (a), (b) and (c) to calculate the refractive index of water with respect to glass.

2 a A ray of light incident on one surface of a 60° glass prism makes an angle of 20° with that surface. Calculate the angle of refraction at this surface, then determine the deviation of the ray produced by the prism. (Use $n = 1.6$.)

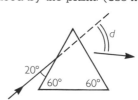

figure 22

b (i)   Calculate the minimum deviation produced by this 60° prism.
(ii)   Determine the angle of incidence at which minimum deviation occurs.

c The refractive index of glass increases as the wavelength of the light decreases. Red and violet are the two extremes of the visible spectrum.
(i)   Use this information to explain the appearance of the spectrum produced by the prism.
(ii)   Determine the angular spread of the spectrum when the red light passes through at minimum deviation.
($n_{\text{red}} = 1.585$ and $n_{\text{violet}} = 1.609$.)

3 This question is about haloes which are sometimes seen around the Sun, particularly by observers in the Arctic and Antarctic. These are due to the refraction of light by tiny hexagonal ice crystals in the atmosphere. We shall simplify the discussion by assuming that the line between the observer and the Sun is horizontal (as is usually the case in the Arctic and the Antarctic), and that the ice crystals are falling with their axes vertical.

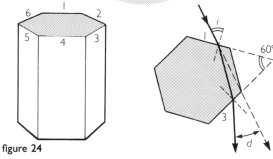

figure 23

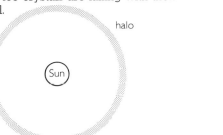

figure 24

Consider a ray of light entering face 1 and leaving via face 3 of the hexagonal crystal shown in figure 24. The crystal deviates the ray through an angle d. Faces 1 and 3 can be thought of as part of a 60° triangular prism.

a Calculate the minimum deviation, D, produced by faces 1 and 3, i.e. by a 60° prism of ice. (Refractive index of ice with respect to air = 1.31.)

b Draw a sketch graph to show how the deviation d produced by a 60° prism (or faces 1 and 3 of these hexagonal crystals) varies with the angle of incidence i.

c Calculate the angle of incidence at which minimum deviation occurs.

d Rays from the Sun will hit the crystals with all possible values of the angle of incidence. Hence, the crystals must refract the light into a large range of angles. Yet the brightest emerging light is at the angle that least deviates the sunlight from its original direction. Use your answer to part (b) to explain this fact.

e If, as we have assumed, all the ice crystals are falling with their axes vertical, will a halo be observed? If not what will be observed? Under what conditions will a halo be observed?

f Now consider what happens if the Sun is well above the horizon, but all the ice crystals are still falling vertically (figure 25). Will the brightest emerging light from the millions of crystals still be deviated by the same angle that you determined in part (a)? Say what happens to the angle of deviation and explain your answer.

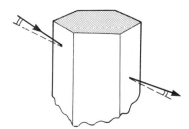

figure 25

4 Consider a horizontal ray of light incident on the surface of a vertical glass rod, as shown in figure 26. The refractive index of glass with respect to air is 1.5.

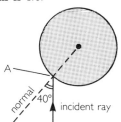

figure 26

a Calculate the angle of refraction at A. Then copy the diagram and draw on it the refracted ray inside the glass rod.

b When the ray inside the glass rod strikes the surface of the rod it will emerge from the glass and be refracted. Calculate the angle by which the emergent ray has been deviated from its original direction. Draw this emergent ray on your copy of figure 26.

5 *Rainbows*
Rainbows are formed by the passage of sunlight through the spherical water droplets of falling rain. Consider figure 27, which shows a ray of

light incident, at an angle i, on the surface of a spherical droplet. Every time the ray strikes an air–water interface, some light will be reflected and some will be refracted. Some of the incident light will be refracted at W, reflected at X, and will emerge along YZ after refraction at Y.

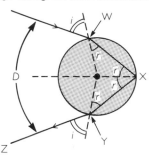

figure 27

a The incident ray is deviated through the angle $(i-r)$ at W, and through $(180° - 2r)$ at X.
(i) Express the angle D in terms of i and r.
(ii) What will be the value of D when the angle of incidence i is zero, as is the case for the ray hitting point P in figure 28?

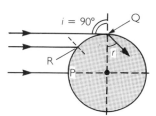

figure 28

b The formation of a rainbow in a fine mist of a liquid which has a refractive index of 1.414 is easier to explain than in the case of water, which has a refractive index of 1.33.
(i) Figure 28 shows a droplet of this liquid. Calculate the angle of refraction r when the angle of incidence i is 90°, as it is at Q. Hence, determine the value of the angle D in this case.
(ii) For the same liquid drop, calculate the angle of refraction r and the angle D when the angle of incidence $i = 45°$, as it is for the ray at R.
(iii) Make a copy of figure 28 and on it mark the paths followed by the three rays P, Q and R as they pass through the drop and return towards the left. Use your diagram to explain that D has a maximum value.
c Now let us consider a parallel beam of sunlight falling on a spherical drop of water (figure 29). We see that all values of i between 0° and 90° will occur. (Refractive index of water = 1.333.)
(i) For different values of i calculate the corresponding values of r and D, and complete the following table:

Angle of incidence, $i/°$	45	50	55	60	65	70	90
Angle of refraction $r/°$							
Angle $D/°$							

(ii) Draw a sketch graph of these results.

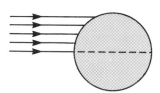

figure 29

Figure 30 shows a beam of parallel sunlight, AB, falling on a spherical water droplet at B. The diagram shows the directions in which the various rays leave the droplet, having been reflected inside.

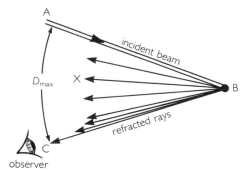

figure 30

(iii) What is the approximate value of D_{max}?
(iv) Explain why we have shown the eye at C, in figure 30, receiving many rays, whereas if it were placed elsewhere, for example at X, it would receive very few.

Now the maximum value of D depends on the refractive index of the water forming the droplets, and this in turn will depend on the wavelength of the light. For water in air the refractive index for red light is 1.331 and for violet light it is 1.343.
d Use this information and your answers to questions (a) and (b) to explain the formation of a rainbow. Use the data to explain why we see red at the top of the rainbow, and explain the direction in which the observer must look.

6 *Apparent depth*
A glass block measures $12 \, cm \times 6 \, cm \times 3 \, cm$. The block is placed on top of a sheet of paper, on which a dot has been made.

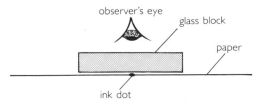

figure 31

a Draw a ray diagram to show the paths taken by light rays that move upwards from the dot at 20° to the vertical.

b If the refractive index of the glass is 1.5, what will be the apparent thickness of the glass block when viewed from above, as in figure 31?

c Explain what differences would be observed by a person looking into a completely calm swimming pool if the refractive index of the water suddenly changed from 1.33 to 1.50.

7 Figure 32 shows a boy observing a goldfish which is swimming around inside a cylindrical

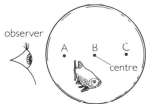

figure 32   Fish bowl

fish bowl. For each of the positions A, B and C, draw ray diagrams to show whether the fish at that point will appear to be (i) nearer to the observer than it really is; or (ii) the correct distance from the observer; or (iii) further from the observer than it really is.

8 Figure 33 shows a drinking straw in a cylindrical glass of lemonade. The drinking straw is straight and has the same diameter all along its length. In the diagram the bottom of the straw, E, rests against the front of the tumbler. The top of the straw, AB, stands out of the lemonade and rests against the rear of the glass. The section AB and the base E lie on one straight line. The features to notice about the diagram are:

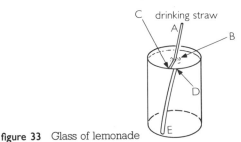

figure 33   Glass of lemonade

(i) the straw appears to bend on entering the lemonade at B, i.e. BC is not in line with AB;

(ii) the section of the straw seen through the curved side of the glass appears to curve away from the observer as he raises his gaze from E to D;

(iii) the diameter of the straw at E appears to be slightly larger than the diameter between A and B;

(iv) the diameter of the straw appears to increase as the observer raises his gaze from E to D.

Take each of these points in turn, and explain with the aid of diagrams how each one comes about.

9 Manufacturers of thin lenses use the following formula

$$\frac{1}{f} = (n - 1)\left[\frac{1}{r_1} - \frac{1}{r_2}\right]$$

where f is the focal length of the lens, r_1 and r_2 are the radii of curvature of its surfaces, and n is the air-to-glass refractive index of the glass used.

   The sign convention for each radius of curvature is illustrated in figure 34. Each radius is measured from the surface concerned; the radius is given a *positive* magnitude if it is measured to the *right*, i.e. in the same direction as the incident light is assumed to be travelling.

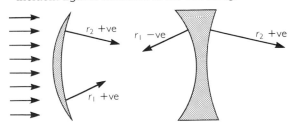

figure 34

   The lenses illustrated in figure 35 are all made from glass of refractive index 1.50. The magnitude of the radius of each face is given below each diagram.

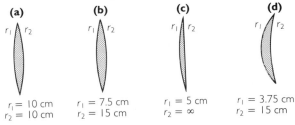

figure 35

   Find the focal length of each lens. State in each case if it is converging or diverging.

10 Read the article about fibre optics below, then answer the questions which follow it.

Fibre optics

Fibre-optic bundles

If light enters the end of a solid glass rod so that the light transmitted into the rod strikes the side of the rod at an angle θ, exceeding the critical angle, then total internal reflection occurs (see figure 1 below). The light continues to be internally reflected back and forth in its passage along the rod, and it emerges from the other end with very little loss of intensity.

This is the principle of fibre optics, in which long glass fibres of very small cross-sectional area transmit light from end to end, even when bent, without much loss of light through their side walls. Such fibres can then be combined into 'bundles' of dozens to thousands of fibres for the efficient conveyance of light from one (often inaccessible) point to another.

Coherent and incoherent bundles

An ideal fibre transmits light independently of its neighbours, so if a bundle of fibres is placed together in an orderly manner along its length, with the relative positions remaining unchanged, actual images may be transmitted along the fibre (see figure 2 below). Such an arrangement is called a coherent bundle, and consists of fibres of very small diameter, about $10\,\mu$m. The ends of the bundle are cut square and polished smooth to prevent distortions. Each fibre transmits a small element of the image which is seen at the other end of the coherent bundle as a mosaic. The eye has to 'look through' the fragmented structure to appreciate a clear image.

In contrast, a bundle of fibres arranged at random is known as an incoherent bundle (or sometimes simply a light guide) and is suitable only for the transport of light not of images. The fibres of such a bundle are relatively large, having diameters of about 50–$100\,\mu$m.

Transmission efficiency and resolution

A light beam inevitably suffers some attenuation in its passage along a fibre core: a 50 per cent loss for every 2 m travelled is typical. In addition, there are 'end losses', which are light losses at the end faces due to partial reflection and incidence on the cladding material. Thus, light sources need to be very powerful, and even then problems can arise when viewing coloured images since different wavelengths have different transmission efficiencies.

The thinner and more numerous the fibres, the greater should be the resolution. However, when the core diameter falls below about $5\,\mu$m diffraction starts to occur and transmission efficiency drops. Hence, although fibres with core diameters down to about $1\,\mu$m have been used, typical diameters are nearer $10\,\mu$m.

(From Jean A. Pope, *Medical Physics* (Heinemann, 1984))

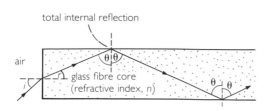

figure 1   Glass fibre

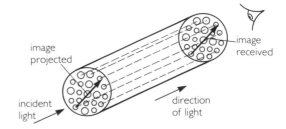

figure 2   Coherent fibre-optic bundle

a The critical angle c (line 2) is the angle of incidence within a medium which would make the angle of the ray, in the air outside the medium, 90°: see figure 36.
Angle θ must be greater than c for total internal reflection to occur.
(i) Apply Snell's law to figure 1 to derive a relationship between c and n.
(ii) Write down Snell's law to relate i, r and n (see figure 1).
(iii) How are r and θ related?

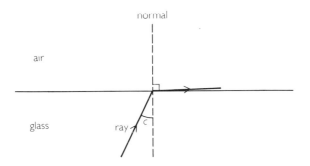

figure 36

(iv) Explain qualitatively why i must be less than a certain value for total internal reflection to occur as shown.
(v) Use the equations from (i), (ii) and (iii) to show that i must be restricted by the equation

$$\sin^2 i \leqslant n_2 - 1$$

b (i) Assume that the separation of a fibre in a coherent bundle is the same size as the fibre diameter. Estimate how close you would have to place your eye to the end of the bundle to be able to see that the image consists of separate points of light (rather like the dots that make up newspaper photographs).
(ii) You should have found, in answering b(i), that you would have to be very close in order for your eye, by itself, to distinguish the separate fibres. In practical uses, such as an endoscope for examining a patient's digestive system from the inside, the end of the bundle is quite small – of the order of millimetres across. The operator therefore views the image through magnifying equipment. Will this make it more or less likely that the operator will be able to detect individual fibres? Is it desirable for the operator to detect individual fibres?
(iii) What does the author mean when she writes: 'The thinner and more numerous the fibres, the greater should be the resolution'?
(iv) The author writes: '... when the core diameter falls below about 5 µm diffraction

starts to occur...' Where does this diffraction occur? Show, using a suitable wavelength for light, that diffraction starts to be significant at about this diameter. Explain why it is a bad thing for the resolution of the bundle.
c Give a reason why it is acceptable for the fibres of an incoherent bundle to be relatively large; why do you think they are in fact made larger?
d '... different wavelengths have different transmission efficiencies.' Suppose the light entering a bundle is white, but the bundle is least efficient at transmitting in the green region of the spectrum. How will the image be affected?
e (i) What does 'attenuation' mean?
(ii) What fraction of the original energy remains after 4 m of travel through the fibre, if it loses 50% for every 2 m?
(iii) In what manner does the energy decrease with distance travelled?

11 a Explain what is meant by (i) a virtual image, (ii) a real image. Illustrate your answers by drawing ray diagrams showing the passage of *two* rays through a converging lens. In each case the two rays should originate from an object which is not on the principal axis.
b A point source of light is placed 50 cm from a thin lens of focal length 20 cm, as illustrated in figure 37.

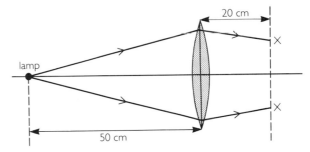

figure 37

(i) Calculate how far from the lens the image is formed. Is the image real or virtual?

Now consider what happens when a diverging lens of focal length 20 cm is placed at XX in the above diagram. The image of the light bulb formed by the first lens acts as a *virtual object* for the second lens.
(ii) Explain what is meant by *virtual object* in this context.
(iii) Calculate the position of the final image produced by this second lens. Is the final image real or virtual? (To be consistent with our sign convention, the distance u between a virtual object and a lens has to be taken as negative.)

12 a Consider a point source of light placed 50 cm from a thin converging lens of focal length 20 cm. Calculate the image distance v for this object distance u, and determine the separation s of the object and the image. Repeat the calculation for different values of u and complete the following table.

Object distance u/cm	23	25	30	40	50	60	100	160
Image distance v/cm								
Separation s/cm								

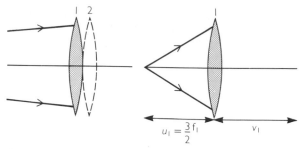

figure 38

b Use the above values to plot a graph of separation s against object distance u. Draw a smooth curve through the points and from your graph determine the least value for the separation of object and image for this lens. What is the object distance when the separation is least?

c Use the values in your table to plot a graph of image distance v against object distance u. Draw a smooth curve through the points, and mark on your graph the object distance u, from part (b), which gave the least separation of object and image. What do you notice about the image distance in this case?

13 Suppose we have two thin lenses of focal lengths f_1 and f_2 placed in contact with one another as shown in figure 38. We shall assume that the thickness of the lenses is very small compared with their focal lengths; therefore we can assume that the distance between the lenses is negligible.

a Draw a ray diagram to show the formation of an image by lens 1, when an object is placed at an object distance equal to $1.5f_1$. Calculate the image distance v_1.

b If the second lens is now placed in position, the image formed by the first lens will act as a *virtual object* for the second lens. Draw a second ray diagram to show how the two lenses form a final image. Calculate the final image distance v.

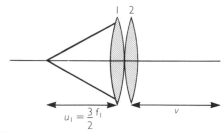

figure 39

c The usual equation for a converging lens is

$$\frac{1}{f} = \frac{1}{u} + \frac{1}{v}$$

Use your answers to parts (a) and (b) to express the effective focal length of the combination of two thin lenses in contact, in terms of f_1 and f_2.

CHAPTER 35

Optical instruments

Introduction

This chapter covers the simple ray treatment of two instruments: a simple telescope, consisting of two lenses, and a microscope. In one of the questions we also consider a reflecting telescope.

GCSE knowledge

We will assume that you are able to:

☐ recall that the paths taken by light rays as they pass through the centre of thin lenses are almost undeviated;

☐ recall that all rays which are parallel to the principal axis as they approach a thin converging lens will be deviated so that they pass through the principal focal point, F.

figure I

Telescope (refracting version)

This consists of two converging lenses. The first one collects and concentrates light from a distant object to produce a real image. The second one behaves as a magnifying glass. This is illustrated in figure 2.

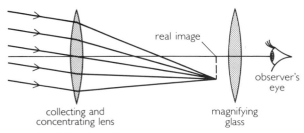

figure 2   Refracting telescope

There are various other designs of telescopes, but they all operate on similar lines. That is, there are two sections:

☐ a light gathering and concentrating system, which can be a lens, lenses or a mirror; followed by

☐ a magnifying glass.

Normal adjustment and near point adjustment

Any optical instrument is said to be in *normal adjustment* when it is emitting *parallel* light. The relaxed eye of a *normal* person focuses parallel light. Therefore, in this arrangement the telescope can be used for long periods without causing eye-strain.

Maximum magnification is produced when the telescope is adjusted so that the final image is formed as near to the observer's eye as possible, that is, at the least distance of distinct vision. This is about 25 cm for normal eyes. This arrangement is known as *near point adjustment*.

Astronomical telescope (ray diagram)

Figure 3 shows the essential features of an astronomical telescope in *normal* adjustment.

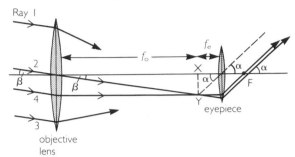

figure 3 Astronomical telescope

Light from any distant object, e.g. from the Moon, will be effectively parallel when it enters the large collecting lens on the front of the telescope. This large lens, which is the lens nearest to the object, is called the *objective lens*. The lens nearest the eye is called the *eye-lens*, or *eyepiece*.

A real image of the distant object is formed by the objective lens in its focal plane, illustrated as XY in figure 3. In *normal* adjustment the eye-lens will emit parallel light, therefore XY must be the focal plane of the eye-lens. Thus, when the telescope is in normal adjustment, the length of the telescope, from objective lens to eyepiece, is equal to the sum of the two focal lengths, i.e. $(f_o + f_e)$.

To construct the ray diagram for the telescope, we start with rays 2 and 4 in figure 3. Ray 2 coming from the top of the distant object strikes the middle of the objective lens, therefore it goes straight through. In the space between the two lenses, ray 4 is parallel to the principal axis, and because ray 4 was originally parallel to ray 2 they must intersect in the focal plane at Y. When ray 4 approaches the eyepiece it is travelling parallel to the principal axis. Therefore ray 4, after passing through the eyepiece, will pass through the principal focal point of the eyepiece, F. The remainder of the diagram can now be constructed by drawing several more rays parallel to rays 2 and 4 as they approach the objective lens, and again parallel to 2 and 4 when they leave the eyepiece. They all intersect at Y.

By inspecting figure 3 we can see that rays entering the objective lens from the top of the object being viewed are moving upwards when they leave the eyepiece. Thus, the image formed by this simple telescope is upside down. For astronomical observation this is not a disadvantage. For everyday use it would be a problem; therefore in terrestrial telescopes we introduce an extra lens between the objective lens and the eyepiece. This is called the *erecting lens*.

Magnification of astronomical telescope in normal adjustment

The important parameter for a telescope is the *angular magnification* (magnifying power), M, it produces, where

$$M = \frac{\text{angle subtended at the eye by the image formed by the telescope}}{\text{angle subtended at the eye by the object when viewed directly}}$$

From figure 3 we see that:

angle subtended by image $= \alpha \approx \tan\alpha$ (because α is small)

$$= \frac{YX}{f_e}$$

and angle subtended by object $= \beta \approx \tan\beta$

$$= \frac{YX}{f_o}$$

$$\therefore \quad M = \frac{\alpha}{\beta}$$

$$= \frac{YX/f_e}{YX/f_o}$$

$$= \frac{f_o}{f_e}$$

$$\therefore \quad \text{Angular magnification,} \quad M = \frac{f_o}{f_e}$$

Example A refracting telescope has an objective lens of focal length 2.5 m. The diameter of the objective lens is 20 cm.

a What will be the magnifying power of the telescope when it is fitted with an eyepiece of focal length 5 cm?

b Where is the best place for the observer's eye to be placed?

a Angular magnification, $\quad M = \dfrac{f_o}{f_e}$

$$= \frac{250 \text{ cm}}{5 \text{ cm}}$$

$$= 50$$

b Redrawing figure 3 will help to answer this question. See figure 4.

All the light entering the telescope passes through a disc-shaped space at EE'; therefore the observer's eye should be at EE'. This disc is known as the *eye-ring*. It is the best position for the eye. If you trace back the dotted and full rays which meet at E, you will find that they both come from B. Hence E is the image of B formed by the

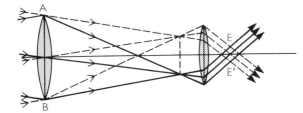

figure 4

eyepiece. In other words, the eye-ring is the circular image of the objective lens formed by the eyepiece.

We may use the equation $1/f = 1/u + 1/v$ to work out the distance v between the eye-ring and eyepiece. In the equation, f is the focal length of the eyepiece, 5 cm.

The objective lens is the object, therefore

$$u = f_o + f_e$$
$$= 255 \text{ cm}$$

Thus

$$\frac{1}{f} = \frac{1}{u} + \frac{1}{v}$$

$$\Rightarrow \quad \frac{1}{5 \text{ cm}} = \frac{1}{255 \text{ cm}} + \frac{1}{v}$$

$$\Rightarrow \quad \frac{1}{v} = \frac{51}{255 \text{ cm}} - \frac{1}{255 \text{ cm}}$$

$$\frac{50}{255 \text{ cm}}$$

$$= \frac{10}{51 \text{ cm}}$$

$$= \frac{1}{5.1 \text{ cm}}$$

$$\therefore \quad v = 5.1 \text{ cm} \approx 5 \text{ cm}$$

That is, the eye should be almost at the focal point of the eyepiece. The size of the eye-ring can be worked out from figure 5.

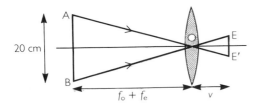

figure 5   Calculating the size of the eye-ring

Similar triangles ABO and E'EO give us

$$\frac{EE'}{AB} = \frac{v}{f_o + f_e}$$

$$\Rightarrow \quad \frac{EE'}{20 \text{ cm}} = \frac{5.1 \text{ cm}}{255 \text{ cm}} = \frac{1}{50}$$

$$\Rightarrow \quad EE' = \frac{20 \text{ cm}}{50} = 0.4 \text{ cm}$$

That is, all the light from the object passes through the small circular eye-ring at EE', the diameter of which is 4 mm.

The increase in intensity of the image is given by the ratio of the area of the objective lens divided by the area of the eye-ring.

$$\text{intensity increase} = \frac{\pi \times (10 \text{ cm})^2}{\pi \times (0.2 \text{ cm})^2}$$

$$= 2500$$

Microscope

A single converging lens can be used as a magnifying glass. In principle the angular magnification, M, can be enormous, but in practice major distortion arises. A better result, with greater usable M, can be achieved by making a 'compound' microscope using two converging lenses.

The lens nearest the eye, called the eyepiece, acts as a magnifying glass, as it did in the telescope. Figure 6 shows the ray diagram of a simple version of the compound microscope in *normal adjustment*, and figure 7 shows *near-point adjustment*.

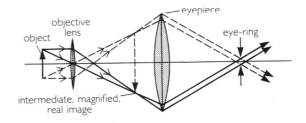

figure 6   Compound microscope in *normal* adjustment

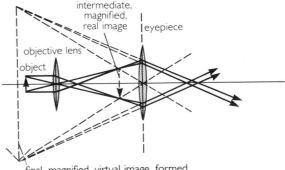

figure 7   Compound microscope in *near-point* adjustment

Summary

1 The refracting telescope
   Normal adjustment (emits parallel light)
   Angular magnification

$$M = \frac{f_o}{f_e}$$

   (See figure 3.)

2 The compound miscroscope
   a Normal adjustment (emits parallel light)
   b Near point adjustment (image formed at least distance of distinct vision)

   (See figures 6 and 7.)

Questions

1 An astronomical telescope consists of two converging lenses: an objective lens of focal length 2.5 m and an eyepiece lens of focal length 2.5 cm.
   a Draw a ray diagram to show the formation of a real image (of a very distant object) by the objective lens.
   b Explain how the eyepiece lens is used as a magnifying glass for viewing the real image formed by the objective lens.
   c State what is meant by *normal adjustment* in the case of an astronomical telescope.
   d (i)   What will be the separation of the lenses when the telescope is in normal adjustment?
   (ii)   For normal adjustment, where is the final image located, and is it real or virtual, erect or inverted?
   (iii)   Explain the term 'angular magnification' as related to an astronomical telescope, and determine its value for this telescope when it is being used in normal adjustment.
   (iv)   Explain where the lens of the observer's eye should be placed in order that he has the best view through the telescope.
   e No one should ever look through a telescope at the Sun because it will result in instantaneous eye damage. When teachers want to show sun-spots to a class, they set an astronomical telescope to project a large image of the Sun onto a screen. Explain how this is done, and say what adjustments must be made to the telescope if it had previously been set in normal adjustment.

2 Two thin converging lenses, of focal lengths 50 cm and 5 cm, are used to construct an astronomical telescope. The telescope is adjusted to form the final virtual image of a distant star at 25 cm from the eyepiece lens.
   a What kind of adjustment is the telescope said to be in when arranged like this?
   b Calculate the distance between the lenses.
   c Calculate the angular magnification of the arrangement.

   d What would the angular magnification be if the telescope were adjusted to give a final image at infinity?
   e Explain why the *angular* magnification (sometimes referred to as the magnifying power) is more appropriate when considering optical instruments such as the telescope, than is *linear* magnification.

3 Refracting astronomical telescopes produce inverted images; therefore they are not used for viewing earthbound objects. In a terrestrial telescope, a third lens placed between the objective lens and the eyepiece lens inverts the image produced by the objective lens, so that the final image is the correct way up. This third lens has no effect on the magnification; it only inverts the image.

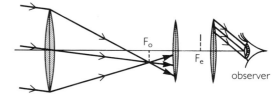

figure 8   Terrestrial telescope

   a One of the problems with terrestrial telescopes is that those which produce high magnifications are too long for hand use. Therefore, when designing a terrestrial telescope you should arrange the erecting lens so that it produces the minimum increase in the length of the telescope: that is, F_o and F_e should be as close together as possible. If f_i is the focal length of the intermediate third lens, what is the minimum distance F_o to F_e?
   b What will be the overall length of this terrestrial telescope when used in normal adjustment?
   c Copy figure 8 and complete it to show how the rays pass through the lenses into the observer's eye. Explain why the observer sees a *magnified erect* image.
   d Discuss the disadvantages of the erecting lens.

4   Consider a compound microscope made of two
    converging lenses: an objective lens of focal
    length 12 mm, and an eyepiece of focal length
    5.0 cm. If an object is placed 13 mm from the
    objective lens, the instrument produces an image
    at infinity.
    a   Sketch a ray diagram to show the paths of two
        off-axis light rays as they pass from one end of
        the object, through the microscope, and into the
        observer's eye. Mark on your diagram the focal
        points of the lenses, and all the known
        distances.
    b   How far must the intermediate image be from
        the objective lens? Calculate the separation of
        the lenses.
    c   By what factor is the intermediate image
        magnified with respect to the object?
    d   Calculate the magnifying power of the
        instrument when used in this way, i.e., the
        angular magnification.

5   It is dangerous to view the Sun directly, and even
    more dangerous to view it via a concave mirror. A
    student knowing this decides to use a concave
    mirror and a plane mirror to project an image of
    the Sun on to a flat white screen as shown in
    figure 9.

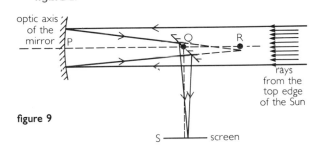

figure 9

a   What angle should the plane mirror make with
    the principle axis of the concave mirror?
b   Where is the principal focal point of the
    concave mirror?
c   What do you know about the distances PQ and
    QS?
d   Describe the effects of making the plane mirror
    smaller.
e   Is the image produced on the screen real or
    virtual?
f   Calculate the size of the image seen on the
    screen, given the following data:
        diameter of Sun = 1.5×10^6 km,
        radius of curvature of concave mirror
        is 50 cm,
        distance from Earth to Sun = 1.5×10^8 km.

The student is familiar with refracting telescopes;
in these a large objective lens concentrates the
light and produces a small, real, inverted image of
the object, which is then magnified by an
eyepiece lens acting as a magnifying glass.

g   Explain where the student should place such an
    eyepiece lens to convert this system of mirrors
    into a telescope, and explain what adjustments
    he should make if he wishes to project a large
    image of the Sun onto the wall of the laboratory.
h   Explain how this system of a concave mirror,
    plane mirror and eyepiece is used as a
    telescope to view the Moon.
i   What will be the angular magnification of this
    telescope when used (with an eyepiece of focal
    length 1 cm) in normal adjustment, i.e. when the
    image is formed at infinity? This type of
    telescope is known as a Newtonian reflector.

SECTION F

NUCLEAR PHYSICS

Einstein: the man who recognised the possibility.

The reality he dreaded.

36   Radioactivity and decay   381
37   Nuclear physics   396

CHAPTER 36

Radioactivity and decay

Introduction

If you flick quickly through this chapter you might feel that there is a certain repetition of what you learnt for your GCSE exam. To some extent that is true but here we look at the topic in more detail and more mathematically. In addition to learning about the nature and properties of α, β and γ radiation, we look at the experimental evidence that allowed the physicists of the 1900s to work out the amount of radioactive material left in a sample after any given time.

GCSE knowledge

We assume that you have a working atomic vocabulary, that is to say that you know the terms nucleus, proton, neutron, isotope and electron. In addition you probably know these points about elementary nuclear theory.

☐ The nucleus is very, very small indeed, in comparison with the atom, being only about 10^{-14} m in diameter. The nucleus is made up of protons and neutrons.

☐ The nucleus can emit energetic α, β or γ radiations. The α-particle is a fast-moving helium nucleus; the β-particle is a fast-moving electron; and the γ-ray is electromagnetic radiation.

☐ We describe the nucleus in terms of its nucleon and proton numbers. The nucleon number A is the combined total of protons and neutrons. The proton number Z is the number of protons. It is the proton number that determines which element the nucleus will form. For example, the common isotope of carbon is described thus: $^{12}_{6}C$; carbon-12 has 6 protons and 6 neutrons in its nucleus.

☐ The emission of an α- or β-particle from the nucleus of an element transforms it to a different element.

☐ Radioactive decay is characterised by a certain *half-life*. If we say an element has a half-life of 1 hour, after 1 hour half the material will have decayed; when 2 hours have elapsed then a quarter of the material will be left.

Radioactivity

If a small amount of radium is placed on top of a light-proof envelope which contains some photographic paper and left for half an hour or so, it is discovered that when the paper is developed it has been affected by the presence of the radium. Our explanation for this is that radium emits some energetic radiation which can pass through the envelope and activate the photographic emulsion. α, β and γ radiations are all emitted by radium. It is possible to distinguish between these because of the substantial differences in their abilities to ionise gases, their penetrating powers and their deflections in magnetic fields.

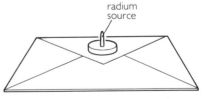

figure 1 Light-proof envelope containing photographic paper

Detection of radiations

Figure 2 shows a diffusion cloud chamber. Solid carbon dioxide (dry ice) at a temperature of $-78\,^{\circ}C$ is held in place underneath a base plate by a sponge. Just above the plate there is a cork ring which is soaked in ethanol. The very low temperature of the base plate causes the ethanol to form a supersaturated vapour at the bottom of the chamber. When a vapour is supersaturated a small local increase in the concentration of the vapour will cause it to condense and form visible droplets.

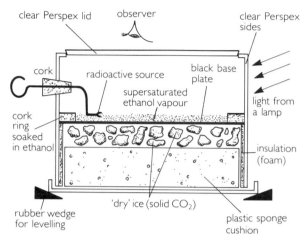

figure 2   A cloud chamber

When an α- or β-particle travels through a gas or vapour, it produces ions. The presence of an ion in the ethanol vapour attracts ethanol molecules to the region around it; the supersaturated ethanol condenses into droplets, and so a visible track is formed showing the path of the ionising radiation. A cloud chamber can usually be made to work better by rubbing the plastic top with a duster; this action produces an electric field which will remove stray ions, and thus the formation of new tracks is encouraged.

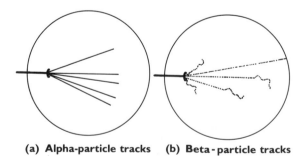

(a) **Alpha-particle tracks**   (b) **Beta-particle tracks**

figure 3   Cloud chamber tracks

Figure 3 shows typical tracks in a cloud chamber for α- and β-particles. The track of an α-particle is very thick and straight, showing that the particle is strongly ionising and massive because it is not deflected very much by air molecules. The tracks are typically 5 cm long. The β-particle tracks are not as

pronounced as the α-particle tracks, indicating that considerably less ionisation is produced. The β-particle's path is far from straight near the end, which would suggest that the β-particle is considerably less massive and therefore more easy to deflect than the α-particle. Most γ-rays leave no trace at all in cloud chambers, revealing that they are only very weakly ionising.

Early workers in radioactivity detected ionising particles by the scintillations that they caused on fluorescent screens. Figure 4 shows a more modern device called a Geiger–Müller tube. A particle entering the tube can trigger a pulse of current which can then be detected by an electronic counter.

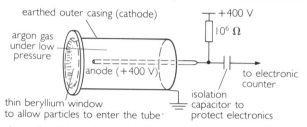

figure 4   Geiger–Müller tube

A particle entering the tube will cause some ionisation, producing positively charged gas ions and electrons (and some negatively charged gas ions). The electrons, having little mass, are rapidly accelerated by the electric field and soon arrive near the anode. In the vicinity of the anode the electric field is very strong, because the anode is a fine rod. When an electron collides with a gas molecule it can ionise it; this process will be repeated several times by a single electron on its way to the anode because the electric field is so strong that the electron does not have to travel very far before it has regained enough energy to ionise another gas molecule. In this way an avalanche of electrons is produced which triggers a large pulse of current to activate the electronic counter.

The positive ions travel more slowly than the electrons because of their greater mass, and so arrive at the cathode some time later. It is quite likely that the positive ions will have sufficient energy to knock more electrons out of the cathode, which would keep the current going. This feature is most undesirable because we want the pulse of current to be very brief so that the tube is ready to detect the next particle. To prevent the ejection of secondary electrons, the tube is filled with a halogen, either chlorine or bromine. On collision with an ion the halogen molecules dissociate, which means that the diatomic molecule is split into two separate atoms. Consequently the gas ions in the tube have a means of losing their energy before reaching the cathode. This process is called *quenching*.

Despite the quenching, the Geiger–Müller tube has a 'dead time' of about 10^{-4} s, which is roughly the time taken by the positive ions to reach the cathode.

This means that if two particles enter the tube within this time they cannot be resolved and so only a single count is recorded. At relatively low count rates this is not important but a correction ought to be made for very high count rates. Although Geiger–Müller tubes are used extensively in schools, they are not used for research purposes because their dead time is too long. Instead, solid state detectors are used which have dead times of the order of 10^{-8} s.

The electronvolt

In atomic and nuclear physics one often encounters the unit of the electronvolt, eV, or mega-electronvolt, MeV. The electronvolt is the quantity of energy gained by an electron when it is accelerated through a p.d. of 1 volt.

Thus $1\,\text{eV} = 1.6 \times 10^{-19}\,\text{J}$, since the charge on an electron is $1.6 \times 10^{-19}\,\text{C}$.

Example How much energy does an α-particle acquire when it is accelerated through 2×10^6 V?

Kinetic energy gained $= qV$

$$= (2 \times 1.6 \times 10^{-19}\,\text{C}) \times (2 \times 10^6\,\text{J C}^{-1})$$

$$= 6.4 \times 10^{-13}\,\text{J}$$

$$= 4\,\text{MeV}$$

The nature and properties of radiation

In this section, a brief summary of the nature and properties of each of the radiations, α, β and γ, is given and then the more important experimental evidence which has given rise to our knowledge is presented.

α-particles are the nuclei of helium atoms. They travel at speeds of the order of $10^7\,\text{m s}^{-1}$ and have energies in the range 4–10 MeV. They can be deflected slightly by a powerful magnet, in such a direction as to confirm their positive charge: see figure 5. They penetrate only a few centimetres through air; their exact range depends on their energy, e.g. an 8 MeV α-particle has a range of 6 cm. They are unable to penetrate any substantial thickness of matter. A piece of paper can effectively stop low energy α-particles, although a thin piece of aluminium foil (0.1 mm) may be needed to stop the most energetic α-particles. α-particles are very strongly ionising, producing about 10^5 ion pairs per centimetre of path length travelled through air.

Our first evidence for the ionising powers of α-particles was provided by the thickness of their cloud chamber tracks. More detailed evidence can be provided by an ionisation chamber and a picoammeter, see figure 6.

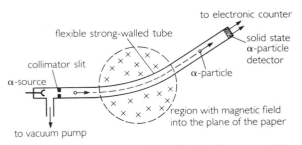

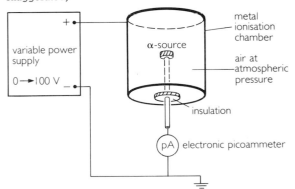

figure 5 Fleming's left hand (motor) rule shows α-particles to be positively charged (the deflection here is exaggerated)

figure 6 An ionisation chamber

Typically you might find that a certain α-source produces an ionisation current of 400 pA. The source is known to have an activity of 4500 Bq. [1 Bq (1 bequerel) = 1 disintegration per second.]

If we assume that each ion produced by the α-particles is only singly charged, then each time an ion pair is produced 1.6×10^{-19} C of charge flows through the picoammeter. Now we can write that the number of ion pairs produced per second, n, is:

$$n = \frac{I}{q} = \frac{4 \times 10^{-10}\,\text{A}}{1.6 \times 10^{-19}\,\text{C}}$$

$$= 2.5 \times 10^9 \text{ ion pairs per second}$$

Since the α-source emits 4500 particles per second it follows that each α-particle produces

$$\frac{2.5 \times 10^9}{4500} = 5.6 \times 10^5 \text{ ion pairs}$$

This also enables us to make a rough estimate of the particle's energy, since the energy required to singly-ionise an air molecule is about 10 eV. If the α-particle can produce 5.6×10^5 ion pairs it must have had initially about 5.6 MeV of energy.

The ionisation chamber can easily be adapted to measure the *range* of α-particles, see figure 7. A gauze can be fitted into the lid to let particles into the ionisation chamber from the outside. The range can be determined by measuring the distance x between the source and the top of the chamber when the ionisation current just drops to zero.

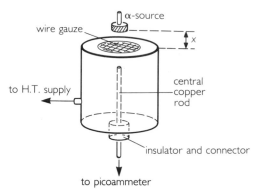

figure 7 Measurement of the range of α-particles using an ionisation chamber

That α-particles are helium nuclei was determined by the Rutherford–Royds experiment of 1909. A sample of radon gas, which emits α-particles, was enclosed in a thin-walled tube, figure 8. The tube was thin enough to allow α-particles to pass through it but thick enough to keep the radon gas inside. Another tube surrounded the radon sample, but the second tube was made with thick glass walls so that the α-particles could not escape. After a week, a gas had collected in the outer tube. The level of mercury was then raised to compress the gas, so that an electrical discharge could be passed through it between the electrodes A and B. The spectrum produced was that of helium, confirming that α-particles were ionised helium. Measurements of the charge-to-mass ratio of α-particles (see below for details) had already shown that this ratio was exactly half that of the proton. This evidence, taken with the Rutherford–Royds result, shows that α-particles are doubly ionised helium ions or simply helium nuclei.

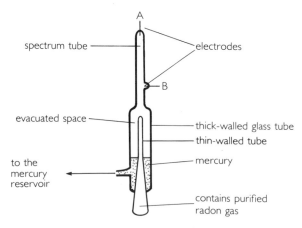

figure 8 Rutherford and Royds' identification of α-particles with helium nuclei

β-*particles* are fast-moving electrons. The range of energies of β-particles emitted by different sources is considerable. Some sources emit particles having

energies as high as 3 MeV, so that they are travelling at about 99% of the speed of light. Also each β-source itself emits particles with widely differing energies, unlike α or γ sources which emit particles of well-defined energy. However, as a very rough guideline it would be fair to say that the energies of β-particles are of the order of 0.5 MeV. On account of the vast range of energies it is difficult to be specific about the range of β-particles; energetic particles can travel as far as 1 m or so through air but will be stopped by several millimetres thickness of aluminium. The number of ions produced per centimetre of path length for a β-particle is typically 10^3, much less than the rate of ion production by an α-particle (10^5/cm of path). However, an α-particle and a β-particle initially carrying the same energy will eventually ionise the same number of atoms, since they both lose their energy by repeated collisions with atoms; but the β-particle will have travelled about 100 times as far in the process. The reasons for the difference in powers of ionisation are:

☐ the β-particles carry less charge;

☐ the β-particles are travelling more quickly.

The more slowly a charged particle travels, the longer it spends near the electrons of a particular atom, with the result that the impulse ($F \times t$) imparted to an electron is bigger and so ionisation is more likely to occur. It is interesting to note that a charged particle produces more ions per centimetre of path length at the limit of its range, because it will be travelling more slowly (figure 9).

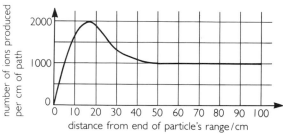

figure 9 Graph showing the number of ions produced by a β-particle per cm of path against distance from the end of its path

β-particles are easily deflected in a magnetic field, and their deflections are such that they confirm that the particles carry a negative charge. In addition, measurements show that the charge-to-mass ratio of β-particles is the same as that for electrons. Figure 10 shows the principle of an experiment to calculate the charge-to-mass ratio of a particle. First, between the points W and Y along its path, the β-particle is subjected to the effects of uniform magnetic and electric fields B_1 and E. The strengths of these fields (acting at right angles to each other and to the line taken by the particles) is adjusted so that the β-particle reaches Y without being deviated from its original path. This means that the force on the particle due to the electric field is exactly equal and opposite to the force due to the magnetic field. So we can write that

$$qE = B_1qv$$

or

$$v = \frac{E}{B_1}$$

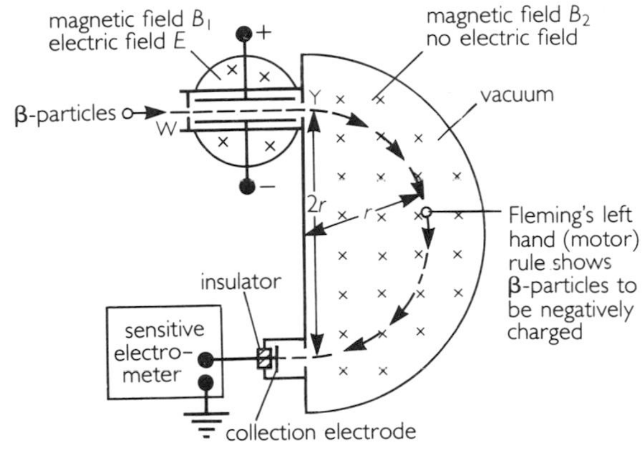

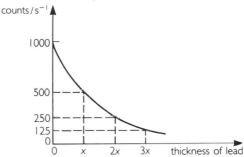

figure 10 Beta-particle spectrometer

Next, the β-particle passes into a second region in which a uniform field B_2 acts, so that the particle goes into a circle of radius r. The force from the magnetic field provides the centripetal force necessary for the particle to move in a circular path, so we can write

$$\frac{mv^2}{r} = B_2qv$$

$$\Rightarrow \qquad \frac{q}{m} = \frac{v}{B_2r} = \frac{E}{B_1B_2r}$$

Using this method it was discovered that the charge-to-mass ratio q/m (sometimes called 'specific charge') for β-particles and electrons was the same. However, early experiments revealed that for energetic β-particles q/m was less than expected. The reason for this discrepancy is that as a particle's velocity approaches the speed of light its mass increases. The mass m of a moving particle is actually given by

$$m = \frac{m_0}{(1 - v^2/c^2)^{1/2}}$$

m_0 being its *rest mass*, v its speed and c the speed of light. Thus if a particle travels at 95% of the speed of light its mass becomes

$$m = \frac{m_0}{(1 - (0.95)^2)^{1/2}} = 3.2\,m_0$$

and thus we would measure a charge-to-mass ratio for such a particle about three times smaller than the accepted value for slow-moving electrons.

γ-rays are packets of electromagnetic waves,

having typical wavelengths in the range 10^{-11} m to 10^{-13} m. The energy of each packet or 'particle' corresponding to those wavelengths is between about 0.1 MeV and 10 MeV. Because γ-particles are uncharged they are only weakly ionising and cannot be deflected at all by a magnetic field. Proof that γ-rays are electromagnetic waves is obtained by showing that they can be diffracted by crystals in the same way that X-rays are, that they can be polarised, and they travel at the speed of light. See page 314.

The way in which γ-rays interact with matter is completely different from the interactions shown by α- and β-particles. An α- or β-particle with a particular energy will be stopped completely by an absorber of well-defined thickness, a thicker absorber being required for a more energetic particle. This is not so for γ-rays. While a few millimetres thickness of lead will reduce considerably the intensity of a beam of γ-rays, it is never possible, in theory, to reduce the intensity to zero. Figure 11 illustrates the way in which the intensity of a γ-source reduces as thicker pieces of lead are inserted between it and a Geiger counter.

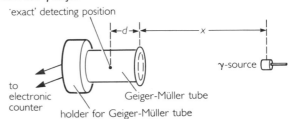

figure 11 Graph of count-rate against thickness of lead

The count rate drops exponentially, showing a characteristic half-thickness. For example, if the half-thickness of lead for a particular γ-source is x, the count rate will be diminished to one-eighth of its original intensity by lead $3x$ thick.

The absorption of γ-rays by air is small. The intensity of γ-radiation at a distance from a small source varies approximately as the inverse square of that distance, in the same way that the intensity of light, or an electric or gravitational field, varies at a distance from a point. Figure 12 shows an experimental means of verifying the inverse square law for γ-rays.

figure 12 Experimental investigation of the inverse square law for γ-rays

The γ-source is placed a distance x from the end of the Geiger tube. However, the exact position inside the tube where the radiation is detected is unknown; we need to correct for this, and shall call it a distance d from the end of the tube. Thus the count-rate C (corrected for background radiation) is expected to be proportional to $1/(d+x)^2$.

$$C \propto \frac{1}{(d+x)^2}$$

$$\Rightarrow \qquad d+x \propto \frac{1}{\sqrt{C}}$$

Thus if a series of count-rates C are taken at different distances x, a graph of x plotted against $1/\sqrt{C}$ should give a straight line with an intercept of $-d$ (figure 13).

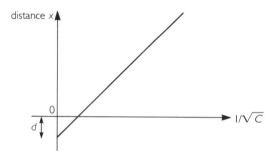

figure 13 Graph from experimental investigation of inverse square law for γ-rays

Chance, probability and decay

If 600 dice are cast we would expect, on average, 100 dice to fall with a one uppermost; but we know enough about statistical fluctuations to realise that we will not see 100 dice with a one on their top face every throw, but some distribution which might be as follows for 25 throws:

110	90	92	98	104
84	123	91	118	88
100	100	103	106	106
92	86	96	104	116
92	75	95	83	102

If we plot the distribution about the mean for a large number of throws, we obtain a graph similar to that illustrated in figure 14, which shows that most of the throws (about two-thirds of the total) lie between 90 and 110, and very few indeed (about one-twentieth of the total) lie outside the range 80 to 120. This form of distribution, which applies to any counting experiment recording random events, is called the *Poisson distribution*. Generally, for an average of N dice showing a one uppermost, two-thirds of the total count will lie in the range $(N-\sqrt{N})$ to $(N+\sqrt{N})$. $\sqrt{N}$ is called the *standard deviation* of the count.

Anyone who has studied the radioactivity of a weak source will know how the count rate is subject to

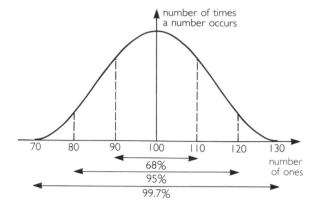

figure 14 Poisson distribution

quite considerable fluctuations. If we were to study a source that emitted on average 100 counts per minute, and the counts for many one-minute periods were recorded, we would find a Poisson distribution identical to the one shown in figure 14. This immediately shows that if we were to take a single measurement of the activity of a radioactive sample, our experiment would be less accurate than if a large number of counts were recorded. If 100 counts were recorded the standard deviation would be 10, so there would be a two-thirds chance of the average count rate lying between 90 and 110. On the other hand if 10 000 counts were recorded the standard deviation would be 100, meaning that the chance of the average count-rate lying between 9900 and 10 100 would be two-thirds. Thus in an experiment in which 100 counts are recorded we say that the count is 100 ± 10, i.e. an uncertainty of 10%, and in an experiment in which 10 000 counts are recorded we say that the count is $10\,000 \pm 100$, i.e. an uncertainty of 1%.

That the fluctuations in radioactive counting experiments fit the Poisson distribution proves that the emission of α, β and γ-rays from a nucleus is a random process. This means that an individual nucleus has a definite fixed probability λ of decaying in 1 second, in the same way that an individual die has a probability of $\frac{1}{6}$ of showing a one each time it is cast. If a die has not shown a one for 99 throws it still has a probability of $\frac{1}{6}$ of showing a one on its hundredth throw. Similarly, if a nucleus has not decayed by emitting a particle over a long period of observation it still has the same probability, λ, of decaying in any one second.

The probability of an individual nucleus decaying per second is λ, and so the fraction $\Delta N/N$ of nuclei out of a sample of N radioactive nuclei decaying per second is $\Delta N/N = \lambda$ (in one throw $\frac{1}{6}$ of the dice turn up a one), and the fraction of nuclei that will decay in a small time Δt is:

$$\frac{\Delta N}{N} = \lambda \Delta t$$

Radioactive decay law

From the last section we can see that the casting of a die is a good analogy to the radioactive decay of a nucleus: both are random events. We will take this analogy further by considering a sample of 600 'active' dice and seeing what happens to our sample after many throws. If a die turns up one, it is said to have decayed and it is therefore removed from our 'active' sample. So if we start with a sample of 600 dice, 100 will 'decay' after the first throw. This will leave 500, and in the next throw a further $\frac{1}{6} \times 500 = 83$ will decay and so on. The table below shows what will happen (on average) after many throws.

Number of active dice	Number of throw	Number of active dice discarded
600	1	100
500	2	83
417	3	70
347	4	58
289	5	48
241	6	40
201	7	

Following our analogy, we would expect a sample of radioactive nuclei to behave in the same way. If the probability of a nucleus decaying per second is $\frac{1}{6}$ then we get the following predictions for a sample of 6000 nuclei:

Number of active nuclei	Time/s	Number of decays per second
6000	1	1000
5000	2	830
4170	3	700
3470	4	580
2890	5	480
2410	6	400

Figure 15 shows these changes graphically. Both the number of active (undecayed) nuclei and the activity of the sample (number of particles emitted per second) decay exponentially with time. A decay is only exponential if the fractional change in the sample per second is the same at all times. This is characterised by a *half-life* (which is observed

experimentally): in a period of just under 4 seconds the activity of the sample is halved, in another 4 seconds it is halved again and so on. Mathematically the constant fractional change per unit time is written as: $\Delta N/N = \lambda \Delta t$, which was predicted by probability arguments earlier.

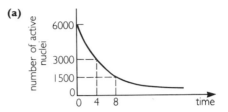

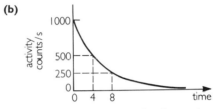

figure 15   Variation of (a) number of undecayed nuclei and (b) activity with time

Mathematics of the decay law

When we consider infinitesimally small time intervals our equation for decay becomes

$$\frac{dN}{N} = -\lambda dt$$

The minus sign was not included earlier, but its significance is that the change dN must be subtracted from N to give the number of remaining nuclei after time dt. The solution to this equation is:

$$N = N_0 e^{-\lambda t}$$

where N is the number of undecayed nuclei remaining after time t, and N_0 is the number of nuclei in the sample originally at $t = 0$.

λ is called the *decay constant*. It is the probability of a single nucleus decaying per second, or it may be related to a large sample of N nuclei by remembering that it is the fractional change in the number of nuclei present per second. The decay constant is related to the half-life $T_{1/2}$ by the equation:

$$\lambda = \frac{0.7}{T_{1/2}}$$

The three important equations governing radioactive decay have been given. We now show the derivation of the last two.
We start with the equation

$$\frac{dN}{N} = -\lambda dt$$

To calculate the number of nuclei left after time t, this equation must be integrated between the limits of N_0 and N for the left hand side and 0 and t for the right hand side.

Thus
$$\int_{N_0}^{N} \frac{dN}{N} = \int_0^t -\lambda \, dt$$

$$\Rightarrow \quad [\ln N]_{N_0}^{N} = -\lambda t$$

$$\Rightarrow \quad \ln N - \ln N_0 = -\lambda t$$

$$\Rightarrow \quad \ln \frac{N}{N_0} = -\lambda t$$

$$\Rightarrow \quad \frac{N}{N_0} = e^{-\lambda t} \text{ or } N = N_0 e^{-\lambda t}$$

After one half-life $T_{1/2}$, $\dfrac{N_0}{2}$ nuclei will be left.

Then
$$\frac{N_0}{2} = N_0 e^{-\lambda T_{1/2}}$$

$$\Rightarrow \quad \tfrac{1}{2} = e^{-\lambda T_{1/2}}$$

$$\Rightarrow \quad \ln \tfrac{1}{2} = -\lambda T_{1/2}$$

$$\Rightarrow \quad -0.7 = -\lambda T_{1/2}$$

$$\Rightarrow \quad \lambda = \frac{0.7}{T_{1/2}}$$

Example Calculate the activity of 1 g of the β-emitter strontium-90, which has a half-life of 28 years.

The number of atoms of strontium in 90 g is 6×10^{23}.

The number N in 1 g $= \dfrac{6 \times 10^{23}}{90} = 6.7 \times 10^{21}$

The decay constant $\lambda = \dfrac{0.7}{T_{1/2}}$

$$= \frac{0.7}{28 \times 3 \times 10^7 \text{ s}}$$

$$= 8.3 \times 10^{-10} \text{ s}^{-1}$$

(There are 3×10^7 s in 1 year.)

The activity is the number of particles emitted per second:

$$\frac{dN}{dt} = -\lambda N$$

$$= 8.3 \times 10^{-10} \times 6.7 \times 10^{21} \text{ s}^{-1}$$

$$= 5.6 \times 10^{12} \text{ particles per second}$$

or 5.6×10^{12} Bq

Measurement of half-life

The half-life of radon can be measured using the apparatus shown in figure 16.

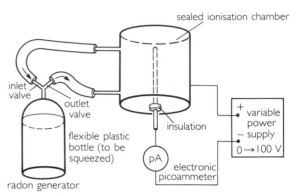

figure 16 Measurement of the half-life of radon gas

Radon gas, often called thoron, is produced in the generator bottle from thorium-232 by a series of decays.

thorium-232 $\xrightarrow{\alpha}$ radium-228
($^{232}_{90}$Th)   ($^{228}_{88}$Ra)
                          $\downarrow \beta$

thorium-228 $\xleftarrow{\beta}$ actinium-228
($^{228}_{90}$Th)   ($^{228}_{89}$Ac)
$\downarrow \alpha$

radium-224 $\xrightarrow{\alpha}$ **radon-220** $\xrightarrow{\alpha}$ polonium-216
($^{224}_{88}$Ra)   ($^{220}_{86}$Rn)   ($^{216}_{84}$Po)

Through lack of space, the half-lives for each decay have not been included. The important ones for this discussion are: thorium-232, 1.4×10^{10} years; radium-224, 3.6 days; radon-220, 52 seconds. All these elements in the decay series are in a radioactive equilibrium, which means that the number of atoms of each element remains almost constant. The equilibrium is dynamic; although radon atoms are continuously decaying into polonium, they are being replaced at exactly the same rate by the decay of radium-224, so the number of radon atoms in the generator bottle remains constant.

Because radon is a gas it is easy to separate from the other elements in the decay series, which are solid. A sample of gas may be squeezed out of the bottle into the ionisation chamber, which is attached to a cell and picoammeter. Because radon emits α-particles, a large ionisation current is produced; we simply record the ionisation current at 10-second intervals, and plot a graph of count-rate against time (figure 17).

Then the half-life may be calculated in the usual fashion. The experiment may be repeated several times quite easily, because in the time taken for the sample of radon in the ionisation chamber to decay, the amount of radon in the plastic bottle will have recovered to its former level.

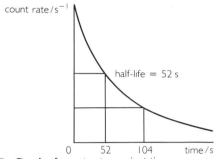

figure 17   Graph of count-rate against time

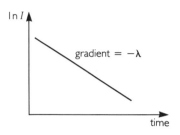

figure 18

The following simple calculation shows that this experiment will be subject to considerable statistical fluctuations. When the elements are in radioactive equilibrium they all have the same activity (emissions per second), so we can write: $\lambda_1 N_1 = \lambda_2 N_2 = \lambda_3 N_3 \ldots$

or

$$\frac{N_1}{T_1} = \frac{N_2}{T_2} = \frac{N_3}{T_3} \text{ etc.}$$

λ_1, T_1, N_1 refer to the decay constant, half-life and number of atoms present for the first member of the decay series, and so on.

Our original sample of thorium (Th) might contain 0.5 g or about 10^{21} atoms, and so the number of radon (Rn) atoms present in the bottle will be given by:

$$N_{\text{Rn}} = \frac{N_{\text{Th}}}{T_{\text{Th}}} \times T_{\text{Rn}}$$

$$= \frac{10^{21}}{1.4 \times 10^{10} \times 3 \times 10^7 \text{ s}} \times 52 \text{ s}$$

$$\simeq 10^5 \text{ atoms}$$

One squeeze of the bottle will not by any means put all 10^5 atoms into the ionisation chamber; perhaps we might only succeed in putting 10^4 atoms into the chamber. The activity of such a number of radon atoms is:

$$\text{activity} = \lambda N = \frac{0.7}{52 \text{ s}} \times 10^4 \simeq 130 \text{ Bq}$$

We know from our previous work on the standard deviation of a radioactive count that the uncertainty in the count can be expressed as $130 \pm \sqrt{130}$, that is, about 9%. As the experiment progresses the count-rate gets smaller and the error bigger.

We can improve our accuracy of the determination of the half-life by the following mathematical trick. For the activity we can write:

$$\text{activity} = \lambda N = \lambda N_0 e^{-\lambda t}$$

$$\Rightarrow \quad \ln(\text{activity}) = \ln(\lambda N_0) - \lambda t$$

Thus if we plot a graph of $\ln(\text{ionisation current})$ against time (figure 18) it will have the form: $\ln I = C - \lambda t$; where C is a constant equal to the intercept on the y-axis. Its gradient is $-\lambda$ and so the half-life may be determined using $\lambda = 0.7/T_{1/2}$.

The method described above will clearly only be able to determine short half-lives. The half-life of a long-lived element (this could be 10^9 years or more) may be determined directly from the activity of a known sample of N atoms, for

$$\lambda = \frac{\text{activity}}{N}$$

and

$$T_{1/2} = \frac{0.7}{\lambda} = \frac{0.7N}{\text{activity}}$$

Nuclear transmutations

The nucleus contains practically all of the mass of an atom. The particles that make up the nucleus are protons and neutrons; both particles have approximately the same mass; the proton is positively charged and the neutron is neutral. Outside the nucleus there are as many electrons as protons, and thus atoms are electrically neutral. We describe a nucleus in terms of its nucleon number A, which is the total number of protons and neutrons, and its proton number Z, which is the number of protons. Thus uranium-238 can be described by $^{238}_{92}\text{U}$, so this nucleus contains 92 protons and $238 - 92 = 146$ neutrons. The proton number determines which element a particular nucleus forms.

We often have different *isotopes* of the same element. For example, $^{14}_{6}\text{C}$ and $^{12}_{6}\text{C}$ are both isotopes of carbon; the former has six protons and eight neutrons in its nucleus, and the latter six protons and six neutrons. Both nuclei are carbon nuclei because they have six protons and therefore the atom will have six electrons outside the nucleus (it is the electrons that determine the chemical properties of an atom); but carbon-14 is a heavier isotope because it contains two extra neutrons. Carbon-12 is the most abundant isotope of carbon and is stable; carbon-14 is rare and is radioactive.

The radioactive emission of particles from a nucleus will cause it to change into a different one. An α-particle is a helium nucleus. Thus by the emission of an α-particle, a polonium nucleus turns into a lead nucleus:

$$^{210}_{84}\text{Po} \rightarrow {}^{206}_{82}\text{Pb} + {}^{4}_{2}\text{He}$$

The nucleon number of the nucleus is decreased by 4, and the proton number is decreased by 2.

A β-particle has very little mass but takes away a negative charge and thus the proton number increases by one:

$$^3_1H \rightarrow {}^3_2He + {}^{\ 0}_{-1}e$$

In β-decay a neutron in the nucleus decays into a proton.

γ-rays are emitted by a nucleus after the emission of α- or β-particles. The reason for this is that after the emission of a particle the nucleus is often left in an energetic state; the extra energy is carried away by the emission of a γ-ray, in the same way that the extra energy that an electron has in a high energy level can be taken away by a photon.

The decay of radioactive potassium illustrates two other processes of nuclear change. The decay can occur either by the emission of a positron (a particle of the same mass as an electron but carrying a positive charge) or by *K-capture*. In the latter process the nucleus captures an electron from the lowest-lying energy shell outside the nucleus. The capture of an electron leaves a hole in the electron shell, which is filled by an electron 'falling' from a higher level. This falling electron emits an X-ray which allows the process to be detected.

$$^{40}_{19}K \rightarrow {}^{40}_{18}Ar + {}^0_1e \quad \text{positron emission}$$

$$^{40}_{19}K + {}^{\ 0}_{-1}e \rightarrow {}^{40}_{18}Ar + \text{X-ray} \quad \text{K-capture}$$

Positron emission is sometimes referred to as β^+-emission to distinguish it from β^--emission, which is the usual form of beta-decay when an electron is emitted.

Artificial nuclear transmutations can be caused by bombarding nuclei with energetic particles. Two famous examples are given below. Rutherford used α-particles to bombard nitrogen and found that protons were emitted:

$$^{14}_7N + {}^4_2He \rightarrow {}^{17}_8O + {}^1_1H \qquad \text{(i)}$$

Cockcroft and Walton used a high-energy beam of protons, which had been accelerated through half a million volts, to bombard lithium. They found that helium was produced:

$$^7_3Li + {}^1_1H \rightarrow {}^4_2He + {}^4_2He$$

Notation for artificial transmutations

It is usual to write equations such as (i) above in this form:

$$^{14}_7N \ (\alpha, \ p)^{17}_8O$$

In this case, this means that the original nitrogen-14 nuclide is bombarded with α-particles; and that oxygen-17 nuclei are formed, protons being emitted in the process.

Summary

1 The typical properties of the three major radiations are listed in the table below.

	α-particles	β-particles	γ-rays
Nature	He-4 nucleus	electron	electromagnetic radiation
Ionising power	10^5 ions cm^{-1}	100 ions cm^{-1}	very slight
Speed	5% of the speed of light	95% of the speed of light	speed of light
Deflection by magnetic field	slight	considerable	zero
Penetration	stopped by thick paper	stopped by sheets of aluminium (3 mm)	reduced in intensity by lead

2 Examples of radioactive decay:

$$^{238}_{92}U \rightarrow {}^{234}_{90}Th + {}^4_2\alpha$$

$$^3_1H \rightarrow {}^3_2He + {}^0_{-1}\beta$$

$$^9_4Be \rightarrow {}^9_4Be + {}^0_0\gamma$$

3 The electronvolt (eV) is the energy gained by an electron when it is accelerated through a p.d. of 1 volt.

$$1\,eV = 1.6 \times 10^{-19}\,J$$

4 The activity of a radioactive sample decays by a constant fraction in a given time; this is characterised by a decay constant λ.

$$\lambda = \frac{dN/N}{dt} = \text{the fractional change per unit time}$$

The activity is given by:

$$\frac{dN}{dt} = -\lambda N$$

5 The number of radioactive nuclei left from an initial number N_0 may be calculated after a time t using:

$$N = N_0 e^{-\lambda t}$$

6 The half-life $T_{1/2}$ is linked to the decay constant as follows:

$$T_{1/2} = \frac{0.7}{\lambda}$$

7 Notation for artificial transmutations:

$$^{14}_7N\,(\alpha, p)\,{}^{17}_8O$$

means

$$^{14}_7N + {}^4_2He \rightarrow {}^1_1H + {}^{17}_8O$$

Questions

1 A radioactive nucleus Y is known to decay to a stable nucleus by the emission of either an α-particle or a β-particle, with a half-life of between 2 and 4 hours. A small sample containing Y is available with an initial activity of about 2000 Bq. Describe how you would determine (i) whether α or β emission occurs (ii) the half-life of Y.

2 A radioactive source is placed close to the window of a Geiger counter. Various absorbers are placed in front of the tube and the following measurements are made:

Absorber	Thickness/mm	Count rate/min^{-1}
none		5000
paper	1.0	3280
Al	0.1	3070
Al	0.5	1940
Al	1.0	230
Pb	1.0	7
Pb	2.0	8

Identify the radiations emitted.

3   A radioactive source is placed close to the window of a Geiger counter. Some thin sheets of metal foil are placed between the source and counter and the measurements below are made. Account for these results.

Number of sheets	0	1	2
Count-rate/min^{-1}	10 000	9977	9953

Number of sheets	4	6	8	16
Count-rate/min^{-1}	9915	4067	1832	1680

Number of sheets	32	64	128	256
Count-rate/min^{-1}	1413	1000	500	125

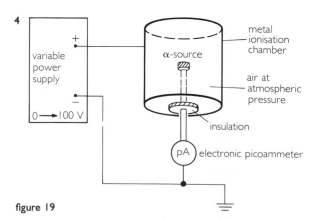

figure 19

The purpose of the experiment shown in figure 19 is to determine the number of ion pairs produced by each α-particle, and hence their energy. An α-source is placed inside an ionisation chamber with a p.d. of about 100 V across it. The picoammeter indicates that the ionisation current is 80 pA.

a   Assuming that the α-particles produce only singly-charged ions in the chamber, work out the number of ion pairs produced per second.
b   The source is then removed from the chamber and is placed a distance of 2 cm from a solid state detector of area 0.25 cm². The count-rate detected by the solid state device is 10 Bq. Deduce the total number of disintegrations produced per second by this source, that would have entered the ionisation chamber. (Assume that the source emits α-particles in all directions and so obeys an inverse square law.)
c   The source is marked as having an activity of 0.075 μCi. How does your activity calculated in part (b) compare with its nominal activity? Can you account for the discrepancy? (1 Ci = 3.7 × 10¹⁰ Bq.)
d   Now calculate the number of ion pairs produced per α-particle. Make also an estimate of the number of ion pairs produced per centimetre of path travelled by the α-particles.
e   The energy required to ionise an air molecule is about 14 eV. Estimate the energy of the α-particles emitted from this source.

5   a   Figure 20(a) shows the path of some β-particles passing through a magnetic field. Explain carefully why this experiment shows that the particles have a negative charge.

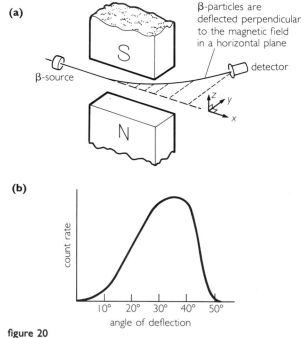

figure 20

b   Figure 20(b) shows how the count rate of the deflected β-particles varies with the angle through which they have been deflected. Explain what this tells us about the energies of β-particles from a particular source. What can you say about the relative energies of particles being deflected by 20° and 40°?

6 A radioactive source is enclosed in a lead container. At a distance of 20 cm from the source, a Geiger tube detects a count rate of 4020 counts per second (c.p.s.); the background count is 20 c.p.s.
   a What radiation does the Geiger tube detect?
   b At what distance away from the source will the detector measure a count rate of 60 c.p.s.?

7 a It has been suggested that the intensity of γ-radiation from a source decreases exponentially through a layer of absorbing material. Below are some results of an experiment, in which a γ-source was placed at a constant distance from a Geiger tube, but a varying thickness of lead was placed in front of the source.

Count-rate/s^{-1}	3000	2700	2430	2190	1970	1770
Thickness of lead/mm	0	1	2	3	4	5

   Test these results for an exponential law.
   b Estimate the thickness of lead that would reduce the intensity of radiation to one-half of its value.
   c What would the count-rate be for a thickness of 15 mm of lead?
   d The intensity of the γ-rays can be described by the equation $I = I_0 e^{-\mu x}$, where I_0 is the original intensity and I is the intensity after absorption by a thickness x of lead. What might μ be called? Find a value for μ.

8 Figure 21 shows an apparatus designed to measure the charge-to-mass ratio of a β-particle. A β-particle enters the tube WY in which two

fields act: a magnetic field, B, acting into the plane of the paper and an electric field E.
   a Explain how it is possible to arrange a combination of fields B_1 and E so that a β-particle passes undeflected through the two slits at the ends of the tube WY.
   b It is found that, for $E = 10^5\,\mathrm{V\,m^{-1}}$ and $B_1 = 0.001$ T, some β-particles pass through WY without deviation. Calculate the speed of these particles.
   c The particles now emerge into a tube YZ, of radius of curvature $r = 0.1$ m. The β-particles are bent around this tube by the action of a second magnetic field B_2 of flux density 6×10^{-3} T. Show that the specific charge for the β-particles is given by:
   $$\frac{q}{m} = \frac{E}{B_1 B_2 r}$$
   Hence evaluate q/m.
   d How does your value of q/m compare with the accepted value of $1.76 \times 10^{11}\,\mathrm{C\,kg^{-1}}$ for an electron? Can you suggest a reason for this discrepancy?
   e Comment on the significance of this experiment.

9 Some physicists at the high-energy laboratory at CERN, Geneva, wished to study the rare isotope fermium-246, which has a half-life of 1.2 seconds. Unfortunately, the nuclear reactor where the isotope of fermium was produced is 800 m away from the target chamber where the experiment was to be performed. With the aid of a fast car and some athletic research assistants, the transfer from reactor to target room took 24.0 seconds.
   a When the sample of fermium left the reactor its activity was estimated to be 10^{12} Bq. What was its activity by the time it reached the target chamber?
   b What mass of fermium-246 was left by the time it reached its destination?

10 The activity of a radioactive sample was found to decay by 20% in 1990. By what per cent will its activity decay in 1991? How will its activity at the beginning of 1993 compare with its activity at the beginning of 1990?

11 Two isotopes of barium, $^{125}_{56}$Ba and $^{123}_{56}$Ba, both decay by the emission of a positron, with half-lives of 6 minutes and 3 minutes respectively. We start with a sample of each isotope, each containing the same number of atoms.
   a Initially which source has the greater activity?
   b How do the activities of the sources compare after (i) 6 minutes, (ii) 12 minutes, (iii) 18 minutes?
   c Sketch graphs, on the same axes, to show how the activities of the two samples change with time.

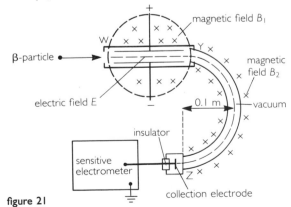

figure 21

12 The concentration of the radioactive isotope carbon-14, $^{14}_{6}C$, present in the remains found in an archaeological dig may be used to estimate the age of the site. $^{14}_{6}C$ does not occur naturally, but is produced in the atmosphere by the action of cosmic rays. Such rays can produce neutrons, and the following reaction occurs:

$$^{14}_{7}N + ^{1}_{0}n \rightarrow ^{14}_{6}C + ^{1}_{1}p$$

a Cro-Magnon man is thought to be one of our ancestors. Five adult skeletons of this species have been found in the Dordogne near Les Eyzies, in France. A sample of charcoal was taken from this site, of mass 0.1 g, and its activity was found to be 0.025 counts min^{-1}; the activity of 2 g of charcoal produced from modern wood was found to be 32 counts min^{-1}. Each activity was corrected for background count. The half-life of carbon-14 is 5730 years. Estimate the age of these Cro-Magnon relics.

b Radioactive decay is random, and therefore a count is subject to errors. The error in a count N is about $\pm\sqrt{N}$. In the dating process described above, the counting was done over a 24-hour period. Estimate roughly the error involved in your previous estimate of the age.

13 Caesium-140 is thought to have a half-life of about 1 minute. A sample of caesium-140 was placed in front of a Geiger tube and the following results were recorded:

Count-rate/s^{-1}	34	26	20	16	12	10
Time/s	0	30	60	90	120	150

Count-rate/s^{-1}	8	7	6	4	5
Time/s	180	210	240	270	300

Do these results confirm the previous statement about the half-life of caesium?

14 A radioactive sample is known to contain two different radioactive elements. The sample was placed in front of a Geiger tube connected to a counter and the following results were obtained; a correction has already been made for background count.

Count-rate/s^{-1}	8000	5090	2980	1640	1080	750	370
Time/min	0	2	5	10	15	20	30

By plotting a graph of ln (count-rate) against time, deduce the half-life of each of the elements present in the sample.

15 a Elements with a proton number higher than uranium (92) can be made artifically in nuclear

reactors. Such a radioactive nuclide is nobelium-257, 257102No. It is highly unstable and decays rapidly. The decay series is shown below:

Element	Symbol	Emission	Half-life
Nobelium	$^{257}_{102}$ No	α	23 seconds
Fermium	Fm	?	3 days
Einsteinium	$^{253}_{99}$ Es	α	20.5 days
Berkelium	Bk	?	314 days
Californium	$^{249}_{98}$ Cf	?	360 years
Curium	$^{245}_{96}$ Cm	?	9300 years
Plutonium	$^{241}_{94}$ Pu	β^- then α	13.2 years
Neptunium	Np	?	2.14×10^6 years
Protactinium	$^{233}_{91}$ Pa	β^-	27 days
Uranium	U	α	1.6×10^5 years
Thorium	Th	?	7300 years
Radium	$^{225}_{88}$ Ra	β^-	15 days
Actinium	Ac	?	10 days
Francium	$^{221}_{87}$ Fr	?	4.8 minutes
Astatine	$^{217}_{85}$ At	α	0.03 seconds
Bismuth	Bi	?	47 minutes
Polonium	$^{213}_{84}$ Po	α	4×10^{-6} seconds
Lead	Pb	β^-	3 hours
Bismuth	Bi	stable	

Copy out the symbols of the elements in the series and fill in the missing mass and atomic numbers of the particles that are emitted where appropriate. (Some of these elements emit γ-rays but these need not be included.)

b After a long time (about 10 million years) radioactive equilibrium will have been achieved. What does this mean?

c Explain why, when equilibrium has been reached, the most abundant nuclides in the decay series will be uranium-233 and neptunium-237.

d If the sample now contains U uranium-233 atoms and N neptunium-237 atoms, what will the value of U/N be?

16 An isotope of plutonium, $^{244}_{94}$Pu, decays by the emission of α-particles with a half-life of 8×10^7 years. The daughter nucleus, $^{240}_{92}$U, then decays by the emission of β-particles, with a half-life of 14 hours.

a Show that the increase in number of uranium atoms may be described by the equation:

$$\frac{dN}{dt} = A - \lambda N$$

where N is the number of uranium atoms, A is the activity of the plutonium sample and λ is the decay constant of uranium-240.

b In an experiment, a Geiger tube that detects only β-particles is placed near to a freshly-prepared sample of plutonium-244 of mass 10^{-5} g. Show that $N = A/\lambda(1 - e^{-\lambda t})$ is the solution to the equation above, then sketch a graph showing, in as much detail as possible, the count-rate detected by the Geiger tube over the first five days of the experiment.

c After a while the count-rate reaches a constant value. After what time will the experimenter deduce that a constant rate has been reached, if he takes a series of 10-second sample counts?

17 An isotope of mendelevium, $^{248}_{101}$Md, has a half-life of 6 seconds, and decays by the emission of 8.3 MeV α-particles.

a Express 8.3 MeV in joules.

b Show that the initial activity of 10^{-5} g of $^{248}_{101}$Md is 2.8×10^{15} Bq.

c Work out the power supplied initially by the radioactive source.

d Work out the energy released by the sample over a period of 1 minute.

e If such a sample, 10^{-5} g of $^{248}_{101}$Md, were placed in a copper can of mass 100 g, deduce the temperature rise of the can after 1 minute. Assume that all the energy released by the α-emission is absorbed by the copper. (The specific heat capacity of copper is $420 \, \mathrm{J\,kg^{-1}\,K^{-1}}$.)

18 Americium-241 has a half-life of 458 years. It decays by α-emission. Deduce the number of α-particles emitted per second by a 1 g sample of $^{241}_{95}$Am.

19 A sample of gold (Au) is exposed to a neutron beam, such that 10^{10} neutrons per second are absorbed in the reaction ^{197}Au(n, γ) ^{198}Au. The nucleide ^{198}Au decays by the emission of a β-particle to ^{198}Hg with a half-life of 3 days.

a What is the equilibrium number of gold atoms?

b How many gold-198 atoms will be present after six days? How many atoms of mercury (Hg) will be present at that time?

20 Scientists have used the radioactive nuclide $^{40}_{19}$K to date samples of rock. The decay process of potassium-40 is complicated. About 90% of the atoms decay by β-emission to form the stable isotope calcium-40. 10% of the atoms decay by a process known as K-capture to form the stable

isotope argon-40. K-capture occurs when an electron from the lowest energy level of the atom is captured by the nucleus, thus turning a proton into a neutron. A further 0.001% of $^{40}_{19}$K atoms decay by the emission of a positron, β$^+$ decay.

a What effect on the mass number of a nucleus do the following processes have: β$^+$ decay, β$^-$ decay, K-capture?

b What effect on the proton number of a nucleus do each of the above processes have?

c Hence write down the proton numbers of calcium and argon.

d When rocks were formed on the Earth they were molten. Why is it reasonable to assume that there was no argon gas in a rock when it was formed?

e A scientist made the following measurements on a rock sample:
   Mass of potassium-40 in sample
   $= 1.2 \times 10^{-3}$ g
   Mass of argon-40 in sample $= 0.36 \times 10^{-3}$ g
   What mass of potassium has decayed?

f Deduce the age of the rock, given that the half-life of potassium-40 is 1.3×10^9 years.

21 A physicist places a 1 mg sample of uranium-238 at a distance of 1 cm from a solid state α-detector (figure 22). The area of the detector is 0.5 cm^2. The recorded count-rate on the counter is 0.5 countss^{-1}

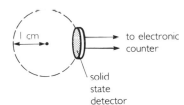

figure 22

a The decay of $^{238}_{92}$U by α-emission clearly occurs in all directions, not just in the direction of the detector. Given that the surface area of a sphere is $4\pi R^2$, show that the fraction of the disintegrations recorded by the counter is about 0.04 of the total.

b Deduce the activity of the sample of uranium.

c Given that 238 g of uranium contain 6×10^{23} atoms, deduce the number of atoms in 1 mg of uranium.

d Using the equation $(dN/dt) = -\lambda N$, show that the decay constant λ for uranium-238 is $5.0 \times 10^{-18} \, \mathrm{s^{-1}}$.

e Hence show that the half-life of uranium-238 is 4.5×10^9 years.

f Explain carefully the problems that arise in trying to measure a count-rate of 0.5 countss^{-1}.

g Approximately how long would the experiment have to last to ensure an error in counting of less than 1%?

CHAPTER 37

Nuclear physics

Introduction

In this chapter we study the stability of the nucleus and the energy involved in binding it together. We encounter Einstein's famous equation $E = mc^2$, and use it to help us understand the processes of fission and fusion.

The text and questions explore some of the social implications of nuclear physics; these include neutron bombs, the hazards of radiation and producing electricity from nuclear power.

GCSE knowledge

You should be familiar with the terms proton, neutron, nucleus, isotope and electron. Also you need to understand isotope symbols such as $^{12}_{6}C$, where 6 is the number of protons in the nucleus and 12 is the number of nucleons (protons and neutrons) in the nucleus.

The discovery of the neutron

Chadwick (1932) discovered that a very penetrating radiation was emitted when beryllium was irradiated with α-particles from a polonium source. This radiation was capable of passing through metal foils, and it interacted strongly with substances containing hydrocarbons, such as paraffin wax.

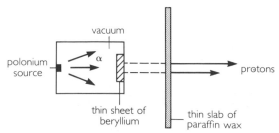

figure 1   Chadwick's experiment

Protons with energies of up to 5.7 MeV were knocked out of a sample of paraffin wax. The radiation was also capable of knocking out the nuclei of nitrogen atoms from nitrogen gas. The maximum recorded energy of such a nitrogen nucleus was found to be 1.5 MeV. By the application of the principles of conservation of energy and momentum to these data (see below), Chadwick showed that the mass of these particles was the same as that of a proton. These particles were called neutrons, and it was suggested that they were produced by the reaction:

$$^{9}_{4}Be + ^{4}_{2}He \rightarrow ^{12}_{6}C + ^{1}_{0}n$$

Neutrons are very penetrating because they are not charged and therefore they produce very little ionisation. The existence of neutrons had been suggested before their discovery, because the relative atomic mass of each element had long been known to be greater than its proton number; for example, the relative atomic mass of nitrogen is 14 but its proton number is 7. The discrepancy could now be explained by having a nucleus containing 7 protons and 7 neutrons.

Mathematical digression on the discovery of the neutron

The maximum energy transfers to the stationary protons and nitrogen nuclei from neutrons will occur for head-on collisions. Applying the laws of conservation of momentum and kinetic energy (we assume elastic collisions) to the collision shown in figure 2, we get:

$$m_1 u = m_1 v_1 + m_2 v_2$$

$$u = v_2 - v_1$$

(coefficient of restitution $e = 1$.)
From these two equations we get

$$\frac{v_2}{u} = \frac{2m_1}{m_1 + m_2} \qquad \text{(i)}$$

If we let m_1 be the mass of a neutron and m_2 be the mass of a proton, then $v_{2(p)}/u$ gives the ratio of the speed of the proton after the collision to that of the neutron before.

figure 2

Similarly, we may write an equation for a collision between a neutron and a stationary nitrogen (N) nucleus (which was known to have a mass, $m_N \simeq 14m_2$):

$$\frac{v_{2(N)}}{u} = \frac{2m_1}{m_1 + 14m_2} \qquad \text{(ii)}$$

Eliminating u from equations (i) and (ii) gives:

$$\frac{v_{2(p)}}{v_{2(N)}} = \frac{m_1 + 14m_2}{m_1 + m_2} \qquad \text{(iii)}$$

The speed of a 5.7 MeV proton is $3.3 \times 10^7 \, \text{ms}^{-1}$ and the speed of a 1.5 MeV nitrogen nucleus is $4.5 \times 10^6 \, \text{ms}^{-1}$

$$\Rightarrow \qquad \frac{v_{2(p)}}{v_{2(N)}} = \frac{3.3 \times 10^7 \, \text{ms}^{-1}}{4.5 \times 10^6 \, \text{ms}^{-1}} = 7.3$$

Whence equation (iii) can be solved to give $m_1 = 1.05m_2$ or, within experimental error, $m_1 \approx m_2$.

Stability of the nucleus

It is necessary to invoke a new force to account for the stability of a nucleus. We call this force the *strong nuclear force*. This force does not distinguish between protons and neutrons, but acts between all nucleons present to bind the nucleus together. (A nucleon is a collective name for a neutron or a proton.)

Figure 3 shows a graph of the number of neutrons plotted against the number of protons for stable nuclei. The following points emerge from a careful study of the graph.

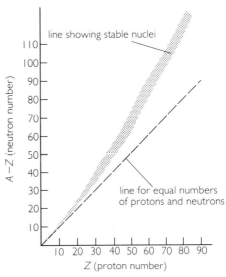

figure 3   Stable nuclei

☐ At low mass numbers the numbers of protons and neutrons are roughly equal, e.g. $^{12}_{6}\text{C}$, $^{16}_{8}\text{O}$; but at high mass numbers the number of neutrons outweighs the number of protons. A likely explanation for this is that for a large nucleus there are very strong repulsive electrostatic forces tending to push the nucleons apart. The effect of a larger number of neutrons is to increase the nuclear binding forces, to outweigh the repulsive electrostatic forces.

☐ Nuclei with too many neutrons tend to decay by β-emission, and heavy nuclei with too many protons tend to decay by α-emission.

☐ The most stable nuclei tend to be the ones which contain both an even number of protons and an even number of neutrons.

The best known equation in physics, $E = mc^2$

$E = mc^2$ is an equation of which most of us have heard. However, unlike a lot of relationships that we have met, it is difficult to derive, being a result of Einstein's theory of special relativity; its implications are very surprising and they have had a dramatic effect on world history.

The equation states the equivalence of mass and energy. If matter disappears then energy is released; to create matter energy has to be provided. In the equation $E = mc^2$, E is the energy released (or provided), m is the disappearing (or created) mass, c is the speed of light $(3 \times 10^8\,\text{ms}^{-1})$. Thus if 1 kg of matter disappears, the energy released is: $E = mc^2 = 1\,\text{kg} \times (3 \times 10^8\,\text{ms}^{-1})^2 = 9 \times 10^{16}\,\text{J}$. This is enough to supply the world's energy requirements for a day or so.

Does Einstein's equation mean that a hot cup of tea is more massive than a cold cup of tea? Yes it does, but not by an amount that we are likely to notice. A hot cup of tea contains thermal energy given by:

$$Q = mc\Delta\theta \quad \text{(using the usual notation)}$$

Making suitable estimates we arrive at:

$$Q = 0.1\,\text{kg} \times 4200\,\text{Jkg}^{-1}\text{K}^{-1} \times 60\,°\text{C} = 25\,200\,\text{J}$$

The mass lost from the tea in cooling down is $m = E/c^2$

$$\Rightarrow \quad m = \frac{25\,200}{9 \times 10^{16}}\,\text{kg} = 3 \times 10^{-13}\,\text{kg}$$

There is of course no way that we can detect such a small change in mass in a cup of tea. But it is possible to detect small changes in the mass of nuclei by the use of sensitive mass spectrometers. This helps us to explain how energy is released in nuclear decay processes.

A common radioactive isotope found in school laboratories is americium-241, which decays into neptunium-237 by the emission of an α-particle. This process may be described by the equation:

$$^{241}_{95}\text{Am} \rightarrow\ ^{237}_{93}\text{Np} + ^4_2\text{He}$$

The mass of the americium nucleus is greater than the combined mass of the neptunium and helium nuclei, which means that energy has been released. We will look later at this decay in greater detail to explain where the energy goes to.

Unified atomic mass unit (u)

The unified atomic mass unit, u, is defined as 1/12 of the mass of the carbon-12 atom. That is,

$$1\,\text{u} = 1.66 \times 10^{-27}\,\text{kg}$$

The energy released by a loss in mass of 1 u would be:

$$E = mc^2$$
$$= 1.66 \times 10^{-27}\,\text{kg} \times (3.00 \times 10^8\,\text{ms}^{-1})^2$$
$$= 1.49 \times 10^{-10}\,\text{J}$$

We may express this energy in MeV, using

$$1\,\text{MeV} = 1.60 \times 10^{-13}\,\text{J}$$

$$\Rightarrow \quad E = \frac{1.49 \times 10^{-10}}{1.60 \times 10^{-13}}$$

$$\Rightarrow \quad \text{Energy equivalent of 1 u}, E = 931\,\text{MeV}$$

Energy of disintegration

Returning to the emission of an α-particle from americium-241, we now examine the masses of the particles in detail. The nuclear masses are:

americium-241	241.0568 u
neptunium-237	237.0482 u
helium-4 (α)	4.0026 u

The combined mass of the neptunium-237 and the α-particle is 241.0508 u. Therefore the mass lost, m, in the disintegration is given by

$$m = 241.0568\,\text{u} - 241.0508\,\text{u}$$
$$= 0.0060\,\text{u}$$

We can now calculate the energy released in the disintegration:

$$E = 0.0060\,\text{u} \times 931\,\text{MeV}\,\text{u}^{-1}$$
$$= 5.6\,\text{MeV}$$

A careful examination of the decay products shows us that emitted from the nucleus of $^{241}_{95}\text{Am}$ are an α-particle of kinetic energy 5.5 MeV and a γ-ray of energy 0.1 MeV. So the loss in mass may be accounted for in the kinetic energy of the ejected α-particle and the energy of the γ-ray ($E = hf$), as they leave the nucleus.

Mass per nucleon

From an elementary working knowledge of physics one would think that the mass of a helium atom is four times the mass of a hydrogen atom, since the former contains four nucleons (two protons and two neutrons) to the latter's one. In fact accurate determination of the masses of these nuclei shows that they have masses of 1.0078 u for hydrogen and 4.0026 u for helium. So helium is less than four times as massive as hydrogen.

This result shows us that the helium nucleus is a stable entity. In stars, thermonuclear reactions cause protons to fuse into helium nuclei (see question 2). In the process of creating helium there is a loss of mass, so energy is radiated as predicted by Einstein's mass–energy equivalence equation. The energy emitted by the newly-formed helium nucleus is called the *binding energy* of the nucleus. The fact that energy is emitted in the process shows that the nucleons in a helium nucleus are in a lower energy state than when they are isolated; they are held together by strong nuclear forces.

Figure 4 shows graphically how the mass per nucleon in a nucleus, measured in atomic mass units (u), varies with the number of nucleons in the nucleus. The points to notice are:

☐ Iron and nickel have the least mass per nucleon; they have, therefore, the greatest binding energy and are the most stable nuclei.

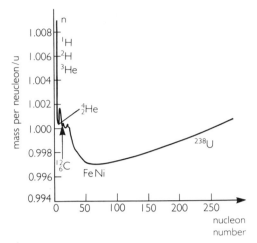

figure 4

☐ The heavy elements have an increasingly greater mass per nucleon as the atomic number rises. They are therefore increasingly unstable, and so we can begin to understand why nuclear decay series occur (question 15 of chapter 36), and why heavy nuclides may undergo fission.

☐ Very light nuclides ($_1^1H$, $_1^2H$, $_2^3He$) have high mass per nucleon. Therefore, if these nuclides combine to form heavier nuclides, energy will be released. This is called *fusion*, and happens spontaneously in the Sun. See question 2.

☐ Pronounced dips in the graph show that some nuclides are particularly stable, e.g. $_2^4He$, $_6^{12}C$.

Nuclear fission

We have already discussed how a nucleus may become more stable by emitting, for example, an α-particle. Another way for a heavy nucleus to stabilise itself is to split into two parts. This process is called nuclear fission. There are quite a number of nuclei that undergo fission: the two most common examples are uranium-235 and plutonium-239. We shall consider the latter as an example.

Plutonium (Pu) is an element that does not occur naturally but is produced artificially in nuclear reactors. Uranium-238, which is the most abundant isotope of that element, is bombarded by energetic neutrons. The following nuclear reactions occur:

$$_{92}^{238}U + _0^1n \rightarrow _{92}^{239}U$$

$$_{92}^{239}U \rightarrow _{93}^{239}Np + _{-1}^0\beta$$

$$_{93}^{239}Np \rightarrow _{94}^{239}Pu + _{-1}^0\beta$$

$_{92}^{239}U$ has a half-life of 23 minutes and $_{93}^{239}Np$ has a half-life of 2.3 days. $_{94}^{239}Pu$ decays by the emission of an α-particle, with a half-life of 24 000 years.

However, if a plutonium nucleus is now struck by a *slow-moving* neutron it can split into two parts with a large release of energy. Figure 5 shows a possible sequence of events. The plutonium nucleus absorbs the neutron (a), but it is left so unstable that it splits into two parts (b). The two nuclei left are also unstable (c), and decay rapidly by the emission of β-particles (d) to leave two stable nuclei, xenon and rubidium (e).

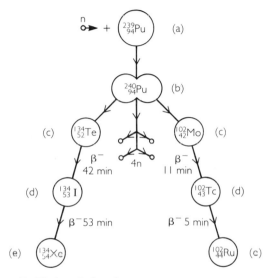

figure 5   Fission of plutonium

Again by using the equation $E = mc^2$, we may work out the energy released on fission.

$$\text{mass of } _{94}^{240}Pu = 240.0538 \text{ u}$$

$$\text{mass of } _{52}^{134}Te = 133.9064 \text{ u}$$

$$\text{mass of } _{42}^{102}Mo = 101.9098 \text{ u}$$

$$\text{mass of } 4 \times _0^1n = 4.0347 \text{ u}$$

So the total mass after fission is 239.8509 u. The mass lost is 0.2030 u, or a mass of $0.203 \times 1.66 \times 10^{-27}$ kg $= 3.36 \times 10^{-28}$ kg. Thus the energy liberated on fission is:

$$E = mc^2$$
$$= 3.36 \times 10^{-28} \text{ kg} \times 9 \times 10^{16} \text{ m}^2\text{s}^{-2}$$
$$= 3.02 \times 10^{-11} \text{ J or } 189 \text{ MeV}$$

This energy of 189 MeV will appear as kinetic energy of the fission fragments. In fact the total energy will be a little greater than this since some mass–energy will be generated in the β-decay processes that follow the fission.

The nuclear reactor

In most power stations in Britain coal is burnt to raise steam. High-pressure steam then drives turbines, which in turn drive the generator that produces electricity. In a nuclear power station a nuclear reactor replaces the coal-burning stage, but after that electricity is produced in the same way (see figure 6).

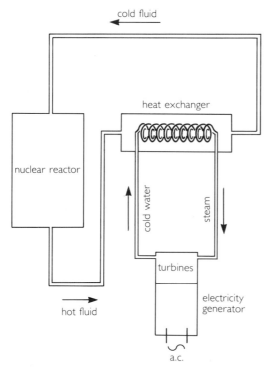

figure 6 Using nuclear power to produce electricity

The nuclear reactor makes use of the energy liberated in the fission of uranium-235. In each fission process, an average of three neutrons are released in the fission products. This allows the fission process to be self-sustaining: the neutrons released in one fission can be absorbed by other nuclei which then undergo fission themselves.

The nuclear reactor is carefully designed so that the rate of energy production is controlled. The nuclear fuel is placed in the reactor in the form of long rods: some or all of these rods may be withdrawn from the reactor at any time. When there is a large number of rods in the reactor, energy is produced at a great rate. With such a high density of rods in the reactor a large fraction of the neutrons produced in fission is absorbed by other fissile nuclei.

On the other hand, if the density of rods in the reactor is low, then a large fraction of the neutrons produced in any fission escape and do not hit another fissile nucleus; thus energy is produced at a slower rate. This principle is used in the manufacture of the plutonium bomb, the only difference being that when the bomb is activated the density of the nuclear fuel is so great that the production of energy is completely unchecked.

The production of electricity from nuclear reactions has several attractive features. These are listed below.

1 A large amount of energy can be produced from a small amount of uranium-235. 1 kg of uranium-235 contains $6 \times 10^{26}/235 = 2.6 \times 10^{24}$ atoms. Each fission releases about 5×10^{-11} J of energy. So 1 kg releases $5 \times 10^{-11} \times 2.6 \times 10^{24}$ J $= 1.3 \times 10^{14}$ J. A typical power station produces energy at a rate of 200 MW. So the energy produced by a power station in 1 year is $2 \times 10^8 \times 3 \times 10^7$ J $= 6 \times 10^{15}$ J. (There are approximately 3×10^7 s in a year.) So 1 kg of fuel would supply a power station for 1 week! This is more attractive than handling millions of tonnes of coal.

2 The natural resources of uranium are sufficient to provide us with energy for tens of thousands of years. Fossil fuels will last us for only a few hundred years.

3 A nuclear reactor can be used to power our major deterrent weapon, the Polaris submarine. Nuclear-powered submarines can stay under water for an indefinite period (provided there is enough food). In a conventional submarine, power is supplied by a diesel generator. Such a generator uses oxygen to burn the fuel and so the submarine has to surface periodically; they also make a lot of noise. Obviously both these features are undesirable for submarines.

On the other hand, nuclear reactors have their problems. There are the dangers to life caused by radiation (see question 9). Neutrons are particularly dangerous to living tissues, and all reactors must be properly shielded. Neutrons are effectively stopped by concrete in nuclear power stations or by polystyrene in nuclear-powered submarines. Although nuclear power is now a matter of some controversy in Britain and elsewhere, France has decided that by the year 2000, 95% of its electricity will be generated by nuclear fission reactors. It is understandable that a government would wish to avoid losing votes by switching to nuclear power, but the British government 200 years hence may be forced into taking that difficult decision. In 200 years time there will be no fossil fuels and so only nuclear power will provide our energy needs, unless some considerable progress is made with alternative means such as wave, wind or hydroelectric power.

Nuclear fusion power

It is possible that within this span of 200 years, power may become available from nuclear fusion. This is the process in which light nucleides combine to form heavier ones, with the release of considerable energy. The energy from the Sun is created by this process – see question 2.

Man-made nuclear fusion has already occurred, in the form of the explosion of the hydrogen bomb. To obtain electrical power from fusion is technically much more difficult, for various reasons:

☐ the reaction has to occur at a controlled rate;

☐ a very high temperature (about 10^8 K) is needed before the reaction will occur;

☐ the reactants have to be contained at this very high temperature;

☐ the energy (in the form of heat) has to be extracted from the reactants.

The technology is being developed by at least three teams of scientists and engineers: in the USA, in the USSR and by a consortium of European countries at the JET project sited at Culham, Oxfordshire. See figure 7.

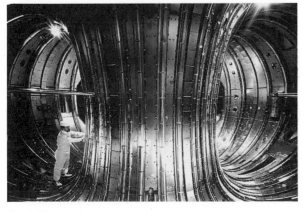

figure 7

Present estimates suggest that nuclear fusion power may be available to us in about 50 years, if all the difficulties can be solved.

Fusion power has several advantages over fission power:

☐ there is virtually limitless fuel available (hydrogen isotope of mass number 2, now called deuterium, occurs naturally in sea water);

☐ the reactants are themselves not dangerous;

☐ there are few harmful by-products, so no danger of a Chernobyl-type accident;

☐ there is very little waste to dispose of.

Several of these points may make this option more palatable to the general public than fission power – if and when the decision-time arrives.

New particles: $E = mc^2$ again

The nuclear physicists of today have a lot of fun making new particles. They claim that the production of new particles tells them something about nuclear forces, but we think that it is a hobby akin to stamp collecting! The principle underlying the production of a new particle is again Einstein's statement of the equivalence of mass and energy.

In modern particle accelerators it is possible to produce a proton with several hundred MeV of kinetic energy. In a collision this excess energy may be used to create mass. Such a process involving a collision between two high-energy protons is described by the equation below.

$$p + p \rightarrow p + \Lambda + K^+$$

In words this equation says that two protons collide to make a proton (p), a lambda (Λ) and a positive kaon (K^+). The lambda is an uncharged particle of mass 1.19 times that of the proton, the K^+ has a mass 0.53 times that of a proton and a charge equal to that of a proton. In adding up masses of both sides of the equation we find that the mass of the left hand side is two proton masses and on the right hand side it is 2.72 proton masses. So an extra mass of 0.72 proton masses has been created, which requires an energy input of about 675 MeV. This energy must originate from kinetic energy of the protons: in such a collision the protons would be travelling at speeds in excess of 90% of the velocity of light.

Summary

1. Chadwick identified neutrons ($^1_0$n) in 1932 by observing their collisions with protons in paraffin wax.

2. Nucleons (neutrons and protons) are held together in the nucleus by the strong nuclear force.

3. Electrostatic repulsive forces between protons can make a nuclide unstable.

4. $E = mc^2$: the equivalence equation relating mass and energy.

5. The atomic mass unit (u) is defined as 1/12 the mass of the carbon-12 atom. $1\,u = 1.66 \times 10^{-27}$ kg.

6. The energy produced by a nuclear reaction is the result of a small net decrease in the mass of the particles involved.

7. The mass per nucleon of a nuclide gives us clues about its stability.

8. Nuclear fission provides some of our electrical power at present.

9. Nuclear fusion may provide electrical power in the future.

Questions

1. Both α- and β-particles lose energy rapidly as they travel through air, because they are charged particles and therefore they interact strongly with electrons in atoms and cause ionisation. Neutrons, on the other hand, are not charged and so only lose energy by hitting the nucleus of another atom.

   a Explain why neutrons are such a very penetrating radiation (more so in fact than γ-radiation).

   b Which would you expect to be a more efficient absorber of neutrons, lead or water? Explain your answer.

   c Figure 8 shows a neutron of mass m hitting a heavier nucleus of mass M with a speed u, and recoiling with speed v_1. By consideration of the conservation of momentum and energy, show that if the collision is elastic the recoil energy of the neutron, $\frac{1}{2}mv_1^2$, is equal to:

   $$\frac{1}{2}m\left(\frac{M-m}{M+m}\right)^2 u^2$$

   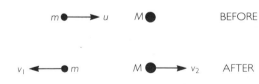

   figure 8

   d Use the formula above to calculate the fraction of energy transmitted to a lead nucleus, relative mass 208, when a neutron (relative

   mass 1) hits it directly. Hence explain why lead tends only to scatter neutrons rather than to absorb them.

   e Repeat the calculation of part (d) for a neutron hitting a stationary proton head-on. Explain what happens to the energy of the knocked-on proton, and hence explain why water is a good absorber of neutrons.

   f One of the newer nuclear weapons is the neutron bomb. Such a bomb releases an intense flux of neutrons which will kill soldiers inside a tank but leave the vehicle itself completely unharmed. Can you explain why?

2. The temperature inside our Sun is thought to be about 15×10^6 K. At these temperatures all the electrons have been stripped off the nuclei, thus leaving a state of matter that is known as a *plasma*. Positively charged nuclei move in stellar interiors with great velocities; they collide violently so that despite the presence of repulsive forces (like charges repel each other) the particles may fuse together, due to the presence of strong but short-ranged nuclear forces, to form new nuclei. The thermonuclear reaction that occurs in our Sun is the conversion of hydrogen to helium. This process is thought to occur as follows:

   (i) $^1_1$H + $^1_1$H → $^2_1$H + $^0_1$e (positron)

   (ii) $^2_1$H + $^1_1$H → $^3_2$He + released energy (γ-rays)

   (iii) $^3_2$He + $^3_2$He → $^4_2$He + $^1_1$H + $^1_1$H + released energy (γ-rays)

a Explain why you would not expect nuclear fusion to occur at low temperatures (a few thousand K).

b How does Einstein's mass equivalence relation ($E = mc^2$) help to explain the release of energy in the fusion process? Would you expect the mass of the constituents of the left hand side of each equation to be greater or smaller than the mass of the constituents of the right hand side of each equation?

c The unified atomic mass unit (u) is defined as 1/12 the mass of the carbon atom $^{12}_{6}C$. Below are listed the masses of some of the particles concerned in the nuclear fusion process.

> mass of proton   1.0073 u
> mass of helium-4 4.0026 u
> mass of positron 0.0005 u

Use this data to show that the production by fusion of one helium-4 nucleus results in a mass loss of about 0.026 u.

d The Sun has a mass of 2×10^{30} kg; it is at the moment in the process of converting hydrogen to helium. What mass will the Sun have lost by the time it has fused all its hydrogen into helium? Assume that when the Sun was formed it was 100% hydrogen.

e Use your answer to (c) to calculate how much energy the Sun will liberate in its lifetime.

f The Sun radiates energy at a rate of 4×10^{26} W. Hence estimate its lifetime. Given that the composition of our Sun is now roughly 25% helium, 75% hydrogen, what does this suggest about the age of our solar system? (More accurate estimates give the age of our solar system to be about 4.5×10^9 years).

3 There are people today who call themselves creationists. They believe that God created the world in six days as described in Genesis; they further deduce from the Bible that the Earth can only be six thousand years old. Argue briefly for or against this concept of the creation. If you argue against the creationists, suggest an alternative means by which we arrived on this planet.

4 Radioactive materials are both useful and hazardous. Write an essay to support this view.

5 Imagine that you have recently been appointed as an adviser to NASA. The director of Space Exploration at NASA needs advice on which radioisotopes would be most suitable to use for power sources for two purposes:

(i) to power a lunar 'buggy' for a short exploration of the Moon by astronauts;

(ii) to power cameras, computer and radio transmitters for a *Voyager* spacecraft intended to explore the outer planets in the solar system.

Explain which isotope you would choose for each purpose. Use the tables below to do this.

Isotope	Half-life of isotope	Main radiations emitted
^{242}Cm (curium)	163 days	α-particle
^{241}Am (americium)	458 years	α-particle
^{90}Sr (strontium)	28 years	β-particle
^{60}Co (cobalt)	5.3 years	β-particle γ-ray
^{160}Tb (terbium)	72 days	β-particle γ-ray

Isotope	Energy of radiation emitted/MeV
^{242}Cm (curium)	6.1
^{241}Am (americium)	5.5
^{90}Sr (strontium)	0.5
^{60}Co (cobalt)	0.3 1.2
^{160}Tb (terbium)	1.7 1.0

6 It is believed that one of the processes that occur in the formation of young stars is this reaction:

$$^{1}_{1}H + ^{1}_{1}H \rightarrow ^{2}_{1}H + e^{+} + \text{energy release}$$

Nuclear forces have a range of about 10^{-15} m, so before this reaction can occur the two protons have to approach to within that distance.

a Calculate the potential energy acquired by two protons approaching to within a distance of 10^{-15} m.

b This increase in potential energy must be provided by an equal loss in kinetic energy of the protons. Assuming that the kinetic energy of a proton in a star is of the order of kT, where k is the Boltzmann constant, estimate the temperature T inside a young star before

the fusion reaction can occur.
($k = 1.4 \times 10^{-23}$ JK^{-1}.)

c Explain where the energy comes from originally to warm up the inside of a growing star.

7 To describe nuclear reactions, the notation A(a,b)B is often used. A(a,b)B means $a + A \rightarrow b + B$ or in words, particle a strikes nucleus A creating nucleus B and emitting particle b; the Q-value for a reaction means the energy released. Listed below are some nuclear reactions.

	Reaction	Q-value/MeV
(i)	^{238}U $(n,\gamma)^{239}$U	4.8
(ii)	$^{7}_{3}$Li$(p,\alpha)^{4}_{2}$He	17.4
(iii)	$^{23}_{11}$Na$(n,\beta)^{24}_{12}$Mg	12.5

a Write out an equation for each of these reactions.

b Explain whether the reaction ^{239}U$(\gamma,n)^{238}$U would work if the incident γ-ray had an energy of 2 MeV. What is the Q-value for this reaction? What is the significance of its negative value?

c What does the second reaction '$^{7}_{3}$Li$(p,\alpha)^{4}_{2}$He', tell us about the stability of the helium nucleus? Why are α-particles often ejected by unstable nuclei?

d $^{24}_{11}$Na is an unstable isotope of sodium; it decays to $^{24}_{12}$Mg by the emission of a β-particle, energy 6 MeV, and two γ-rays of energy 1.4 MeV and 2.8 MeV. The Q-value for the third reaction is 12.5 MeV; how do you account for the missing energy?

8 Read the passage below, then answer the questions which follow it.

Biological effects of radiation
Animal cells are composed mainly of water, and therefore it is the effects of radiation on water that are of vital importance. It now seems that when ionising radiations are present, highly reactive ions are produced:

$$H_2O + \text{absorbed energy} \rightarrow H_2O^+ + e^-$$

$$H_2O + e^- \rightarrow H_2O^-$$

These reactive ions may interact with other molecules such as RNA or DNA (which may not actually have been hit by radiation), thus causing permanent damage to a cell.

The most common unit used to describe the interaction of radiation with tissue is the sievert (Sv), which is the quantity of radiation that will result in an energy absorption of 1 Jkg^{-1} of

animal tissue. However, it is not just the quantity of radiation that is important but the nature of it. Densely ionising particles, such as α-particles which produce very large numbers of ions along each millimetre of their tracks, are more likely to cause permanent damage to a molecule, because they will lose their energy in a highly localised area. It is the linear energy transfer (LET) that determines the radiation damage; the LET is the energy deposited per metre of particle path length. LET $\propto q^2/v^2$, where q is the particle's charge and v its speed.

The LET determines the RBE (relative biological effectiveness) of a radiation. The RBEs of γ- and X-radiation are about equal, and are defined as 1. The RBE of fast β-particles is 1, but slow β-particles have an RBE of 2. Neutrons have an RBE of 3 and α-particles an RBE of 10.

Even having taken into account the RBE of different radiations, it is not possible to state exactly the effect of 1 Sv. This is because some animals are more radiosensitive than others, and different tissues in the same animal vary. In general, the most radiosensitive cells are those which have the highest division rate or those which retain the capacity for division for the longest time. The total body dose (X- or γ-rays) that will kill 50% of a certain population (the LD$_{50}$)is 25 Sv for guinea pigs, 6.5 Sv for man and 30 Sv for newts. (The figure for man was calculated from the Hiroshima and Nagasaki data.)

Figure 9 shows how the percentage of deaths in a population increases with radiation dose.

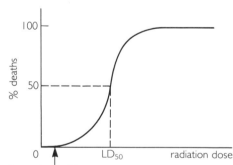

figure 9   threshold

The problem is still not yet solved, because the effect of a dose depends on how it is administered. The table on the next page shows how the dose necessary to produce a certain intensity of erythema (reddening of the skin) in man varies with the dose rate.

Radiations not only cause gross damage to tissues, they cause mutations. While there is a lower limit below which radiation death is most unlikely, no such limit appears to exist for mutations. Careful experiments on mice have shown that the number of radiation-induced mutations is proportional to the dose. Thus any exposure to radiation might be genetically harmful. In Britain, the background radiation is 1 mSv per year, and there do not appear to be any genetic mutations in regions where the background radiation level is twice as high as normal. See figure 10.

Dose rate/ mSv min^{-1}	Irradiation time required/min	Total dose required/mSv
50	1	50
0.5	15.5	78
0.5	260	130
0.05	4500	225

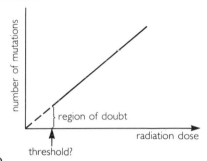

figure 10

a Explain carefully why the RBE of α-particles is so much higher than that of γ-rays, and why the RBE is greater for slow β-particles than it is for fast-moving β-particles.

b Despite having a low RBE, why is a source of γ-rays potentially more dangerous than a source of α-particles?

c Explain carefully why you think that it is unwise to expose a pregnant woman to X-rays.

d A 5 MeV α-particle will travel about 5 cm through air. Calculate its LET in Jm^{-1}.

e Use the information in part (d) and in the passage to estimate the range of a 5 MeV proton through air.

f A man of mass 70 kg is exposed to an average dose of 5×10^{-5} Sv during a series of X-ray photographs taken at a hospital. Given that the energy of the X-ray photons was 10 keV, estimate the total number of photons that passed through his body. ($e = 1.6 \times 10^{-19}$ C.)

g In an accident at a nuclear reactor, a worker was exposed to a dose of 4 Sv over a period of 1 minute. Use the data in the passage to estimate the probability that he will die.

Another worker received the same dose but spread over a period of 15 minutes. What can you say about his chances of survival? Explain your answer.

h Discuss whether you think that the advantages of nuclear power outweigh any possible disadvantages.

9 *Nuclear cross-section*
A typical nucleus has a diameter of about 10^{-14} m and therefore a cross-sectional area of about 10^{-28} m^2. Nuclear physicists talk in terms of *absorption cross-sections* for various nuclear processes, e.g. we might say that a particular nucleus has an absorption cross-section of 10^{-28} m^2 for the absorption of neutrons. However the absorption cross-section, σ, is very often substantially different from the real size or geometrical cross-section of the nucleus. It would appear that nuclear physiscists have a sense of humour: the unit of cross-section, 10^{-28} m^2, is called a *barn*, because, so I am told, a physicist once described a nuclear cross-section as being 'as big as a barn door'.

a This part of the question enables us to work out the fraction of an incident beam of neutrons that will be absorbed by a material. The material has a number density n (atoms per m^3) and has a thickness t. So the number of nuclei in the path of an incident beam of neutrons is nt (nuclei per m^2). Thus the total cross-section for absorption is an effective area $nt\sigma$, out of an area of incidence for the neutron beam of 1 m^2; and $nt\sigma$ is therefore the fraction of incident neutrons that will be absorbed.

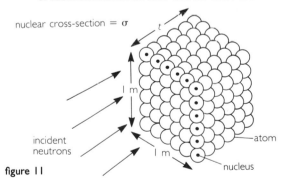

figure 11

A nuclear reactor produces a neutron flux density of 10^8 neutrons per second per m^2. The neutrons fall on a sample of uranium of cross-sectional area 10^{-2} m^2 and thickness 1 mm. The nuclear cross-section for the absorption of the neutrons is 10 barns; uranium contains 5×10^{28} atoms per m^3. Calculate the fraction of neutrons absorbed per second, and hence the total number.

b In America at the present there is an experiment in progress which is designed to monitor small particles called neutrinos which come from the Sun. They are produced in the reaction:

$$^1_1p + ^1_1p \rightarrow ^2_1d + ^0_1\beta + \nu$$

ν is a neutrino and d is a deuteron, i.e. a nucleus consisting of one proton plus one neutron.

These neutrinos arrive in great abundance; the neutrino flux at the Earth's distance from the Sun is about 4×10^{13} per m^2 per second. However, neutrinos hardly ever interact with matter, and most of them go straight through the Earth. The neutrinos are detected in a vast tank of cleaning fluid, tetrachloroethene (C_2Cl_4), which is buried in a disused mine. The tank is roughly a cube measuring $100\,m \times 100\,m \times 100\,m$.

The neutrinos interact with ^{37}Cl in the fluid to produce argon in the reaction

$$^{37}_{17}Cl + \nu \rightarrow {}^{37}_{18}Ar + {}^{0}_{-1}\beta$$

The scientists detect an average production of 9 argon atoms per month.

Use the information given here to calculate the absorption cross-section for the neutrinos by ^{37}Cl, given that there are approximately 10^{34} atoms of ^{37}Cl in the tank. Compare your answer for the cross-section with the size of the chlorine nucleus and comment on the difference.

c Do you think we should take the results of the above experiment at all seriously?

10 *Moderation in a nuclear reactor*
a In a nuclear reactor energy can be produced by the typical chain reaction:

$$^{235}_{92}U + {}^{1}_{0}n \rightarrow {}^{144}_{56}Ba + {}^{90}_{36}Kr + 2{}^{1}_{0}n$$

The chain reaction is maintained because the two neutrons emitted can cause other fissions to occur. However, the neutrons emitted in the fission process have large energies, of the order of 2 MeV. Explain, with the aid of

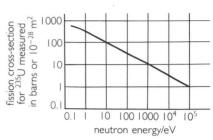

figure 12

figure 12, why such neutrons are not very likely to trigger off further fissions.

b In between the fuel rods there is a material such as graphite or lithium, which is called a *moderator*. The purpose of the moderator is to slow the neutrons down.

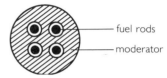

figure 13

Explain (i) why it is necessary to slow the neutrons down to maintain the chain reaction, (ii) why the materials mentioned are chosen as moderators, as opposed to anything more massive.

c When a neutron is emitted in a fission reaction, is it more likely to cause another fission to occur in its own fuel rod or in a neighbouring one?

ATOMIC PHYSICS

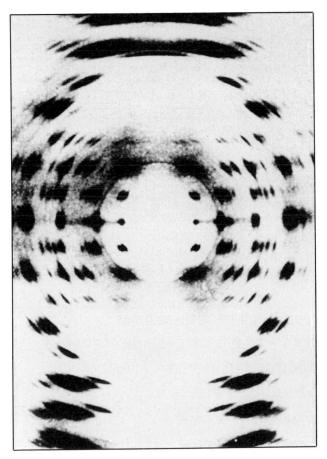

X-rays are used to reveal to us the structure of materials. This photograph shows us the structure of DNA...

... and James Dewey Watson (together with
Francis Crick) earned the Nobel Prize for
unravelling the double helical structure of the DNA
molecule.

38   Discovery of the nucleus   409
39   Light – wave or particle?   413
40   Electrons in atoms   420
41   Electrons – waves or
      particles?   428

Consider the following paragraph:
  'We know that the radius of an atom is about 10^{-10} m, and that the nucleus of an atom is yet smaller, having a radius of the order of 10^{-14} m. Furthermore, we know that electrons orbit the nucleus in certain fixed energy levels, and that electrons can jump from one level to another.'
  No doubt you will either have read or will have heard your teacher make a statement similar to the one in the previous paragraph. The purpose of the next four chapters is to describe the experimental evidence that allows us to make such statements. In addition we will be looking at some theoretical models that arise as a result of the experiments. It is important that you appreciate throughout this section that the theoretical ideas about atoms are based firmly on experimental evidence.

CHAPTER 38

Discovery of the nucleus

Introduction

In this short chapter we consider just one experiment, the scattering of α-particles by gold foil. This experiment was first performed in 1909 by Geiger and Marsden, under the direction of Rutherford. The results were crucial to the development of the nuclear model of the atom. Today we take this model for granted; at that time it had barely occurred to anybody that the structure of an atom might be like this.

GCSE knowledge

We assume you know that:

☐ atoms are very small; they are too small to see, and have a radius of about 10^{-10} m;

☐ they have a nucleus which is small in comparison with the size of the atom itself;

☐ the nucleus contains protons and neutrons (figure 1);

☐ the protons and neutrons are of similar mass, and make up most of the atom's mass;

☐ the protons carry one positive charge, the neutrons are neutral;

☐ in orbit around the nucleus there are electrons;

☐ these electrons carry one negative charge, but their mass is very small in comparison with a proton or neutron;

☐ the charge on a proton is the same size as that on an electron;

☐ there are as many protons as electrons in a neutral atom.

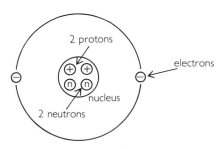

figure 1   A helium atom, not to scale

The size of the nucleus: Rutherford scattering

In 1909 Geiger and Marsden, guided by Rutherford, did a series of experiments in which α-particles were scattered from a variety of metal foils, as illustrated in figure 2. They detected the scattered particles by observing scintillations on a fluorescent screen, viewed through a microscope. (A scintillation is an individual tiny flash of light, which occurs each time an α-particle strikes the fluorescent screen.) This was a laborious task, since they counted about 100 000 such scintillations in each experiment. Nowadays we have the luxuries of a Geiger tube and an electronic counter.

Geiger and Marsden discovered that most of the α-particles were deviated through only small angles; some were deflected through larger angles, e.g. to B in figure 2(b); and a very small fraction of the incident particles was even scattered backwards through very large angles indeed, e.g. to C and D. These results were a surprise, since at the time it was thought that the atom was a homogenous sphere with a uniform density. An α-particle in collision with such an atom might perhaps undergo a small deviation, which for the sake of argument we will take to be 1° or so; thus, in passing through a foil several hundred atoms thick, deviations through larger angles would be possible on this model. However, to be deflected through a large angle such as 150°, an α-particle would have to experience 150 deviations successively in the same direction. This is very unlikely indeed, as the following calculation shows. (The discussion is taken further in question 4.)

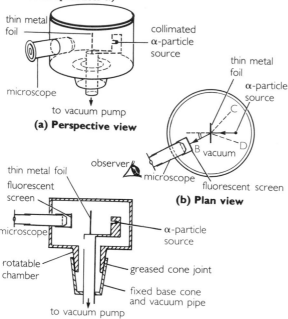

(a) Perspective view

(b) Plan view

(c) Sectional view

figure 2   Geiger–Marsden experiment on α-particle scattering

Taking a simplistic model, we will imagine a particle to be deflected only to the left or the right (in practice of course we are dealing with a three-dimensional problem). Then the probability of being deflected one way 150 times is $(\frac{1}{2})^{150} \approx 10^{-45}$, which means that 1 in 10^{45} particles will be deflected through an angle of 150° if the atoms are of uniform density. In their experiment Geiger and Marsden discovered that for a gold foil of thickness 6×10^{-7} m, 1 in 8000 particles were deflected through an angle of 150° or more. Although this is still a small fraction of the total incident particles, it is far more than expected on the old homogenous-sphere model of the atom.

These results led Rutherford to suggest that the atom consisted of a very small, though massive, positively charged nucleus. Thus practically all the mass of the atom is concentrated at the centre, leaving only electrons outside the positively-charged nucleus. This hypothesis allows the scattering data to be satisfactorily explained. The α-particle (which is positively charged) experiences a repulsive force from the nucleus. However, because the nucleus is so very small, most α-particles experience only a small deviation, because they do not approach the nucleus very closely. On the other hand, a few do approach the nucleus closely and thus are deflected through very large angles.

A rough estimate of the nuclear radius may be made. From X-ray diffraction work it can be

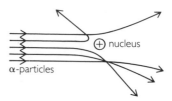

figure 3   Scattering of α-particles by a nucleus

established that the diameter of a gold atom is 3×10^{-10} m. Thus in this experiment the gold foil was $(6 \times 10^{-7} \text{ m})/(3 \times 10^{-10} \text{ m}) = 2000$ atoms thick. We will assume that if an α-particle is scattered through an angle of 150° or more then it has made a head-on collision with the nucleus. So in the Geiger–Marsden experiment 1 in 8000 incident particles did that. However, had the foil been only 1 atom thick then the fraction scattered through this large angle would have been 2000 times smaller, i.e. 1 in 16 000 000. This ratio is also, therefore, the ratio of the cross-sectional area of the nucleus to the cross-sectional area of the whole atom. Since the area of a circle is proportional to the square of the diameter, it follows that the diameter of the gold nucleus is roughly 4000 times smaller than that of the atom, i.e.

$$\frac{3 \times 10^{10}}{4000} \text{ m} = 7.5 \times 10^{-14} \text{ m}$$

In fact it turns out that this estimate is a little too big, because the α-particles are repelled by the large electrostatic forces without actually reaching the edge of the nucleus. (The questions that follow suggest more accurate ways of measuring nuclear radii.)

By assuming that the scattering was caused by electrostatic repulsion from a very small nucleus, Rutherford showed that the number of particles N scattered per unit area through an angle θ to a small detector a distance r from the metal foil would be:

$$N = \frac{N_0 n t e^4 Z^2 \operatorname{cosec}^4(\theta/2)}{16 \pi^2 \varepsilon_0^2 r^2 m^2 v^4}$$

where N_0 is the incident number of particles, n the number of nuclei per unit volume, t the thickness of the foil, Z the proton number of the nucleus, e the electronic charge, m the mass of the α-particle and v its speed. Geiger and Marsden verified this formula by measuring N for various scattering angles θ.

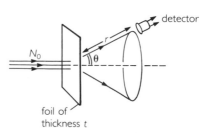

figure 4   Geometry of experiment on α-particle scattering

Questions

1 An upper limit for the diameter of a carbon nucleus may be obtained from data similar to Geiger and Marsden's. In an experiment, about 1 in 20 000 α-particles was scattered by an angle of more than 150°; this is taken to mean that the particle had 'scored a direct hit'. The thickness of the carbon foil was 2 μm.

   a Given that a carbon atom has a diameter of about 2×10^{-10} m, deduce how many atoms thick the foil was.

   b Assuming that the probability of an α-particle hitting a nucleus is proportional to the thickness of the foil, deduce what fraction of α-particles would have been back-scattered by an angle greater than 150° had the foil been 1 atom thick.

   c Hence deduce the ratio of the cross-sectional area of a carbon atom to the cross-sectional area of its nucleus.

   d Finally deduce the diameter of the carbon nucleus.

2 The previous question only leads to an upper limit for the nuclear diameter, because the incident α-particles were of relatively low energy (5 MeV) and so may not actually have reached the nucleus itself (see figure 5).

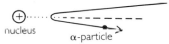

figure 5   Nucleus scattering low energy α-particle

   In α-scattering experiments, it is found generally that the number of particles scattered through a given angle can be predicted by the Rutherford scattering formula (page 410). This formula is derived by considering the electrostatic repulsion between nucleus and α-parpticle. However, for high-energy α-particles the Rutherford scattering formula breaks down, because such a high-energy α-particle can approach the nucleus sufficiently closely for the strong nuclear forces to act. Thus the deflection is no longer governed solely by the electrostatic forces (figure 6).

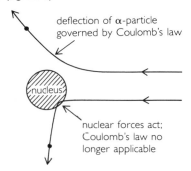

figure 6

Figure 7 summarises the results of a scattering experiment using 30 MeV α-particles, with uranium as the target nucleus. The graph shows that the observed scattering is as predicted by the Rutherford scattering formula up to scattering angles of 120°.

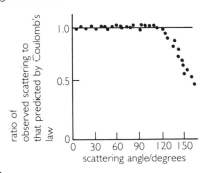

figure 7

   a Explain why the predictions of the Rutherford formula hold for small angles of scattering but not for large angles.

   b Write down an expression for the electrical potential energy between two particles carrying charges q_1 and q_2 when separated by a distance r.

   c Estimate the closest distance of approach, r_0, between a 30 MeV α-particle and a uranium nucleus, by assuming that at the separation r_0 all the α-particle's original kinetic energy has been converted into potential energy.

   d Is your answer to part (c) the radius of a uranium nucleus? Explain your answer.

3 Figure 8 shows an α-particle being scattered through an angle of about 90°. In an experiment it is found that n α-particles per unit time are scattered through an angle of 90° off a thin foil of zinc.

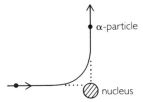

figure 8   α-particle being scattered at 90° by a nucleus

   a What will happen to n if the foil is replaced by a zinc foil of twice the thickness?

   b What will happen to n if the foil is replaced by a sheet of graphite of the same thickness?

   c What will happen to n if the original piece of zinc foil is irradiated with α-particles with greater kinetic energy?

You are *not* expected to provide quantitative answers to these questions, but you should explain the physics which leads to your answers.

4 *The plum-pudding model of the atom*
An early model of the atom was called the 'plum-pudding' model. The idea was that the main substance of the atom was a positively-charged 'pudding' with a few negative plums (electrons) embedded inside the positive charge. People were uncertain as to what the pudding was – protons and neutrons had yet to be identified as part of the nucleus. The idea seems fairly sensible. After all, why should atoms not be solid? Furthermore an atom *without* a great concentration of charge would explain why α- and β-particles pass through matter. (Until Geiger and Marsden's experiments no one had observed large-angle scattering of α-particles.)

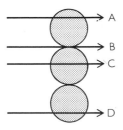

figure 9   Plum pudding model of the atom

This question is about the size of deviation an α-particle might expect on passing through an atom, if the plum-pudding model were correct.
a Imagine an α-particle passing through an atom which is part of a long line of such atoms (figure 10). Explain on which of the paths A, B, C, D, and α-particle might be deflected. For an atom like gold, there would be about 80 electrons dotted around inside the atom.

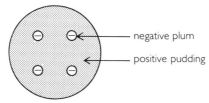

figure 10

Part (a) is difficult to answer, but you might have decided that the α-particle would be deflected a little if it passed close to an electron. We will now make a very rough estimate as to the size of this deflection. Suppose an α-particle passes within about 10^{-14} m of an electron as shown in figure 11. It will experience a sideways force which will pull it towards the electron.

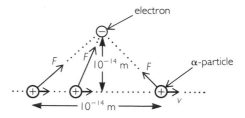

figure 11

b Show that the maximum value of this force is about 4.6 N. ($e = 1.6 \times 10^{-19}$ C)
c The effect of the electron/α-particle 'collision' is complicated to work out, since the force is changing in size and direction. We will approximate it roughly by saying that a sideways force acts over a distance of about 10^{-14} m as shown. If the α-particle is travelling at 10^7 ms^{-1}, how long does the force act for?
d Use your answers to (b) and (c) to calculate the deflecting impulse on the α-particle.
e Given that an α-particle has a mass of 6.8×10^{-27} kg, calculate the momentum of the α-particle before it meets the electron.
f Now estimate the deflection of the α-particle, using the vector diagram in figure 12.

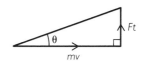

figure 12

g Comment on the size of deflection, and criticise any approximations that we have made.

5 Physicists talk confidently about the diameters of an atom, and even about the diameter of a nucleus. Write an essay to explain to an intelligent non-scientist why physicists believe atoms and nuclei are the size that they claim. Your essay should include descriptions of relevant experimental evidence.

CHAPTER 39

Light: wave or particle?

Introduction

Your earlier work on light will have convinced you that light is a wave. Light is diffracted when it passes through a small slit and two sources of light may be made to interfere. We also know that light waves carry energy.

Things turn out not to be quite so simple. Experiments show that electrons can be removed from a metal surface by light. This might not surprise you, since light carries energy, so an electron can gain this energy and therefore escape from the attractive forces of the atoms in the metal. However, for some metals (zinc for example), it is possible to remove electrons with a weak source of ultraviolet light, but not with a bright white light. From such experiments comes the idea that light comes in packets of energy, which behave like particles.

GCSE knowledge

We shall assume you know that:

☐ electrons have negative charges;

☐ there are 'free' electrons which can move through metals;

☐ in metals there are positive ions which vibrate about fixed positions, but cannot translate through the metal;

☐ like charges repel one another.

Photoelectricity

In the 1880s an interesting phenomenon was discovered. It was found that a weak source of ultraviolet light, when held near to a negatively-charged electroscope, caused it to discharge, whereas a bright white light source did not. However, the ultraviolet light source could not discharge the electroscope when it was positively charged. See figure 1. The explanation that weak ultraviolet light can knock electrons out of the metal, but the more intense white light source cannot is curious, since the wave theory of light predicts that the more intense light source should transfer more energy to the metal.

Energy is required to remove electrons from a metal surface, and we might therefore expect the bright white light source to be the more likely to discharge the electroscope. After all, in thermionic emission processes it is well known that the hotter the filament, the more electrons are emitted from it.

More detailed experiments on the photoelectric effect may be performed using the apparatus illustrated in figure 2. The evacuated photoelectric cell contains a cathode and an anode; a suitable cathode would be potassium, since visible light can remove electrons from its surface. The potential divider allows the anode to be made either positive or negative with respect to the cathode.

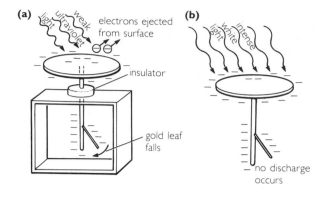

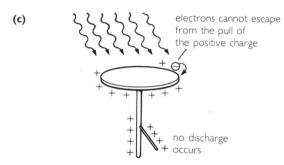

figure I Gold-leaf electroscope used to demonstrate the photoelectric effect

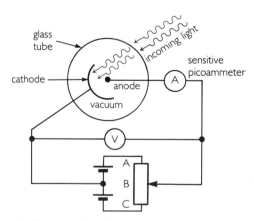

figure 2 Photoelectric cell

The following important results can be obtained:

I Provided that the anode is at a potential of 10 V or so above the cathode, all the emitted electrons will be collected by the anode. It is found

that the current is proportional to the intensity of the incoming light. Figure 3 shows the variation of current with anode potential for two sources of light with different intensities but the same wavelength.

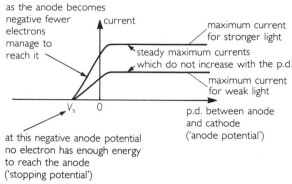

figure 3 Graph of current against anode potential for the photoelectric cell

2 Even for vanishingly small intensities of light, electrons are emitted by the cathode. Moreover, they are emitted as soon as a light source is switched on, with no time delay.

3 A particular cathode material will not emit electrons if the incident light is below a certain frequency, no matter how intense the source.

4 The kinetic energy of the emitted electrons depends on the frequency of the light, but not on the intensity. The energy of the electrons may be measured by making the anode negative with respect to the cathode; this may be achieved by sliding the contact S into the region BC of the potential divider (figure 2). Then as the reverse p.d. is increased the current decreases, until at a certain p.d., V_s, which is called the stopping potential, no current passes at all. This is because the initial kinetic energy of the electrons is not enough to do the necessary work against the electric field to get them to the anode. By applying the principle of conservation of energy, we may write that $\frac{1}{2}mv^2 = eV_s$; thus the initial kinetic energy of a photoelectron may be calculated. A typical graph of stopping potential against frequency is shown in figure 4.

The first result above is to be expected from the wave theory of light. However, we would not expect electrons to be emitted for vanishingly small intensities (result 2), because, if the energy is spread uniformly over a metal surface, it would take a long time for an area the size of an atom to accumulate enough energy for an electron to be emitted. Also, a wave theory of light predicts no relationship between energy and frequency, such as that of results 3 and 4.

To explain these results Einstein suggested in 1905 that light was behaving as a particle rather than as a wave. This leads to the idea that when light is

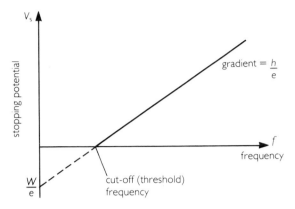

figure 4 Graph of stopping potential against incident frequency for a particular cathode material

emitted from a source, the energy is not spread uniformly over a volume of space but is carried by a finite number of particles which we call *photons*. Further, it was proposed that the energy E of a photon is given by $E = hf$, where h is called the Planck constant and f is the frequency of the radiation. h can be measured using the graph of figure 4 (see above); it has the value 6.6×10^{-34} J s.

This hypothesis enables us to explain why electrons are emitted even at vanishingly low intensities of light. One photon will interact with one electron, figure 6, and provided the photon carries sufficient energy the electron is knocked out of the metal. However, if the photon's energy were spread over a larger area (as in the wave theory) this would not occur, figure 5.

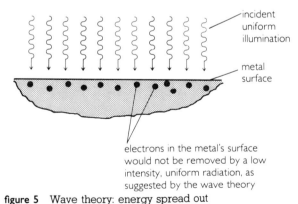

figure 5 Wave theory: energy spread out

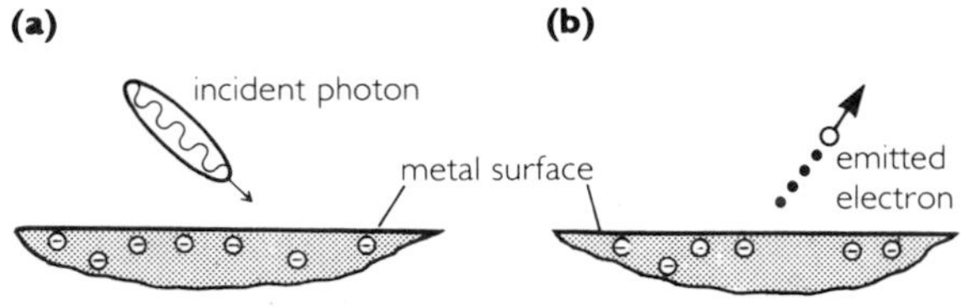

(a)   **(b)**

figure 6 (a) Incident energy concentrated in one photon; (b) the photon has been absorbed and an electron has been emitted

A more intense source of light will cause more photons to arrive at a metal surface, and therefore more electrons will be emitted. On this model it is easy to see that if light of very long wavelength (low frequency) is used to illuminate the cathode, the incident photons will carry too small an amount of energy to knock out any electrons.

We may draw an analogy: if I throw a 1 kg sandbag out of a window into a crowded street I will knock someone out; but if I pour the same quantity of sand out of the bag I will shower a lot of people with sand but render them no harm.

This model may also be used to explain the graph of stopping potential V_s against frequency, figure 4. The energy of the photon is used up in two ways: first in doing work W to remove the electron, and second in giving the electron kinetic energy. This may be expressed mathematically:

$$hf = W + \tfrac{1}{2}mv^2 = W + eV_s$$

The electron's kinetic energy is converted to electrical potential energy eV_s, since V_s is just large enough to stop the electron. Thus we can rearrange the above equation to give

$$V_s = \frac{h}{e}f - \frac{W}{e}$$

So the gradient of the graph is h/e and the intercept $-W/e$. W is called the *work function* of the cathode metal.

Crooke's radiometer: a demonstration of light particles

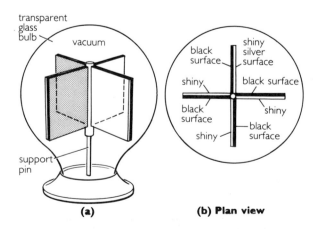

figure 7 Crooke's radiometer

Crooke's radiometer consists of four vanes in the form of a cross, which are delicately balanced on a pivot so that they are free to rotate. They are enclosed in a completely evacuated bulb.

Each vane has a black and a shiny side. The vane rotates with the black side leading. We may explain this by the theory that light consists of particles. Each particle, or photon, has momentum. The photons hitting the black side are absorbed, and therefore impart a certain impulse I to the vane (figure 8). On the shiny side the photons are reflected and so impart twice the impulse, $2I$, to the vane. Thus the shiny side experiences a bigger force than the dull side, which explains the observed direction of rotation.

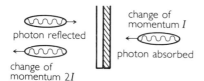

figure 8   Photons being reflected by the shiny side of a vane and absorbed by the black side

It is possible to buy similar radiometers in shops, but these will rotate the other way, because they have air in them at low pressure! It is left as an exercise for you to explain why.

Wave/particle duality

That light behaves as a wave is clearly illustrated by diffraction and interference experiments; and yet the photoelectric effect clearly illustrates that light can also behave as a particle which carries energy and momentum. This at first sight is rather alarming, since we cannot totally reject one theory and accept the other. So we are led to believe that light exhibits both wave and particle properties.

The concept of wave/particle duality is a difficult one to accept because we have preconceived ideas about particles and waves. We think in terms of billiard balls and ripples on a pond. In the macroscopic world, particles and waves are totally separate entities, but in the microscopic world of photons and electrons they overlap.

An example of how difficult it can be to reconcile the two models follows. Figure 9 shows a typical Young's slit experiment. Light passes through the double slit, and we can use the wave theory of light to explain the regions of constructive interference marked X and destructive interference marked O. We say that the waves from the two slits arrive in phase at X and out of phase at O.

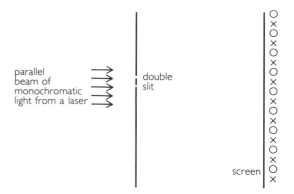

figure 9   Young's slits experiment

Taylor (1909) carried out such an experiment, but with a light source so distant and feeble that only one photon at a time could pass through the apparatus. (This is not difficult to achieve – light from a distant star will pass into our eyes one photon at a time.) He discovered that photons reached only the parts marked X but not those marked O. If we were to repeat Taylor's experiment today, and replace his fluorescent screen with a bank of photomultipliers and scintillation counters, we would discover an intensity distribution similar to the one shown in figure 10, if we detected photons for a long time. Each photomultiplier would periodically detect a single photon, but the pattern would gradually build up to produce the well known double-slit intensity pattern. Thus we can say that a single photon is more likely to arrive at a point X, but has no chance at all of arriving at a point O.

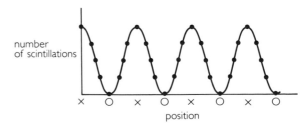

figure 10   Each dot represents the count from a given photomultiplier tube

We cannot predict what will happen to a single photon as it passes through the apparatus, but we can predict what will happen to a lot of photons by using the wave model of light. Thus in talking about individual photons, e.g. in the photoelectric effect, it is convenient to use a particle model of light; but a wave model must be retained to predict what will happen to a large number of photons in an interference or diffraction experiment.

Summary

1 The important experimental results of the photoelectric effect are these:
  a Electrons are removed from a metal surface by light; the number removed is proportional to the intensity of the light.
  b The energy of each ejected electron gets bigger as the frequency of light increases.
  c Below a cut-off or threshold frequency of light (different for different metals), electrons are not emitted.

2 A theory to explain these results is:
  a Light comes in packets of energy called photons. Their energy E is given by:
$$E = hf$$
  b One photon interacts with one electron in the metal.

c The energy E is used in two ways: first to remove the electron from the metal – this uses energy W (W is called the work function of the metal); secondly the remaining energy of the photon gives the ejected electron kinetic energy.

3 a If the energy (and hence frequency) of a photon is too low, no electrons will be ejected.
  b As the energy of a photon (and hence frequency) increases, so does the kinetic energy of the ejected electrons.
  c As the intensity of light rises, more photons hit the metal surface so more electrons are emitted.

Questions

$h = 6.6 \times 10^{-34}\,\mathrm{J\,s}$ $c = 3 \times 10^8\,\mathrm{m\,s^{-1}}$

1 a What is the frequency of red light of wavelength 700 nm?
  b How much energy is carried by a photon of this wavelength (i) in joules? (ii) in electronvolts?
  c How much energy is carried by each photon that leaves a radio aerial broadcasting on a frequency of 3.0 MHz?
  d What is the wavelength of a 15 keV X-ray?

2 A substance, when heated in a Bunsen flame, emits radiation of wavelengths 240 nm, 400 nm and 600 nm. Calculate the energy of a photon of each of these wavelengths, and hence sketch a diagram for the electron energy levels in an atom of this substance.

3 A laser produces a beam of light of wavelength 500 nm; the power of the laser is 1 mW and the light beam has a cross-sectional area of 5 mm^2.
  a What is the colour of the laser light?
  b What is the energy of each photon?
  c How many photons pass a given point in the beam per second?
  d How does the intensity of the laser beam compare with that of a 60 W light bulb when viewed at a distance of 1 m? Assume that 5% of the energy input to a light bulb is actually converted into light.

4 A 100 W sodium street lamp emits light of average wavelength 5×10^{-7} m. Estimate the number of photons per second that enter the eye of an observer at a distance of 100 m.

5 Newtonian mechanics led us to believe that matter could be neither created nor destroyed. However, Einstein's theory of relativity showed that there was an equivalence between mass and energy. According to this theory, if a quantity of matter, mass m, disappears then energy is created instead. The quantity of energy is $E = mc^2$, where c = speed of light, 3×10^8 m s^{-1}.
  Figure 11 shows a γ-ray entering a cloud chamber along AB (but leaving no track). At B the γ-ray disintegrates into an electron–positron pair. A magnetic field of intensity 0.05 T acts out of the plane of the paper. Initially the radius of curvature of both paths is 5 mm.

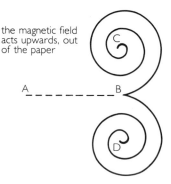

the magnetic field acts upwards, out of the paper

figure 11   Pair production

  a Why does the γ-ray leave no path in the cloud chamber?
  b Which is the path of the positron?

c Why do the paths spiral inwards?
d Estimate the initial kinetic energy of the electron and the positron.
   ($m_e = m_p = 9.1 \times 10^{-31}$ kg; $e = 1.6 \times 10^{-19}$ C)
e Estimate the wavelength of the incoming γ-ray, stating clearly any assumptions in the calculation. (Don't forget $E = mc^2$.)
f When a positron–electron pair is annihilated in a collision, why must two γ-rays be produced?

6 A gold-leaf electroscope was made with a freshly cleaned zinc plate on top of it, figure 12. The following experiments were carried out.

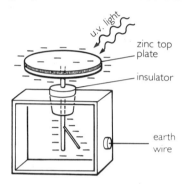

figure 12

(i)   The plate was negatively charged and illuminated by a weak ultraviolet light. The electroscope discharged in a few seconds.

(ii) The electroscope was recharged and the ultraviolet lamp was replaced by a very strong white light source (100 W bulb). The electroscope did not discharge.

(iii) The plate was positively charged and again the weak ultraviolet light was used to illuminate it. The electroscope did not discharge.

An explanation for the discharge of the electroscope is that electrons are knocked out of the zinc plate by the ultraviolet light.

a Explain why these experiments show that positive particles are not emitted from the electroscope.
b Light shows both diffraction and interference phenomena; this is evidence that light behaves as a wave. Explain carefully why wave theory cannot adequately account for the photoelectric effect. (If in doubt do the calculation in part (c)).
c Suppose the light from the ultraviolet source (in experiment (ii)) falls on the zinc plate with intensity 0.4 mW m^{-2}. It is reasonable to suppose that an electron can only absorb energy from an area of several atomic spacings. Use this assumption, together with the information that it requires 4 eV to remove an electron from zinc, to estimate the time for such an electron to absorb sufficient energy to

escape from the metal (diameter of zinc atoms = 2×10^{-10} m). Do we notice such a time delay?
d Now outline a theory that can account satisfactorily for the photoelectric effect.

7 In photoelectric experiments on sodium it was discovered that only light of a wavelength shorter than 5.4×10^{-7} m will cause electrons to be emitted.
a Will red light cause electrons to be emitted from sodium?
b What is the work function of sodium?
c What is the maximum kinetic energy that an electron could have if it were knocked out of the metal by a photon of wavelength
   (i) 2.7×10^{-7} m, (ii) 3.5×10^{-7} m?

8 Two metal plates X and Y are contained in an evacuated container, and are connected into a circuit as shown in figure 13. Graph A in figure 14 shows the current I through the microammeter as a function of the p.d. applied across XY when monochromatic light is allowed to fall on plate X.

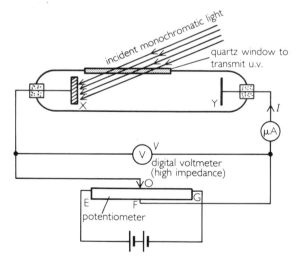

figure 13

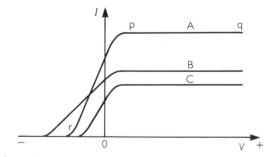

figure 14

a Where would the sliding contact O have to be to give the part of the graph pq, in EF or in FG? Explain your answer.
b Why does the current in graph A reach a steady value along pq, and become no bigger even if the p.d. is increased still further?
c Explain the decrease in current along the region pr of graph A.
d What changes have been made in the experiment to produce the graphs B and C?

9 You may well already have seen the demonstration in which a ping pong ball is held suspended in a fast-flowing stream of air. The air hitting the ball causes an upwards force which counterbalances the ball's weight. The ball stays in the stream of air because the pressure in a fast-flowing stream of air is less than that of stationary air. So if the ball moves to the edge of the stream it is pushed back into the moving air by the higher-pressure stationary air. See figure 15.

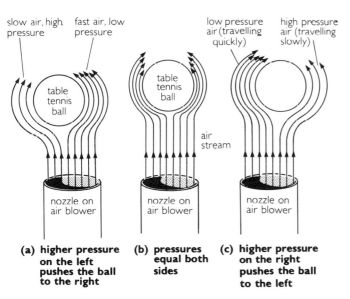

(a) **higher pressure on the left pushes the ball to the right**   (b) **pressures equal both sides**   (c) **higher pressure on the right pushes the ball to the left**

figure 15   Balancing ball demonstration

In a similar way, light from a powerful laser has been able to lift small glass spheres of radius 10^{-5} m. This question examines how the lift occurs, and why the balls remain stable in the laser beam.
a Figure 16 shows the passage of two light rays through a small glass sphere. A photon of light carries momentum. Draw a vector diagram to show the change of momentum of photons following the paths AA' and BB', and hence explain why a force is exerted on the sphere to lift it.

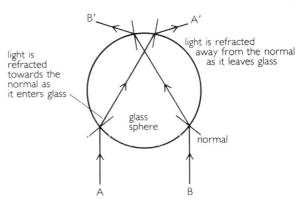

figure 16

b Now suppose that the sphere reached the edge of the beam so that light fell only on the left hand side and photons could follow path AA' but not BB'. Explain why in addition to a lifting force the sphere is pushed to the left.
c Could a small steel sphere be supported *with stability* in the same way? The light would be reflected off the surface.

CHAPTER 40

Electrons in atoms

Introduction

In the last chapter we learnt that light exists as packets of energy called photons. In this chapter we will see that the idea of photons, taken together with the experimental observation that only a few particular wavelengths of light are emitted from atoms, leads us to the conclusion that electrons can only exist in a limited number of energy levels.

GCSE knowledge

In this chapter we assume that you are already familiar with the following ideas.

☐ Electrons move around a small positively-charged nucleus.

☐ Electrostatic forces are responsible for keeping the electrons close to the nucleus.

☐ It is possible to remove electrons from atoms. This can be done by (i) rubbing a material, (ii) heating it, (iii) bombarding it with radiation (α-particles, for instance).

☐ When electrons are removed from atoms positive ions are produced.

☐ When electrons attach themselves to atoms negative ions are produced.

Emission spectra

Figure 1 shows a simple apparatus for investigating the spectrum emitted by an ordinary white light source. A collimated beam is allowed to fall onto a diffraction grating; the spectrum thus produced shows the familiar range of colours stretching continuously from red through to violet.

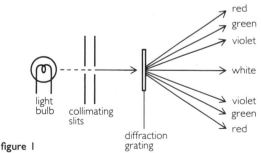

figure 1

If the white light source is replaced by a sodium vapour lamp a different picture emerges. Since the light emitted from the sodium lamp appears to be yellow, we might expect to see a continuous band of colour centred on the yellow part of the spectrum. Figure 2 illustrates what we actually do observe. Instead of a continuous band of colours, we see a few very well defined lines of light. (We see lines because the light has passed through a narrow slit.) The brightest line is yellow light of wavelength 590 nm.

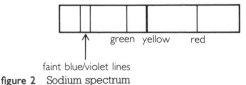

figure 2   Sodium spectrum

Bearing in mind the photoelectric effect, which showed us that light comes in packets of energy, the sodium spectrum shows us that the sodium atoms emit only photons of certain energies. The energy of a photon is:

$$E = hf = \frac{hc}{\lambda}$$

So for instance, the energy of a yellow photon is

$$E = \frac{6.6 \times 10^{-34} \times 3.0 \times 10^{8}}{590 \times 10^{-9}} \text{ J}$$

$$= 3.4 \times 10^{-19} \text{ J}$$

$$= 2.1 \text{ eV}$$

The theory that we put forward to explain our observations is that electrons in sodium may only occupy certain well-defined energy levels. This result is surprising. After all, it is perfectly possible for us to put a satellite into orbit around the Earth at any chosen height. Since the action of a gravitational field on a mass is in very many ways analogous to the action of an electric field on a charged particle, we would expect an electron to occupy any chosen orbit around a nucleus. But it appears that the electron is confined to certain specific orbits or energy levels. Such levels are repesented by the horizontal lines in figure 3.

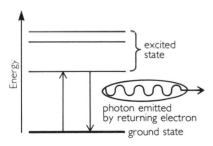

figure 3   Electronic energy levels in an atom

Normally, at room temperature the electrons in sodium occupy the lowest available energy levels. However, at the higher temperatures that the sodium atoms experience in a discharge tube, the outermost electron may receive enough thermal energy to lift it to one of the higher levels (or excited states). The electron does not stay in the higher level for very long, but falls down again to the lower level. The extra energy that the electron had in its excited state is now emitted in the form of a photon.

The hydrogen spectrum and energy levels

Figure 4 shows the appearance of the hydrogen spectrum in the visible region. Photographic plates also reveal a multitude of lines in the ultraviolet and infrared parts of the spectrum. The full spectrum consists of several series of lines. By working out the wavelength λ of a spectral line, we can also work out the corresponding energy level change ($E = hc/\lambda$), and thus deduce a possible scheme of energy levels for the atom. The calculated energy levels are displayed in figure 5, and the corresponding orbital changes are shown in figure 6.

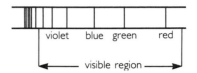

figure 4   Hydrogen spectrum

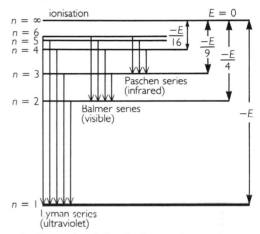

figure 5   Energy levels in a hydrogen atom

E is the ionisation energy of hydrogen, that is, the energy required to take an electron from the lowest energy level ($n = 1$) completely away from the nucleus. The second level lies $E/4$ and the third level $E/9$ below the level at which the electron becomes ionised; in general the nth level lies E/n^2 below the ionised level ($n = \infty$). Thus the higher energy levels lie closer together than the lower ones.

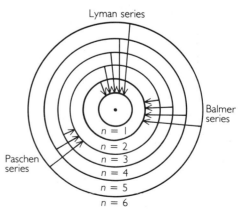

figure 6   Electron transitions in a hydrogen atom

It is important to remember that when the electron is removed, it is a long way from the nucleus and has no electrical potential energy with respect to the nucleus; so when the electron is attached to the nucleus it has a negative quantity of energy. From figure 5 we can see that in general for a transition from level m down to level n, the electron will lose energy equal to $E[(1/n^2) - (1/m^2)]$, which will be the energy of the emitted photon. For hydrogen the ionisation energy E is 13.6 eV. As an example, we use this to calculate the wavelength of the photon emitted in the transition from level 4 to level 2 (Balmer series):

$$hf = E\left(\frac{1}{2^2} - \frac{1}{4^2}\right)$$

$$\Rightarrow \quad \frac{hc}{\lambda} = E \times \frac{3}{16}$$

$$\Rightarrow \quad \lambda = \frac{16hc}{3E}$$

$$= \frac{16 \times 6.6 \times 10^{-34} \times 3.0 \times 10^8}{3 \times 13.6 \times 1.6 \times 10^{-19}} \text{ m}$$

$$= 4.9 \times 10^{-7} \text{ m}$$

This transition will appear as a green line in the visible part of the spectrum.

X-ray spectra

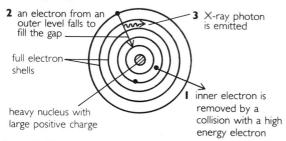

2 an electron from an outer level falls to fill the gap

full electron shells

heavy nucleus with large positive charge

3 X-ray photon is emitted

1 inner electron is removed by a collision with a high energy electron

figure 7 How an X-ray spectrum is produced

A heavy atom has a nucleus with a large positive charge surrounded by several full electron shells. Because of the large charge on the nucleus the electrons in the innermost shells are very strongly bound to the nucleus; the energy required to move such a low-lying electron completely away from the nucleus can be 10keV or more, as opposed to 13.6 eV for hydrogen. Electrons can be removed from the inner shells of heavy atoms by bombarding a target with a stream of very high energy electrons. When an electron is removed from a low-lying level, it is replaced by an electron falling from one of the higher energy levels. As the electron falls from the higher level its extra energy is emitted in the form of a photon, but as the energy change is large the wavelength of the photon is very small, and is in the X-ray region of the electromagnetic spectrum ($\lambda \approx 10^{-10}$ m). The principle of an X-ray tube is shown in figure 8.

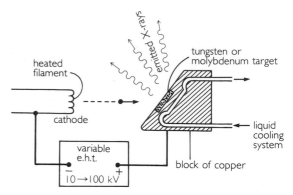

heated filament

cathode

variable e.h.t.

$\overline{10 \rightarrow 100 \text{ kV}}$

emitted X-rays

tungsten or molybdenum target

liquid cooling system

block of copper

figure 8 X-ray tube

A typical X-ray spectrum is illustrated in figure 9. The two sharp lines are caused by electrons being removed from low-lying levels as described above. In addition, there is a continuous background of X-rays called the *bremsstrahlung*, which means 'radiation from slowing-down processes'. This results from electrons striking the target and suffering repeated small energy losses by interacting with atoms. Each small loss of the electron's kinetic energy is radiated as a single quantum of X-radiation. Because any energy loss is possible there is a continuous bremsstrahlung spectrum. The X-ray spectrum does however have a minimum possible wavelength, which will be emitted when an incident electron loses all its energy in a single collision. If V is the p.d. between cathode and anode, λ_{min} may be calculated from the equation

$$eV = \frac{hc}{\lambda_{min}}$$

$$\Rightarrow \quad \lambda_{min} = \frac{hc}{eV}$$

$$= 6.2 \times 10^{-11} \text{ m for } V = 20\,000 \text{ V}$$

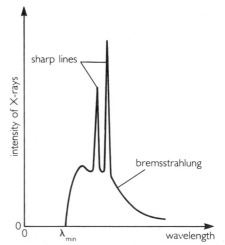

intensity of X-rays

sharp lines

bremsstrahlung

λ_{min}

wavelength

figure 9 Typical X-ray spectrum

Collisions between electrons and atoms

Normally when a low-energy electron collides with an atom the collision is elastic, and the electron bounces off the atom with virtually no loss of energy. This is similar to throwing a ping-pong ball at a billiard ball; the very much lighter ping-pong ball bounces back with its original energy, and the massive billiard ball hardly recoils at all.

Figure 10 shows an apparatus in which it is possible to detect inelastic collisions between electrons and atoms.

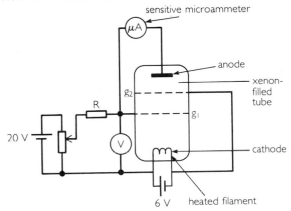

figure 10

A glass tube is filled with xenon under low pressure. Electrons are accelerated by a potential difference V between a grid, g_1, and the cathode. A second grid, g_2, is at the same potential as the cathode and therefore slows down the electrons, but provided they only collide elastically in the region betwen g_1 and g_2 they can still reach the anode. As the p.d. between g_1 and cathode rises, g_1 attracts a larger fraction of the electrons emitted by the cathode, so the anode current gradually rises (figure 11).

However, at an accelerating p.d. of 8 V the anode current falls, suggesting that the number of electrons passing g_2 is falling. This indicates that some electrons are now losing energy in collisions in the region between g_1 and g_2. The explanation for this is

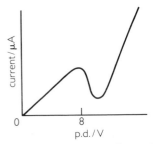

figure 11   Graph of anode current against p.d. between grid and cathode

that xenon has an energy level which lies about 8 eV above the ground state energy level. Until the colliding electrons have acquired 8 eV of kinetic energy it is not possible to excite one of the electrons in xenon to this higher level. This experiment lends support to the idea of electrons being fixed in certain energy levels, because it is evident that there is a gap of 8 eV between the ground state and the first excited energy level in xenon.

Bohr theory of the hydrogen atom

Following the ideas outlined in the previous sections, Bohr made the following postulates about the atom.

☐ Each electron in an atom revolves around a small nucleus in a fixed circular orbit.

☐ An electron in an orbit has a definite amount of energy, so since only certain orbits are permitted there must also be only certain allowed energy levels.

☐ When an electron is in an orbit it does not emit any energy, but an electron can jump from one level to another. When the electron falls to a lower level it emits energy $E = hf$.

So far these statements have introduced nothing that was not previously described in this chapter. However, Bohr made one further postulate which was that the angular momentum of the electron had to be an integral multiple of $h/2\pi$. Expressed mathematically that is:

$$mvr = n\frac{h}{2\pi} \qquad \text{(i)}$$

The arguments that Bohr used to justify this equation are very complicated, so we will take it for granted. Now we know that the electron moves in a circular orbit around the proton, so the electrostatic force of attraction between the two particles must provide the centripetal force necessary to keep the electron in its circular path. Thus we may write:

$$\frac{mv^2}{r} = \frac{e^2}{4\pi\varepsilon_0 r^2} \qquad \text{(ii)}$$

From equation (ii) it follows that:

$$v^2 = \frac{e^2}{4\pi\varepsilon_0 mr} \qquad \text{(iii)}$$

but from equation (i),

$$mr = \frac{nh}{2\pi v} \qquad \text{(iv)}$$

Substitution of equation (iv) into equation (iii) yields the result:

$$v = \frac{e^2}{2\varepsilon_0 nh} \qquad \text{(v)}$$

Next we consider the energy of an electron in orbit. Its energy is made up of two parts, potential and kinetic. Thus the electron's energy E is:

$$E = -\frac{e^2}{4\pi\varepsilon_0 r} + \tfrac{1}{2}mv^2 \qquad \text{(vi)}$$

But from equation (ii) we see that $mv^2 = e^2/4\pi\varepsilon_0 r$ so the total energy of the electron is:

$$E = -mv^2 + \tfrac{1}{2}mv^2 = -\tfrac{1}{2}mv^2 \qquad \text{(vii)}$$

Finally, using the result of equation (v), the energy of the nth level in the Bohr atom is:

$$E = -\frac{me^4}{8\varepsilon_0^2 n^2 h^2} \qquad \text{(viii)}$$

Thus the Bohr model of the atom has predicted the $1/n^2$ rule for the hydrogen levels. We may also use equation (viii) to predict the hydrogen atom's ionisation energy which will be:

$$\text{ionisation energy} = \frac{me^4}{8\varepsilon_0^2 h^2}$$

$$= \frac{(9.1 \times 10^{-31}) \times (1.6 \times 10^{-19})^4}{8 \times (8.85 \times 10^{-12})^2 (6.6 \times 10^{-34})^2} \, \text{J}$$

$$= 2.2 \times 10^{-18} \, \text{J}$$

$$= 14 \, \text{eV}$$

The agreement between experiment and theory is impressive. We can now also use the theory to predict the radius of the hydrogen atom. From equation (iv) and equation (v) we can show that

$$r = \frac{n^2 h^2 \varepsilon_0}{\pi m e^2} \qquad \text{(ix)}$$

So for the lowest level $n = 1$

$$r = \frac{h^2 \varepsilon_0}{\pi m e^2} \simeq 5 \times 10^{-11} \, \text{m}.$$

Despite the obvious successes of the Bohr model, it is limited in its application since it only works for a single electron in orbit around a nucleus. It fails for the helium atom. The Bohr theory breaks down because it neglects one very important aspect of electron behaviour: electrons behave not only as particles but as waves, a feature discussed in the next chapter.

Summary

1 The photoelectric effect (chapter 39) showed us that light exists as photons, each of which has a specific energy E given by $E = hf$, where h is the Planck constant and f the frequency of the light.

2 When vapours are heated light is emitted. However, the light emitted corresponds to a small number of specific wavelengths. Since each wavelength seen is carried by photons of a specific energy, this means that atoms only lose energy in certain very specific amounts. We explain this by the theory that electrons exist in only a few energy levels. When an atom is heated its electrons can gain energy and move into higher levels. When an electron falls back to a lower level the extra energy that it had in the higher level is carried away by a photon.

3 The theory of electron energy levels is given further confirmation by electron collision experiments and X-ray spectra.

4 X-ray spectra:
   a continuous due to bremsstrahlung.
   b line due to level changes. See figure 9.

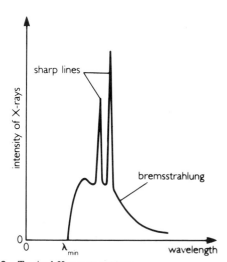

figure 9   Typical X-ray spectrum

5 Bohr theory of the hydrogen atom:
   a The energy of the nth level is

$$E = -\frac{me^4}{8\varepsilon_0^2 n^2 h^2} \qquad \text{(viii)}$$

   b The radius of the hydrogen atom:

$$r = \frac{n^2 h^2 \varepsilon_0}{\pi m e^2} \qquad \text{(ix)}$$

Questions

1 Figure 12(a) shows an experiment similar to that of Hertz and Franck, in which electrons may be accelerated from a hot filament by the p.d. across GF. Helium under low pressure is contained inside the valve. Figure 12(b) shows how the current detected by the microammeter varies with the accelerating p.d. across GF.

(a)

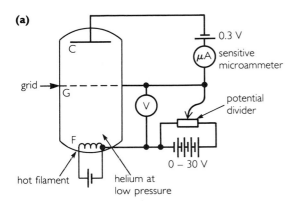

(b)

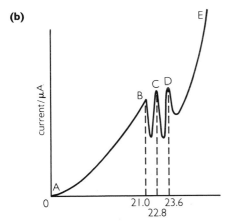

figure 12

a Explain why, as the p.d. V_{GF} increases, the current rises steadily on the region AB of the graph.
b Explain why, in an elastic collision between a small object like an electron and a much larger object like an atom, the electron loses very little kinetic energy. Hence explain why, despite many collisions with helium atoms, electrons arrive at C and then pass through the ammeter.
c If electrons were to undergo inelastic collisions in the region FG, what would happen to the current measured by the microammeter? By what mechanism could such an inelastic collision occur?

d What is the significance of the peaks B, C and D in figure 12(b)?
e What differences would you see in figure 12(b) if the p.d. V_{GC} were increased from 0.3 V to 2 V? Explain your answer carefully.
f What has happened to the gas at point E on the graph?
g Explain what you would expect to observe if the helium were replaced by sodium vapour and the experiment were repeated.
h Calculate the wavelength of one spectral line, in the *visible* region of the electromagnetic spectrum, which you think should be emitted by hot helium.

2 Figure 13 shows the energy levels in a sodium atom. The arrows show the transitions that are most likely to occur, although others are possible.
a What is the minimum energy required to ionise singly a sodium atom in its lowest energy state?
b The wavelengths of the sodium D-lines are 589.6 nm and 589.0 nm respectively. These transitions are the most common, and therefore give the strongest spectral lines. What colour do you expect a sodium discharge tube to be?
c Use the diagram to determine the wavelength of light emitted for the transitions 4s→3p; 3d→3p; 4f→3d.
d Which transition gives rise to a spectral line of wavelength 620 nm?

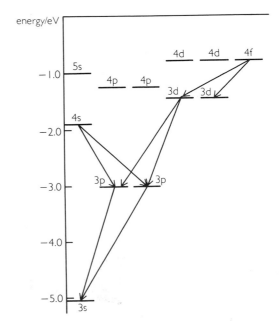

figure 13

3   The apparatus shown in figure 14 was set up to
examine the spectrum of hydrogen in the
ultraviolet. The lens forms an image of the slit in
the plane of the photographic paper, and the
diffraction grating splits up the spectrum. The
diffraction grating has 3000 lines mm^{-1}.

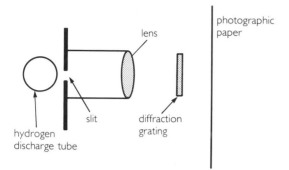

figure 14

a   When the film is developed, four lines are
discovered in the ultraviolet end of the
spectrum. They have been diffracted through
angles of 16.7°, 17.1°, 18.0°, 21.5°. Work out the
wavelength of each line.
b   Work out the energy changes in each case.
Assuming that each electron transition involved
is from an excited state to the ground state,
draw a diagram to show the energy levels of
hydrogen.
c   Calculate two other possible energy changes
that could occur, one in the visible part and one
in the infrared part of the electromagnetic
spectrum.
d   Would your lines chosen in part (c) appear on
the photographic paper? If so, through what
angles will the radiations have been diffracted?

4   Figure 15(a) shows the essential features of an
X-ray beam tube. Electrons are removed from a
filament by thermionic emission, and are then
accelerated through a potential difference of
12 kV before hitting a metal target. Figure 15(b)
shows a typical X-ray spectrum.

(a)

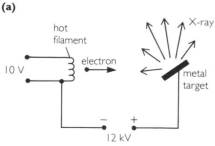

figure 15(a)

(b)

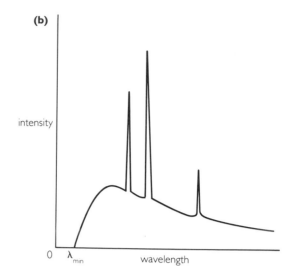

figure 15(b)

a   Figure 16 shows a high-energy electron striking
a target atom and removing an electron in a low
energy level. An outer electron then falls to the
lower energy level, thus filling the gap created.
Explain how this process accounts for the sharp
lines seen in the X-ray spectrum.
b   Account for the background continuous X-ray
spectrum, and explain why the X-rays cannot
have less than a particular wavelength λ_{min}.
c   For this particular X-ray tube with the
accelerating voltage set at 12 kV, λ_{min} is found
to be 1.0×10^{-10} m. Given that $e = 1.6 \times 10^{-19}$ C
and that $c = 3 \times 10^8$ ms^{-1}, make an estimate of
the Planck constant.

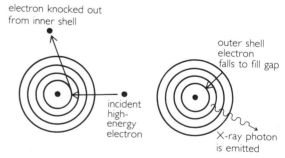

figure 16

5   Figure 17 shows an experiment to observe the
absorption of light by sodium. A Bunsen flame is
impregnated with sodium vapour using brine-
soaked chalk, and a white light source is viewed
through a diffraction grating after the light has
passed through the sodium flame.

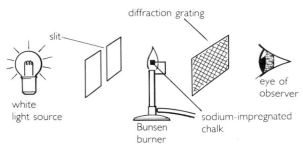

figure 17

a Figure 18 shows the observed spectrum, with a dark line in the yellow part of the spectrum. Give an explanation of the presence of this line.

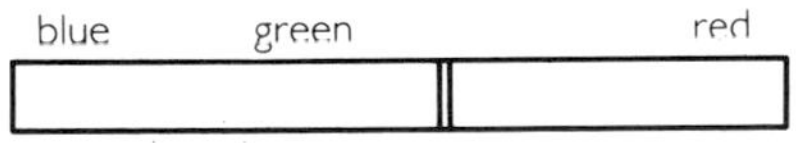

blue	green	red

figure 18

b In the experiment described above, the white light source is now replaced by a sodium lamp. What would be the effect on the intensity of sodium light seen by the observer, if the sodium flame were now removed from in front of the slit? Explain.

6 An early model of the hydrogen atom consisted of a sphere with positive charge uniformly distributed throughout it. An electron was free to move inside the sphere, acted on only by electrostatic forces.

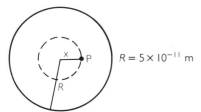

figure 19

a At a point P, the electron is only acted on effectively by the positive charge contained within the radius x. Explain briefly why this is so.

b Calculate the net force acting on the electron at P, and hence show that, provided the electron stays within the charged sphere, it executes simple harmonic motion. Calculate the frequency of this motion.

c Calculate the energy required to move the electron away from the centre of the sphere to infinity. Comment on your answer; how does it compare with the experimentally measured ionisation energy of 13.6 eV?

d Discuss briefly why physicists no longer believe this model to be the correct representation of the atom.

7 Write an essay to explain why physicists believe that electrons exist in discrete energy levels. Your answer should include accounts of any relevant experimental evidence.

CHAPTER 41

Electrons – waves or particles?

Introduction

Up to now you have probably thought of an electron as a tiny particle with mass 9.1×10^{-31} kg and charge 1.6×10^{-19} C. You are now about to learn that an electron can also behave as a wave. As a result of this discovery we will be able to understand from a theoretical point of view why electrons can only exist, inside atoms, in a few well-defined energy levels.

GCSE knowledge

The background for this chapter is not GCSE but work covered in the earlier chapters of this book. In particular chapter 21 (Electrical potential), chapter 8 (Resonance and standing waves) and chapter 31 (Interference).

Electron waves

Figure 1(a) illustrates an apparatus that can be used to demonstrate the wave properties of electrons. An electron gun is used to accelerate electrons so that they pass through a thin graphite target. Then the electrons move on to strike a fluorescent screen.

The screen reveals a series of concentric rings, thus showing that the electrons only strike the screen in certain positions: figure 1(b). This phenomenon is similar to the one of photons passing through a double slit and being diffracted to certain allowed positions. We use a wave theory to explain how

photons behave when passing through a slit, and so it would now be appropriate to use a wave theory to describe the behaviour of electrons as they pass through the layers of atoms in the graphite target. In the past we might have treated electrons like particles, but now they are behaving as waves since we see bright and dark rings on a screen corresponding to places of constructive and destructive interference.

When a narrow beam of monochromatic light passes through a diffraction grating, we see a series

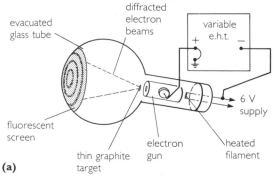

(a)

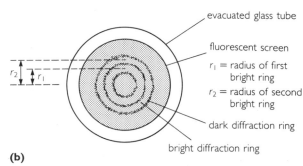

(b)

figure 1   (a) Electron diffraction tube; (b) front view of the fluorescent screen showing bright and dark rings of dots, corresponding to those angles for which the path difference between adjacent slits is a whole number of wavelengths. If, however, we make the light pass through lots of gratings which have their slits lying along a large number of different directions, then we see a series of rings. Graphite consists of many different crystals oriented randomly, and so we see these electron diffraction rings in the experiment described above.

When we carry out further experiments, we find that the diameter of the rings gets smaller if the accelerating voltage is increased. To be precise, the radius of the first ring is inversely proportional to the square root of the p.d., or

$$r \propto \frac{1}{\sqrt{V}} \qquad \text{(i)}$$

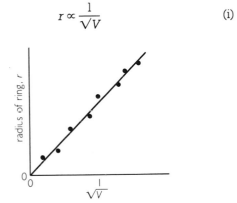

figure 2 Graph showing how the radius of the rings varies with the accelerating p.d.

If the separation of the planes of atoms responsible for the diffraction is d, we may write $\lambda = d \sin\theta$, since the planes are acting like a diffraction grating. However, $\sin\theta \approx r/D$ (see figure 3), so

$$\lambda = \frac{dr}{D} \qquad \text{(ii)}$$

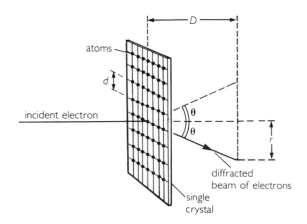

figure 3 Diffraction of electrons by a single crystal

Therefore, the wavelength of the electrons, λ, is proportional to the radius of the diffraction ring, r. The electrons have been accelerated through a p.d. V. So, applying the principle that the electron's potential energy eV is converted into kinetic energy, $\frac{1}{2}mv^2$, we may write:

$$eV = \tfrac{1}{2}mv^2$$
$$\Rightarrow \qquad V = \frac{m^2v^2}{2me} = \frac{p^2}{2me} \qquad \text{(iii)}$$

where p is the momentum of the electron.

Now we can link the wavelength and momentum of the electron. From equation (iii) we see that:

$$p \propto \sqrt{V} \qquad \text{(iv)}$$

But from expressions (ii) and (i), $\lambda \propto r$ and $r \propto 1/\sqrt{V}$, so it follows that

$$\lambda \propto \frac{1}{\sqrt{V}} \text{ or } \frac{1}{\lambda} \propto \sqrt{V} \qquad \text{(v)}$$

From (iv) and (v) we now get the important result that the wavelength of the electron λ is inversely proportional to the momentum, p, or

$$\lambda \propto \frac{1}{p}$$

Note that in this argument we treated the electron first as a wave, then as a particle. So, like photons, electrons can show both wave and particle properties.

Finally, we would like to know the constant of proportionality in the equation $\lambda = \text{constant}/p$. This can be determined by measurement of a pair of values for λ and p. The following data is taken from a typical electron diffraction experiment using a graphite target: the separation of the layers d responsible for a diffraction ring of radius 17 mm is 1.7×10^{-10} m (this was determined in a separate experiment using X-ray diffraction techniques). The distance D from the target to the screen is 13 cm. The electrons were accelerated through a potential of 3000 V.

First we calculate the momentum:

$$eV = \tfrac{1}{2}mv^2 = \frac{p^2}{2m}$$
$$\Rightarrow p = \sqrt{2meV}$$
$$= \sqrt{2 \times 9.1 \times 10^{-31} \times 1.6 \times 10^{-19} \times 3000} \text{ kg m s}^{-1}$$
$$= 3.0 \times 10^{-23} \text{ kg m s}^{-1}$$

The wavelength is given by:

$$\lambda = \frac{dr}{D}$$
$$= \frac{1.7 \times 10^{-10} \times 0.017}{0.13} \text{ m}$$
$$= 2.2 \times 10^{-11} \text{ m}$$

Thus the constant of proportionality is given by:

$$\text{constant} = \lambda p = 6.6 \times 10^{-34}\,\text{Js},$$

which is the Planck constant.

So we now have the De Broglie equation, named after the man who first proposed it:

$$\lambda = \frac{h}{p}$$

The De Broglie equation does not just refer to electrons, but to all particles. However, for a particle to show its wave properties, its wavelength must be of comparable size to the aperture through which it is passing. Thus electrons of wavelength 2×10^{-11} m showed diffraction and interference effects when passing through an aperture about ten times bigger. On the other hand, we would not expect to observe wave properties for an α-particle travelling at $10^7\,\text{ms}^{-1}$; its wavelength is

$$\lambda = \frac{h}{p}$$
$$= \frac{6.6 \times 10^{-34}}{6.8 \times 10^{-27} \times 10^7}\,\text{m}$$
$$\approx 10^{-14}\,\text{m}$$

Heavier particles (atoms for example) will only reveal their wave-like nature if they are travelling very slowly. The reason that we do not observe diffraction as we pass through a doorway is that we are very massive particles, so our momentum is large and our wavelength very small indeed (about 10^{-36} m).

Electron standing waves: energy levels

Figure 4 shows two possible forms of standing waves on a string, the upper one being the fundamental mode and the lower one the second harmonic. The standing wave patterns occur only when an integral number of half wavelengths can fit into the string.

(a)

(b)

figure 4   Standing waves on a string

We know that electrons behave like waves; further, we know that electrons are confined to a limited space by the attractive forces of the nucleus. Taking these two ideas together suggests that electrons will behave like standing waves when they are 'trapped inside an atom'. Exactly what an electron standing wave looks like is impossible to say; but it would be

reasonable to suggest that an electron in its lowest level has one 'loop', and higher levels have successively more 'loops'. Here the idea of standing waves in musical instruments is helpful. If an organ pipe is blown gently we only hear the fundamental note, i.e. one half wavelength occupies the length of the pipe. If on the other hand, we blow very strongly, there is enough energy to cause overtones to sound. Overtones are notes with a higher frequency, and thus shorter wavelengths. There might now be three half wavelengths to the length of the pipe – analogous to our three 'loops' in the higher electron energy level.

Let us now adopt a very simple view of a hydrogen atom, and assume that an electron is confined to a spherical box of diameter x, and that regardless of the amount of energy that it has got the atom does not change its size. Figure 5 illustrates the first three standing waves that will fit across the diameter of the spherical box.

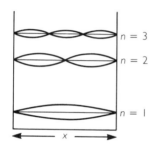

figure 5   Standing waves in a box

Using the de Broglie equation we know that the momentum of an electron is $p = h/\lambda$. Thus the kinetic energy of an electron is

$$E = \tfrac{1}{2}mv^2 = \frac{p^2}{2m} = \frac{h^2}{2m\lambda^2}$$

The potential energy of the electron is fixed, because the diameter of the atom is constant. Only certain wavelengths will fit into the atom, and so only certain values of kinetic energy are allowed, because the kinetic energy is governed by the wavelength of the electron. Thus a simple wave model of electrons predicts that electrons in atoms will be confined to a certain finite number of levels.

Before we leave this topic, we can at least check whether our wave model predicts a reasonable size for a hydrogen atom. Available experimental data suggests a radius of about 10^{-10} m for the hydrogen atom. We will adopt this radius and see whether it allows a viable atom to exist.

The total energy of an electron bound in an atom is negative. This energy is made up of two parts, kinetic and potential. The kinetic energy is inevitably positive; but the potential energy is negative, because work is done to take the electron away from the nucleus; and so the atom can only exist if the potential energy is numerically bigger than (though of the opposite sign to) the kinetic energy.

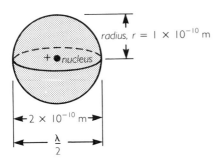

radius, $r = 1 \times 10^{-10}$ m

2×10^{-10} m

$\dfrac{\lambda}{2}$

figure 6   An atom

Figure 6 represents our atom. The average potential energy of the electron V is:

$$V = \frac{-e^2}{4\pi\varepsilon_0 r}$$

$$= -2.3 \times 10^{-18}\,\text{J}$$

$$= -14.4\,\text{eV}$$

Half a wave fits into the diameter, therefore the wavelength of the electron is 4×10^{-10} m. We can calculate the kinetic energy, E_k, from:

$$E_k = \frac{p^2}{2m} = \frac{h^2}{2m\lambda^2}$$

$$= 1.5 \times 10^{-18}\,\text{J}$$

$$= 9.3\,\text{eV}$$

The net energy of the electron is thus -5.1 eV; therefore the atom *can* exist. In fact the energy will be a little more negative than this. The electron, being a wave, can occupy any position within the atom (even inside the nucleus), and on average it will be closer to the nucleus than 10^{-10} m, and so its average potential energy will be more negative than we have allowed for.

We will now consider what happens if we choose a different size of atom. If the radius were much bigger than 10^{-10} m, the electron would have little kinetic energy and so would tend to fall back towards the nucleus. Now let us see what happens if we try to make the atom smaller. Suppose now that the radius is about 5×10^{-11} m. Since the potential energy of the electron is inversely proportional to the distance from the nucleus, if r is halved V is doubled. So in our new atom, $E_p = -28.8$ eV. However, the kinetic energy of the electron is inversely proportional to λ^2, so if r is halved, E_k is quadrupled, and the new kinetic energy would be $+37.2$ eV. The electron's kinetic energy now outweighs its potential energy, so it would escape from the atom. Thus with these arguments our model of electron waves has enabled us to predict a value for the size of a hydrogen atom which is consistent with the experimentally determined value.

Summary

1 Electrons when fired through layers of regularly arrayed atoms show diffraction effects. This experimental result leads us to believe that electrons show wave properties as well as particle ones.

2 The wavelength (λ) and momentum (p) of the electron are linked by the equation

$$\lambda = \frac{h}{p} \quad (h \text{ is the Planck constant.})$$

The wavelength, λ, is sometimes referred to as the *de Broglie* wavelength.

3 When electrons are confined to small volumes (such as atoms) they behave as standing waves. The result of this is that only certain wavelengths of electron can fit into an atom. Since the energy of an electron depends on its wavelength, it follows that only certain energy levels can exist in an atom. This line of reasoning, stemming from the idea of electron waves, provides a theoretical answer to why electrons exist in discrete energy levels in atoms. We have already seen the experimental evidence for the existence of these levels, but we can now begin to see why these levels exist, from a theoretical point of view.

Questions

Planck's constant, $h = 6.6 \times 10^{-34}$ Js;
mass of proton, $m_p = 1.7 \times 10^{-27}$ kg;
charge on an electron, $e = 1.6 \times 10^{-19}$ C;
rest mass of an electron, $m_e = 9.1 \times 10^{-31}$ kg;
speed of light, $c = 3 \times 10^8$ ms^{-1}.

1 Calculate the wavelength of an electron with a velocity of 10^7 ms^{-1}. Would you expect such an electron to be diffracted considerably when made to pass through a crystal with atomic spacing 3×10^{-10} m? Would the electron be more likely to show diffraction if it were made to travel faster?

2 What is the wavelength of a 5 MeV α-particle?

3 Interference experiments show that light is a wave, and yet there is a theory that light is a particle. Experiments with cathode ray tubes show that electrons are particles, yet there is a theory that they can also be waves.

 Faced with such a paragraph, a student who was just beginning an A-level course would be confused. Write an essay to explain the concept of *wave-particle duality* to such a student. In your discussion you should describe briefly any experiments which support the ideas in the paragraph above.

4 Figure 7 shows a graphical representation of the experimental results obtained when some very energetic electrons are scattered by the nuclei of $^7_3$Li. The electrons travel close to the speed of light and have a momentum of 3.3×10^{-19} kgms^{-1}.

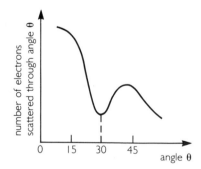

figure 7

 a Calculate the wavelength of one of the electrons using the equation $p = h/\lambda$.
 b The scattering of the electrons is very similar to the scattering of light waves by small particles. This allows us to use the equation $\sin\theta = 1.22\lambda/d$ to calculate the size of the nucleus. Here θ is the angle at which a scattering minimum occurs, λ is the de Broglie wavelength of the electrons and d is the nuclear diameter. Hence calculate the nuclear radius of $^7_3$Li.

 c A rough guide to the radius of a nucleus is given by the formula $r = 1.2 \times 10^{-15}A^{1/3}$ m, where A is the nucleon number. What radius does this formula predict for $^7_3$Li? How does it compare with your previous value?

5 Figures 8(a) and (b) show two possible standing wave modes of vibration for a string.

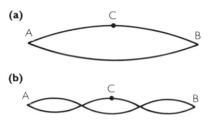

figure 8

 a Explain what conditions must exist for standing waves to arise, and sketch two other possible modes for this string.
 b How does the frequency of the waves in figure 8(b) compare with that of the waves in figure 8(a).
 c If the waves both have the same amplitude; use your answer to (b) to calculate the ratio of the average speeds of the string at C for the two cases.
 d Hence calculate the ratio of the kinetic energies of the string in the two cases.
 e Your answer to (d) will have shown that the string is in a higher energy state in figure 8(b). Now explain why if a flute is blown gently a low note may be produced, and a high note can be produced by blowing hard (without changing the position of your fingers on the flute).
 f What relevance has this question to atoms?

6 When a proton and an electron are separated by an infinite distance, their electrical potential energy, V, is zero. When separated by a distance r, $V = -e^2/4\pi\varepsilon_0 r$.

figure 9

 a Explain how V can be a negative quantity.
 b Explain why the electron's kinetic energy, E_k, is always a positive quantity.

c (i) If the electron is free (i.e. not attached to an atom), explain why $(E_k + V)$ is positive.
(ii) If the electron is attached to an atom, explain why $(E_k + V)$ is negative.

We will now use energy arguments to calculate roughly the size of a hydrogen atom. In our simple model of the atom we will assume that the electron is confined to a volume of radius 10^{-10} m.

d Calculate V for an electron at a distance of 10^{-10} m from the nucleus.

e Assuming that the trapped electron behaves as a standing wave as shown in figure 10, calculate its de Broglie wavelength.

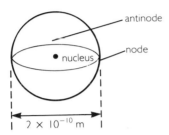

figure 10 An atom

f Calculate the electron's kinetic energy.
g Use your answers to (d) and (f) to explain why a stable hydrogen atom of radius 10^{-10} m may be formed.
h Suppose that the atomic radius were reduced now to 5×10^{-11} m. Calculate the new values for V and E_k.
i Explain why it is not possible to have a hydrogen atom as small as the one in part (h).
j Is it possible to have a hydrogen atom with a radius greater than 10^{-10} m? Explain.

7 Figure 11 shows a simple attempt to calculate energy levels in a hydrogen atom. The electron is trapped in a box with inflexible sides, but may occupy a variety of energy levels; if the electron acquires enough energy it may eventually escape out of the top of the box, leaving the atom ionised.

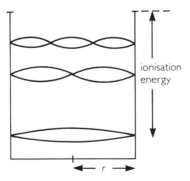

ionisation energy

r = radius of atom

figure 11

a What is the de Broglie wavelength for the electron in its lowest energy level?
b What is the momentum of the electron in the lowest level?
c Hence show that the kinetic energy of the electron in its lowest level is

$$E = \frac{h^2}{32mr^2}$$

d Deduce expressions for the kinetic energy of the electron in its second and third levels. Show that your answers are consistent with the formula $E = n^2h^2/32mr^2$ where n is the 1st, 2nd, 3rd etc. energy level.

e Does this scheme of energy levels fit in with that observed experimentally?

f If an electron were to fall from level n_2 to level n_1, a photon of energy $E = (n_2^2 - n_1^2)h^2/32mr^2$ would be emitted. Using the formula $E = hc/\lambda$, calculate the wavelength of such a photon for an electronic transition from the second to the first level. Take the radius of the atom to be 10^{-10} m. To what part of the electromagnetic spectrum does this photon belong? Comment on your answer.

8 Figure 12 shows a slightly more sophisticated model of the hydrogen atom than the one discussed in the previous question. Here we take some account of the fact that the electron's potential energy V varies with its distance from the nucleus. The diagram also shows that the radius of the atom increases as the electron moves into a higher energy level.

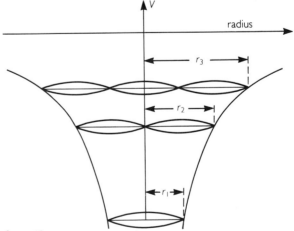

figure 12

Figure 13 shows that the potential energy V of the electron is proportional to $1/r$, where r is the radius of the atom. Suppose that the maximum distance that the electron can go from the nucleus is $2r$; then at this distance the electron's kinetic energy E_k is zero. If the maximum distance that the electron goes from the nucleus is $2r$, then a rough approximation of its average distance from the nucleus is r. Let the total energy of the electron be E, and the average potential and kinetic energies be $-V$ and E_k respectively. Then

$$E = -V + E_k \qquad (i)$$

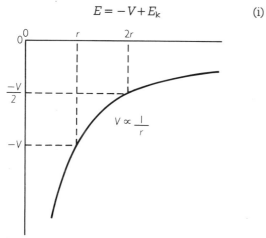

figure 13

a Explain why $E = -V/2$ (ii) and $E_k = +V/2$ (iii) and $E = -E_k$ (iv).

b Refer to figure 12 and show that the formula $n(\lambda/2) = 2r$ relates the de Broglie wavelength of the electron to the radius of the atom for the first three energy levels.

c Use the equation $p = h/\lambda$ together with $n(\lambda/2) = 2r$ to show that the kinetic energy of an electron in the nth energy level is:

$$E_k = \frac{n^2 h^2}{32 m r^2}$$

d Using equations (ii) and (iv), show that the total energy E satisfies the following:

$$E \propto \frac{n^2}{r^2}$$

$$E \propto \frac{1}{r}$$

e Thus explain why $E \propto 1/n^2$. Comment on this result in the light of the experimental data from the hydrogen spectrum.

f Comment on any simplifying assumptions that are made in this model.

9 This question is about the design of an experiment to produce a finely collimated beam of silver atoms. The atoms emerge from a furnace at 1500 K; the beam then travels to a detector placed 10 m away from the collimating slit (figure 14).

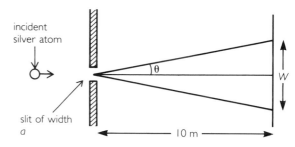

figure 14

a The kinetic energy of the emerging atoms is of the order of kT, where k is the Boltzmann constant, and T the Kelvin temperature. Hence show that the de Broglie wavelength of the atoms is about $\lambda = h/(2mkT)^{1/2}$; and that $\lambda \approx 8 \times 10^{-11}$ m. ($k = 1.4 \times 10^{-23}$ J K^{-1}; mass of a silver atom $= 1.8 \times 10^{-25}$ kg).

b If $a \gg \lambda$, what is the size of W?

c If $\lambda \gg a$, what is the approximate size of W?

d Combine your results for (b) and (c) to show that in general, for small θ, $W = a + 10m \times (2\lambda/a)$.

e Show that the smallest value of W occurs when $10m \times 2\lambda = a^2$. Hence show that the smallest width W of the beam of atoms is about 8×10^{-5} m.

SECTION H

HEAT

Energy can be converted from one form to another...

. . . but it is always a one-way process

42   The kinetic theory of gases   437

43   Thermodynamics   449

44   Heat transfer   468

CHAPTER 42

The kinetic theory of gases

Introduction

We start with a brief summary of the experimental background to the gas laws, which are the experimental results that the kinetic theory has to explain. We follow this with a very short historical background to the concept of *molecules* and the rules that they are assumed to obey. Then we continue with the elementary theory of gases based on billiard-ball type molecules. These ideas are then developed into the gas laws (i.e. Boyle's law, Charles' law, etc.). The molecular interpretation of temperature follows. We finish off by considering the limitations of the kinetic theory.

GCSE knowledge

It will be assumed that you are able to recall:

☐ that we consider matter to consist of molecules;

☐ the particle model of gases, where the particles are assumed to be minute molecules which are similar to billiard balls or steel ball-bearings;

☐ the evidence for the model, which assumes there are such things as gas molecules (Brownian motion and bromine diffusion);

☐ the gas laws, which hold for a fixed mass of ideal gas. Representing pressure, volume and temperature in kelvin by p, V and T respectively, they are:
  (i)   Boyle's law: pV = constant if T = constant;
  (ii) Charles' law: V/T = constant if p = constant;
  (iii) Pressure law: p/T = constant if V = constant;

☐ that absolute zero (0 K) is at $-273\,°C$ and that $0\,°C$ is 273 K;

☐ that momentum = mass × velocity;

☐ that impulse = force × time = change in momentum;

☐ that kinetic energy = $\frac{1}{2}mv^2$.

The gas laws

The experimental relationships between pressure, volume and temperature of a gas were investigated by Boyle (1660), Amontons (1702) and Charles (1787). They discovered that for temperatures which are not too low, and pressures which are not too high, real gases obeyed the three gas laws.

Boyle's law Figure 1 shows a simple form of apparatus which can be used to verify Boyle's law. A fixed mass of gas is trapped above the mercury in the left hand tube. The volume of the trapped gas is proportional to the length ℓ. The pressure p is atmospheric pressure plus that due to the column of mercury of height h.

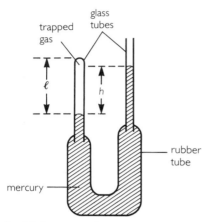

figure 1 Boyle's law apparatus

A graph of pressure against volume is shown in figure 2(a). The curve labelled T_1 is obtained if the temperature of the apparatus is kept constant at temperature T_1. If a higher temperature, T_2, is maintained then the curve labelled T_2 is obtained. A curve of constant temperature is called an *isothermal* curve, or *isotherm*.

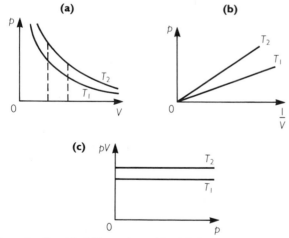

figure 2 Graphical illustrations of Boyle's law

Boyle discovered that:
  'For a fixed mass of gas at constant temperature, the pressure is inversely proportional to its volume.'
In equation form:

$$pV = \text{constant if } T \text{ is constant.}$$

The law is illustrated graphically in figures 2(a), (b) and (c).

Pressure law The apparatus illustrated in figure 1 can be modified by surrounding the glass tube containing the trapped air with a hot water jacket. The volume of the trapped air can be kept constant by maintaining the mercury level in the left

hand tube at a fixed mark. When the temperature of the water bath surrounding the trapped air is increased, the level of the mercury in the right hand tube will have to be raised in order to keep the volume of trapped air constant.
Amontons discovered that:

  'For a fixed mass of gas at constant volume, the pressure is proportional to the kelvin temperature.'

In equation form:

$$p \propto T \text{ if } V \text{ is constant.}$$

The law is illustrated graphically in figure 3.

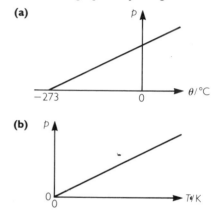

figure 3 Graphical illustrations of the pressure law

Charles' law By using a slightly modified version of the apparatus, Charles discovered that:

  'For a fixed mass of gas at constant pressure, the volume is proportional to the kelvin temperature.'

In equation form:

$$V \propto T \text{ if } p \text{ is constant.}$$

The equation of state The three gas laws can be summarised as follows.
  For a fixed mass of gas

$$pV = \text{constant if } T = \text{constant} \qquad \text{(i)}$$
$$p/T = \text{constant if } V = \text{constant} \qquad \text{(ii)}$$
$$V/T = \text{constant if } p = \text{constant} \qquad \text{(iii)}$$

These three laws are clearly special cases of the general equation

$$\frac{p_1 V_1}{T_1} = \frac{p_2 V_2}{T_2} = \text{constant}$$

which relates an initial state p_1, V_1 and T_1, for a fixed mass of gas, with a final state p_2, V_2 and T_2. The equations (i), (ii) and (iii) are obtained from this general equation by considering what happens when
(i) $T_1 = T_2$

(ii) $V_1 = V_2$
(iii) $p_1 = p_2$

The magnitude of the constant depends on how many molecules there are in the trapped gas. If we are dealing with *one mole* of gas the constant is represented by the symbol R. Two moles of gas will occupy twice as much volume, therefore if there are n moles of gas we can write

$$\frac{pV}{T} = nR \qquad \text{(iv)}$$

This is known as the equation of state for an ideal gas, or simply as the *ideal gas equation*. R is called the (universal) molar gas constant.

1 mole (1 mol) of a substance is the amount of that substance that contains the same number of elementary units as there are atoms in 12 grams of carbon-12. This number is known as the Avogadro constant, N_A. Its measured value is $N_A = 6.022 \times 10^{23}\,\text{mol}^{-1}$. Therefore, one mole of a gas will contain 6.022×10^{23} molecules.

The molar gas constant, R

The value of R can be obtained from the following experimental data and definitions.

☐   Standard atmospheric pressure $= 1.013 \times 10^5\,\text{Nm}^{-2}$.

☐   The density of hydrogen is $8.99 \times 10^{-2}\,\text{kg m}^{-3}$, when it is at standard temperature and pressure.

☐ $0\,°C$, or $273\,K$, is called standard temperature.

☐   The relative molecular mass of hydrogen is 2.016, i.e. 1 mole of hydrogen molecules has a mass of 2.016 g.

The volume occupied by one mole of hydrogen at standard temperature and pressure is given by

volume = mass/density

$$= (2.016 \times 10^{-3}\,\text{kg})/(8.99 \times 10^{-2}\,\text{kg m}^{-3})$$

$$= 22.4 \times 10^{-3}\,\text{m}^3$$

(The molar volume V_m is the same for all gases. At standard temperature and pressure,

$$V_m = 22.4 \times 10^{-3}\,\text{m}^3\text{mol}^{-1})$$

Substituting for these in equation (iv) gives

$$R = \frac{pV}{nT}$$

$$= \frac{(1.013 \times 10^5\,\text{Nm}^{-2}) \times (22.4 \times 10^{-3}\,\text{m}^3)}{(1\,\text{mol}) \times (273\,\text{K})}$$

$$\Rightarrow R = 8.3\,\text{Jmol}^{-1}\text{K}^{-1}$$

R turns out to have the same value for all gases if it is determined at low pressures and high temperatures, i.e. temperatures and pressures at which the molecules are well separated, and well above the boiling point of the liquid form.

The origins of the theory

Two Greek philosophers, Leucippus and Democritus, proposed in 450 BC that we think of matter as made of minute, hard, invisible, indivisible particles. They gave them the name ατομνς, which means indivisible. The 'atoms' of these Greeks correspond to the molecules of modern science. The Roman poet-philosopher Lucretius (55 BC) extended this model by considering the atoms to be in ceaseless, random motion, colliding and rebounding from one another.

Much later, Hooke ($\sim$1680) tried to explain Boyle's law in terms of the molecules undergoing elastic collisions with the walls of the container. Isaac Newton ($\sim$1686) then *invented* his laws of motion, from which the modern kinetic theory was derived.

The kinetic theory of gases

It is important to form as clear an idea as possible of the *model* of the gaseous state on which the kinetic theory is based. The kinetic theory rests essentially upon two hypotheses:

☐   that there are such things as molecules, and that a gas is a collection of molecules;

☐   that these molecules are in constant random motion, and that heat (sometimes called 'internal energy') is a manifestation of this molecular motion.

The most direct visual evidence for the second hypothesis is Brownian motion, where smoke particles, viewed through a microscope, are seen to move around randomly as they are struck by air molecules.

When setting out to develop a theory it is desirable to have as concrete a representation as possible to visualise. We shall start by associating the idea of a molecule with that of a dense spherical body of great elasticity and rigidity, something like an ideal billiard-ball or steel ball-bearing. Before deriving a mathematical expression for the pressure in such an ideal gas, we need to list the additional assumptions we shall be making.

☐   The molecules are of infinitesimal size, i.e. the volume of the gas molecules is negligible compared with the volume occupied by the gas.

☐   The intermolecular forces are negligible except during a collision, i.e. molecules do not attract or repel one another except when they collide.

☐   The duration of a collision is negligible compared with the time spent in free motion between collisions.

☐   A molecule moves with uniform velocity between collisions, i.e. the effect of gravity is taken to be negligible over these small distances.

☐   The collisions between molecules, and between molecules and walls, are perfectly elastic.

☐   All collisions obey Newtonian mechanics.

The pressure exerted by an ideal gas

In order to derive a mathematical expression for the gas pressure let us imagine a rectangular box containing ideal gas. We shall start by imagining that the box contains only a single gas molecule, and that this molecule is moving parallel to the x-axis with speed c_x, as illustrated in figure 4.

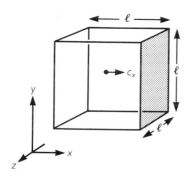

figure 4

The molecule bounces elastically off the shaded face of the box. Therefore its change of momentum at the shaded face is given by

change in momentum = momentum before
$\qquad$ − momentum after

$\qquad$ = (mc_x to the right) − (mc_x to the left)

$\qquad$ = $mc_x - (-mc_x)$

$\qquad$ = $2mc_x$ (remember momentum is a vector)

The molecule will travel a distance 2ℓ before it collides again with the same face. Hence, the time between collisions with the right hand face will be ($2\ell/c_x$). Therefore, the frequency of collisions by the molecule with this face is ($c_x/2\ell$).

Now, average force F on molecule (to the left) = change in momentum/time

$\Rightarrow$ F = (change in momentum/collision) $\times$ (no. of collisions/s).

$\Rightarrow$ $F = (2mc_x) \times (c_x/2\ell)$

$\qquad$ = mc_x^2/ℓ (to the left)

This is the push by the face on the molecule.

Thus, by Newton's third law of motion, the push of the molecule on the face is (mc_x^2/ℓ) to the right.

Therefore the pressure p exerted on the face by this one molecule is

$$p = \frac{\text{force}}{\text{area}}$$

$$= \frac{mc_x^2/\ell}{\ell^2}$$

$$= \frac{mc_x^2}{\ell^3}$$

If the box contains N molecules all travelling with different velocities c_1, c_2, c_3, etc., then the pressure exerted on the shaded face will be:

$$p = \frac{m}{\ell^3}[(c_x^2)_1 + (c_x^2)_2 + (c_x^2)_3{}^2 + \ldots + (c_x^2)_N]$$

The velocity of each molecule is written with a subscript, such as $(c_x)_1$ or $(c_x)_2$, because we are only considering their components of velocity parallel to the x-axis. The mean value of these c_x^2-terms is given by

$$\text{mean value} = \overline{c_x^2} = \frac{[(c_x^2)_1 + (c_x^2)_2 + (c_x^2)_3 + \ldots + (c_x^2)_N]}{N}$$

Substituting for this in the above equation gives

$$p = \frac{m}{\ell^3}[N \cdot \overline{c_x^2}]$$

$\Rightarrow$ $\qquad\qquad$ $$p = \frac{Nm}{\ell^3}[\overline{c_x^2}]$$ $\qquad$ (i)

Molecules do not only travel parallel to the x-axis, as we have been assuming. They travel in random directions, therefore their velocities can be resolved into x-, y- and z-components as depicted in figure 5.

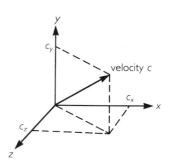

figure 5

By applying Pythagoras' theorem in three dimensions we can write for the velocity c:

$$c^2 = c_x^2 + c_y^2 + c_z^2$$

This relationship will also hold for the mean square values. Hence we may write

$$\overline{c^2} = \overline{c_x^2} + \overline{c_y^2} + \overline{c_z^2}$$

Because we are dealing with a very large number, N, of molecules in the box, and they are all travelling randomly in direction, it follows that the mean values of the x-, y- and z-components are equal. That is

$$\overline{c_x^2} = \overline{c_y^2} = \overline{c_z^2}$$

Hence the *mean square speed* of the molecules is

$$\overline{c^2} = 3(\overline{c_x^2})$$

Substituting for this value of $(\overline{c_x^2})$ in equation (i) above, along with V for the volume of the box ($V = \ell^3$) we obtain

$$p = \frac{Nm}{V}[\tfrac{1}{3}\overline{c^2}] \qquad \text{(ii)}$$

$$= \tfrac{1}{3} \cdot \frac{N}{V} \cdot [\tfrac{1}{2}m\overline{c^2}] \cdot 2$$

$$= \tfrac{2}{3} \cdot \frac{N}{V} \cdot [\tfrac{1}{2}m\overline{c^2}]$$

i.e. $pV = \tfrac{2}{3}N \times$ (mean kinetic energy of the molecules)

or

$$pV = \tfrac{2}{3}N\overline{E_k}$$

In equation (ii) the quantity Nm is the total mass of gas in the box. Therefore (Nm/V) gives the mass per unit volume, or density, ρ. So we may write

$$p = (Nm/V) \cdot [\tfrac{1}{3}\overline{c^2}]$$

i.e

$$p = \tfrac{1}{3}\rho\overline{c^2} \qquad \text{(iii)}$$

Root-mean-square (r.m.s.) speed, $c_{\text{r.m.s.}}$

We can now substitute some known values of ρ and p into equation (iii) and calculate $\overline{c^2}$. For example, at $0\,°C$ and normal atmospheric pressure $(1.0 \times 10^5\,\mathrm{Nm^{-2}})$ the density of helium is $0.18\,\mathrm{kg\,m^{-3}}$.

$$\Rightarrow \overline{c^2} = 3 \times (1.0 \times 10^5\,\mathrm{Nm^{-2}})/(0.18\,\mathrm{kg\,m^{-3}})$$
$$= 1.7 \times 10^6\,\mathrm{m^2\,s^{-2}}$$

The square root of $\overline{c^2}$ is called the root-mean-square speed, $c_{\text{r.m.s.}}$.

$$c_{\text{r.m.s.}} = \sqrt{\overline{c^2}} = 1300\,\mathrm{ms^{-1}}$$

The root-mean-square speed of molecules is the same order of magnitude as the mean speed, but they are not the same quantity. This simple example makes the point.

Example What are the mean speed and the root-mean-square speed of three molecules travelling at $300\,\mathrm{ms^{-1}}$, $400\,\mathrm{ms^{-1}}$ and $500\,\mathrm{ms^{-1}}$?

The mean speed $\overline{c} = \tfrac{1}{3}(300 + 400 + 500)\,\mathrm{ms^{-1}}$

$$\Rightarrow \overline{c} = 400\,\mathrm{ms^{-1}}$$

The r.m.s. speed $c_{\text{r.m.s.}} = \sqrt{\overline{c^2}}$

$$= \sqrt{\frac{300^2 + 400^2 + 500^2}{3}}\,\mathrm{ms^{-1}}$$

$$= 408\,\mathrm{ms^{-1}}$$

Distribution of molecular speeds

For a quantity of gas at constant temperature, the values of $\overline{c}$ and $\sqrt{\overline{c^2}}$ remain constant. Individual molecules of course have a very wide range of different speeds. At a given instant, the proportions of molecules with particular speeds are illustrated by the graphs of figure 6.

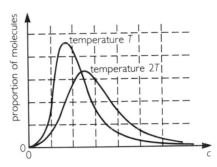

figure 6

Though the speed of an individual molecule changes each time it makes a collision, the distribution of speeds throughout the gas, as illustrated in figure 5, remains constant. Graphs such as these can be predicted using advanced theory, and confirmed by experiment. (See question 5, page 446).

Relationship between temperature and mean kinetic energy

If we compare the equation

$$pV = \tfrac{2}{3}N \cdot (\overline{E_k})$$

with the equation of state

$$pV = nRT$$

we see that

$$\tfrac{2}{3}N \cdot (\overline{E_k}) = nRT$$

Now $N = nN_A$, where N_A is the Avogadro constant.

$$\Rightarrow \quad \tfrac{2}{3}N_A \cdot (\overline{E_k}) = RT$$

therefore

$$\overline{E_k} = \tfrac{3}{2} \cdot \frac{R}{N_A} \cdot T$$

$$\Rightarrow \quad \overline{E_k} \text{ of the molecule} = \tfrac{3}{2} \cdot k \cdot T$$

where $k = R/N_A$ is called the Boltzmann constant. Hence we have shown that the mean kinetic energy of the molecules is proportional to the kelvin temperature.

The Boltzmann constant

At the start of this chapter we obtained the value $8.3\,\mathrm{J\,mol^{-1}\,K^{-1}}$ for the molar gas constant R, and we saw that the Avogadro constant, N_A, has the value $6.022 \times 10^{23}\,\mathrm{mol^{-1}}$. Thus

the Boltzmann constant $k = \dfrac{R}{N_A} = \dfrac{8.3\,\mathrm{J\,mol^{-1}\,K^{-1}}}{6.02 \times 10^{23}\,\mathrm{mol^{-1}}}$

$\Rightarrow \qquad k = 1.38 \times 10^{-23}\,\mathrm{J\,K^{-1}}$

We can now express the equation of state in terms of the Boltzmann constant k, instead of the molar gas constant R. Substituting $R = kN_A$ in equation (iv) on page 439 gives

$$pV = n(kN_A)T = nN_A kT$$

But the total number of molecules in the gas is $N = nN_A$.
Therefore, $\qquad\qquad pV = NkT$

Deductions from $pV = \frac{2}{3}N(E_k$ of the molecules)

1 *Boyle's law* This relates the pressure to the volume of a fixed mass of gas, when the temperature is kept constant. We have associated temperature with the mean kinetic energy E_k of the gas molecules. Therefore, if the temperature is constant then the mean kinetic energy of the molecules is constant.
  Now since $pV = \frac{2}{3}N \times (\overline{E_k}$ of the molecules)

$\Rightarrow \qquad\qquad pV = \text{constant}$

i.e. $p \propto 1/V \qquad$ if T is constant

2 *Charles' law* This relates the volume to the temperature of a fixed mass of gas when the pressure is kept constant.
Starting from

$$pV = \frac{2}{3}N \times (\overline{E_k}\ \text{of the molecules})$$

we deduced earlier that

$$pV = NkT$$

i.e. $V \propto T \qquad$ if p is constant

3 *Pressure law* This relates the pressure to the temperature of a fixed mass of gas when the volume is kept constant. Starting from $pV = \frac{2}{3}N\overline{E_k}$, we deduce

$$p \propto T \qquad \text{if } V \text{ is constant}$$

4 *Dalton's law of partial pressures* Consider a mixture of two different gases in a volume V. Each gas will occupy the whole volume.

for gas 1,

$$p_1 V = \frac{2}{3}N_1(\tfrac{1}{2}m_1\overline{c_1^2})$$

for gas 2,

$$p_2 V = \frac{2}{3}N_2(\tfrac{1}{2}m_2\overline{c_2^2})$$

where p_1 and p_2 are the partial pressures exerted by each gas. They are the pressures that the gases would exert if each alone occupied the volume. When the gases have been mixed for some time they will have acquired equal temperatures, i.e. the mean kinetic energies of the molecules of gases 1 and 2 will be equal.

$\Rightarrow \qquad \tfrac{1}{2}m_1\overline{c_1^2} = \tfrac{1}{2}m_2\overline{c_2^2} = \overline{E_k}$

If we add the above two expressions for $p_1 V$ and $p_2 V$ we obtain for the mixture

$$p_1 V + p_2 V = (p_1 + p_2)V = \frac{2}{3}(N_1 + N_2) \times \overline{E_k}$$

But

$$p_{\text{total}} \times V = \frac{2}{3}(N_1 + N_2) \times \overline{E_k}$$

Hence, total pressure p_{total} is given by

$$p_{\text{total}} = p_1 + p_2$$

This is known as Dalton's law of partial pressures and is written as
  'When two or more gases, which do not react chemically, are present in the same container, the total pressure is the sum of the partial pressures.'

5 *Avogadro's law* Consider two different ideal gases.

For gas 1

$$p_1 V_1 = \frac{2}{3}N_1(\tfrac{1}{2}m_1 c_1^2)$$

For gas 2

$$p_2 V_2 = \frac{2}{3}N_2(\tfrac{1}{2}m_2 c_2^2)$$

If their pressures, volumes and temperatures are equal, then

$$\frac{p_1 V_1}{\tfrac{1}{2}m_1 c_1^2} = \frac{p_2 V_2}{\tfrac{1}{2}m_2 c_2^2}$$

which means that $N_1 = N_2$.
This is known as Avogadro's law and is stated as:
  'Equal volumes of ideal gases at the same temperature and pressure contain equal numbers of molecules.'

Behaviour of real gases

We can look at this problem in two ways:

1 we can experimentally compare how real gases behave with the way in which our theory predicts an ideal gas will behave, or

2 we can look at our theory for an ideal gas and see what arises when we remove some of our simplifying assumptions.

  Figure 7 shows graphs which are predicted by the

kinetic theory for an ideal gas. The graphs shown in figure 8 indicate how real gases behave.

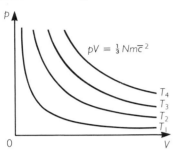

$$pV = \tfrac{1}{3} Nm\overline{c}^{\,2}$$

T_4
T_3
T_2
T_1

figure 7

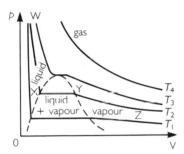

gas
W
liquid
X — Y
liquid + vapour   vapour   Z
T_4
T_3
T_2
T_1

figure 8

If we compare the isothermals in figure 7 with those in figure 8, we immediately see several areas where the ideal gas theory breaks down. All the isothermal curves in figure 7 are smooth, rectangular hyperbolae, whereas in figure 8 we see that real gases only give rectangular hyperbolae at high temperatures (T_4).

The effects of finite molecular size

One of the assumptions in the kinetic theory for an ideal gas is that the size of the molecules is negligible. This assumption will become unreasonable as the gas is subjected to greater and greater pressures. If the pressure were continually increased all the molecules would end up touching one another as illustrated in figure 9.

The kinetic theory effectively assumed that all of the volume V was available for free movement. However, if the volume occupied by the molecules themselves is b, the volume available for free movement is actually $(V - b)$. At low pressure this is nearly equal to V, but at higher pressures a more accurate version of the equation of state is required. It is:

$$p(V - b) = nRT.$$

The isothermals corresponding to this equation are plotted in figure 10.

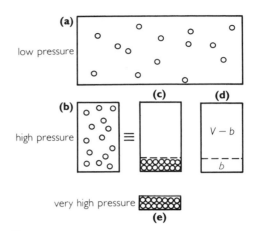

(a) low pressure

(b) high pressure

(c)   (d) $V - b$ b

very high pressure

(e)

figure 9

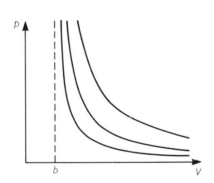

b V

figure 10

If the curves in figure 10 are compared with those in figure 8, it will be seen that there is a lower limit b to the volume of the real gas as the pressure is increased.

The effect of intermolecular forces

In deriving the ideal gas equation, we assumed that the intermolecular forces were zero, except at the moment of collision. Any form of actual intermolecular force is likely to fall off with increasing distance between the molecules. Therefore, the effect of an intermolecular force will be negligible if the molecules are well separated and travelling at high speed, i.e. at low pressure and high temperature. Under opposite conditions, at high pressures and low temperatures, the molecules will be very close together and travelling slowly. Attractive intermolecular forces will cause them to bond together, or to slow down as they separate.

Both of these effects will result in the gas exerting a lower pressure on its surroundings than kinetic theory predicts. This effect can be seen by comparing the kinetic theory curves, labelled T_3 and

T_4 in figure 7, with the real gas curves, labelled T_3 and T_4 in figure 8.

At high temperatures and fairly low pressures the curves labelled T_4 are almost identical. But at a lower temperature, T_3, the curve for the real gas indicates a lower pressure than the kinetic theory predicts; this difference becomes more noticeable as the volume of the gas is reduced and the molecules become closer together. (See figure 11.)

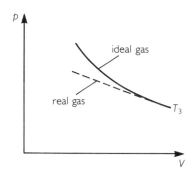

figure 11

In fact it is these intermolecular forces which, at low temperatures and high pressures, cause the atoms to bond together to form liquid rather than gas. In figure 8 the isothermal for T_2 has a portion, labelled XY, which is parallel to the V-axis. Consider what this represents. At large volumes the isotherm has the general shape of a rectangular hyperbola, which is how ideal gases behave. Therefore, in the region YZ the material behaves as a gas.

At high pressures the volume does not change much, i.e. the material appears to be incompressible. Liquids are not very compressible, therefore it would be reasonable to suppose that the region WX represents the liquid phase.

Now let us consider the horizontal section XY. As the volume of a container is gradually increased, as illustrated in figure 12, molecules will evaporate to maintain a constant pressure. But once all the liquid has evaporated the pressure will start to fall.

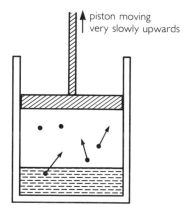

figure 12

Therefore, it would be reasonable to suspect that the horizontal section XY in figure 8 represents a mixture of liquid and gas. If the experiment is carried out using a transparent cylinder this is seen to be the case.

It should be noted that in figure 8 the isotherms with $T > T_3$ do not have horizontal sections. When $T > T_3$ a gas does not turn into liquid even at very high pressures. When a gas has $T < T_3$, i.e. it will turn into a liquid when it is compressed, we call it a *vapour*. T_3 is known as the *critical temperature* of the substance.

Summary

1 Experimental evidence for Boyle's law, Charles' law and the pressure law.

2 The equation of state for an ideal gas: $pV = nRT$.

3 Kinetic theory: hypotheses and assumptions
   a There are such things as molecules, and a gas is a collection of molecules.
   b The molecules are in constant, random motion.
   c The molecules are of infinitesimal size, i.e. the volume of the gas molecules is negligible compared with the volume occupied by the gas.
   d The intermolecular forces are negligible except during a collision.
   e The duration of a collision is negligible compared with the time spent in free motion between collisions.
   f A molecule moves with uniform velocity between collisions, i.e. the effect of gravity is taken to be negligible.
   g The collisions between molecules, and between molecules and walls, are perfectly elastic.
   h All collisions obey Newtonian mechanics.

4 Kinetic theory: deductions
   a Derivation of $p = \frac{1}{3}\rho\overline{c^2}$
   b Derivation of $pV = \frac{2}{3}N\,(\overline{E_k}$ of the molecules)
   c Deduction that $\frac{1}{2}m\overline{v^2} = \frac{3}{2}kT$ and $pV = NkT$

5 Deductions from $pV = \frac{2}{3}N\,(\overline{E_k})$
   a Boyle's law: $p \propto 1/V$ if T = constant
   b Charles' law: $V \propto T$ if p = constant
   c Pressure law: $p \propto T$ if V = constant
   d Dalton's law: $p_{total} = p_1 + p_2 + p_3$ etc.
   e Avogadro's law: $N_1 = N_2$ if $p_1 = p_2$, $V_1 = V_2$ and $T_1 = T_2$

6 Behaviour of real gases
   a The effect of finite molecular size.
   b The effects of intermolecular forces. See figure 8

7 S.T.P.
   a Standard Temperature is 273 K.
   b Standard Atmospheric Pressure is $1.01 \times 10^5\,\mathrm{Nm^{-2}}$.

Questions

1 a What is the equation of state for an ideal gas?
   b What is the numerical value of standard temperature?
   c What is the numerical value of standard atmospheric pressure measured in Pa? Convert your answer into millimetres of mercury (mm Hg), given that the density of mercury is $13.6 \times 10^3\,\mathrm{kgm^{-3}}$ and the gravitational field strength on the surface of the Earth is $9.8\,\mathrm{Nkg^{-1}}$.
   d What is meant by s.t.p.?
   e Calculate the molar volume of oxygen, V_m, that is, the volume occupied per mole, given that the relative molecular mass of oxygen is 32 and its density at s.t.p. is $1.43\,\mathrm{kgm^{-3}}$.
   f Use your answers to part (a), (b), (c) and (e) to calculate the value of the molar gas constant, R.

2 a Explain what is meant by an ideal gas.
   b What assumptions are made for the model of an ideal gas in deriving the expression
   $$p = \tfrac{1}{3}\rho\overline{c^2}$$
   where p is the pressure, ρ is the density and $\overline{c^2}$ is the mean-square-speed?
   c Explain what is meant by the root-mean-square speed, $c_{r.m.s.}$.
   d Calculate the molecular speed (r.m.s. value) for oxygen at $0\,°C$, if the density is $1.43\,\mathrm{kgm^{-3}}$ when the pressure is $1.0 \times 10^5\,\mathrm{Nm^{-2}}$, i.e. at s.t.p.

   e Calculate the r.m.s. speed of hydrogen molecules at s.t.p., if the molar mass of hydrogen is $2 \times 10^{-3}\,\mathrm{kgmol^{-1}}$ and the molar volume is $2.24 \times 10^{-2}\,\mathrm{m^3mol^{-1}}$. (Standard pressure = $1.01 \times 10^5\,\mathrm{Nm^{-2}}$.)
   f Imagine that a fellow student of yours is having difficulty in understanding why the speed of sound through air at s.t.p. $(331\,\mathrm{ms^{-1}})$ is less than the root-mean-square speed at s.t.p. $(485\,\mathrm{ms^{-1}})$, or the mean speed $(447\,\mathrm{ms^{-1}})$. Write a short explanation which will clear up these difficulties for him.
   g The measured velocities of sound at s.t.p. in hydrogen and oxygen are $1255\,\mathrm{ms^{-1}}$ and $315\,\mathrm{ms^{-1}}$ respectively. Would you expect these values, given the r.m.s. values that you obtained in answering parts (d) and (e)? Explain your answer.

3 a Define pressure.
   b Discuss qualitatively how the pressure exerted by a gas may be explained in terms of the movement and collisions of gas molecules.
   c The simple kinetic theory of gases derives the equation $p = \frac{1}{3}\rho\overline{c^2}$ where p is the gas pressure, ρ is the density and $\overline{c^2}$ is the mean-square speed. Explain what is meant by mean-square speed.
   d Show how the equation $pV = \frac{2}{3}N(\overline{E_k}$ of the

molecules) may be derived from $p = \frac{1}{3}\rho\overline{c^2}$, where V is the volume of the gas and N is the number of molecules trapped within volume V.

e Establish the relationship between T and $\overline{E_k}$ for an ideal gas.

f By reference to an ideal gas, explain what the Boltzmann constant is.

4 a What are the relative molecular masses of helium, nitrogen and carbon monoxide?

b Explain why the speed of sound in helium ($895\ ms^{-1}$ at s.t.p.) is greater than the speed of sound in nitrogen ($337\ ms^{-1}$ at s.t.p.).

c If, for carbon monoxide at $0\ °C$, $c_{r.m.s.} = 493\ ms^{-1}$, what would you expect the speed of sound to be in carbon monoxide at s.t.p.?

5 Figure 13 illustrates a form of apparatus that has been used to study the distribution of molecular speeds in gases and vapours. The apparatus is mounted inside a vacuum chamber.

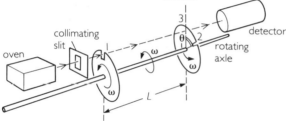

figure 13

The oven maintains the gas or vapour at a steady temperature. Molecules of gas emerge from the oven, and some of them will travel through the collimating slit which produces a parallel beam of molecules. The two discs attached to the rotating axle have small slits cut into their edges. These are labelled 1 and 2. Molecules which take the same time to pass from position 1 to position 3 as it takes the slit in the second disc to rotate from 2 to 3 will pass through the system and on into the detector. The rotating discs thus act as a velocity selector. In this apparatus the speed of rotation of the discs is maintained at a steady value while a reading is taken, then the speed is increased.

a Explain how the rotating discs behave as a velocity selector.

b Derive an expression for the velocity selected when the axle rotates at angular speed ω (use the marked values in the diagram).

The result of such an experiment done on nitrogen at 290 K is illustrated in figure 14.

c The word 'average' is a vague term used by many people, when really they mean one of the following more exact words: mode, median, mean or root-mean-square. Associate each of these names with one of the dotted lines

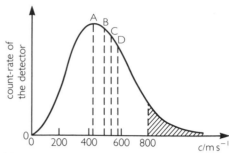

figure 14

labelled A, B, C, D on the graph and explain your choice.

d The space under the curve and beyond $800\ ms^{-1}$ has been shaded in. Explain what this shaded area signifies.

6 a Each of the graphs in figure 15 shows how the number of molecules with a given speed (in a given sample of gas) depends on that speed. Use the two graphs to explain why the Earth's atmosphere is rich in nitrogen but contains very little helium.

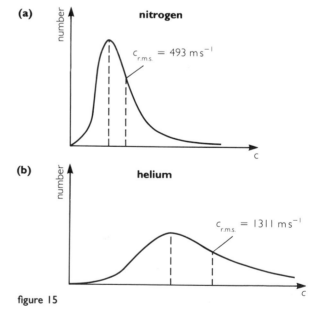

figure 15

b Use Charles' law ($V/T =$ constant at constant pressure) to calculate the molar volume at standard pressure and 6000 K, the temperature on the surface of the Sun. The molar volume at s.t.p. is $2.24 \times 10^{-2}\ m^3 mol^{-1}$. What will be the density of hydrogen at 6000 K and standard pressure? Use the equation $p = \frac{1}{3}\rho\overline{c^2}$ to determine $c_{r.m.s.}$ for hydrogen at 6000 K.

c The mass of the Sun is 2.0×10^{30} kg and its diameter is 150×10^6 km. Calculate the escape velocity from the surface of the Sun (see page 213). Do you think there is significant 'solar wind' of positively-charged protons striking the Earth? Why does this cosmic ray shower from the Sun not reach the *surface* of the Earth, where it would cause skin cancer among other things? ($G = 6.67 \times 10^{-11}$ Nkg^{-2}m^2.)

d Comment on the graphs in figure 16, which show the speed distributions for three different samples of hydrogen.

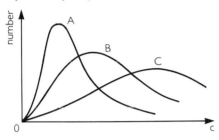

figure 16

7   This question is about Graham's law of diffusion. The law states that the rate at which a gas of density ρ diffuses through another gas, or escapes through a hole, is proportional to $1/\sqrt{\rho}$, provided other conditions, such as temperature and pressure, are fixed.

This law was discovered experimentally by Graham. It can also be deduced from the kinetic theory, thus:

rate of diffusion $\propto \bar{c}$ (seems sensible)

⇒   rate $\propto \sqrt{\overline{c^2}}$ ($\bar{c} \propto \sqrt{\overline{c^2}}$)

⇒   rate $\propto \dfrac{1}{\sqrt{\rho}}$ ($p = \frac{1}{3}\rho\overline{c^2}$)

The apparatus in figure 17 is sometimes used in schools to demonstrate Graham's law.

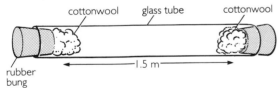

figure 17

It consists of a long, clean glass tube containing air at room pressure and temperature. One ball of cotton wool is soaked in concentrated hydrochloric acid (HCl) and the other ball is soaked in concentrated ammonia (NH$_4$OH). The two balls of cotton wool are introduced simultaneously into opposite ends of the long glass tube, then the tube is sealed with rubber bungs. After about 10 minutes a white cloud starts to develop nearer the end containing HCl. (The

atomic masses of hydrogen, nitrogen and chlorine are 1, 14 and 35.5 respectively.)

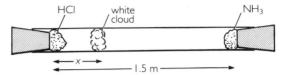

figure 18   Sectional view of figure 17

a Explain why the white cloud forms nearer the HCl end.

b Explain why it is several minutes before the cloud starts to appear.

c Use the information given to estimate the value of x (see figure 18). Explain your reasoning fully.

d What differences would it make to the results of this experiment if all the apparatus were placed in an oven at about 300 °C and the experiment then performed at this higher temperature?

8   The apparatus shown in figure 19 is sometimes used in schools to demonstrate the high speed at which bromine molecules travel, and the slow speed at which they diffuse.

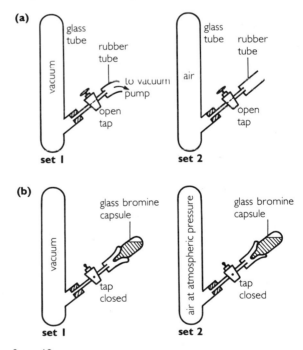

figure 19

Two identical sets of apparatus are placed side by side. One of the large glass tubes is evacuated (set 1); the other one is full of air at atmospheric pressure (set 2). Each set of apparatus has a bromine capsule sealed into the rubber tube above its tap.

a The density of nitrogen at s.t.p. is $1.25\,\mathrm{kg\,m}^{-3}$. Calculate the r.m.s. speed of nitrogen molecules at s.t.p., given that standard atmospheric pressure is $1.01 \times 10^5\,\mathrm{N\,m}^{-2}$.

b Use your answer to part (a) to determine the r.m.s. speed of nitrogen molecules at room temperature (20 °C).

c Use your answer to part (b) to calculate the r.m.s. speed of bromine molecules at room temperature. The relative molecular masses of nitrogen and bromine molecules are 28 and 160.

d Consider the above experiment. When the bromine capsule is broken and the tap then opened, the glass tube immediately fills with brown vapour in the case of the evacuated tube (set 1). On the other hand, in the air-filled tube (set 2) the brown vapour rises from the bottom of the tube very slowly. The situation after about 8 minutes (500 s) is illustrated in figure 20. Explain, in some detail, each of these observations.

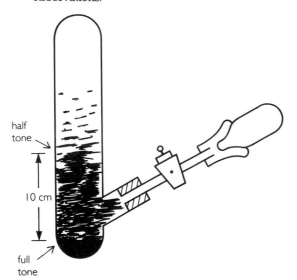

figure 20

e A bromine molecule will progress randomly through the air molecules as it undergoes repeated collisions. Although the distance travelled between collisions will vary enormously, it is useful to introduce the idea of an average distance between collisions. We call this average distance the *mean free path* λ. If a bromine molecule undergoes N collision in 500 s, then the total distance travelled by the molecule will be $N\lambda$.

What is the numerical value of $N\lambda$?

f Statistical analysis of the random motion of molecules reveals that the average distance travelled by a molecule undergoing N collisions is $\sqrt{N}.\lambda$ (figure 21).

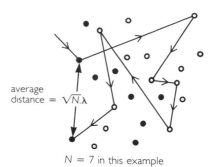

average distance = $\sqrt{N}.\lambda$

$N = 7$ in this example

figure 21

The height at which the half tone (figure 20) appears after 500 s indicates that the average distance travelled by the bromine molecules in 500 s was 0.1 m. Use this information, and your answer to part (e), to determine the value of N, i.e. the average number of collisions suffered by a bromine molecule in 500 s.

g Substitute your value of N into your answer to part (f) to determine the value of λ. That is, determine a value for the mean free path of the bromine molecules.

9 Density of oxygen at s.t.p. = $1.43\,\mathrm{kg\,m}^{-3}$
Density of liquid nitrogen = $1.04 \times 10^3\,\mathrm{kg\,m}^{-3}$
Avogadro constant $N_A = 6.02 \times 10^{23}\,\mathrm{mol}^{-1}$
Relative molecular masses of oxygen and nitrogen are 32 and 28 respectively.

a Determine the density of nitrogen at s.t.p.

b Calculate the volume occupied by 1 mole of nitrogen at s.t.p.

c Calculate the volume occupied by 1 mole of liquid nitrogen. Assume that the molecules in liquid nitrogen are packed as shown in figure 22, when d is the diameter of a molecule.

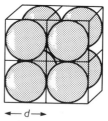

figure 22

d Determine the mean separation of nitrogen molecules in nitrogen gas at s.t.p.

e In question 8 you determine the mean free path of bromine molecules in air. Compare your answer to part (e) with your answer to question 8.

CHAPTER 43

Thermodynamics

Introduction

The everyday word *temperature* is our starting point. Its meaning is discussed, and then real scales of temperature are introduced. This is extended into the ideal-gas scale, and the absolute thermodynamic scale of temperature. The interrelationships between the *macroscopic* quantities pressure, temperature, volume and internal energy are then explored; this topic is called *thermodynamics*. This leads on to the relationship between molar heat capacity at constant volume, C_V, and molar heat capacity at constant pressure, C_p. Finally the crucial second law of themodynamics is developed; and we consider why it is inevitable that power stations and motor cars are inefficient converters of energy.

GCSE knowledge

It will be assumed that you are able to recall:

☐ that numerical values of temperature are used to indicate whether or not one object is hotter or colder than another object;

☐ that most objects expand when their temperatures rise;

☐ the gas laws:
  (i)   Boyle's law $p \propto 1/V$ if T is constant
  (ii) Charles' law $V \propto T$ if p is constant
  (iii) Pressure law $p \propto T$ if V is constant

☐ that heat is energy that flows from a body to its surroundings, or vice versa, because of a temperature difference between them;

☐ that work is done on a system when an external force moves its point of application; and that this involves energy being transferred into the system;

☐ that the efficiency of a system is given by

$$\text{efficiency} = \frac{\text{useful work out}}{\text{energy in}}$$

Temperature

Temperature is an everyday word which refers to the way in which the senses of the human body respond to the hotness or coldness of any system; e.g. an ice cube feels cold, therefore we say that it has a low temperature. For scientific use, temperature must have a numerical scale. The temperature scale we choose must be consistent and reproducible. If we are going to introduce a scale of temperature then we need to have a concept of two objects having the same temperature as each other. Consider two objects placed in good thermal contact with one another, one object being hotter than the other. We find that the hotter object cools down and the colder object warms up. We explain these facts by saying that heat energy flows from the higher temperature (the hot object) to the lower temperature (the cold object) as in figure 1. When two bodies are in good thermal contact and there is no net flow of heat energy between them, we say that they are in thermal equilibrium.

The idea that heat is a form of energy can be justified in many ways; for example, rubbing one's hands together is doing work and this results in heat being generated.

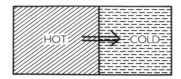

figure I   Heat flows from a hot object to a cold object

The temperature of a system is the property that determines whether or not a system is in thermal equilibrium with other systems.

Scale of temperature

Thermometers are used for measuring temperature. A thermometer uses a particular property of a substance which is sensitive to temperature changes. Such a property is known as a 'thermometric' property. Once a thermometric property has been chosen, an agreed set of procedures for assigning numerical values to temperatures has to be applied.

The best known thermometer is the mercury-in-glass one. The thermometric property chosen in this case is the length of the mercury thread in a capillary tube. See figure 2.

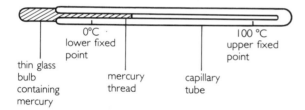

figure 2   A mercury-in-glass thermometer

The internationally-agreed procedures are as follows.

1 Mark the two fixed points.
   a) The *lower fixed point* is known as the *ice point* and is given the temperature of 0 °C (°C stands for degrees Celsius). It is the temperature at which ice melts when under standard atmospheric pressure.
   b) The *upper fixed point* is known as the *steam point* and is given the temperature 100 °C. It is the temperature of steam above water which is boiling under standard atmospheric pressure.

2 Divide the interval between the two fixed points into 100 equal divisions, and say that each division represents a change in temperature of 1 Celsius degree.

All thermometers set up in this way will agree with one another at only two points on their scales, namely the lower and upper fixed points of 0 °C and 100 °C.

Temperatures are defined by this method as follows. The values X_{100} and X_0 of the thermometric property are those taken at the steam point and ice point respectively. If X_θ is the value at another temperature θ, which we wish to measure, then

$$\theta = \left\{\frac{X_\theta - X_0}{X_{100} - X_0}\right\} \times 100 \text{ °C}$$

There are two different reasons for discrepancies between thermometers. Firstly, two mercury thermometers may disagree with one another when placed side by side in a fluid at, say, about 50 °C. This is because the bores of the capillaries may not be exactly uniform; also the glasses used to make each thermometer may expand by different amounts, due to the different constituents of each. Secondly, even an 'ideal' mercury thermometer would probably not agree with a resistance thermometer (or any other type) at 50 °C, due to the differing variations in their thermometric properties with temperature.

To explain this more fully we need to say what is meant by an ideal mercury thermometer. We mean one that is made from a glass which does not expand with temperature, and which has an exactly uniform bore, but has real mercury in it. When the length of the mercury thread in such a thermometer is, for example, half-way between its 0 °C mark and its 100 °C mark, we say the temperature is 50 °C on the mercury-in-glass scale. But we would find that the resistance of a platinum resistance thermometer, placed in the same water bath, is not exactly half-way between its 0°C and 100°C values (see page 452 for further details). The disagreement between scales is small in the range 0 °C to 100 °C, but it does mean that we need to say on which type of thermometer any particular temperature has been measured.

We have associated temperature with the idea of heat energy flowing from a hot body to a colder body. Therefore, an ideal temperature scale should be based on a property which varies linearly with the transfer of heat energy. That is, if twice as much energy flowed into a body, then its temperature would rise twice as much. If this were the case for the mercury-in-glass thermometer, the amounts of energy taken in by the thermometer in going from the 0 °C mark to the 50 °C mark, and from the 50 °C to the 100 °C mark would be equal. Unfortunately this is not quite the case. This non-linearity is explored in more detail in question 2.

The ideal-gas scale of temperature

A constant-volume gas thermometer approaches the ideal of a linear relationship between the temperature indicated and the quantity of heat energy taken in by the thermometer. We define the temperature scale by using the equation $pV = nRT$, that is, the ideal gas equation. Therefore the temperature indicated is proportional to the pressure in the gas. The essential details of a constant-volume gas thermometer are illustrated in figure 3.

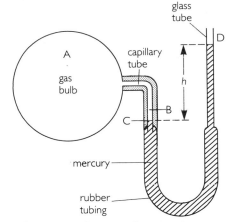

figure 3   A constant-volume gas thermometer

A gas is trapped in bulb A, which is connected to the mercury manometer through a capillary tube. The volume of the gas is kept constant by adjusting the height of the mercury at D, until the mercury just touches the level indicator at C. The pressure P of the gas in A is given by:

$$p = \text{atmospheric pressure} + h\rho g$$

This apparatus has many disadvantages, such as

☐   taking a long time to reach equilibrium,

☐   having a large heat capacity, therefore removing heat from the object being measured,

☐   needing many corrections in calculating the final temperature, such as allowing for the colder gas in space B, and the expansion of bulb A.

But, because it approaches the ideal of a linear relationship between temperature and heat energy, it is used as a standard in National Physical Laboratories throughout the world.

The thermometer illustrated in figure 3 is full of real gas. It should be a gas which approaches as nearly as possible the behaviour of an ideal gas. Helium at low pressure does this well. An ideal gas is one which obeys the equation $pV/T = \text{constant}$, for a fixed mass of gas trapped in a sealed vessel.

The basic concept behind the constant-volume gas thermometer is that, if the pressure p is plotted against temperature θ, the graph is a straight line through the coordinates of the lower and upper fixed points (figure 4).

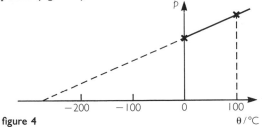

figure 4

This plot of p against temperature is a straight line that extrapolates back to give zero pressure at $-273\,°C$. This leads to the concept of *absolute temperature*, and the *Kelvin scale of temperature*. The divisions on the Celsius and Kelvin scales are equal in magnitude. To convert degrees Celsius (°C) into Kelvin (K) we therefore use this equation:

$$T/K = \theta/°C + 273$$

Thus, for example, 0 °C becomes 273 K.

The absolute thermodynamic scale of temperature

The ideal-gas scale of temperature is independent of the properties of any gas in particular. Temperatures on the ideal-gas scale are calculated by taking repeated measurements with a constant-volume gas thermometer at lower and lower pressures, and then extrapolating these measured values to find out what you would obtain if you could achieve zero pressure. This is a long and tedious process, which does in general depend on the properties of the gas being used. In order to be independent of the properties of a particular gas, an *absolute thermodynamic scale of temperature* has been defined. This temperature scale is defined in terms of heat energy flowing into, or out of, a body. If twice as much energy flows out then the fall in temperature will be twice as great. The absolute thermodynamic scale of temperature, also called the *Kelvin temperature scale*, is essentially identical to the ideal-gas scale.

The platinum resistance thermometer

This relies on the fact that the electrical resistance of a wire of pure platinum increases with temperature. On the platinum resistance scale the temperature is defined as

$$\theta = \left(\frac{R_\theta - R_0}{R_{100} - R_0}\right) \times 100\,°C$$

where R_0 and R_{100} are the resistances of the wire at the ice and steam points. R_θ is the resistance of the wire at the temperature required.

It is usual to measure the resistance of any type of resistance thermometer, be it platinum wire or carbon thermistor, by maintaining a known current through it and measuring the p.d. across it using a sensitive potentiometer.

The thermocouple thermometer

Whenever two dissimilar metals are in good electrical contact, the electrons migrate across the boundary between the two metals. This will result in a potential difference between the two metals. This contact potential difference, which is always present at a junction of dissimilar metals, varies with temperature.

The arrangement of figure 5 forms a 'thermocouple'. If both junctions are at the same temperature, the two contact p.ds cancel out, since they act in opposite directions round the circuit, and there is zero output. But if the junctions are at different temperatures, the two contact p.ds are not equal in magnitude, and there is a net output e.m.f. This is called a thermal e.m.f., ε.

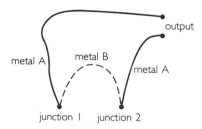

figure 5 A thermocouple

An arrangement for using this effect as a thermometer is illustrated in figure 6. The reference junction is kept at 0 °C (in a mixture of ice and water), and the thermocouple is calibrated by measuring the thermal e.m.f. ε_{100} produced when the test junction is at the steam point temperature, 100 °C.

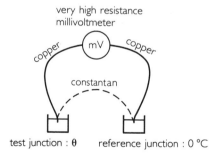

figure 6

On the thermocouple scale the temperature is defined as

$$\theta = \left(\frac{\varepsilon_\theta - \varepsilon_0}{\varepsilon_{100} - \varepsilon_0}\right) \times 100\,°C$$

where ε_θ is the thermal e.m.f. at the temperature required.

However, when the test junction is at 0 °C, there is zero output: so

$$\varepsilon_0 = 0$$

Thus

$$\theta = \frac{\varepsilon_\theta}{\varepsilon_{100}} \times 100\,°C$$

Disagreement between scales of temperature

We mentioned earlier in this chapter that when the length of the mercury thread of an ideal mercury-in-glass thermometer was half-way between the 0 °C mark and the 100 °C mark then we say that the temperature is '50 °C on the mercury-in-glass scale'. We went on to say that a platinum resistance thermometer, placed in the same water bath, would not give a resistance exactly half-way between its fixed point values of R_0 and R_{100}. Nor would the pressure of a constant-volume gas thermometer be half-way between p_0 and p_{100}, nor the e.m.f. of a thermocouple be half-way either.

This disagreement is due to the fact that all these temperature scales have been defined by the equation

$$\theta = \frac{X_\theta - X_0}{X_{100} - X_0} \times 100\,°C$$

which arbitrarily defines a linear relationship between the thermometric property being measured and the temperature on that particular scale. Unfortunately, as we have seen, if the property X is measured against the constant-volume gas scale or the absolute thermodynamic scale, then a different, non-linear graph is likely to result.

As an illustration, figure 7 shows the variation of resistance R with absolute thermodynamic temperature θ. This curve is approximately given by the equation

$$R_\theta = R_0(1 + A\theta + B\theta^2)$$

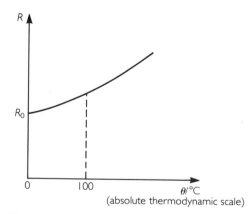

figure 7

The constants R_0, A and B can be determined by measurements at the ice-point (0 °C), steam point (100 °C) and sulphur-point (444.6 °C), which is the boiling point of sulphur at standard atmospheric pressure.

Example

Temperature on a mercury scale/°C	0	30	60	90	100	150	200	250	300
e.m.f. of thermocouple/mV	0	0.90	1.69	2.36	2.56	3.36	3.84	4.00	3.86

A student is asked to compare the thermometer scales of a mercury-in-glass thermometer designed for the range $-10\,°C$ to $+360\,°C$, and a copper–iron thermocouple (see page 452). He places the two thermometers side by side in a metal block and raises the temperature slowly. He obtains the data above.

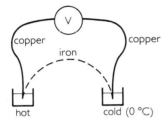

figure 8

a Plot a graph of e.m.f. against mercury temperature.
b What is the *fundamental interval*, i.e. $X_{100} - X_0$, on the thermocouple scale?
c When the mercury thermometer reads $50\,°C$, what temperature does the thermocouple indicate?
d What does the thermocouple indicate the temperature to be when the mercury thermometer indicates a temperature of $300\,°C$?

a

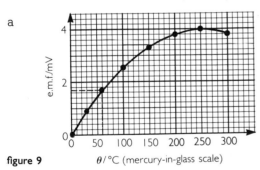

figure 9

b The fundamental interval is

$$\varepsilon_{100} - \varepsilon_0 = (2.56 - 0)\,mV = 2.56\,mV$$

c From the graph, when the mercury-in-glass thermometer reads $50\,°C$,

$$\varepsilon = 1.44\,mV$$

$$\Rightarrow \theta\ (\text{thermocouple scale}) = \frac{\varepsilon_\theta - \varepsilon_0}{\varepsilon_{100} - \varepsilon_0} \times 100\,°C$$

$$= \frac{1.44}{2.56} \times 100\,°C$$

$$= 56\,°C$$

d $\quad \theta\ (\text{thermocouple scale}) = \frac{\varepsilon_\theta - \varepsilon_0}{\varepsilon_{100} - \varepsilon_0} \times 100\,°C$

$$= \frac{3.84}{2.56} \times 100\,°C$$

$$= 150\,°C$$

i.e. when the mercury thermometer indicates $300\,°C$ this thermocouple indicates $150\,°C$.

The international scale of temperature

The thermodynamic scale is only open to direct realisation through using gas thermometers. Because of the experimental difficulties involved in doing this, a practical scale of temperature called the *International Scale of Temperature* was introduced. The scale conforms with the thermodynamic scale as closely as possible. In order to make use of it, certain fixed points had to be set up. The sulphur-point $(444.6\,°C)$ is one of these internationally-accepted values.

Zeroth law of thermodynamics

This very basic law was proposed long after the first law had become an accepted part of physics, hence its unusual name. Consider figure 10, which shows three bodies separated by either good conducting walls or good insulating walls.

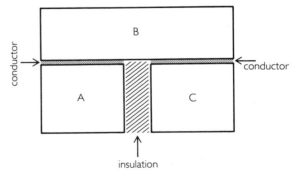

figure 10

The zeroth law states that after some time body A will be in thermal equilibrium with body B, and body C will also be in thermal equilibrium with body B. Therefore body A must be in thermal equilibrium with body C, from which it is insulated.

The first law of thermodynamics

Consider a fixed mass of gas at pressure p trapped inside a cylinder, one end of which is sealed by a frictionless piston (see figure 11). If a force is applied to the piston causing it to move, then the gas will be compressed, and work will be done on the gas. That is, energy is given to the gas by the force doing the work.

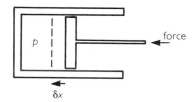

figure 11

In order to understand this, consider an individual molecule of the gas as it collides with the moving piston. The relative velocity of the molecule and piston before the collision is $(v + u)$. Because the collision is perfectly elastic, the relative velocity after the collision must be $(v + u)$. Therefore the speed of the molecule increases from v before the collision to $(v + 2u)$ after the collision.

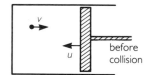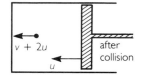

figure 12

Consider the piston moving a very small distance δx, due to the external force. If the external force causes the piston to move very slowly, then the applied external force must very nearly equal the force on the face of the piston due to the pressure of the gas, i.e. force = pressure × area. We are concerned with the work done on the gas, δW, by the piston when it moves a small distance δx.

work done = force × distance moved

$$\Rightarrow \quad \delta W = (pA).\delta x$$

$$= p.\delta V$$

As the piston moves, the pressure in the gas changes. Therefore the work done on the gas when the piston moves a finite distance is given by

$$\Delta W = \int_{V_1}^{V_2} p \, .dV$$

Note that it is often convenient to evaluate ΔW from the area beneath the appropriate p-against-V graph.

If, instead of pushing the piston in, the gas pushes the piston outwards, then we say that the gas does work on its environment.

Consider what can happen when heat, ΔQ, is supplied to an ideal gas.

| 1 | The mean kinetic energy of the gas molecule can increase. In the present discussion we shall say that the *internal energy U* of the gas increases by ΔU.

| 2 | Some of the energy may be used to do *external work*, ΔW, on the environment. That is, the gas pushes back the environment.

The *first law of thermodynamics* expresses the conservation of energy in this situation. Namely

| Heat energy supplied to the system | = | Increase in internal energy of the system | + | Work done on the environment by the system |

This equation is usually written as

$$\Delta Q = \Delta U + \Delta W$$

The quantity Q is sometimes referred to as the *enthalpy* of the system.

There are four methods of producing changes in gases which it is useful to consider:

☐   isothermal processes: this means T = constant;

☐   isovolumetric processes: this means V = constant;

☐   isobaric processes: this means p = constant;

☐   adiabatic processes: this means ΔQ = zero, i.e. no heat energy is supplied to it, or extracted from it.

These four processes are illustrated in the diagrams in figure 13.

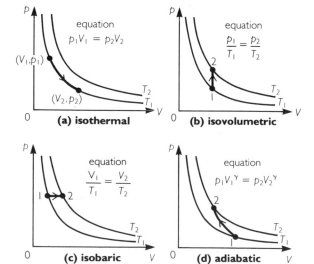

figure 13

The first three are quite straightforward and were studied earlier in chapter 42 as Boyle's law, the pressure law and Charles' law. The adiabatic case needs further explanation. The equation is derived below, after we have introduced molar heat capacities in the next section.

The molar heat capacities of gases

It is usual to consider only two different molar heat capacities, namely,

☐ the molar heat capacity when the gas remains at constant volume, C_V;

☐ the molar heat capacity when the gas remains at constant pressure, C_p.

Constant volume   We define molar heat capacity at constant volume, C_V, as

$$C_V = \frac{\Delta Q}{n \Delta T}$$

where

n = amount of gas present (in mole);
ΔQ = the heat energy given to the gas;
ΔT = the temperature rise produced in the gas.

But the first law of thermodynamics states that $\Delta Q = \Delta U + \Delta W$, therefore

$$C_V n \Delta T = \Delta U + \Delta W$$

But at constant volume we know that the external work $\Delta W = p \Delta V =$ zero.

Therefore $n C_V \Delta T = \Delta U$ (i)

Constant pressure   We define molar heat capacity at constant pressure, C_p, as

$$C_p = \frac{\Delta Q}{n \Delta T}$$

At constant pressure the external work $\Delta W = p \Delta V$. Substituting for ΔW and ΔQ in $\Delta Q = \Delta U + \Delta W$ gives

$$C_p n \Delta T = \Delta U + p \Delta V$$ (ii)

We have already seen that the equation of state for an ideal gas is

$$pV = nRT$$

Therefore, if the pressure is constant, then for a small change in temperature we shall have

$$p \Delta V = nR \Delta T$$

If we substitute this and equation (i) into equation (ii) we obtain

$$C_p n \Delta T = n C_V \Delta T + nR \Delta T$$

This simplifies to

$$C_p - C_V = R$$

This is true for any ideal gas, i.e. one for which there is no intermolecular potential energy. It is sometimes known as Mayer's equation.

Adiabatic processes in ideal gases

This derivation is only included for completeness. Knowledge of the final equation is all that is needed for A-level courses.

In adiabatic changes heat energy neither flows out of, nor into a gas. That is $\Delta Q = 0$. From the first law of thermodynamics we can see that this implies
$$\Delta Q = \Delta U + \Delta W = 0$$

i.e. $$\Delta U = -\Delta W$$ (iii)

If we are dealing with very small changes then we may write $\Delta W = p \delta V$
From equation (i) we see that

$$\Delta U = n C_V \delta T$$

If we substitute for ΔW and ΔU in equation (iii) above we obtain

$$n C_V \delta T = -p \delta V$$ (iv)

Since we are dealing with an ideal gas we can make use of the equation of state, namely

$$pV = nRT$$

Using differential calculus, we can differentiate this to write

$$p \delta V + V \delta p \approx nR \delta T$$

Rearranging this equation gives

$$\delta T = \frac{1}{nR}[p \delta V + V \delta p]$$

Substituting for this in equation (iv) gives

$$n C_V \left\{ \frac{1}{nR}[p \delta V + V \delta p] \right\} = -p \delta V$$

i.e. $$C_V [p \delta V + V \delta p] = -Rp \delta V$$

Rearranging this and substituting for $(C_V + R)$ gives

$$C_V V \delta p = -(C_V + R)p \delta V = -C_p p \delta V$$

i.e. $$\frac{C_p}{C_V} \cdot \frac{\delta V}{V} = -\frac{\delta p}{p}$$

Putting $C_p/C_V = \gamma$, the ratio of the molar heat capacities, the equation becomes

$$\gamma \int_{V_1}^{V_2} \frac{dV}{V} = -\int_{p_1}^{p_2} \frac{dp}{p}$$

If we assume that C_p and C_V do not vary with temperature, which is the case for an ideal gas, then we can integrate this equation:

$$\gamma\int_{v_1}^{v_2}\frac{\mathrm{d}V}{V} = -\int_{p_1}^{p_2}\frac{\mathrm{d}p}{p}$$

to give

$$\gamma.\log_e\left(\frac{V_2}{V_1}\right) = -\log_e\left(\frac{p_2}{p_1}\right)$$

i.e.

$$p_1V_1^{\gamma} = p_2V_2^{\gamma} = \text{constant}$$

Thus the equation for adiabatic changes in an ideal gas is

$$p_1V_1^{\gamma} = p_2V_2^{\gamma} = \text{constant}$$

Such an adiabatic change is illustrated in figure 13(d). The final temperature in an adiabatic change can always be calculated by remembering that this is an ideal gas, therefore the ideal gas equation can always be applied.

The value of γ

Earlier, in chapter 42, we saw that for an ideal gas the average kinetic energy of the molecules is $\frac{3}{2}kT$. We also defined the molar heat capacity at constant volume as

$$C_V = \frac{\Delta Q}{n\Delta T} = \frac{\Delta U}{n\Delta T}$$

Therefore we can write

$$C_V = \frac{\Delta U}{n\Delta T} = \frac{nN_A[\frac{3}{2}k\Delta T]}{n\Delta T}$$

$$\Rightarrow \qquad C_V = \frac{3}{2}kN_A = \frac{3}{2}R$$

Also we derived Mayer's equation
$$C_p - C_V = R$$

$$\Rightarrow \qquad C_p = R + C_V = \frac{5}{2}R$$

$\Rightarrow$ For an ideal monatomic gas,

$$\gamma = \frac{C_p}{C_V} = \frac{\frac{5}{2}R}{\frac{3}{2}R} = \frac{5}{3} = 1.67$$

Example Consider an ideal gas at a pressure of 2.0×10^5 Pa. If it is expanded adiabatically until its volume is doubled, what will be its new pressure? (Assume $\gamma = 1.67$.)

$$p_1V_1^{\gamma} = p_2V_2^{\gamma}$$

$$\Rightarrow \qquad p_2 = p_1\left(\frac{V_1}{V_2}\right)^{\gamma}$$

$$= 2.0 \times 10^5 \text{ Pa } (\tfrac{1}{2})^{1.67}$$

$$= 0.63 \times 10^5 \text{ Pa}$$

Equipartition of energy

So far we have assumed that our ideal gas consists of infinitesimally small, hard, spherical molecules; and that these molecules can only have translational kinetic energy,

i.e.   total E_k of a molecule $= \frac{1}{2}mc_x^2 + \frac{1}{2}mc_y^2 + \frac{1}{2}mc_z^2$

We averaged the x-, y- and z-components separately, and assumed that there was no preferred direction for molecular motion. This resulted in the mean kinetic energy associated with each degree of freedom (x, y or z direction in this case) being

$$\tfrac{1}{2}m\overline{c_x^2} = \tfrac{1}{2}m\overline{c_y^2} = \tfrac{1}{2}m\overline{c_z^2} = \tfrac{1}{3}(\tfrac{1}{2}m\overline{c^2}) = \tfrac{1}{2}kT$$

This is known as the *principle of the equipartition of energy*.

This assumption led to $\gamma = 1.67$ for ideal gases, and resulted in the adiabatic equation

$$p_1V_1^{\gamma} = p_2V_2^{\gamma}$$

It turns out that the monatomic gas helium obeys this equation well when we take $\gamma = 1.67$. But in experiments diatomic hydrogen gas does *not* obey this equation, if 1.67 is taken as the value for γ.

The diatomic molecule illustrated in figure 14 can have rotational kinetic energy about the x-axis and the y-axis as well as translational kinetic energy.

i.e. total kinetic energy E_k of a molecule
$$= \tfrac{1}{2}mc_x^2 + \tfrac{1}{2}mc_y^2 + \tfrac{1}{2}mc_z^2 + \tfrac{1}{2}I_x\omega_x^2 + \tfrac{1}{2}I_y\omega_y^2$$

where ω_x is the angular velocity of the molecule about the x-axis, and I_x is the moment of inertia of the molecule about the x-axis. (See chapter 4 for a discussion of ω_x and I_x.)

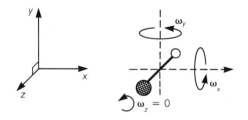

figure 14   Rotational states of a diatomic molecule

If we assume that kinetic energy is randomly shared between all *five* terms in this equation, and that each term has a mean value of $\frac{1}{2}kT$ as we saw earlier, then

$$\overline{E_k} \text{ of the molecules} = 5(\tfrac{1}{2}kT)$$

This means that an increase in the internal energy ΔU of this ideal gas of diatomic molecules will be given by

$$\Delta U = nN_A\{\tfrac{5}{2}k\Delta T\}$$

Earlier we saw that the molar heat capacity at constant volume C_V was defined as $C_V = \Delta U/n\Delta T$. Therefore substituting for ΔU gives for a monatomic gas:

$$C_V = \frac{\Delta U}{n\Delta T}$$
$$= \frac{nN_A[\frac{3}{2}k\Delta T]}{n\Delta T}$$
$$= N_A[\tfrac{3}{2}k]$$
$$= \tfrac{3}{2}R$$

By comparison a diatomic ideal gas will have

$$C_V = \frac{\Delta U}{n\Delta T}$$
$$= N_A[\tfrac{5}{2}k]$$
$$= \tfrac{5}{2}R$$

Also $\qquad C_p - C_V = R$

$\Rightarrow \qquad\qquad C_p = C_V + R$
$$= \tfrac{5}{2}R + R$$
$$= \tfrac{7}{2}R$$

So for a diatomic ideal gas,

$$\gamma = \frac{C_p}{C_V} = \frac{\frac{7}{2}R}{\frac{5}{2}R} = 1.40$$

Hydrogen obeys the adiabatic equation if γ is given the value 1.4.

The triatomic molecule illustrated in figure 15 can have rotational kinetic energy about its z-axis as well as about its x- and y-axes. This would also be true for any molecule with more than three atoms.

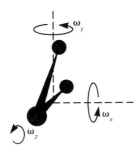

figure 15 Rotational states of a triatomic molecule

Now we shall assume that kinetic energy is randomly shared between all *six* degrees of freedom, and that each term has a mean value of $\frac{1}{2}kT$. For the total kinetic energy E_k we shall now have

$$E_k = \tfrac{1}{2}mc_x^2 + \tfrac{1}{2}mc_y^2 + \tfrac{1}{2}mc_z^2 + \tfrac{1}{2}I_x\omega_x^2 + \tfrac{1}{2}I_y\omega_y^2 + \tfrac{1}{2}I_z\omega_z^2$$

and $\qquad\qquad C_V = N_A[\tfrac{6}{2}k] = 3R$

Also $\qquad\qquad C_p = C_V + R$
$$= 3R + R$$
$$= 4R$$

Therefore for a polyatomic ideal gas we have

$$\gamma = \frac{C_p}{C_V} = \frac{4R}{3R} = 1.33$$

Methane obeys the adiabatic equation if γ is given the value 1.33.

Example Consider a container full of hydrogen at low pressure. The temperature of the system is 27 °C. If the volume of the container is rapidly doubled what will be the new temperature of the gas?

Hydrogen is diatomic, so $\gamma = 1.40$. Rapid expansion means that there is not enough time for heat to flow into the system. Therefore we can assume the change is adiabatic.

$\Rightarrow \qquad p_1V_1{}^\gamma = p_2V_2{}^\gamma$

$\Rightarrow \qquad \dfrac{p_2}{p_1} = \dfrac{V_1{}^\gamma}{V_2{}^\gamma} \qquad\qquad\qquad$ (i)

Also $\qquad \dfrac{p_1V_1}{T_1} = \dfrac{p_2V_2}{T_2} \qquad$ (ideal gas equation)

$\Rightarrow \qquad \dfrac{p_2}{p_1} = \dfrac{T_2V_1}{T_1V_2} \qquad\qquad\qquad$ (ii)

Combining (i) and (ii) leads to

$$\frac{V_1{}^\gamma}{V_2{}^\gamma} = \frac{T_2V_1}{T_1V_2}$$

$\Rightarrow \qquad T_2 = T_1\left(\dfrac{V_1}{V_2}\right)^{\gamma-1}$
$$= (27 + 273)\,\text{K} \times (\tfrac{1}{2})^{1.40-1}$$
$$= 227\,\text{K}$$
$$= -46\,°\text{C}$$

The second law of thermodynamics

This law was first proposed by Sadi Carnot in 1824, 25 years before Lord Kelvin invented what is now known as the first law of thermodynamics, and long before Fowler formulated the zeroth law in 1931. The second law of thermodynamics enables us to predict in which direction a process will proceed. This idea is illustrated in figure 16, which shows a beaker containing marbles of two different colours, but of the same size; (a), (b), (c) and (d) represent the situation at different times. Between each drawing the beaker was shaken a little.

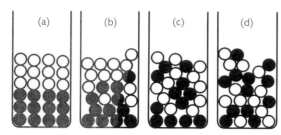

figure 16   Marbles in a beaker

Consider the question, 'Which order is correct? Did (a) occur first followed by (b), (c) and then (d); or was (d) first, ending up with (a)?' From your past experience you would probably reason that chaotic systems do not, left to themselves, become ordered, whereas ordered systems do become disordered. Therefore, the order must be (a), (b), (c) then (d). You deduced this answer because you know that *there is a tendency on the part of nature to proceed towards a state of greater chaos.* You have just applied the second law of thermodynamics.

Here are two other examples illustrating this movement from order to disorder.

1 Two isolated, different gases in connecting chambers diffuse into one another when the partition is removed; see figure 17. At any later time you will never find the two gases separated into their initial chambers. Order becomes disorder, but disorder does not become order (without outside interference).

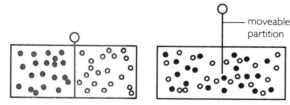

figure 17   Mixing gases

2 A car travelling at speed has kinetic energy. When the brakes are applied the road, the tyres, the brake-pads and the brake-drums become hot; the energy is spread out into the surrounding molecules. The heat energy once generated in this way never turns itself back into kinetic energy of the car.

In all the above cases there is a slight probability that the reverse order might occur. That is, the marbles might become ordered, the gases might separate. But all of these reverse processes are highly unlikely: so unlikely, in fact, that we are justified in using the word 'never'. You will now see that the second law of thermodynamics is a statistical law: a law of probability.

Today it is far more usual for scientists to think of the second law in terms of *entropy*:

'Any spontaneous process (change) in an isolated system always results in an increase in the entropy of that system.'

From what we have said so far, this *entropy* must be a measure of the chaos that results from a process. We now need to explain just what we mean by entropy.

Entropy

In the nineteenth century the second law of thermodynamics was invented by heat engineers in their quest to improve steam engines and steam turbines. Consequently they thought of entropy in terms of heat energy exchanges and temperature. Today we emphasise the statistical nature of the second law. The two approaches can be related by considering heat passing from a hot body to a cold one. In order to simplify the ideas we shall consider a model with two solids, of just three atoms each; see figure 18(a). The hot body, at temperature T_1, has four quanta of heat energy, and the cold body, at temperature T_2, has only two quanta. The quanta of heat energy can be arranged in different ways between the three atoms in each solid; some of the other possible arrangements are illustrated in figure 18(b).

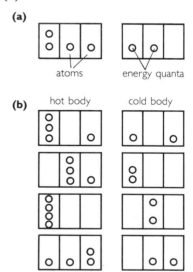

figure 18

You should be able to convince yourself that there are 15 different ways of sharing four quanta between three atoms, and 6 different ways of sharing two quanta between three atoms. Because each arrangement of quanta in the hot body is independent of the arrangement in the cold body, there are $6 \times 15 = 90$ different ways of arranging the quanta with the bodies as they are.

When the two bodies are placed in thermal contact with one another, heat energy will flow from the hot to the cold body. Here we shall impose a known experimental result on our model; the two bodies will achieve equal temperatures, so they will share the quanta of energy: 3 quanta each, as shown in figure 19(a). There are 10 different ways of sharing three quanta between three different atoms; some of the other possible arrangements are shown in figure 19(b). Because the arrangements in the two bodies are independent, there are $10 \times 10 = 100$ different ways of arranging the quanta.

(a)

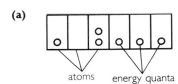

atoms   energy quanta

(b)

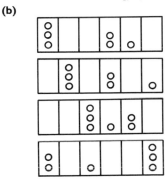

figure 19

What we have shown is that putting a hot body in contact with a cold body results in an *increase* in the possible ways of sharing the quanta. Increasing the possible arrangements means increasing the possible disorder of a system.

We need to generalise this analysis to the case where we have realistic solids of many atoms, sharing many heat quanta.

Let W_1 = the number of different arrangements of quanta in the hot body,
W_2 = the number of different arrangements of quanta in the cold body,
and W_1' and W_2' = the values when the bodies have exchanged some heat energy.

The number of different ways, W, of arranging the quanta in the two separate bodies, before thermal contact is made, is

$$W = W_1 W_2$$

Now consider the entropy.

Let S_1 = the entropy of the hot body,
S_2 = the entropy of the cold body.

Therefore the entropy S of the combined system of two bodies will be given by

$$S = S_1 + S_2$$

assuming that entropies of systems are additive.

If S is a measure of the chaos, then this equation has to be related to

$$W = W_1 W_2$$

This suggests that we should choose $S = k \ln W$, where k is a positive constant.

When we do this: $S = S_1 + S_2$

becomes $k \ln W = k \ln W_1 + k \ln W_2$

$\Rightarrow$ $k \ln W = k \ln (W_1 W_2)$

where k is a constant.

Earlier we quoted the second law of thermodynamics as 'Any spontaneous process (change) in an isolated system always results in an increase in the entropy of that system.' Let us see if this is true in the simple three-atom solids we considered. The changes in entropy of the hot and cold bodies are given by:

hot: $\Delta S_1 = k \ln W_1' - k \ln W_1$

$$= k \ln \left(\frac{W_1'}{W_1} \right)$$

$$= k \ln (10/15)$$

$$= -0.41 \, k$$

cold: $\Delta S_2 = k \ln \left(\frac{W_2'}{W_2} \right)$

$$= k \ln (10/6)$$

$$= +0.51 \, k$$

Therefore, the total entropy change ΔS is given by

$$\Delta S = \Delta S_1 + \Delta S_2$$

$$= -0.41 \, k + 0.51 \, k$$

$$= +0.10 \, k$$

This is an increase in overall entropy, as it should be. Notice that the entropy of the hotter body decreases ($\Delta S_1 = -0.41 \, k$); but the entropy of the colder body increases by a greater numerical amount ($\Delta S_2 = +0.51 \, k$). This net increase in entropy is what makes the change likely to occur.

Earlier we said that we would relate the statistical approach to entropy with that adopted by the heat engineers of the last century. The easiest way to do this is to deal with the expansion of an ideal gas.

We shall start by considering another simple model: a game in which counters marked 1 to 6 are shuffled between the two halves of a board; see figure 20(a). The game proceeds by throwing a six-sided die. The counter whose number comes up is then moved onto the opposite half of the board. If a class of twenty students plays this game, each student throwing the die 1000 times and keeping a running tally of how many counters are in each half, an average over the whole class, for the whole session, will reveal three counters in each half. The random, chaotic behaviour of nature, *pure chance*, produces a definitie average equilibrium situation: 3 in each half.

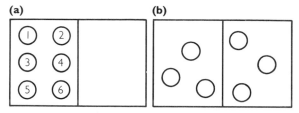

(a)   **(b)**

figure 20

Now imagine a bigger board, with each half subdivided into W different positions, and with lots of counters. There are W different possible positions for a counter on the left half of the board. There is an equal number for the same counter on the right half of the board. Hence, if we are dealing with one counter only, the total number of different possible arrangements (positions) on both boards, W', would equal $2W$. That is $W'/W = 2$ for one counter when it has both halves to choose from. Each additional counter can be placed in two ways, left or right. Since the two ways for any additional counter combine with all previous possibilities of the previous counters, adding an extra counter *doubles* the previous number of arrangements. It follows that doubling the volume available to N molecules of an ideal gas increases the number of possible arrangements of the molecules by the factor $W'/W = 2^N$.

Let us now consider what happens if the volume of the gas is only increased by a very small amount ΔV, as illustrated in figure 21.

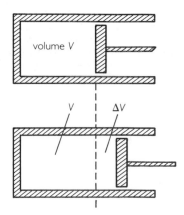

volume V

V ΔV

figure 21   Gas trapped in cylinder by moveable piston

When doubling the volume.

$$W'/W = 2^N$$

By extension, increasing the volume from V to $(V + \Delta V)$ will increase the number of possible arrangements by

$$W'/W = [(V + \Delta V)/V]^N$$
$$= (1 + \Delta V/V)^N$$

Earlier we showed that the change in entropy ΔS was given by

$$\Delta S = k(\ln W' - \ln W)$$

i.e. $$\Delta S = k\ln(W'/W)$$

Therefore in this case

$$\Delta S = k\ln[(1 + \Delta V/V)^N]$$
$$= kN\ln(1 + \Delta V/V)$$

If $\Delta V/V$ is small then $\ln(1 + \Delta V/V) \simeq \Delta V/V$ (check this on your calculator for $\Delta V/V = 0.01$). This gives

$$\Delta S \simeq kN\Delta V/V$$

For an ideal gas (see chapter 42)

$$pV = kNT$$
$$\Rightarrow \qquad kN = pV/T$$

Substituting for kN in $\Delta S = kN\Delta V/V$ gives

$$\Delta S = (pV/T)(\Delta V/V)$$
$$= p\Delta V/T$$

In this case, the first law of thermodynamics states

$$\Delta Q = \Delta U + p\Delta V$$

Any temperature change will be negligible if $\Delta V/V \ll 1$. Therefore $\Delta U \simeq 0$, and $\Delta Q \simeq p\Delta V$

Hence $$\Delta S \simeq \Delta Q/T$$

This was the expression derived by the heat engineers in the last century. Note that we have identified the constant k in

$$\Delta S = k\ln(W'/W)$$

with the Boltzmann constant. The definition $\Delta S = \Delta Q/T$ indicates that the units for entropy are JK^{-1}.

Thus the change in entropy is defined, from a statistical point of view, as

$$\Delta S = k(\ln W' - \ln W)$$

and from a heat-transfer point of view as

$$\Delta S = \Delta Q/T$$

An application of the second law: the efficiency of a heat engine

Many forms of engine depend on extracting mechanical work from heat. The turbine/generator sets in power stations and the motor-car engine are two examples. In both cases a fuel is burnt to produce heat, and a fluid – steam in the power station, exhaust gases in the car – is made to do work on its way from a hot region, at temperature T_h, to a cooler region at T_c. Figure 22 illustrates the process.

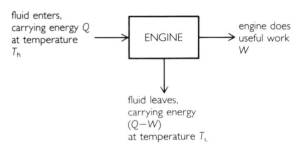

figure 22   Basic principle of a heat engine

Now the furnace which provides the heat to the fluid suffers an entropy loss. If its change in entropy is ΔS_f, then

$$\Delta S_f = -\frac{Q}{T_h}$$

But the fluid passes on heat $(Q - W)$ to the heat 'sink', which has an entropy gain ΔS_s given by

$$\Delta S_s = +\frac{(Q - W)}{T_c}$$

Thus the net entropy change of the universe ΔS is given by

$$\Delta S = \Delta S_f + \Delta S_s$$
$$= -\frac{Q}{T_h} + \frac{Q - W}{T_c}$$

The second law says

$$\Delta S \geqslant 0$$

$$\Rightarrow \quad \frac{Q - W}{T_c} - \frac{Q}{T_h} \geqslant 0$$

A few lines of algebra leads to this result:

$$\frac{W}{Q} \leqslant \frac{T_h - T_c}{T_h}$$

Now W/Q is the efficiency of the system: the fraction of the initially available energy Q which has become useful work W. So we have, for a heat engine,

$$\text{efficiency} \leqslant \frac{T_h - T_c}{T_h}$$

η is the usual symbol used for efficiency.

Notes

1   This expression sets a maximum possible value for the efficiency. For given values of T_h and T_c, there is a maximum attainable efficiency, which is inevitably less than 1, or 100%. In practice an engine is unlikely to reach even this efficiency.

2   As an example, consider a typical power station. T_h is the steam temperature – say 500 °C. T_c is the cooling-tower temperature – roughly the atmospheric temperature, say 20 °C.

$$\text{Then efficiency} \leqslant \frac{(500 + 273)\,\text{K} - (20 + 273)\,\text{K}}{(500 + 273)\,\text{K}}$$

$$\leqslant \frac{480}{773} \text{ or } 62\%$$

In practice, power stations rarely attain efficiencies greater than about 30%. But the point is that a large wastage of heat is *inevitable*, however good the design.

3   The best efficiencies are obtained using engines where T_h is as high as possible, as you can see from the equation above. Thus, for example, car engines are being designed to work at higher and higher temperatures, which in turn demands better materials; jet engines operate at such high temperatures that the metal is continuously red-hot; and power-station engineers aim for steam temperatures as high as they can design pressure containers to accommodate.

Summary

1 a The temperature of a system is the property that determines whether or not a system is in thermal equilibrium with other systems.
   b The *lower fixed point*, 0 °C, is the melting point of ice at standard atmospheric pressure.
   c The *upper fixed point*, 100 °C, is the boiling point of pure water at standard atmospheric pressure.
   d Temperature is defined by

$$\theta = \frac{X_\theta - X_0}{X_{100} - X_0} \times 100 \,°C$$

   e Mercury-in-glass thermometer, its construction and limitations.
   See figure 2.
   f The constant-volume gas thermometer.
   See figure 3.
   g The ideal-gas scale of temperature is defined by $pV = nRT$.
   h The absolute thermodynamic scale of temperature.
   i The platinum resistance thermometer, and its improved equation $R_\theta = R_0(1 + A\theta + B\theta^2)$.
   j The thermocouple thermometer.
   See figure 6.
   k Disagreement between scales of temperature.
   l The international scale of temperature.

2 Zeroth law of thermodynamics:
   'If two systems are separately in thermal equilibrium with a third, then they must also be in thermal equilibrium with each other.'

3 a The first law of thermodynamics:
$$\Delta Q = \Delta U + \Delta W$$

   b The molar heat capacity at constant volume
$$C_V = \frac{\Delta Q}{n\Delta T} = \frac{\Delta U}{n\Delta T}$$

   c The molar heat capacity at constant pressure
$$C_p = \frac{\Delta Q}{n\Delta T} = \frac{\Delta U + \Delta W}{n\Delta T}$$

   d Mayer's equation: $(C_p - C_V) = R$
   e Adiabatic processes: $p_1 V_1{}^\gamma = p_2 V_2{}^\gamma$

4 Equipartition of energy
   a For a monatomic gas $\overline{E_k} = \frac{3}{2}kT$
   b For a diatomic gas $\overline{E_k} = \frac{5}{2}kT$
   c For a polyatomic gas $\overline{E_k} = \frac{6}{2}kT$
   d Values of the ratio $\gamma = C_p/C_V$:
   (i) Monatomic gas $\gamma = 1.67$
   (ii) Diatomic gas $\gamma = 1.40$
   (iii) Polyatomic gas $\gamma = 1.33$

5 The second law of thermodynamics:
   a 'Nature tends towards a state of greater chaos.'
   b A spontaneous change in a system results in

$$\Delta S > 0 \text{ for the system.}$$

   Both (a) and (b) are statements of the second law.

   c
$$\Delta S = k(\ln W' - \ln W)$$

   d
$$\Delta S = \frac{\Delta Q}{T}$$

   e The maximum theoretical efficiency η of a heat engine is given by

$$\eta \leqslant \frac{T_h - T_c}{T_h}$$

Questions

1 a What is meant by temperature?
  b Explain what is meant by a scale of temperature, and discuss the principles of setting up such a scale.

In the following table, column one lists examples of thermometers, column two indicates the thermometric properties on which the instruments depend, and column three gives the

details of the measurements which must be made.

Fill in the blank spaces labelled (c) to (h) in the following table.
  c For each of the thermometers in the table indicate the approximate range of temperatures over which it is usable.

Thermometer	Thermometric property	Measurements
Mercury-in-glass	change in length of the mercury thread	the position of the end of the mercury thread relative to the two fixed points on the scale
Constant-volume gas thermometer	(c)	(d)
Thermocouple	(e)	(f)
Platinum resistance thermometer	(g)	(h)

2   This question explores how the accuracy and sensitivity of a mercury-in-glass thermometer depends on various properties of the materials used, and on the external environment.

The essential features of a mercury-in-glass thermometer are illustrated in figure 23. A commercial mercury-in-glass thermometer has two marks etched on the outside of its glass tube, corresponding to the internationally agreed lower and upper fixed points. The distance between these two etched marks is then divided into 100 equal divisions.

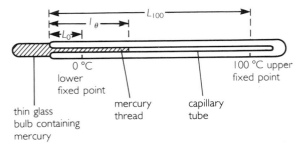

figure 23

a   Explain what is meant by lower and upper fixed points.
b   What thermometric property is used in the mercury-in-glass thermometer?
c   How is the temperature θ on the Celsius scale defined for these thermometers?

Let us now consider an ideal thermometer made from glass which does not expand, and which has a perfectly uniform bore to its capillary tube. Figure 24 shows how mercury expands with respect to the ideal-gas temperature.

d   Explain what is meant by 'ideal-gas temperature'.
e   Use the graph to explain whether a mercury-in-glass thermometer reads too high or too low

compared with an ideal-gas thermometer when measuring a temperature (i) about 50 °C, and (ii) about 120 °C.
f   Real thermometers do not have exactly uniform bores to their capillary tubes. Discuss what effect this will have on the readings.

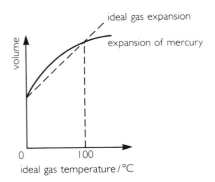

figure 24

The following two graphs (figure 25) show how two different glasses expand with respect to the ideal gas temperature. The shallow slope of these graphs is to indicate that the cubic expansivity of glass is less than that of mercury, which was illustrated in figure 24.

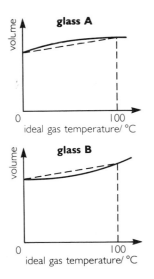

figure 25

g   Compare and contrast the readings you would obtain on two real mercury-in-glass thermometers, one made from glass A, and the other one made from glass B.
h   Why does the mercury reservoir at the end of the thermometer have very thin glass walls?
i   A student wishing to know the temperature at which water boils in a domestic pressure

cooker drills a hole in the lid and seals a mercury-in-glass thermometer through the hole. (It would be *extremely dangerous* to try this.) He sets the pressure control to 2 atmospheres. Explain why the temperature reading that he obtains is not an accurate representation of the temperature inside the chamber.

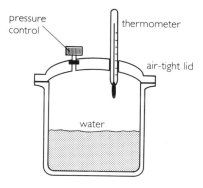

figure 26

The thermometers used by a group of students have scales which are 20 cm long. The divisions etched on the glass are 1 mm apart. Therefore the thermometer has been designed to be read to the nearest 0.5 °C. The students are not satisfied with this sensitivity. One student suggests that if the thermometers had been made with capillary tubes of finer bore then the mercury expansion could be 5 mm for a temperature rise of 0.5 °C, which would mean that the thermometer could be read to the nearest 0.1 °C. Another student says that it would be better to use a capillary tube of the same bore, and obtain the extra expansion by having a bulb of mercury which is five times larger.

j　Compare and contrast their ideas, which both give the same sensitivity, and explain which one gives the greater accuracy.

3　a　The words 'heat' and 'temperature' are often confused in everyday speech. Write a short discussion, aimed at non-scientists, to distinguish clearly between the two.

　　b　Explain how you would set up a scale of temperature.

　　c　Describe the essential details of a constant-volume gas thermometer, and discuss the choice of gas.

　　d　Why is it not convenient to use a constant-volume gas thermometer in most practical situations? If it is usually not convenient to use it, then to what use is it put?

Discuss the relative merits of (i) a mercury-in-glass thermometer, (ii) a platinum resistance thermometer, (iii) a thermocouple thermometer, for measuring:

e　the temperature inside an oven maintained at about 300 °C;

f　the temperature inside a domestic pressure cooker.

4　a　Explain why some thermometric properties are better than others for measuring temperature.

　　b　Explain why the constant-volume gas thermometer is used as a standard, and describe how it is used to realise the ideal-gas scale of temperature.

　　c　Compare and contrast the advantages and the disadvantages of the constant-volume gas thermometer with those of the thermocouple.

　　d　Discuss the factors which affect the sensitivity of a mercury-in-glass thermometer.

　　e　Describe what factors affect the accuracy of a mercury-in-glass thermometer, and then explain why a mercury-in-glass thermometer is not a suitable instrument for temperature measurement by a skin diver operating on the sea bed.

5　a　Explain what are meant by the lower fixed point and the upper fixed point.

Figure 27 shows a mercury-in-glass thermometer, where L_0, L_{100} and L_θ are the lengths of the mercury thread at 0 °C, 100 °C and θ.

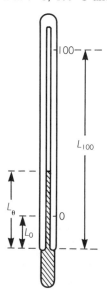

figure 27

b　Define the Celsius scale of temperature for the mercury-in-glass thermometer.

The following table of data shows the values of the thermometric properties of four different thermometers when they are all used to measure the temperature (at the same place and time) in a factory. The table also contains the thermometric values at the steam point and the ice point.

Thermometer	Thermometric value at steam point	Thermometric value at ice point	Thermometric value at factory temp.	Factory temp.
Platinum resistance thermometer Resistance/Ω	18.30	16.80	17.41	
Thermocouple p.d./mV	4.10	0.000	1.61	
Mercury-in-glass Length L/cm	27.9	2.00	10.4	
Constant volume Gas pressure/10^5 Pa	1.343	0.983	1.127	

c The final column of the table has been left blank. Use the definition of the Celsius scale of temperature to determine the temperature of the factory according to each thermometer.

d What do you think the temperature in the factory is on the absolute thermodynamic scale of temperature?

e Sketch graphs to indicate the variation with absolute thermodynamic temperature of (i) the resistance of the platinum resistance thermometer, (ii) the e.m.f. of the thermocouple.

6 a Define the Celsius scale of temperature.

The International Practical Temperature Scale was defined in 1968. The standard instrument used for measuring temperatures between 630 °C and 1064 °C is a thermocouple made with one wire of platinum and the other a platinum(10%)–rhodium alloy. The e.m.f.–temperature relationship is represented by a quadratic equation of the form

$$E = a + bt + ct^2$$

where t is the International Practical Temperature.

The e.m.f.s generated by such a thermocouple at certain fixed points are given in the table below.

International practical temp/°C	0	100	200	700	800	1000
e.m.f./μV	0	645	1440	6274	7345	9585

b Sketch graphs to indicate how the e.m.f. of this thermocouple varies with temperature over the range (i) 0–100 °C, (ii) 700–1000 °C.

c The equation defining the Celsius scale of temperature arbitrarily defines a linear relationship between thermometric property and temperature. Calculate the Celsius temperature corresponding to an e.m.f. of 1440 μV. Compare your answer with the International Practical Temperature.

d Determine the constants a, b and c for the equation $E = a + bt + ct^2$ in the range 0–200 °C.

e Now use your values of the constants a, b and c to find the disagreement between the fixed point value of 800 °C and that value determined from the equation $E = a + bt + ct^2$.

7 a The first law of thermodynamics can be written in the form

$$\Delta Q = \Delta U + \Delta W.$$

Explain what the terms ΔQ, ΔU and ΔW represent.

b In each of the following changes, state and explain whether the values of ΔQ, ΔU and ΔW are positive, negative or zero.

(i) A gas in an insulated, non-expanding chamber is heated from 293 K to 323 K.

(ii) A gas in a metal chamber is slowly compressed in volume from 1 ℓ to 0.1 ℓ, so that its temperature remains constant.

(iii) A party balloon in a room deflates slightly during the night when the central heating goes off.

(iv) A gas in an insulated piston chamber is rapidly expanded from 0.1 ℓ to 1 ℓ.

(v) A liquid in an insulated bowl which is covered by an insulating lid is stirred by an electric beater which is driven by an electric motor above the lid.

(vi) An *ideal* gas in an insulated chamber occupies only half of the chamber (figure 28). The other half of the chamber is evacuated. When the partition is removed the gas expands to fill the whole of the chamber.

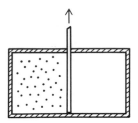

figure 28

(vii) Consider a *real* gas occupying the half chamber in the above question. What are the answers this time when the partition is removed?

(viii) A soap film formed between two pieces of wire, as illustrated in figure 29, is extended by pulling the piece of wire AB back to position CD. (Regard the soap film as the system.)

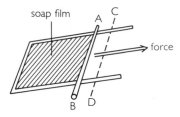

soap film

force

figure 29

(ix) A small amount of petrol is burned in a sealed copper chamber which is full of oxygen. (Regard the mixture of petrol and oxygen as the system.)

8 a Explain what is meant by the first law of thermodynamics.
   b Explain why the molar heat capacity at constant pressure, C_p, is greater than the molar heat capacity at constant volume, C_V, as is indicated by Mayer's equation, $C_p - C_V = R$, for an ideal gas. What does R represent in this equation?
   c What is meant by equipartition of energy?
   d Explain why the mean energy of the molecules is greater for diatomic molecules, such as oxygen or nitrogen, than it is for monatomic molecules, such as helium or argon, when they are at the same temperature.
   e Use Mayer's equation, from part (b), to show that, for an ideal monatomic gas, the ratio $\gamma = C_p/C_V$ of the molar heat capacities should be equal to 5/3.

9 a State the first law of thermodynamics.
   b What evidence is there to substantiate this law?

c Why, when quoting the molar heat capacity of a gas, is it essential to give the physical conditions under which the temperature rise occurs? In most books only two molar heat capacities for gases are quoted. What are the physical conditions for these two heat capacities?
d Explain what is meant by (i) an adiabatic change, (ii) an isothermal change, (iii) an isovolumetric change, (iv) an isobaric change.

Consider helium gas, at 1 atmosphere pressure and a temperature of 300 K, in a piston chamber of volume 2 ℓ (2000 cm³). Call this state A.
  The piston is *slowly* pushed in until the pressure has increased from 1 atmosphere to 2 atmospheres (state B). Then it is pushed in *rapidly* until it reaches 5 atmospheres pressure (state C).
e Explain what type of change each pressure change represents.
f Determine the temperature and volume (i) at the end of the slow change, at state B; (ii) at the end of the rapid change, at state C. Sketch these changes on a p-against-V graph.

Now consider the helium expanding *slowly* from state C to a further state, D; then *rapidly* expanding back to the conditions of state A.
g Determine the temperature, pressure and volume that state D must have if the final state, A, is to be achieved.
h Sketch these changes onto your previous graph. The final picture should form a closed circuit of four sides, each side being curved.
i State, with reasons, which is greater: the work done on the helium during compression, or the work done by the helium during expansion.

10 This question is about the change in entropy ΔS which occurs when a gas expands isothermally to twice its original volume. We will obtain expressions for ΔS using both the statistical and the heat-transfer points of view, and show that they are equivalent.
  Consider 1 mole of an ideal gas. The equation of state $pV = nRT$ applies, therefore, with $n = 1$ mole. The gas expands isothermally from volume V_0 to volume $2V_0$. See figure 30.

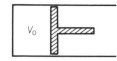

figure 30

Statistical view Imagine the new volume to be like two halves of a box: initially all the molecules were in the left hand half, now they are randomly distributed in both halves.

a What is the number of ways W for them all to be in the left hand half?

b There are N_A molecules. What is the number of ways W' that they could be randomly distributed in both halves.

c Write down the statistical-view expression for ΔS.

d Put your values for W and W' into this expression. Evaluate it as far as you can. You should be able to express it as $kN_A\ln 2$, and show that this has the value 5.8 J K^{-1}.

Heat-transfer view   For this view, we begin with the first law: $\Delta Q = \Delta U + \Delta W$.

e Explain why $\Delta U = 0$ for the expansion which occurs.

f ΔW can be evaluated using the equation

$$\Delta W = \int_{V_0}^{2V_0} p\,dV$$

Substitute for p using the equation of state, and hence evaluate ΔW. It should come out as $nRT\ln 2$.

g Write down the heat-transfer-view expression for ΔS.

h Use the information now available to evaluate ΔS using this heat-transfer expression. You should arrive at

$$\Delta S = R\ln 2 = 5.8 \text{ J K}^{-1}$$

which agrees with part (d) above.

11 Find the entropy changes which occur when the following changes take place, at constant temperature, using helium gas.

a One mole expands from volume 20 ℓ to volume 40 ℓ.

b Two moles expand from volume 20 ℓ to volume 60 ℓ.

c Three moles are compressed from pressure 1 atmosphere to pressure 10 atmospheres.

12 Estimate the entropy change which occurs for the substance mentioned when

a a cup of tea cools from its initial high temperature to room temperature;

b a cube of ice at 0 °C turns to water at 0 °C;

c a 1 kg block of wood falls from the bench to the floor.
(Specific latent heat of fusion of ice = 3.4×10^5 J kg^{-1}.)

13 This question is about why ice changes into water at 0 °C, and not at any other temperature. When ice changes to water, the material gains entropy of 22 J K^{-1} mol^{-1}.

$$\Delta S = +22 \text{ J K}^{-1}\text{mol}^{-1}$$

This value can be worked out statistically by using the known facts about the structures of solid ice and liquid water. Separate experimentation shows that the latent heat required for this change is 6000 J mol^{-1}. This heat must come from the surroundings, and must result in an entropy decrease in these surroundings.

a Why do you think the material gains entropy when ice changes to water?

b Suppose the ice were to melt at -10 °C (263 K). What would be the entropy change in the surroundings, ΔS_s? Would it be positive or negative?

c The net entropy change of the universe, ΔS, is given by $\Delta S = \Delta S_w + \Delta S_s$. What is the value of ΔS in the supposed case of part (b)? Will this change in fact occur?

d Now suppose the ice were to be in surroundings at $+10$ °C (283 K). Calculate ΔS_s and hence ΔS for this case. Will this ice melt? Explain your answer.

e Now calculate the temperature at which the ice would have to melt so that $\Delta S = 0$. Comment on your answer.

CHAPTER 44

Heat transfer

Introduction

This chapter starts with a small section on specific heat capacities and specific latent heats. It then has an extensive coverage of thermal conduction, finishing off with a section on radiation.

GCSE knowledge

It will be assumed that you are able to recall:

- □   that heat energy flows from regions of higher temperature to regions of lower temperature;
- □   that heat energy passes through all solids by conduction;
- □   that heat energy passes through a vacuum by radiation: for example, from the Sun to the Earth;
- □   that convection occurs only in liquids and gases;
- □   that metals are in general much better conductors of heat energy than non-metals.

Heat capacity and specific heat capacity

The *heat capacity C* of an object is defined by the equation:

$$Q = C\Delta\theta$$

where Q is an amount of heat supplied to the object, and $\Delta\theta$ is its rise in temperature. Unit: $J\,K^{-1}$.

The *specific heat capacity, c,* of a material is defined by the equation:

$$Q = mc\Delta\theta$$

where Q is an amount of heat supplied to an object of mass m made of that material and $\Delta\theta$ is the temperature rise of the object. Unit: $J\,kg^{-1}K^{-1}$.

Typical values of specific heat capacities are given in the following table.

Substance	$c/J\,kg^{-1}\,K^{-1}$
Aluminium	910
Copper	380
Lead	130
Glass	700
Air (20 °C)	1010
Ice	2100
Rubber	1100–2000
Water	4200
Paraffin oil	2100

Measurement of specific heat capacity

Questions 1, 2 and 6 examine in detail three different methods for measuring specific heat capacity c. Essentially, each method involves supplying a measurable amount of heat Q to an object of mass m made from the substance being investigated. If the temperature rise $\Delta\theta$ can be measured, then all the quantities in the defining equation except c are known. Thus c can be calculated. For more information, we recommend that you study the questions referred to above.

Latent heat and specific latent heat

When a substance changes state, that is, melts, vaporises or sublimes, it does not change its temperature until all of the material has changed state. Yet the substance absorbs heat. We call this *latent* (or hidden) *heat* because it does not produce a change in temperature.

The *specific latent heat*, ℓ, of a material is defined by the equation

$$Q = m\ell$$

where Q is the amount of heat required to change the state of a mass m of the material.

In general, any particular material will have two values for ℓ, one for fusion or melting (ℓ_f), and one for vaporisation (boiling or subliming), ℓ_v. The same defining equation applies for both quantities. The value for vaporisation is usually much bigger than the value for fusion, since in the vaporisation case a large change in volume takes place: the material must do work during its expansion.

Methods for measuring ℓ_f and ℓ_v are described in detail in questions 3 and 4 respectively.

Newton's law of cooling

Newton's law applies when a body is cooling under conditions of *forced convection* (for example, when a fan blows a steady draught over the object). It states that the rate of loss of heat, dQ/dt, to the surroundings is proportional to the excess temperature $(\theta - \theta_0)$, where θ is the temperature of the hot body, and θ_0 is the temperature of the surroundings.

i.e.

$$\frac{dQ}{dt} = -K(\theta - \theta_0) \qquad \text{(i)}$$

K is the constant of proportionality, whose value depends on the area of the body's surface and the nature of its surface. The minus sign indicates that K has a positive value when heat energy is lost.

This law turns out to be a good approximation for cooling under conditions of *natural convection* also, provided that the excess temperature is not too high.

In practice this means roughly $(\theta - \theta_0) \leqslant 50$ K.

The experimental verification of this law is illustrated in question 5.

Conduction of heat

Many houses today have double-glazed windows, and sheets of expanded polystyrene in the cavities of their outer walls. Both of these are ways of reducing the heat energy being conducted from the hot inside of the house to the cold outside, particularly during the winter.

Not all solids conduct equally well. A common way of illustrating this point is illustrated in figure 1.

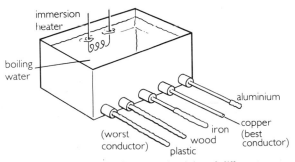

figure 1 Comparing the heat conductivity of different materials

Solid rods of different materials, but having equal lengths and diameters, are dipped into molten paraffin wax and withdrawn to allow a coating of wax to solidify on them. They are then inserted in the holes in the tank of water as shown. The boiling water is in contact with the ends of the rods. After a few minutes the wax will have melted to different distances along the rods, as illustrated in the diagram. This indicates the differences in their thermal conductivities.

Thermal conductivity k

In order to study conduction in more detail let us consider a metal rod of uniform cross-sectional area. The analysis is easier if the temperatures at each end of the rod are fixed. This can be achieved by soldering the ends of the metal rod into the walls of metal beakers which contain boiling water and melting ice respectively – see figure 2.

After some time (perhaps one hour), steady-state conditions will be achieved. That is, the temperature at any point on the bar will have reached its final steady value. The temperature variation with position for an unlagged bar is shown in figure 2(b), and the flow of heat energy along, and out of the sides of, the bar is also shown. It is much easier to analyse what is happening if no heat is lost through the sides of the bar. This case is illustrated in figure 2(c), where all the heat that flows into the left hand end of the bar must flow out of the right hand end, because the bar is in our imagination lagged with perfectly-insulating material.

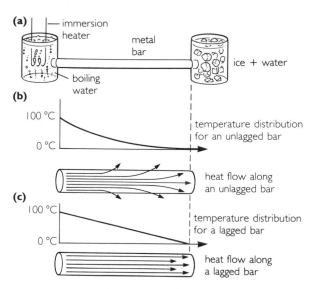

figure 2 Thermal conduction in a metal bar

Let us consider the bar in figure 2(c). It is redrawn in figure 3. In order to derive an equation for the flow of heat energy through this bar, it will help if we draw an analogy between the flow of heat along this bar, and the flow of electricity through a similar metal conductor.

figure 3

It may also be helpful if an analogy drawn earlier, between water flowing through a pipe and electricity flowing through a wire, is reinforced here. Figure 4 draws the analogies between the corresponding quantities in the different branches of physics.

We now derive the equations governing the situations in figure 4 by an intuitive approach. We start off by making some proportional statements which seem rather obvious.

1 *Dependence on cross-sectional area* If a second identical pipe carried water out of the large tank, then it would carry an identical amount of water per unit time, if it were at the same vertical depth below the surface of the water. That is, having

two identical pipes allows twice the rate of water flow out of the tank. Similarly, if a wire of twice the cross-sectional area is used in the electrical case, then the quantity of electrical charge flowing per unit time will double. Hence, we should expect that the quantity of heat energy flowing per unit time along a metal bar of twice the cross-sectional area will be double that which flowed along the original bar. (Note that doubling the area of the pipe in fact more than doubles the rate of flow of water, because there is less viscous drag from the walls. No such effects occur in either electricity or heat flow.)

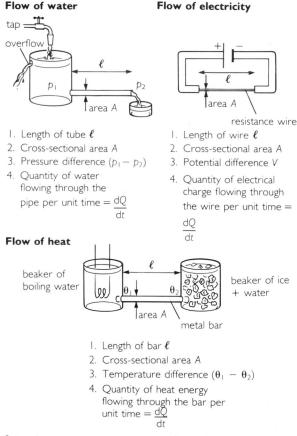

Flow of water

1. Length of tube ℓ
2. Cross-sectional area A
3. Pressure difference $(p_1 - p_2)$
4. Quantity of water flowing through the pipe per unit time $= \dfrac{dQ}{dt}$

Flow of electricity

1. Length of wire ℓ
2. Cross-sectional area A
3. Potential difference V
4. Quantity of electrical charge flowing through the wire per unit time $= \dfrac{dQ}{dt}$

Flow of heat

1. Length of bar ℓ
2. Cross-sectional area A
3. Temperature difference $(\theta_1 - \theta_2)$
4. Quantity of heat energy flowing through the bar per unit time $= \dfrac{dQ}{dt}$

figure 4

2 *Dependence on pressure gradient* In the electrical case the current (the quantity of electric charge flowing in unit time) will be doubled if either the potential difference across the resistance wire is doubled, or the length of the resistance wire is halved. Either of these changes will double the potential gradient along the wire. Doubling the pressure gradient or doubling the temperature gradient will result in similar changes in the rates of flow for water or heat energy respectively. Steps 1 and 2 are summarised in the table on the next page.

Flow of water	Flow of electrical current	Flow of heat
1. $\dfrac{dQ}{dt} \propto A$	1. $\dfrac{dQ}{dt} \propto A$	1. $\dfrac{dQ}{dt} \propto A$
2. $\dfrac{dQ}{dt} \propto \left[\dfrac{P_1 - P_2}{\ell}\right]$	2. $\dfrac{dQ}{dt} \propto \dfrac{V}{\ell}$	2. $\dfrac{dQ}{dt} \propto \left[\dfrac{\theta_1 - \theta_2}{\ell}\right]$
The equation is not given because viscosity complicates the problem	$\therefore I = \dfrac{dQ}{dt}$ $= \sigma A \dfrac{V}{\ell}$	$\dfrac{dQ}{dt} = kA\left[\dfrac{\theta_1 - \theta_2}{\ell}\right]$

Hence the equation which defines the thermal conductivity k of a material is

$$\frac{dQ}{dt} = kA\left(\frac{\theta_1 - \theta_2}{\ell}\right) \qquad \text{(ii)}$$

This definition applies provided the lines of heat flow are parallel, as in figure 2(c).

Experimental determination of k for a good conductor

The apparatus, usually referred to as Searle's Bar after its originator, is illustrated in figure 5.

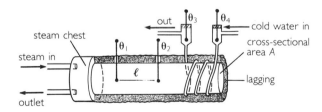

figure 5   Searle's bar

The bar being studied is heavily insulated and is heated at one end by steam. A copper pipe is soldered around the far end, and cold water flows through the pipe. The water carries away the heat energy which arrives at the right hand end of the metal bar. A steady state has been achieved when all four thermometers give steady readings. This may take an hour. In order to use equation (ii) to determine k we need to work out dQ/dt, that is, the rate of heat flow along the bar. This heat is taken away from the cold end of the bar by the mass $\dot{m}$ of cold water flowing through the cooling system in unit time. (The dot over the m indicates a 'rate'.)

If the specific heat capacity of water is denoted by c then we may write for this cooling water:

$$Q = (\text{mass}) \times c \times (\text{temperature change})$$

$$\Rightarrow \quad \frac{dQ}{dt} = (\text{mass/unit time}) \times c \times (\theta_3 - \theta_4)$$

$$\Rightarrow \quad \frac{dQ}{dt} = \dot{m}c(\theta_3 - \theta_4)$$

$\dot{m}$ is determined by collecting the water; dQ/dt can now be calculated and substituted into equation (ii) to give the value of k.

For accurate determinations:

☐   the insulation must be very good,

☐   the temperatures must be steady,

☐   the flow of water must be constant,

☐   the holes drilled into the bar for the thermometers must be of negligible size compared with the cross-section of the bar.

The last condition could be achieved by drilling very fine holes and inserting thermocouples, or using a bar of large cross-sectional area and using mercury thermometers.

Units and magnitudes of thermal conductivities

Equation (ii) enables us to see what the units of k must be:

$$k = \frac{dQ}{dt} \bigg/ A\left(\frac{\theta_1 - \theta_2}{\ell}\right)$$

$$\Rightarrow \text{ units of } k \text{ are } Js^{-1}/m^2\left(\frac{K}{m}\right) = \frac{Js^{-1}}{Km}$$

$$= Wm^{-1}K^{-1}$$

The thermal conductivities of some common substances at room temperature are given in the following table:

Substance	$k/Wm^{-1}K^{-1}$
Silver	420
Copper	380
Aluminium	230
Iron	60
Mercury	8
Glass	1
Brick	1
Rubber	0.2
Asbestos	0.13
Felt	0.04
Air	0.03

Models of thermal conduction

Non-metallic solids The model used here considers the solid to be made up of molecules, all of which are locked into positions within a lattice. The molecules are thought of as billiard balls suspended by springs (electron bonds). This is illustrated in figure 6.

figure 6

When extra heat energy is given to molecule A it will start to vibrate more violently about its mean position. The violence of its oscillations will be transmitted along the bond connecting it to B. Therefore, the oscillations of B will become larger. Hence, we say that heat energy has been conducted from A to B. This approach is extended to the whole of the solid. This process will be slow because B will resonate in sympathy with the oscillations of A, but it will take many oscillations to build up to large amplitudes in B. This means that non-metals are expected to have small thermal conductivities according to this model.

Solid metals The model of a metal assumes that there are free electrons which can move through the whole of the lattice, and that the lattice consists of positive ions which are bound into position. It is assumed that these free electrons undergo collisions with the bound ions of the lattice, but on average they travel past many bound ions before they undergo their next collision; see figure 7.

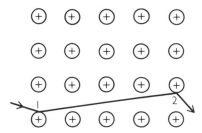

figure 7

Suppose that one end of a metal bar has been given extra heat energy; then the ions in that region will be vibrating more violently, as was the case of molecules in the non-metals. If a free electron collides with a violently vibrating ion (labelled 1 in figure 7), then it is highly likely that the free electron will bounce off with an increased kinetic energy. The free electron, travelling at about $10^6 \, \text{ms}^{-1}$, will rapidly travel past perhaps ten or more ions before colliding with another ion. The violence of this collision (labelled 2 in figure 7) will most probably cause this ion to oscillate more violently than before. This means that our model predicts that solid metals should have much greater thermal conductivities than do solid non-metals, as is in fact the case.

Thermal radiation

Thermal radiation is the radiation emitted by a body by virtue of its temperature. The radiation consists of electromagnetic waves. When thermal radiation is dispersed by a suitable prism, or diffraction grating, a continuous spectrum is obtained. The distribution of the energy over the wavelengths depends on the temperature of the emitting body. At temperatures below about 700 °C most of the energy is associated with infrared waves. At higher temperatures some visible radiation is emitted, and the total amount of energy radiated is greater.

Thermal detectors

Thermal radiation can be detected by thermometers or thermopiles. An ideal thermal detector will absorb all the thermal radiation incident upon it. Lamp-black, a paint consisting of very finely divided carbon, approaches this ideal very closely. It absorbs 95% of the incident radiation in the range 0–350 °C. Therefore the detecting surface of your detector should be painted with lamp-black.

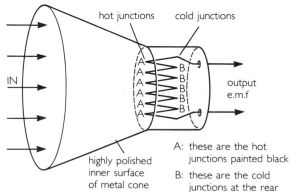

figure 8 A thermopile

The essential features of a simple *thermopile* are illustrated in figure 8. It consists of many thermocouples connected in series. The reason for this construction is that each individual thermocouple will only generate a fraction of a millivolt; but several hundred in series will generate a measurable e.m.f.

The temperature of the hot junction rises as its blackened surface absorbs the incident radiation. As its temperature rises the blackened surface will radiate more thermal energy. The temperature of the blackened surface will become steady when the rate of gain of heat energy from the surroundings is equal to the rate of loss of heat energy to the surroundings. When this has been achieved the e.m.f. is measured with a voltmeter of high resistance.

Emission of thermal radiation by different surfaces

An experiment designed to study the rate at which a surface emits thermal radiation is illustrated in figure 9. The metal tank has four different faces: one painted with lamp-black (dull black), one painted with black gloss paint, one painted dull white, and the fourth is a highly polished metal surface. The tank is filled with boiling water. This tank is known as a Leslie cube after the man who first investigated the problem in this way.

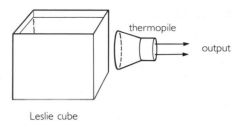

figure 9   Experiment to study emission of thermal radiation by surfaces

The thermopile is positioned so that only radiation from the face being investigated will travel down the cone and on to the blackened hot junction. The e.m.f. produced by the thermopile is greatest when the thermal detector is facing the dull black surface, and least when it is facing the highly polished surface. Hence, the highly polished surface is the worst emitter of thermal radiation, and the dull black surface is the best emitter.

Summary

1   The heat capacity C of an object is defined by
$$Q = C\Delta\theta$$

2   The specific heat capacity c of a material is defined by
$$Q = mc\Delta\theta$$

3   The specific latent heat ℓ of a material (fusion or vaporisation) is defined by
$$Q = m\ell$$

4   Newton's law of cooling:
$$\frac{dQ}{dt} = -K(\theta - \theta_0)$$

5   The thermal conductivity k of a material is defined by the heat conduction equation:
$$\frac{dQ}{dt} = kA\left(\frac{\theta_1 - \theta_2}{\ell}\right)$$

6   Analogies for thermal conductivity:
$$\text{electric current } I = \sigma A\frac{V}{\ell}$$
$$\text{fluid flow in a pipe } I \propto A, \frac{p_1 - p_2}{\ell}$$

7   Thermal conductivity k for a metal can be determined using Searle's method.

8   Thermal conduction in non-metals occurs through the intermolecular bonds.

9   Thermal conduction in metals occurs through the bonds and also by the migration of free electrons carrying kinetic energy.

10   Thermal radiation can be detected using a thermopile – with a black-painted surface for best absorption of radiation.

11   Dull black surfaces are the best emitters of thermal radiation.

Questions

I Figure 10 shows a copper beaker, resting on a small piece of felt, inside a slightly larger beaker. The inner beaker is partly filled with water. When a quantity of hot material is stirred into the water, the water and material end up at a common equilibrium temperature. If the specific heat capacities of water (4200 J kg^{-1}K^{-1}) and copper (380 J kg^{-1}K^{-1}) are used then it is possible to determine the specific heat capacity of the hot solid, provided that the initial temperature of the hot material is known.

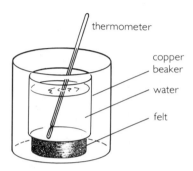

figure 10

A student wishing to find the specific heat capacity c of rubber cut a rubber ring, mass 0.020 kg, into about 100 small pieces. He then placed them in a boiling tube which he immersed in a bath of boiling water. See figure 11. After one hour he rapidly emptied the hot rubber into the cold water in the copper beaker and stirred continuously until the temperature indicated by the thermometer ceased to rise. The final temperature of the mixture was 24 °C.

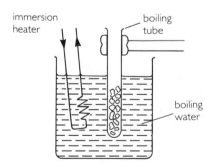

figure 11

Data:
  mass of copper beaker = 0.050 kg
  mass of cold water = 0.15 kg
  initial temperature of cold water = 18 °C

a (i) Calculate how much heat energy was gained by the water as its temperature rose from 18 °C to 24 °C.
(ii) Calculate how much heat energy was gained by the copper beaker as its temperature rose from 18 °C to 24 °C.
(iii) The student assumed that the heat energy gained by the beaker and water was equal to the total amount of heat energy lost by the rubber as it cooled from 100 °C. Determine the value for the specific heat capacity c of rubber that the student obtained.

b Discuss in some detail the following comments made by fellow students.
(i) 'It is highly unlikely that one hour was long enough for the rubber, which is a poor conductor of heat, to reach a uniform temperature of 100 °C.'
(ii) 'It would have been better to keep the rubber bung as a whole, then once it was transferred to the water it could have been held below the surface by the stirrer. This would have prevented heat losses to the atmosphere, which occurred when the rubber pieces floated on the surface.'

2 Nernst developed a method for measuring the specific heat capacity of solids. Its essential features are illustrated in figure 12.

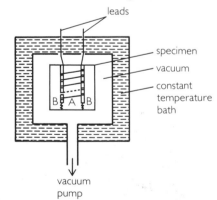

figure 12

A platinum heating coil, covered by a very thin layer of insulating varnish, is wound around a solid cylinder of the metal being studied. The varnish prevents the heating coil being shorted by the block. The cylinder, labelled A in figure 12, and the platinum coil fit closely inside a larger block, labelled B, of the same metal. The thin copper leads of the heating coil suspend the block of metal being studied inside the evacuated vessel. The temperature of the constant-temperature bath is adjusted to be equal to the temperature at which the metal is to be studied.

An electric current passing through the platinum coil will heat the metal block (i.e. A and B). If the p.d. across the platinum coil and the current through it are known, and the length of time measured, then the heat energy transferred to the block can be calculated.

The temperature rise of the metal block can be found by using the platinum resistance wire as a thermometer. Its resistance should be measured immediately before the heating current is switched on, and then after the heating current has been switched off.

Data:
  combined mass of cylinders (A and B)
    $= 0.20\,\text{kg}$
  heating current through coil $I = 1.1 \times 10^{-1}\,\text{A}$
  potential difference across coil $V = 1.2\,\text{V}$
  time for which heater was on $t = 300\,\text{s}$
  resistance of platinum $R_1 = 11.05\,\Omega$
    (immediately before heating)
  resistance of platinum $R_2 = 11.07\,\Omega$
    (immediately after heating)
  temperature coefficient of resistivity of platinum
    $\alpha = 3.8 \times 10^{-3}\,\text{K}^{-1}$

a Calculate the heat energy emitted by the platinum resistance coil during the heating period.

b Calculate the temperature rise of the platinum coil during the experiment. [α, referred to above, is defined by the equation $R_2 = R_1(1 + \alpha\Delta\theta)$.]

c From your answers to parts (a) and (b) and the above data, calculate the specific heat capacity of the metal of which the cylinder is constructed.

d Discuss in some detail the following comments:
  (i)   'Although the metal block started off at the same temperature as the surrounding constant-temperature bath, it ended up at a higher temperature due to the heating. Therefore, heat lost to the surroundings should be corrected for.'
  (ii) 'The final resistance of the platinum was measured *immediately* after the heating current was switched off, therefore it will not represent an *average* temperature for the solid block of A and B.'
  (iii) 'What about the mass of the platinum wire?'

3 Figure 13 shows a copper beaker, resting on a small piece of expanded polystyrene foam, inside a slightly larger plastic beaker. The copper beaker is partly filled with water. In order to determine the specific latent heat of fusion of ice, ℓ_f, a weighed amount of ice is dropped into the water, and the final temperature of the water is taken once all the ice has melted.

A student using this apparatus obtained the following results.

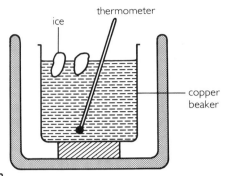

figure 13

A copper beaker of mass 0.050 kg contained 0.20 kg of water which was initially at room temperature (20.0 °C). She took four small ice cubes from the freezer, weighed them (total mass 0.020 kg) and put them straight into the water in the copper beaker. She stirred the mixture with the plastic stirring rod, and after about 10 minutes they had melted and the final temperature of the water was 10.8 °C.

a (i)   Calculate how much heat energy was lost by the 0.20 kg of water as it cooled from 20.0 °C to 10.8 °C, given that the specific heat capacity of water is 4200 $\text{J}\,\text{kg}^{-1}\,\text{K}^{-1}$.
  (ii) Calculate how much heat energy was lost by the copper beaker as it cooled from 20.0 °C to 10.8 °C, given that the specific heat capacity of copper is 380 $\text{J}\,\text{kg}^{-1}\,\text{K}^{-1}$.
  (iii) The student assumed that the heat energy lost by the beaker and water was gained by the ice cubes, first of all as they melted, and then as their temperature rose to the final 10.8 °C. Calculate how much heat energy was taken in by the *melted* ice cubes as their temperature rose from 0 °C to 10.8 °C.
  (iv) From your answer to (i), (ii) and (iii), determine how much heat energy the student assumed was needed to melt 0.020 kg of ice. Hence, determine the value for the specific latent heat of fusion ℓ_f that the student obtained.

b Explain the following points about the apparatus:
  (i)   the choice of copper for the inner beaker;
  (ii) the narrowness of the air gap between the inner and outer beakers,
  (iii) the choice of expanded polystyrene foam for the support pad.

c Discuss in some detail the following comments made by fellow students:
  (i)   'The temperature of the ice from the freezer was less than 0 °C, therefore the calculated value ℓ_f was too small.'
  (ii) 'If the ice had been crushed then dried with blotting paper, before being dropped into the water, a more accurate value of ℓ_f would have been obtained.'

(iii) 'Heat must have been gained from the room, which was at 20 °C, once the mixture started to cool. Therefore, it would have been better to start with water at about 25 °C.'
(iv) 'The apparatus should have had a lid on it.'

4 The apparatus illustrated in figure 14 can be used to determine the specific latent heat of vaporisation, ℓ_v, of liquids.

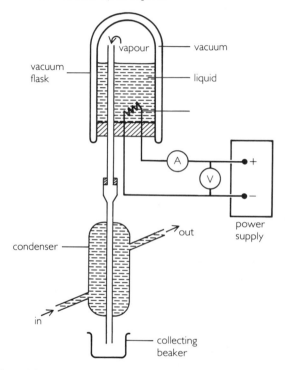

figure 14

An electric current passing through the heating coil causes the liquid to boil. The vapour produced passes down the escape tube, condenses in the condenser, and is collected as liquid in the beaker at the bottom of the apparatus. A student using this apparatus obtained the following results.

Using water (250 cm³), which was initially at 20 °C, when the initial p.d. and current were 12 V and 5.5 A, he found no liquid appeared in the beaker for 25 minutes. When liquid did start to appear in the beaker he noticed that the voltmeter still indicated 12 V, but the ammeter reading had dropped to 5.0 A. During the first 300 s (5 minutes) of collection he obtained 7.7 g of liquid. A quarter of an hour later he collected the condensing liquid again; this time he collected 7.9 g in 300 s.

a (i) Approximately how much electrical energy was delivered to the apparatus in the first 25 minutes? Show your working.

(ii) Use your estimate of the electrical energy to make a rough estimate of the average specific heat capacity of water over the range 20 °C–100 °C. Explain your method clearly.
(iii) State with reasoning whether you think your estimated value for the specific heat capacity is too high or too low.
b (i) From the final data, calculate a value for ℓ_v for water. Show your reasoning clearly.
(ii) Do you think your value is equal to the accepted value of the specific latent heat of vaporisation? Explain your answer.
(iii) Explain why less water was collected during the first 5 minutes, that is, 7.7 g, compared with 7.9 g later.
c The vacuum flask consists of very thin glass walls which enclose a vacuum. All the glass surfaces inside the vacuum are silvered. The flask is sealed with a rubber bung through which the copper leads to the heater coil, and the glass tube to the condenser, pass. Discuss in detail how this apparatus loses heat by (i) convection, (ii) conduction and (iii) thermal radiation. Pay particular attention to how these losses are minimised, and say where you think the greatest losses will occur.
d In answering part b(ii) you were asked to say what was wrong with your estimate. What further experiment could you do to correct this estimated value?

5 An apparatus to verify Newton's law of cooling is depicted in figure 15. It consists of a copper beaker containing hot water. The beaker stands on an insulating surface (e.g. a pad of felt), and is in a steady stream of air from the electric fan. The temperature of the water is recorded at one-minute intervals on a sensitive thermometer. Before each reading the water is gently stirred. The following is a typical set of results. (Room temperature $\theta_0 = 18$ °C.)

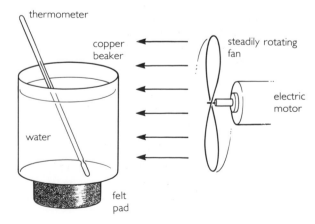

figure 15 Newton's law of cooling

Time/minutes	Temp/°C	Time/minutes	Temp/°C
0	45.0	11	30.6
1	43.2	12	29.8
3	41.5	13	29.0
4	38.5	14	28.2
5	37.1	15	27.5
6	35.8	16	26.9
7	34.6	17	26.3
8	33.5	18	25.8
9	32.5	19	25.2
10	31.5	20	24.8

a Plot a graph of temperature against time and draw an accurate curve through the points.

b The gradient of the graph you drew at any temperature θ will equal the rate of fall of temperature $d\theta/dt$ at temperature θ. Very carefully construct tangents to the curve at the following temperatures and complete the table of data.

θ/°C		41	37	33	30	26
$\dfrac{d\theta}{dt}$/K min^{-1}						

Next, plot a graph of $d\theta/dt$ against the excess temperature $(\theta - \theta_0)$. Does your graph agree with Newton's law? Explain your answer fully.

c If Newton's law of cooling applies, the graph of temperature θ against time is exponential. This can be tested numerically by taking ratios of excess temperatures at fixed time intervals and checking that they are all equal. Carry out this test on the data for the following pairs of results:

d Explain why a constant ratio in this test verifies Newton's law, i.e. show how the constant ratio is related to exponential decay.

6 In an experiment to determine the specific heat capacity, c, of aluminium, a cylindrical block of aluminium, mass 1.0 kg, was electrically heated by a 15 W heater. The heater fitted tightly into the hole in the centre of the block as illustrated in figure 16.

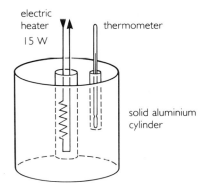

figure 16

The block was suspended by thin nylon thread in a draught-free room at 20 °C. The temperature on the thermometer was recorded at 1-minute intervals. Between 5 minutes and 15 minutes the thermometer reading rose steadily by 8.7 K, i.e. the graph of thermometer reading against time was approximately a straight line over this range (23.0 °C to 31.7 °C). As time passed the thermometer reading rose more slowly, then finally stabilised at 76 °C.

a Explain why the thermometer reading only rose by 3.0 K in the first 5 minutes, whereas it rose by 4.4 K in the second 5 minutes, after the heater was switched on.

b Explain why the temperature of the block eventually stabilised even though the heater was still switched on.

Interval/mins	0–4	4–8	8–12	12–16	16–20
Initial temp./°C	45.0	38.5			
Initial excess temp./K	(45 – 18)				
Final excess temp./K	(38.5 – 18)				
Ratio					

c Assume that Newton's law of cooling applied, i.e. the rate of loss of heat from the block was proportional to the excess temperature of the block (above that of the room). Using this assumption calculate:

(i) the rate of loss of heat from the block when the thermometer registered 27 °C (half-way up the linear section);

(ii) the specific heat capacity of aluminium at 27 °C, using a correction for the heat loss.

(iii) Explain why it may not be reasonable to assume that Newton's law applies in this experiment.

d When a similar experiment was performed with an iron cylinder of mass 1.0 kg, under the same experimental conditions (same heater, 15 W, same room temperature, 20 °C), the final steady temperature was considerably higher than the 76 °C obtained when using the aluminium block. Discuss in some detail the most likely reasons for this.

7 A cast-iron frying pan containing water boiling at 100 °C is standing on an electric hot-plate. The boiling water is reduced from a depth of 1.0 cm to 0.5 cm in 3 minutes. Calculate the temperature of the under surface of the pan, given that the pan has a diameter of 20 cm and its base is 3 mm thick. Explain fully your calculation, and make clear the effect of any assumptions that you make.

(Thermal conductivity of iron may be taken as $66 \, \mathrm{W\,m^{-1}K^{-1}}$, and the specific latent heat of vaporisation of water can be taken as $2.3 \times 10^6 \, \mathrm{J\,kg^{-1}}$.)

8 The material used to lag a hot water tank has a thermal conductivity of $4 \times 10^{-2} \, \mathrm{W\,m^{-1}K^{-1}}$. Copper has thermal conductivity of $400 \, \mathrm{W\,m^{-1}K^{-1}}$. The temperature in the airing cupboard containing the hot water tank is 27 °C, and the dimensions of the tank and its lagging are shown in figure 17.

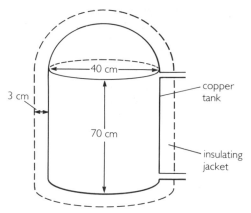

figure 17

a What electrical power must be supplied to the immersion heater in the tank in order that the temperature of the water is maintained at a steady 70 °C? The calculation is considerably simplified if it is assumed that the copper walls have a negligible temperature difference across them. Justify this assumption before you make it.

b How long will it take for the temperature of the water to fall by 1 K once the heater has been switched off? Hence, estimate how long it will take for the temperature to fall to 45 °C. Do you think the time that you have estimated for 45 °C is what actually happens? If not suggest other processes which affect the time.

Data:
density of water = $10^3 \, \mathrm{kg\,m^{-3}}$
specific heat capacity of water
= $4200 \, \mathrm{J\,kg^{-1}\,K^{-1}}$

9 A section of a house wall, two bricks thick, is illustrated in figure 18. Houses built at the start of the century were normally built in this way. Brick and mortar have a thermal conductivity of $0.6 \, \mathrm{W\,m^{-1}K^{-1}}$.

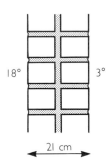

figure 18 Solid wall

a (i) Calculate the rate of flow of heat per unit area (in $\mathrm{W\,m^{-2}}$) through the brick wall if a steady state exists, where the temperature of the wall's surface inside the room is 18 °C and that of the wall's outer surface is 3 °C.

(ii) Explain why, under those conditions, the temperature of the outer surface of the wall must be greater than the ambient air temperature outside. Also explain why the temperature of the wall's inner surface must be less than the general temperature of the room.

Houses built in the 1940s and 50s had cavity walls. The structure of those walls is depicted in figure 19. You will notice that a 5 cm air gap separates two walls of brick. The thermal conductivity of air is $0.03 \, \mathrm{W\,m^{-1}K^{-1}}$.

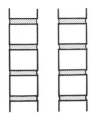

figure 19   Cavity wall

b (i)   Explain why, when steady-state conditions apply, the rate of heat flow across each layer is the same.
(ii)   Draw a sketch graph of temperature against distance through the wall to show how the temperature varies from 18 °C on the surface of the inner wall to 3 °C on the surface of the outer wall.
(iii)   Calculate, showing your working, the thickness of brick equivalent to 5 cm of air. Hence, or otherwise, calculate the rate of loss of heat per unit area (in W m^{-2}) through the cavity wall, and compare your answer with the value that you obtained for the solid brick wall.
(iv)   In actual fact, the improvement obtained in insulation is not nearly as good as your calculations suggest. Explain why this is the case.

A common practice in the 1970s was to inject expanded polystyrene foam into the cavities of earlier houses. The gas in the foam bubbles can be taken to have a thermal conductivity approximately equal to that of air.
c (i)   Explain why the thermal conductivity of the solid foam is approximately equal to that of the gas trapped in its bubbles.
(ii)   Explain why the answer that you obtained to part b(ii) above will apply to a wall that has had its cavity filled in this way.
(iii)   If the outer wall is no longer losing so much heat to the outer air, will its temperature still be 3 °C? Explain your answer fully and extend your discussion to include the temperature of the surface of the inner wall and how it is related to average room temperature.
(iv)   In view of your answer to part c(iii), will the saving in heating bills be as great as your answer to part b(iii) suggested?

10 This question is about an electric fuse, the cartridge type that is used in a 13 A plug. The structure of this fuse is shown in figure 20.
   The electric current in the fuse wire produces a heating effect. This can be understood better if we consider a small section of the fuse wire, of length δx, illustrated in figure 21. The resistance of this section will be $\delta R_x = \rho_\theta \times (\delta_x / A)$, where ρ_θ is the resistivity of the fuse wire when it is at

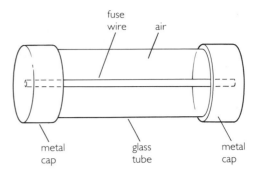

figure 20   Cartridge fuse

temperature θ, and θ is the temperature of the wire at a distance x from the middle.
   Heat can escape from the middle position of the wire by
(i)   conduction along the fuse wire to the large metal caps at the end,
(ii)   radiation to the cooler case of the fuse,
(iii)   convection due to the gas molecules colliding with the hot wire.

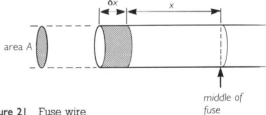

figure 21   Fuse wire

a At which point on the wire do you think the temperature will be greatest? Explain your answer.
b By considering what is happening in the small section of length δx shown in figure 21, i.e. (i) heat flowing in one end, (ii) heat flowing out of the other end, (iii) heat generated within the section, (iv) heat lost from both sides, decide how the temperature θ along the wire varies with position. Make a sketch graph of temperature against position, and explain how you arrived at this shape.
c Where will the fuse melt when the maximum current is exceeded?

11 a Draw two labelled graphs to show the energy distribution as functions of wavelength for a black body which is first of all at 1500 K and then at 6000 K. Your drawings should illustrate reasonably accurately Wien's Law and Stefan's Law. Give numerical ratios to the features of your graphs which illustrate these laws.
b Wien's Law is sometimes used as a basis for measuring high temperatures. The temperature so determined is known as the Colour Temperature. Why is this so?

SECTION I

APPENDICES

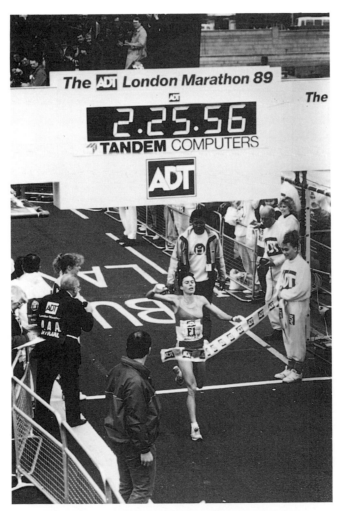

Physicists always need to take accurate measurements of time and displacement. Veronique Marot, winner of the 1989 London Marathon, has her time recorded to an accuracy of 1 part in a million...

*. . . and this seismograph at the California Academy
of Sciences detects some minor Earth tremors*

45   Units and dimensions   483
46   Mathematics   489

CHAPTER 45

Units and dimensions

Physical quantities

Study of the physical universe ('physics') involves description and measurement.

A *physical quantity*, for example velocity, is a property whose meaning can be defined by agreement so that all physicists share the same understanding of the term. A quantity is usually defined by an equation which refers to a specified physical situation; sometimes a verbal definition is used instead.

To measure a physical quantity it is necessary to define a *unit*. This again must be by common agreement. The magnitude of the quantity can then be specified by saying that it is a certain *number* of the defined units.

Table 1 showing all the physical quantities used in this book, together with their standard symbols and units, appears on page 484. Quantities are defined at appropriate places in the text.

SI units

Scientists throughout the world now use a common system, known as SI (Système International d'Unités). The system has seven *base units*, whose definitions are as far as possible reproducible in a laboratory anywhere. All other quantities have units which are expressible in terms of the base units. Some of the units so expressed are given their own names, and are then called *derived units*.

The seven base units are:

☐ **metre, m** The metre is the length equal to 1 650 763.73 wavelengths in vacuum of the radiation from the transition between the levels $2p_{10}$ and $5d_5$ of the krypton-86 atom.

☐ **kilogram, kg** The kilogram is the unit of mass; it is equal to the mass of the international prototype of the kilogram.

☐ **second, s** The second is the duration of 9 192 631 770 periods of the radiation from the transition between the two hyperfine levels of the ground state of the caesium-133 atom.

☐ **ampere, A** The ampere is that constant current which, when passing through two straight parallel conductors of infinite length, of negligible circular cross-section, and placed 1 metre apart in vacuum, would produce between these conductors a force of 2×10^{-7} newton per metre of their length.

☐ **kelvin, K** The kelvin is the unit of thermodynamic temperature. It is the fraction 1/273.16 of the thermodynamic temperature of the triple point of water.

☐ **candela, cd** The candela is the unit of luminous intensity, which is a measure of the 'luminous energy emitted per second per unit solid angle' by a point source such as S in figure 1. The unit of solid angle is called the steradian, sr, so the candela can be expressed in units of watts per steradian. The candela is the luminous intensity of a source that emits monochromatic radiation of frequency 5.4×10^{14} Hz that has radiant intensity of 1/683 watt per steradian.

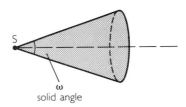

solid angle

figure 1

☐ **mole, mol** The mole is the amount of substance of a system which contains as many elementary particles as there are atoms in 0.012 kilogram of carbon-12.

Note that the kilogram is the only base unit not at present defined using a reproducible situation.

The ampere and the candela include in their definitions the derived units of newton and watt respectively. This at first sight looks odd, since the ampere and candela are supposed to be base units. However, the newton and watt can both be expressed in terms of the fundamental units of the kilogram, metre and second.

Table 1   Physical quantities, symbols and units

Quantity	Symbol	Unit
length	ℓ	metre (m)
height	h	
radius	r	
diameter	d	
displacement, distance	s	
wavelength	λ	
amplitude	A	
mass	m, M	kilogram (kg)
time	t	second (s)
time-period	T	
half-life	$T_{1/2}, t_{1/2}$	
time constant	τ	
current	I	ampere (A)
r.m.s. value of a.c.	$I_{r.m.s.}$	
peak value of a.c.	I_0	
temperature (thermodynamic)	T	kelvin (K)
common temperature	θ	degree Celsius (°C)
amount of substance	n	mole (mol)
area	A	m^2
volume	V	m^3
density	ρ	$kg\,m^{-3}$
speed, velocity	u, v	ms^{-1}
speed of electromagnetic waves	c	
speed of other waves	v	
velocity of a molecule	c	
acceleration	a	ms^{-2}
acceleration due to gravity	g	
force	F	newton (N) $1\,N = 1\,kg\,ms^{-2}$
weight	W	
gravitational field strength	g	Nkg^{-1} $(1\,Nkg^{-1} = 1\,ms^{-2})$
work	W	joule (J) $1\,J = 1\,Nm$
energy	E	J
potential energy	E_p	
kinetic energy	E_k	
heat	Q	
internal energy	U	
electrical energy	W	
work function	W	
power	P	watt (W) $1\,W = 1\,Js^{-1}$
pressure	p, P	pascal (Pa) $1\,Pa = 1\,Nm^{-2}$
moment of a force	M	Nm
torque, moment of a couple	T	Nm
momentum, impulse	p	$Ns = kg\,ms^{-1}$
angle	α, θ, ϕ	radian (rad) *or*
phase angle	ϕ	degree (°)

moment of inertia	I	$\text{kg}\,\text{m}^2$
angular momentum	L	Js
coefficient of friction	μ	no unit
Avogadro constant	L, N_A	mol^{-1}
tensile or compressive stress	σ	Pa
linear strain	ε	no unit
Young modulus	E	Pa
spring constant	k	$\text{N}\,\text{m}^{-1}$
electric charge	Q, q	coulomb (C) $1\,\text{C} = 1\,\text{As}$
electron charge	e	
potential difference (p.d.)	V	volt (V) $1\,\text{V} = 1\,\text{J}\,\text{C}^{-1}$
e.m.f.	E	
electric potential	V	
r.m.s. value of alternating e.m.f.	$E_{\text{r.m.s.}}$	
peak value of alternating e.m.f.	E_0	
resistance	R	ohm (Ω) $1\,\Omega = 1\,\text{V}\,\text{A}^{-1}$
internal resistance	r	
resistivity	ρ	$\Omega\,\text{m}$
temperature coefficient of resistivity	α	K^{-1}
conductivity	σ	$\Omega^{-1}\text{m}^{-1}$
electric current density	J	$\text{A}\,\text{m}^{-2}$
capacitance	C	farad (F) $1\,\text{F} - 1\,\text{C}\,\text{V}^{-1}$
number density	n	m^{-3}
frequency	f	hertz (Hz) $1\,\text{Hz} = 1\,\text{s}^{-1}$
reactance	X	Ω
impedance	Z	Ω
angular velocity speed of rotation	ω	$\text{rad}\,\text{s}^{-1}$
gravitational potential	V	$\text{J}\,\text{kg}^{-1}$
gravitational constant	G	$\text{N}\,\text{m}^2\text{kg}^{-2}$
permittivity of free space	ε_0	$\text{F}\,\text{m}^{-1}$
relative permittivity	ε_r	no unit
electric field strength	E	$\text{N}\,\text{C}^{-1}, \text{V}\,\text{m}^{-1}$
electric charge density	σ	$\text{C}\,\text{m}^{-2}$
radioactive decay constant	λ	s^{-1}
activity (of radioactive source)	A	bequerel (Bq) $1\,\text{Bq} = 1\,\text{s}^{-1}$ Curie (Ci) – obsolete $1\,\text{Ci} = 3.7 \times 10^{10}\,\text{Bq}$
proton number	Z	no unit
nucleon number	A	no unit
neutron number	N	no unit
magnetic flux density (field)	B	tesla (T) $1\,\text{T} = 1\,\text{N}\,\text{A}^{-1}\text{m}^{-1}$
magnetic flux	Φ	weber (Wb) $1\,\text{Wb} = 1\,\text{T}\,\text{m}^2$

self inductance	L	henry (H) $1\,\mathrm{H} = 1\,\mathrm{Wb\,A^{-1}}$
mutual inductance	M	H
permeability of free space	μ_0	$\mathrm{Hm^{-1}} = \mathrm{NA^{-2}}$
relative permeability	μ_r	no unit
number of turns of wire	N	no unit
number of turns per unit length	n	$\mathrm{m^{-1}}$
order of interference order of spectrum	n	no unit
slit separation grating spacing	s	m
number of molecules number of nuclei	N	no unit
molar gas constant	R	$\mathrm{Jmol^{-1}K^{-1}}$
molar volume	V_m	$\mathrm{m^3mol^{-1}}$
Boltzmann constant	k	$\mathrm{JK^{-1}}$
entropy	S	$\mathrm{JK^{-1}}$
Planck constant	h	Js
thermal conductivity	k	$\mathrm{Wm^{-1}K^{-1}}$
heat capacity	C	$\mathrm{JK^{-1}}$
specific heat capacity	c	$\mathrm{Jkg^{-1}K^{-1}}$
specific latent heat	ℓ	$\mathrm{Jkg^{-1}}$

Standard prefixes for SI units

Table 2

Factor	Prefix	Symbol
10^{-12}	pico	p
10^{-9}	nano	n
10^{-6}	micro	μ
10^{-3}	milli	m
10^{-2}	centi	c
10^{3}	kilo	k
10^{6}	mega	M
10^{9}	giga	G
10^{12}	tera	T

Examples:
$1\,\mathrm{GW} = 1 \times 10^9\,\mathrm{W}$
$1\,\mathrm{M\Omega} = 1 \times 10^6\,\Omega$
$1\,\mathrm{nF} = 1 \times 10^{-9}\,\mathrm{F}$

Table 2 gives the prefixes which are commonly used with base and derived units to multiply them by the factor indicated.

Note that the word 'kilogram' consists of 'gram' already prefixed by 'kilo'. In all mass units the standard prefixes are added to the word 'gram' ($1\,\mathrm{g} = 1 \times 10^{-3}\,\mathrm{kg}$).

Conventions for written physics

1 In typeset work, symbols for quantities are in italics, and units are in upright type.

2 Units:
a) Names of units written in full begin with a small letter, even though the unit may be named after a person.
b) The symbol for a unit which is named after a person does begin with a capital letter, e.g. Hz (hertz), N (newton).
c) The letter 's' is not added to units to create a plural form.

d) A full stop is not placed after a unit symbol (except to end a sentence in the normal way).

Examples   5 newton and 5 N are both correct; 5 Newton, 5 newtons, 5 N. and 5 n are all incorrect.

3   A symbol for a quantity denotes its magnitude, that is, both its numerical value and the appropriate SI unit. Thus, for example, '$m = 5$ kg' is a correct statement, but '$m = 5$' and 'mass $= m$ kg' are incorrect. In other words, a unit symbol should always appear after a numerical value, but never after a quantity symbol.

4   In calculations it is correct practice to include units with numerical values in each line of the calculation. For example:

$$\text{acceleration} = \frac{v - u}{t}$$
$$= \frac{12\,\text{ms}^{-1} - 2\,\text{ms}^{-1}}{5\,\text{s}}$$
$$= 2\,\text{ms}^{-2}$$

5   In calculations, provided one uses the correct SI unit for every item substituted into a formula, then the value of the quantity being calculated is always automatically obtained in the correct SI unit.

Dimensions of a quantity

The dimensions of a quantity show the way in which certain base quantities combine to form it. A system of dimensions may be built up in a very similar way to the SI unit system. Base dimensions are chosen arbitrarily; then the dimensions of other quantities are found using their definitions. Dimension symbols are by convention capital letters.

If one chooses as base dimensions the same quantities as the SI base unit quantities, then the systems are almost completely parallel. With this choice of base units, table 3 shows the base dimension symbols conventionally used.

Table 3

Quantity	SI unit symbol	Dimension symbol
mass	kg	M
length	m	L
time	s	T
current	A	I
temperature	K	Θ

Table 4 shows the dimensions of selected other quantities, using the base dimensions in table 3, and the comparison with their SI units.

Table 4

Quantity	SI unit symbols	Dimension symbols
density	$\text{kg}\,\text{m}^{-3}$	ML^{-3}
energy	$\text{J}\,(= \text{kg}\,\text{m}^2\,\text{s}^{-2})$	ML^2T^{-2}
p.d.	$\text{V}\,(= \text{J}\,\text{C}^{-1} = \text{kg}\,\text{m}^2\,\text{s}^{-3}\text{A}^{-1})$	$\text{ML}^2\text{T}^{-3}\text{I}^{-1}$

It would be possible to choose a different set of base dimensions and produce a different, but equally valid, table of dimensions for all quantities. For example, instead of M, L, T, I and Θ one could choose density (D), velocity (V), energy (E), charge (Q) and specific heat capcity (C).

The principle value of the idea of dimensions is in checking a complex expression for consistency. All terms related by addition, subtraction, equality or inequality must have the same units and dimensions. If at any stage in a piece of work this is not so, then an error must have occurred. For example, consider this expression for the energy of an oscillator, in which the symbols have their conventional meaning:

$$E = \tfrac{1}{2}mv^2 + \tfrac{1}{2}Fs^2$$

Ignore the numerical factors of $\tfrac{1}{2}$, and consider the units of each term separately:

quantities:	E	$\tfrac{1}{2}mv^2$	$\tfrac{1}{2}Fs^2$
units:	J	$\text{kg}.(\text{ms}^{-1})^2$	$(\text{kg}\,\text{ms}^{-2}).\text{m}^2$
$\Rightarrow$	$\text{kg}\,\text{m}^2\text{s}^{-2}$	$\Rightarrow \text{kg}\,\text{m}^2\text{s}^{-2}$	$\Rightarrow \text{kg}\,\text{m}^3\text{s}^{-2}$

We can conclude that the original equation *must* have been wrong, because when reduced to base units the three terms are not all the same. Put another way, the equation is dimensionally unsound, because the three expressions have different dimensions of length (L).

Note that the unit of angle, radian, has no dimensions, and can be omitted when it arises in unit-checking operations.

Dimensional analysis

We can take the use of dimensions a stage further, and apply them to help solve very complicated problems that we might otherwise not be able to solve. Here is an example.

We know that energy is produced in stars by a process of thermonuclear fusion. As a result of this energy production the centre of a star is at an extremely high temperature, 10^7 K in the case of the Sun. The particles in the centre of the star move at great speeds and so the pressure at the centre of a star is also very high, 10^8 atmosphere or more. It is this high pressure that prevents the collapse of the star under the influence of gravitational forces.

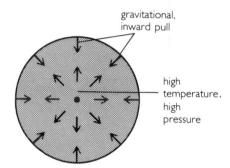

figure 2   Forces inside a star

What we would like to know is this: how long, approximately, would it take for the star to collapse if the thermonuclear process were to stop? Now we have absolutely no idea how to solve this problem by a rigorous method, but we can guess at the factors that will affect it. We think that it will depend on the universal constant of gravitation G, the mass of the star M, and the radius of the star R.

Thus we can write:

$$t = kG^x M^y R^z$$

where t is the time of collapse and k is a dimensionless constant.

Now x, y and z are all unknown, but we do know that the dimensions *must* be the same on each side. So we will express the equation in terms of the base dimensions of mass M, length L and time T.

G is defined in the equation:

$$F = \frac{GM_1 M_2}{R^2}$$

(See chapter 18.)

F is the force between two point masses M_1 and M_2 a distance R apart. Thus G has units $\mathrm{Nm^2 kg^{-2}}$. The newton has units $\mathrm{kgms^{-2}}$, and so G has units of:

$$(\mathrm{kgms^{-2}}) \times \mathrm{m^2 kg^{-2}} = \mathrm{kg^{-1} m^3 s^{-2}}$$

We may now write a dimensional equation for the time of collapse:

$$[\mathrm{T}] = [\mathrm{M^{-1} L^3 T^{-2}}]^x [\mathrm{M}]^y [\mathrm{L}]^z$$

By comparing the dimensions on the left hand side of this equation with those on the right, we can see the following statements must be true if our equation is dimensionally correct:

$$\mathrm{T^1 = T^{-2x}} \Rightarrow x = -\tfrac{1}{2}$$

$$\mathrm{M^0 = M^{-x} . M^y} \Rightarrow y - x = 0 \Rightarrow x = y = -\tfrac{1}{2}$$

$$\mathrm{L^0 = L^{3x} . L^z} \Rightarrow 3x + z = 0 \Rightarrow z = -3x = \tfrac{3}{2}$$

So we arrive at the result $x = -\tfrac{1}{2}$, $y = -\tfrac{1}{2}$, $z = \tfrac{3}{2}$ and so

$$t = k \frac{R^{\frac{3}{2}}}{G^{\frac{1}{2}} M^{\frac{1}{2}}}$$
$$= k \left(\frac{R^3}{GM} \right)^{\frac{1}{2}}$$

The possibility of a numerical constant k appearing in a rigorous analysis need not worry us too much, because it is likely to have an order of magnitude near unity.

Having got this far we should at least try out this formula for our Sun, for which $R = 7 \times 10^8$ m and $M = 2 \times 10^{30}$ kg. G is a constant having a value of $6.7 \times 10^{-11}\,\mathrm{Nkg^{-2}m^2}$. Assume $k = 1$.

Thus

$$t = k \left(\frac{R^3}{GM} \right)^{1/2}$$
$$= \left[\frac{(7 \times 10^8\,\mathrm{m})^3}{(6.7 \times 10^{-11}\,\mathrm{Nkg^{-2}m^2})(2 \times 10^{30}\,\mathrm{kg})} \right]^{1/2}$$
$$= 1600\,\mathrm{s}$$

So this result tells us that we have not got long to live if the Sun's thermonuclear reaction does stop.

Making a measurement

One assumes in making any measurement that there is a *true* value, to which one intends to get as near as possible. No measuring instrument will give this true value exactly. Instruments are limited by their *accuracy* and by their *sensitivity*.

The *accuracy* of an instrument is the degree to which its calibration matches the values which the SI base units would give. An inaccurate instrument gives rise to *systematic* errors, which will not show up even in a repeated set of readings. To guard against this, the instrument must be checked against other instruments.

The *sensitivity* of an instrument is the smallest scale division to which it can be read. For example, a metre rule is usually sensitive to 1 mm. In a more complex situation it may be possible to improve on the simple sensitivity of an instrument. When timing an oscillation with a stop-watch, for example, a number of oscillations should be timed in one measurement; since the value required is then the total time divided by the number of oscillations, the sensitivity is improved by that factor. Also, in many situations a single reading, such as the time of fall of an object through a given height, can be repeated for a number of trials; this helps to reduce the effect of the *random* error built into the experimental design.

CHAPTER 46

Mathematics

Introduction

We have tried to write this book using a low level of mathematics, but there is no doubt at all that to be a good physicist you also need to be a good mathematician. If you are studying A-level maths then you will already know most of the content of this chapter, the purpose of which is to summarise useful pieces of mathematics that you are likely to encounter in an A-level physics course.

Notation

The following symbols are often encountered.

>   greater than, e.g. $x > 1$ means x is greater than 1.

<   less than, e.g. $x < 1$ means x is less than 1.

≫   very much greater than, e.g. $x \gg 1$ means x is very much greater than 1.

≪   very much less than, e.g. $x \ll 1$ means x is very much less than 1.

≈   approximately equal to, e.g. $x = 3.08$ so $x \approx 3$.

∝   proportional to, e.g. $I \propto V$ for an ohmic conductor.

$\langle x \rangle$ the average value of x.

Σx the sum of some variable x. If, for example, x represents the odd numbers between 1 and 10, then

$$\Sigma x = 1 + 3 + 5 + 7 + 9 = 25.$$

Δx a change in x.

δx a very small change in x.

dx/dt the rate of change of x with t. (It is not necessary to know how to differentiate but you need to understand the mathematical shorthand.)

⇒   it follows that

     e.g. $y = x^2 - 7$

     $\Rightarrow x^2 = y + 7$

     $\Rightarrow x = \sqrt{y + 7}$

Errors

In making any measurements in an experiment it is important to have some feel for the size of the errors involved. Here are some examples of measurements obtained and how we can deal with the errors.

1 To measure the time period of a pendulum, we would determine the time interval for 10 swings; this helps to reduce the error, since our timing error is split between 10 oscillations. We get some feel for the probable error in timing by repeating the measurements. Thus we might record the following eight results for the time of 10 swings:

7.6 s, 7.7 s, 8.0 s, 7.7 s, 7.8 s, 7.9 s, 7.8 s, 7.8 s.

The average of these results is 7.8 s. Also we know that the time for 10 swings lies between 7.6 s and 8.0 s. Thus we can express the time for 1 swing as: $T = 0.78\,\text{s} \pm 0.02\,\text{s}$.

For such a simple experiment, eight repetitions is rather excessive; in practice two or three should be enough to give you a feel for the size of the error involved.

2 The measurement of the length of a piece of string is a simple task, but it gives an example of how to add errors. We may decide to measure the distance to each end of the string from some fixed external point; with our ruler we can measure each length to within 0.5 mm. Thus measurements for a string AB may be as follows:

     distance from fixed point to A $= 651\,\text{mm} \pm 0.5\,\text{mm}$
     distance from fixed point to B $= 237\,\text{mm} \pm 0.5\,\text{mm}$
     ⇒   length of AB $= 414\,\text{mm} \pm 1\,\text{mm}$

So the uncertainty in each measurement doubles the error in our length measurement.

3 It was fairly straightforward to deal with two errors in the last example because we were dealing with errors in the same quantity, namely length. Now consider this problem. Suppose we wished to calculate the resistivity of a wire, and we had measured the following quantities to the stated accuracy:

     resistance $R = 20.0\,\Omega \pm 1.0\,\Omega$ (5% error)
     length $\ell = 1.00\,\text{m} \pm 0.01\,\text{m}$ (1% error)
     radius $r = 0.1\,\text{mm} \pm 0.01\,\text{mm}$ (10% error)

To deal with these errors we have to make use of the Binomial Theorem:

$$(1 + x)^n = 1 + nx + \frac{n(n-1)x^2}{1 \times 2} + \frac{n(n-1)(n-2)}{1 \times 2 \times 3}x^3 \ldots$$

If $x \ll 1$ then we can neglect terms in x^2 and higher orders, and then we get the useful approximations:

$$(1+x)^n \approx 1+nx$$

$$(1+x)^{-n} \approx 1-nx$$

Now we will apply these approximations to deal with the errors in our resistivity calculation.

$$R = \frac{\rho\ell}{A} \quad \text{(page 121)}$$

$$\Rightarrow \quad \rho = \frac{RA}{\ell}$$

$$= \frac{R\pi r^2}{\ell}$$

$$= \frac{\pi R r^2}{\ell}$$

Let the error in R be δR, the error in r be δr, and the error in ℓ be $\delta\ell$. Then

$$\rho = \frac{\pi(R \pm \delta R)(r \pm \delta r)^2}{(\ell \pm \delta\ell)}$$

$$= \frac{\pi R r^2 [1 \pm (\delta R/R)][1 \pm (\delta r/r)]^2}{\ell[1 \pm (\delta\ell/\ell)]}$$

Using the binomial approximations it follows that:

$$\rho = \frac{\pi R r^2}{\ell}\left(1 \pm \frac{\delta R}{R}\right)\left(1 \pm 2\frac{\delta r}{r}\right)\left(1 \mp \frac{\delta\ell}{\ell}\right)$$

$$\Rightarrow \quad \rho = \frac{\pi R r^2}{\ell}\left(1 \pm \frac{\delta R}{R} \pm 2\frac{\delta r}{r} \mp \frac{\delta\ell}{\ell}\right)$$

In the last line we have ignored any terms such as $(\delta R/R) \times (\delta r/r)$, since these will be very small.

Using our measurements we now get:

$$\rho = \frac{\pi \times (20\,\Omega) \times (10^{-4}\,\text{m})^2}{1\,\text{m}} \times \left[1 \pm \frac{1}{20} \pm 2 \times \frac{1}{10} \mp \frac{1}{100}\right]$$

$$\Rightarrow \quad \rho = 6.3 \times 10^{-7}\,\Omega\,\text{m}[1 \pm 0.26]$$

When we consider the large errors in this particular experiment we ought to express the resistivity in the form:

$$\rho = (6 \pm 2) \times 10^{-7}\,\Omega\,\text{m}$$

The errors are so large that we can only reliably express our result to 1 significant figure.

To summarise, the way to calculate the percentage error in ρ was to calculate the percentage errors in R, r and ℓ. Then:

percentage error in ρ
= % error in R + 2 × % error in r + % error in ℓ.
The error in r has to be included twice because r appears twice in the formula in the form of r^2.

Example At low temperatures the specific heat capacity c of a substance is given by the formula:

$$c = AT^3$$

A is a constant and T is the kelvin temperature. What

is the percentage error in c if we know the temperature to be $T = (100 \pm 2)$ K?

Answer: the percentage error in T is 2% so the error in c is 6%.

Systematic errors

So far we have only mentioned random errors. Such an error has an equal chance of being either positive or negative. Random errors may be caused by one of the following:

☐   lack of sensitivity in the measuring instrument;

☐   carelessness or imperfection of the person doing the experiment;

☐   the particular measurement in question may not be reproducible because the quantity being measured may fluctuate a little as time passes.

A systematic error will cause a random set of errors to be spread about a value which is not the true value. Such errors could be caused in the following ways.

☐   The operator may consistently make a mistake. For instance, a person using an ammeter which has two parallel scales may use the wrong one.

☐   An instrument could be incorrectly calibrated. For example, a thermometer might read 100 °C when in fact the temperature is really 90 °C.

☐   An instrument might have a zero error. A micrometer screw gauge often has such an error.

As long as we are aware of systematic errors, we can eliminate them or allow for them.

Graphs and errors

Often when we do an experiment we use a graph to present our results. This is a good idea because it allows us to see at a glance how one quantity varies with another. Also we can gain useful information from a graph, for example the gradient of a V/I graph for an ohmic resistor will give us the resistance.

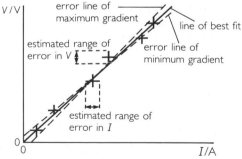

figure 1   Graph showing error bars

In figure 1 such a graph is shown. The points have been plotted to show the estimated error in V and I

for each point. The resistance can now be calculated from the gradient of the line of best fit.

We can get a feel for the size of error involved by estimating the gradients of the two error lines. These are lines that will pass through the error ranges of all the points, one with the maximum possible gradient, the other with the minimum possible gradient.

Labelling axes of graphs

The purpose of a graph is to display numerical data in as convenient a form as possible. The convention that is most widely adopted is to put only numbers on the axes, but to label the axes to show the units. Some examples are shown in figure 2.

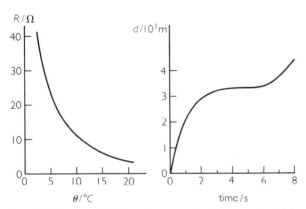

figure 2   (a) A graph of resistance against temperature; (b) a graph of distance against time

Figure 2(a) shows a graph of resistance against temperature for an electrical component. At 5 °C the resistance of the component is 25 Ω. So we can write $R = 25\,\Omega$ or $R/\Omega = 25$, the solidus meaning 'divided by'. In the same way, in figure 2(b) we notice that at time = 2 s, $d/10^3$ m = 3, so $d = 3 \times 10^3$ m. The same notation may be adopted for labelling the headings of columns in tables of readings.

Significant figures

A physicist ought to have some feel for the accuracy to which a result should be expressed. This can be best illustrated by a simple example. What is the volume of a cube of side length $\ell = 2.3$ cm?

Answer: volume $= \ell^3 = (2.3\text{ cm})^3 = 12.167\text{ cm}^3$. But how many significant figures should the answer contain? The length ℓ of the cube was quoted as 2.3 cm; this means that ℓ must lie in the range: $2.25\text{ cm} < \ell < 2.35\text{ cm}$. In other words our stated accuracy is $\ell = (2.3 \pm 0.05)$ cm; this is an error of about 2%. Using the result from the last section, our volume thus has an error of about 6%. Therefore we ought to express the volume as: $V = (12.167 \pm 0.79)\text{ cm}^3$. Clearly, with such a large uncertainty the last two decimal places are certainly not significant, and even

the first decimal place is of dubious value. So our answer to the problem ought to be expressed as $V = 12\text{ cm}^3$.

In A-level exams many marks are thrown away by candidates writing down all the decimal places that their calculator gives them at the end of a calculation. You do not need to worry too much about whether an answer should be expressed to 2 or 3 significant figures but you will certainly lose marks if you fail to show some awareness of the importance of rounding answers down to a sensible number of significant figures. A safe rule is to express your answer to the same number of significant figures as was used in the question.

Estimates

Physicists ought to acquire the ability to make reasoned estimates of virtually any quantity. To do this well it helps if you have learnt many physical constants; you ought to know the values of e, h, c, m_e, ε_0, μ_0, etc. In addition, it is not adequate for a physicist merely to have heard of the Young modulus or resistivity for example; he ought to have some idea of typical values of these and other important quantities. Some worked examples follow.

Example 1   About how many oxygen atoms are there in a bath full of water?

volume of water $\approx 2\text{ m} \times 0.5\text{ m} \times 0.5\text{ m} = 0.5\text{ m}^3$
mass of water = density × volume
$= 1000\text{ kg m}^{-3} \times 0.5\text{ m}^3 = 500\text{ kg}$
relative molecular mass of water $= 18 \approx 20$
$\Rightarrow$ 1 mole has a mass of 0.02 kg
$\Rightarrow$ 500 kg is (500/0.02) mole = 25 000 mole
1 mole contains 6×10^{23} molecules (and the same number of oxygen atoms since there is one oxygen atom per molecule of water).
So the number of oxygen atoms in the bath is:
$$N = 2.5 \times 10^4 \times 6 \times 10^{23}$$
$$= 1.5 \times 10^{28}$$
$$\approx 10^{28}$$

You need to know quite a lot to do that estimate. You needed to know: the Avagadro constant, the density of water, the chemical formula of water (H_2O) and the atomic masses of hydrogen and oxygen. You needed to make a reasonable guess at the size of a bath (ours is a nice large one!).

Example 2   Estimate the greatest length of steel rope that could be suspended vertically from one end.

In figure 3, when the stress at X exceeds the breaking stress for steel S, then the rope breaks. The stress is caused by the rope's weight.

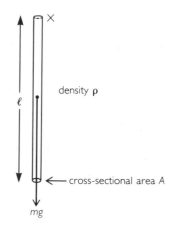

figure 3

Thus when the rope breaks

$$S = \frac{mg}{A}$$

$$\Rightarrow \qquad S = \frac{\rho \ell A g}{A} = \rho \ell g$$

$$\Rightarrow \qquad \ell = \frac{S}{\rho g}$$

Now $S \approx 10^8\,\mathrm{Nm^{-2}}$, $\rho \approx 10^4\,\mathrm{kg\,m^{-3}}$, $g \approx 10\,\mathrm{Nkg^{-1}}$

$$\Rightarrow \qquad \ell \approx \frac{10^8\,\mathrm{Nm^{-2}}}{(10^4\,\mathrm{kg\,m^{-3}}) \times (10\,\mathrm{Nkg^{-1}})} \approx 10^3\,\mathrm{m}$$

Example 3   How many piano tuners are there in Chicago?

You probably feel that you have never really had a burning desire to know the answer to this particular question. But the point of this example is to illustrate that we can make an order-of-magnitude estimate for practically anything.

First we make some estimates.
population of Chicago $\approx 5 \times 10^6$;
there are about 5 people per household;
1 house in 20 has a piano;

a piano gets tuned on average once every year;
a piano tuner can tune 20 pianos a week.

Calculation:

There are 10^6 houses in Chicago and so about 50 000 pianos.
A piano tuner can tune about $50 \times 20 = 1000$ pianos a year.
So Chicago ought to have about 50 piano tuners. Rounding our answer off to an order-of-magnitude figure, we can expect Chicago to have about 10^2 piano tuners. This is a sensible sort of answer; 10 would be too few and there would not be enough work for 10^3.

Straight-line graph

Figure 4 shows a typical straight-line graph. The equation for such a graph is:

$$y = mx + c$$

c is the intercept; m is the gradient, or slope:
$m = a/b$.

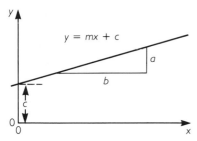

figure 4   A typical straight-line graph

Since a graph deals with pure numbers, m and c must also be pure numbers, without units.

In handling experimental data one often elects to plot a straight-line graph, since useful information may be obtained from the gradient and intercept of such a graph. However, it is not always easy to decide what variables to plot. The table below gives some examples.

Equation	Plot on y-axis	Plot on x-axis	Gradient m	Intercept c
$E = \dfrac{q}{4\pi\varepsilon_0 r^2}$	$E/\mathrm{Vm^{-1}}$	$\dfrac{1}{r^2}/\mathrm{m^{-2}}$	$\dfrac{q}{4\pi\varepsilon_0}/\mathrm{Vm}$	0
$T^2 h = A h^3 + B$	$T^2 h/\mathrm{s^2 m}$	$h^3/\mathrm{m^3}$	$A/\mathrm{s^2 m^{-2}}$	$B/\mathrm{s^2 m}$
$q = q_0 e^{-t/RC}$	$\ln(q/\mathrm{C})$	t/s	$-\dfrac{1}{RC}/\mathrm{s^{-1}}$	$\ln(q_0/\mathrm{C})$
$s = \frac{1}{2} g t^2$	s/m	$t^2/\mathrm{s^2}$	$\frac{1}{2} g/\mathrm{ms^{-2}}$	0

Vectors

You will often need to use vectors in an A-level physics course. You ought to be able to add or subtract vectors, and resolve a vector into two components.

We add vectors according to the parallelogram law. Figure 5 shows how this can be done; we cannot simply add vectors algebraically: we must take account of directions as well as magnitudes. Figure 6 shows how we can take one vector away from another, using the idea that $\mathbf{Q} - \mathbf{P} = \mathbf{Q} + (-\mathbf{P})$.

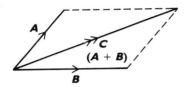

figure 5   Adding vectors

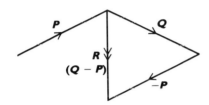

figure 6   Subtracting vectors

Figure 5 shows how we can add two vectors $\mathbf{A} + \mathbf{B}$ to make vector $\mathbf{C}$. The reverse process, in which we replace a vector $\mathbf{C}$ by $\mathbf{A}$ and $\mathbf{B}$, is called finding the *components* of a vector. There is an infinite number of possible parallelograms for which $\mathbf{C}$ could be a diagonal, which means that $\mathbf{C}$ could be split into any components that we choose.

The most useful way of taking components of a vector is to split the vector into two perpendicular parts; this is called *resolving* the vector. Figure 7 shows how vector $\mathbf{Z}$ has been resolved into a vertical component $\mathbf{Y}$ and a horizontal component $\mathbf{X}$. Simple trigonometry may be applied to calculate the magnitudes of these components.

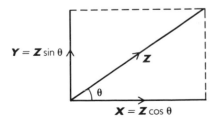

figure 7   Resolving a vector into components

Angles

Angle θ is defined by $\theta = s/r$; this is illustrated in figure 8. s is the distance along a circular arc and r is the radius of that arc.

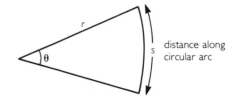

figure 8

The SI unit of angular measure is the radian (rad). One radian is defined as the angle θ subtended by arc s when $s = r$.

In one complete revolution there are 2π radians, since

$$\theta = \frac{2\pi r}{r} = 2\pi \text{ rad}$$

so we may connect radians and degrees by the equation

$$2\pi \text{ rad} = 360°$$

$$\Rightarrow \qquad 1 \text{ rad} = \frac{360°}{2\pi} \approx 57.3°$$

There are some useful approximations that we can make when θ is very small. It is important that you remember that these approximations are only valid if θ is measured in radians. Figure 9 illustrates these approximations. In the triangle OAB, angle OBA is a right angle, and $x \ll D$.

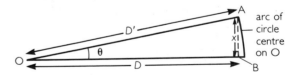

figure 9

We can make the following approximations.

| 1 |

$$\theta \approx \frac{x}{D}$$

because θ is so small that the arc of the circle AB is very nearly the same length as the distance x.

| 2 |

$$\tan\theta = \frac{x}{D} \approx \theta$$

$$\Rightarrow \qquad \theta \approx \tan\theta$$

| 3 |

$$\sin\theta = \frac{x}{D'} \approx \frac{x}{D} \approx \theta$$

$$\Rightarrow \qquad \theta \approx \sin\theta$$

The approximation that $D' \approx D$ can be justified as follows:

$$D' = (D^2 + x^2)^{1/2} \quad \text{(Pythagoras' theorem)}$$

but since $x \ll D$ we can ignore x^2 in comparison with D^2.

4

$$\cos\theta = \frac{D}{D'} \approx 1$$

All these approximations are particularly useful in the field of optics.

Logarithms

A logarithm (log) is defined in the following way.
If $y = 10^x$
then $\log_{10}y = x$; in this case the log is to base 10.
   In science we commonly encounter natural logarithms which are to base e; $e \approx 2.718$.
If $y = e^x$, then $\ln y = x$. $\ln y$ means $\log_e y$.
Useful relationships involving logarithms are:

$$\log(xy) = \log x + \log y$$

$$\log\left(\frac{y}{x}\right) = \log y - \log x$$

$$\log(x)^n = n \log x$$

Algebra and equations

You will find it a great help if you sharpen up your ability to handle algebra. An argument involving algebra is often the neatest way to deal with a problem. You need to be able to handle several equations at once, and to use them to derive a formula for the quantity in which you are interested.
   Here is an example of a fairly typical A-level problem on radioactivity (chapter 36). We show two ways to handle the problem.

Example   A radioisotope has a half-life of 28 days; it emits β-particles of energy 1.0 MeV. The isotope has a nucleon number of 128. What is the maximum power that could be extracted from a sample of 12.8 g of this isotope?

Method 1: several separate calculations.

$$\text{number of moles} = \frac{12.8}{128} = 0.10$$

$$\Rightarrow \text{number of atoms} = 0.10 \times 6.0 \times 10^{23} = 6.0 \times 10^{22}$$

The decay constant for the isotope $= \lambda = \dfrac{0.69}{T_{1/2}}$

$$\Rightarrow \quad \lambda = \frac{0.69}{28 \times 24 \times 3600}\, \text{s}^{-1}$$

$$= 2.9 \times 10^{-7}\,\text{s}^{-1}$$

The number of β-particles emitted per second is given by:

$$-\frac{dN}{dt} = \lambda N$$

$$= 2.9 \times 10^{-7}\text{s}^{-1} \times 6.0 \times 10^{22}$$

$$= 1.7 \times 10^{16}\,\text{s}^{-1}$$

(The minus sign is required because dN/dt has a negative value.)

Power = energy of each particle × number of particles emitted/unit time

$$= 1.6 \times 10^{-13}\,\text{J} \times 1.7 \times 10^{16}\,\text{s}^{-1}$$

$$= 2.7 \times 10^3\,\text{W}$$

Method 2: use of algebra

$$\frac{dN}{dt} = -\lambda N \tag{i}$$

$$\lambda = \frac{0.69}{T_{1/2}} \tag{ii}$$

$$P = -E \cdot \frac{dN}{dt} \tag{iii}$$

$$N = L \times \frac{m}{M} \tag{iv}$$

L = the Avogadro constant, m = mass of sample, M = molar mass of sample
From (i) and (iv):

$$\frac{dN}{dt} = -\lambda \frac{Lm}{M}$$

$$\Rightarrow \quad -\frac{dN}{dt} = \frac{0.69Lm}{T_{1/2}M} \quad \text{from equation (ii)}$$

$$\Rightarrow \quad P = \frac{0.69ELm}{T_{1/2}M} \quad \text{from equation (iii)}$$

$$\Rightarrow$$

$$P = \frac{0.69 \times (1.6 \times 10^{-13}\,\text{J}) \times (6 \times 10^{23}\,\text{mol}^{-1}) \times (12.8 \times 10^{-3}\,\text{kg})}{(28 \times 24 \times 3600\,\text{s}) \times (128 \times 10^{-3}\,\text{kg}\,\text{mol}^{-1})}$$

$$= 2.7 \times 10^3\,\text{W}$$

Both methods are good since they give the right answer. Method 1 is to be recommended for those who find it easier to deal with numbers. Method 2 is probably quicker for those who are used to handling equations. In the second method we only use the calculator once.
   We now give a harder example to illustrate how to solve quadratic and simultaneous equations. (You are only likely to meet mathematics of this difficulty at S-level.)

Example   An object of mass 2 kg is travelling at $2\,\text{ms}^{-1}$ when it collides elastically with a stationary object of mass 1 kg. The two objects travel along the same direction after the collision as the first object did before the collision. What are the speeds of the two objects after the collision?

(a) Before the collision

(b) After the collision

figure 10   (In this example u and v are being used as numbers in order to simplify the equations)

From the conservation of momentum we may write:
$$4 = 2u + v \qquad \text{(i)}$$
(see figure 10(b)).

In an elastic collision the kinetic energy of the objects will be the same before and after the collision:
$$\Rightarrow \qquad 4 = u^2 + \tfrac{1}{2}v^2 \qquad \text{(ii)}$$

From equation (i)
$$v = 4 - 2u$$

Substitution of this into equation (ii) gives:
$$4 = u^2 + \tfrac{1}{2}(4 - 2u)^2$$
$$= u^2 + \tfrac{1}{2}(16 - 16u + 4u^2)$$
$$= u^2 + 8 - 8u + 2u^2$$
$$\Rightarrow \qquad 3u^2 - 8u + 4 = 0$$

To solve this quadratic we use the following result: the equation $ax^2 + bx + c = 0$ has the solution
$$x = \frac{-b \pm \sqrt{b^2 - 4ac}}{2a}$$

Thus
$$u = \frac{8 \pm \sqrt{64 - 4 \times 3 \times 4}}{2 \times 3}$$
$$= \frac{4}{3} \pm \frac{2}{3}$$
$$\Rightarrow \qquad u = 2 \text{ or } \frac{2}{3}$$

The solution in which we are interested is $u = \tfrac{2}{3}$. If $u = 2$ it means that the two objects missed each other and $v = 0$.

In equation (i), if $u = \tfrac{2}{3}$ then:
$$4 = \frac{4}{3} + v$$
$$\Rightarrow \qquad v = \frac{8}{3}$$

Thus the solution to the equation is:

the 1 kg object travels off at $\tfrac{8}{3}\,\mathrm{ms}^{-1}$;

the 2 kg object travels on at $\tfrac{2}{3}\,\mathrm{ms}^{-1}$.

Note that in chapter 3 such problems are solved using the coefficient of restitution. That is a more straightforward method, but we chose not to use it so that we could demonstrate the solving of a quadratic equation.

Calculus

The use of calculus in A-level physics is not essential, but it is useful. We summarise below some of the most commonly encountered derivatives and integrals.

$y = f(x)$	$\dfrac{dy}{dx} = f'(x)$	$\int y \cdot dx$	
$y = x^n$	nx^{n-1}	$\dfrac{x^{n+1}}{n+1} + c$	this is not true however if $n = -1$
$y = \dfrac{1}{x}$	$-\dfrac{1}{x^2}$	$\ln x + c$	
$y = e^x$	e^x	$e^x + c$	
$y = \sin x$	$\cos x$	$-\cos x + c$	
$y = \cos x$	$-\sin x$	$\sin x + c$	

In this book you will find that we have made use of elementary calculus in areas such as simple harmonic motion, a.c. theory and radioactive decay. If you are studying A-level mathematics you should not find our use of calculus too difficult. Below we give some further examples of ways in which calculus may be helpful to us, although you are only likely to meet such examples at S-level.

Example 1   A copper disc has a hole through its centre of radius a. The radius of the disc is b, its thickness t, and its resistivity is ρ. Calculate the electrical resistance of the disc between its inside rim and its outer edge, in terms of ρ, a, b, and t. Figure 11 shows the disc. The resistance δR of a thin section of width δr is given by:
$$\delta R = \frac{\rho \ell}{A}$$
$$= \frac{\rho \, \delta r}{2\pi r t}$$

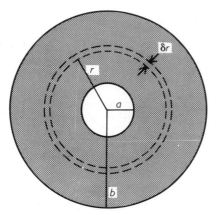

figure 11   Copper disc with hole through centre

so the resistance between inside and outside is given by:

$$R = \int_a^b \frac{\rho \, dr}{2\pi rt}$$

$$= \frac{\rho}{2\pi t} \int_a^b \frac{dr}{r}$$

$$= \frac{\rho}{2\pi t} \ln\left(\frac{b}{a}\right)$$

Example 2   Figure 12 shows two point charges $+q$ at A and B, which are separated by a distance $2a$. At what distance along the line OP would we find the maximum field? OP bisects AB perpendicularly.

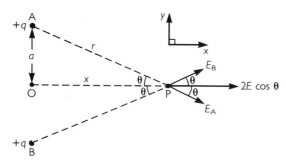

figure 12

At P there are two contributions towards the field, E_A and E_B. By symmetry the two components of E_A and E_B in the y-direction cancel each other out, leaving only the components in the x-direction.
   So the field at P is

$$E_P = 2E \cos\theta$$

where E is the magnitude of the field due to one of the charges.

Now $$E = \frac{q}{4\pi\varepsilon_0 r^2}$$ (page 248)

$$\Rightarrow \qquad E_P = \frac{2q \cos\theta}{4\pi\varepsilon_0 r^2}$$

But $$\cos\theta = \frac{x}{r} \text{ and } r = (a^2 + x^2)^{1/2}$$

$$\Rightarrow \qquad E_P = \frac{2qx}{4\pi\varepsilon_0(a^2 + x^2)^{3/2}}$$

For a maximum value, $\dfrac{dE_P}{dx} = 0$.

Now $$\frac{dE_P}{dx} =$$

$$\frac{2q \times 4\pi\varepsilon_0(a^2 + x^2)^{3/2} - 2qx \times 4\pi\varepsilon_0 \times \frac{3}{2}(a^2 + x^2)^{1/2} \times 2x}{[4\pi\varepsilon_0(a^2 + x^2)^{3/2}]^2}$$

But when $$\frac{dE_P}{dx} = 0$$

$$\Rightarrow \qquad (a^2 + x^2) = 3x^2$$

$$\Rightarrow \qquad x^2 = \frac{a^2}{2}$$

So $$x = \pm\frac{a}{\sqrt{2}}$$

Differential equations

Physicists tend not to meet a great number of different differential equations, but they tend to meet the same sort of ones over and over again. We give examples of the three commonest types of equations.

Example 1   Exponential decay
For example the decay of charge from a capacitor is described by:

$$IR + \frac{Q}{C} = 0 \qquad \text{(page 153)}$$

$$\Rightarrow \qquad \frac{dQ}{dt}R + \frac{Q}{C} = 0 \qquad \text{(i)}$$

The solution is of the form $Q = Q_0 e^{-\lambda t}$. Substitution of this into equation (i) gives

$$-\lambda Q_0 e^{-\lambda t} \times R + \frac{Q_0}{C}e^{-\lambda t} = 0$$

$$\Rightarrow \qquad \lambda = \frac{1}{RC}$$

so $Q = Q_0 e^{-t/RC}$ is the solution.

Example 2   Exponential growth to a maximum value
When an inductor is connected to a battery, current

is governed by this equation:

$$V = IR + L\frac{\mathrm{d}I}{\mathrm{d}t} \qquad \text{(ii)}$$

The solution is of the form

$$I = I_0(1 - e^{-\lambda t})$$

I_0 is the value of the current when $t = \infty$, because when $t = \infty$, $e^{-\lambda t} = 0$. Hence $V = I_0 R$.
Substitution of this solution into equation (ii) gives:

$$V = I_0(1 - e^{-\lambda t})R + \lambda L I_0 e^{-\lambda t}$$

$$\Rightarrow \qquad V = I_0 R - I_0 e^{-\lambda t}R + \lambda L I_0 e^{-\lambda t}$$

$$\Rightarrow \qquad V = I_0 R(1 - e^{-\lambda t} + \lambda\frac{L}{R}e^{-\lambda t})$$

$$\Rightarrow \qquad \frac{V}{R} = I_0\left[1 - e^{-\lambda t}\left(1 - \lambda\frac{L}{R}\right)\right]$$

Comparing this with $I = I_0(1 - e^{-\lambda t})$ and remembering that $V = I_0 R$, we see that $[1 - \lambda(L/R)] = 0$ i.e. $\lambda = R/L$.

So the solution to the equation is:

$$I = \frac{V}{R}(1 - e^{-Rt/L})$$

Example 3 Oscillatory equation
The simple harmonic motion of a mass m acted on by a restoring force of k per unit displacement can be described by:

$$\frac{\mathrm{d}^2 x}{\mathrm{d}t^2} = -\frac{k}{m}x \qquad \text{(page 84)}$$

The solution is of the form $x = x_0 \cos\omega t$ (or $x_0 \sin\omega t$)

$$\Rightarrow \qquad -x_0\omega^2 \cos\omega t = -\frac{k}{m}x_0 \cos\omega t$$

$$\Rightarrow \qquad \omega^2 = \frac{k}{m}$$

Thus $x = x_0 \cos(k/m)^{1/2}t$ is a solution to the equation.

Note that in all the solutions to these equations we proceeded by *knowing* in advance the correct form of the solution.

Index

absolute
  temperature 451
  thermodynamic scale of
    temperature 451
  zero 451
absorption of radiation 472,
  473
  spectra 333
a.c. circuits 165ff
acceleration
  angular 50, 51
  equations of motion 20
  $F = ma$ 22, 23
  in circle 50, 51, 222
  of charge 227, 264, 265
  of gravity 21
accelerometer 19
adiabatic change 454, 455
aerial 305
air wedge fringes 328, 329
alpha-particle 383, 384
  scattering of 409, 410
alternating current
  in capacitors 166
  in inductors 168ff
  power 168, 170
  reactance 170ff, 176
  r.m.s. 166
  smoothing circuit 175, 176
ammeter, d.c. 108
AMONTONS 437
ampere 107, 108, 285, 286,
  483
  circuit law 259, 260
  hour 117
  turns 259
amplification
  transistor 190
amplitude 295, 296
  a.c. 165
AND-gate 179
angular
  acceleration 54
  impulse 54
  magnification 374
  momentum 54
  velocity 52
antinode 327
apparent depth 368, 369
Archimedes' principle 92
armature 291
artificial disintegration 396
artificial nuclear
  transmutations 390
astable circuit 180
astronomical telescope 374
atmospheric pressure 439
atomic energy levels 421,
  430, 431, 433, 434
atomic
  mass number 381
  nucleus 409ff
  spacing 61

  structure 61, 62
  unit 398
atoms (evidence of) 61
Avogadro constant 439
Avogadro's Law 442

B (flux density) 252, 253
  measurement of 255ff
back-e.m.f.
  of induction 273, 282
  of a motor 286, 291
  of polarization 228, 245
Barton's pendula 97
balancing ball
  demonstration 419
Balmer series 421
barrier p.d. 126
beats 303, 334
Becquerel 383, 388, 389
Bernoulli's principle 32, 33,
  103, 419
beta decay
  by electrons 384
  by positron 390
beta-particles 384, 385
  charge-to-mass-ratio 383,
    393
  magnetic deflection
    of 383, 392
betatron 290
binding energy 398
biological effects of
  radiation 400ff
bistable circuit 181
black body radiation 473
black hole 210
Bohr's theory of hydrogen
  atom 240, 423
Boltzmann constant 442
Boyle's law 437, 442
Boys' measurement of G 208
brakes (car) 2
breaking stress (strength) 67
Brewster's law 309
brittle 67
bromine diffusion 447, 448
Brownian motion 437
bubble raft 62, 70

calibration
  of thermometer 450ff
camera
  mirror lens 359
  pinhole 354
  refractor lens 359
candela 483
capacitance 146, 147
  measurement of 148
capacitor
  charging of 146, 147
  discharging of 150, 151
  electrolytic 157
  energy stored in 150

  in parallel 149
  in series 149
  in timing circuits 156
  parallel plate 243, 244
  water analogy 148
  with a.c. 166, 167
car ignition system 290
carbon dating 394
carbon dioxide (solid) 381,
  382
catapult field 255
cathode ray oscilloscope,
  CRO, (use of) 112, 160
cavity walls 479
Celsius temperature
  scale 450, 451
centre of mass 10
centripetal force 50
cepheid variable star 218
Chadwick 396
chain reaction 399, 400, 406
characteristics of sound 101
charge carriers 123
charge of electron 241, 242
charge-to-mass-ratio 385,
  393
Charles' law 438, 442
chromatic aberration 365
circular motion 50ff, 222, 223
circular coil 253
closed pipe 100, 101
cloud chamber 382
circular motion 80
Cockcroft and Walton 390
coefficient
  of friction 3
  of restitution 42, 45
coherent sources 344
collimator 344
collisions 40, 41
  elastic 41ff
  inelastic 42
colour temperature 479
combination of thin
  lenses 370
comparison
  of e.m.f.s 141
  of resistance 133, 141
components
  of force 4, 5
  of velocity 18
composite materials 70
compound microscope 375,
  377
concave
  lens 365
  mirror 356
conduction
  comparison of 469
  of bad conductor 472
  of good conductor 471f
  thermal 469ff
  through gases 423

conductivity 121
conductors
  metallic 122
conservation
  of energy 25
  of angular momentum 55
  of momentum 38, 39
constant volume gas scale (of
  temperature) 450, 451
continuous spectra 343, 347
converging mirror 356
convex (converging)
  lens 363, 364
  mirror 356
cooling, Newton's law of 469
corkscrew rule,
  (Maxwell) 253
corpuscular theory of
  light 451, 452
corrosometer 142
coulomb 109
coulombmeter 154
coulomb's law 229, 411
couple
  on coil 285, 286
  on rigid body 7, 9
critical
  angle 362
  potential 425
  temperature (gas) 444
Crooke's radiometer 415
current against p.d.
  graph 114
current balance 255, 261
curved mirror formulae 356,
  357
cusp 360
cylindrical lens 354
cylindrical mirrors 355ff

Dalton's law of partial
  pressures 442
damped oscillation 87
De Broglie 430
De Broglie's law 430
decay constant 387, 388
decay series 394
degrees of freedom
Democritus 439
deuterium 402
deviation from gas laws 442,
  443
dextro-rotary 313
dichroism 313
dielectric
  constant 245
  polarisation 228, 245
difference of molar heats 455
differentiating circuit 161
diffraction 316ff
  at a hole 319, 323
  at a lens (telescope) 324
  at a slit 318, 330

diffraction – *cont.*
  by an aerial 321
  by loudspeaker 325
  grating 331, 343, 420
  of atoms 434
  of electrons 428
  of microwaves 00
  of X-rays 62
diffuse reflection 354
diffusion
  Graham's law of 447, 448
  of bromine 447
dimensions
  applications of 483, 484
diode (pn junction) 126
diode bridge 118, 175
dislocation 69
dispersion by prism 347
displacement (waves) 78
diverging
  lens 365
  spherical mirror 356
Doppler effect 19, 202, 299, 300, 302
drift velocity (of ions) 110, 111
driving mirror 358
dry ice 381, 382
ductile 67
ductile fracture 69
dynamics 17ff
dynamo 134, 274, 277

Earth
  escape velocity 213
  mass of 223
  vertical component of 276
eddy currents 284
efficiency 28, 284
  electrical 284
  of a heat engine 461
Einstein's mass law 397, 401
elastic deformation 67
elastic
  collisions 41, 42
  limit 67
  modulus 66
electrical interference 292
electric field
  strength 226ff, 231
  lines 228, 236, 247
electric
  flux 247ff
  motor (d.c.) 285, 286, 291
  potential 233ff
  potential flame probe 235
electric fuse 479
electromagnetic
  induction 271ff, 280, 281
  radiation 304ff
  waves 304
electromagnetism 251ff
electron
  charge 241, 242
  diffraction 70, 428, 429

e/m 385
  gun 111, 428
  lens 269, 270
  mass 385, 393
  microscopy 62
  orbit 421
  shells 421, 425
  standing waves 430
  waves 428
electronvolt 383
electroscope (gold leaf) 227, 414, 418
electrostatic fields 228
  shuttling ball
    demonstration of 238
emission spectra 333, 420
emissivity 473
energy
  conservation of 25
  elastic 26, 27, 66, 67
  electrical 113
  gravitational 26
  in capacitor 150
  in collisions 40
  levels (atom) 421, 430, 431
  nuclear 397, 398
  shells 421
  sound (SHM) 86
  vibrational 86
enthalpy 454
entropy 458, 467
equations
  of motion 20
  of state 438
equilibrium conditions 12
  stable 11
  unstable 12
equipartition of energy 456
equipotentials
  electrostatic 236
  gravitational 217
errors
  counting 386
escape velocity 213, 214, 447
excitation potential 425
exclusive OR-gate 187
exponential decay 87, 151ff, 160, 161
eye-ring 375

Faraday's laws of
  induction 271, 272, 276
fibre optics 371
fine beam tube 265, 269
fission (nuclear) 399
fixed points
  (thermometer) 450
Fizeau's method 314
flame probe (electric
  potential) 235, 240
Fleming's left hand rule 255, 265
flux
  electric 247ff
  magnetic 252

flux density, B 252
focal length
  of lenses 363, 364
  of mirrors 356
focussing, electron 269, 270
force and acceleration 22ff
force
  on charges (Bev) 265, 266
  on conductor 255, 256
Foucault's method 309
Franck and Hertz 425
free body diagram 5
friction
  coefficient of 3
fuse (electric) 479
fusion (nuclear) 401ff

G, measurement of 207, 208
galvanometer 263
  sensitivity of, 263
$\gamma = C_p/C_v$ 462
gamma-rays 385, 386
  speed (measurement) 314
gas
  constant 439
  ideal 440
  laws 437, 438, 442
  molar volume 439
  real 442
  thermometer 450, 451
  velocity of sound in 298, 299
gases
  diffusion of 447
  distribution of molecular
    speeds 441, 446
  equation of state 438, 439
Gauss's theorem 248
gear wheels 58
Geiger and Marsden 409
Geiger-Muller tube 382
generators (a.c. and
  d.c.) 274
geostationary orbit 224
geosynchronous orbit 224
germanium 123
gold-leaf electroscope 227, 414, 418
Graham's law of diffusion 447
gravitation (Newton's law
  of) 205
gravitational constant, G 205
  measurement of 207
gravitational
  equipotentials 217
  fields 209, 215, 216
  orbits 222, 223
  potential 213, 214
  potential energy 26, 212
gravity
  acceleration of 21, 209

half-adders 187
half-life 387ff
half-thickness 385, 393

Hall
  effect 266, 270
  voltage 206
haloes 367
harmonics
  of pipes 100
  of strings 98
heat capacity 468, 473
  energy 454
heat engine 461
heat transfer 468
heating effect of current 113
henry, the 161
high pass filter 161
holes 123
Hooke's law 64, 65, 68
Hubble's constant 220, 221
Huygen's construction 345
hydroelectric power 145
hydrogen atom
  Bohr theory of 423
  isotopes 402
hysteresis
  in rubber 70
  in transformers 285

ideal gas
  equation 438
  adiabatic change 454
  isobaric change 454, 455
  isothermal change 454
  isovolumetric change 454
  scale of temperature 450
ignition system (of a car) 290
impedance 165
impulse 38
induced
  charge 228, 235
  current (e.m.f.) 271ff
inductance
  mutual 282, 283
  self 168ff, 280, 281
induction (magnetic) 271ff
inelastic collisions 40ff
inertia (moments of) 54
inertial mass 23
infra-red radiation 472
insulators 124
integrating circuit 161, 193, 199
interference
  light 328, 329
  microwaves 327
  sound 328, 329
  in thin films 349
intensity (electric field
  strength) 228ff
interference (light) 329
  in thin film 349
  in wedge films 328
  of waves 327
interferometer
  Michelson's 336
  radio (four-aerial) 341
  Stellar 338

intermolecular forces
(energy) 68, 74, 76
internal energy of
gases 454ff
internal resistance
of cell 134, 135
of an e.h.t. unit 143
international temperature
scale 453
intrinsic semi-conductors
(pure) 123
inverse square law
electrostatics 229, 230
gravitation 206
gamma rays 385, 386
light 218
ion implantation 127, 128
ionization
current 423, 425
of gases 382, 384
potential 423
isobaric change 454
isothermal change 454
isotopes 389
isovolumetric change 454

joule, the 25
junction diode 126

K-capture 390
kelvin 483
kelvin scale of
temperature 451
Kepler's laws 206, 210, 222
Kerr cell 315
kilowatt-hour 117
kinetic energy 26
kinetic theory (of
gases) 437ff
assumption (ideal gas) 439
behaviour of real
gases 442
calculation of pressure 440
deductions from 442
distribution of molecular
speeds 441, 446
Graham's law of
diffusion 447
root-mean-square (r.m.s.)
speed 441
temperature 441
Kirchoff's laws
(electricity) 135

lamp-black 472
laser 334
latent heat 469
of evaporation 476
of fusion 475
lateral inversion 355
lens
blooming of 327
chromatic aberration 365
converging (convex) 363,
364

diverging (concave) 365,
366
spherical aberration 365
lens combinations 370
lens makers' formula 369
Lenz's law 161, 272
Leslie's cube 473
Leucippus 439
light-dependent resistor
(LDR) 132
light (rectilinear
propogation) 353
line spectra 420, 421, 426,
427
linear air track 39
linear flow (heat) 470
linear momentum 38ff
lines of force
electric 228
gravitational 209
magnetic 252, 253
liquid state 443
Lloyd's mirror 336, 337
logic gates 179
loudness 101
loudspeaker
construction 262
sound emission 298
low pass filter 161
LR circuits 161, 162
Lucretius 439
Lyman series 421

magnetic
circuit 259, 260
fields 251ff
force on charged
particles 264
flux 252
flux density, B 252, 253
in a coil 253
in a solenoid 254
in a straight wire 253,
257
lens 269, 270
monopoles 279
permeability 254
torque 262
magnetohydrodynamic
generator 292
magnetomotive force 259
magnification (angular) of
telescope 374
magnifying glass 373
majority carriers 123
malleable 67
Maltese cross tube 227
Malus's law 308
mass
gravitational 24
inertial 23
number (nuclear
number) 389
of the sun 223

spectrometers 269
mass-energy relation 397,
398
maximum power 135
Maxwell's right hand grip
rule 253
Mayer's equation 455
mean free path 448
mean separation of gas
molecules 448
mean square velocity 441
mercury thermometer 463
metal conductor 122
metre bridge 133, 139, 141
Michelson's
interferometer 336
microscope, compound 375,
377
microwaves
polarisations 307, 312
superposition 327
Millikan's oil drop
experiment 241, 242
minimum deviation by
prism 363
mirror
car 358
concave 355
convex 356
hallof 358
lens for camera 359
parabolic 359
plane 354
modulus of elasticity
(Young's modulus) 65, 66,
68
molar
gas constant 439
heat capacity, C_v 455, 462
heat capacity, C_p 455, 462
mole 483
molecular forces 68, 74, 76
molecular speed
distribution of 446
mean value 445
measurement of 446
root-mean-square
value 441, 445
temperature
dependence 446
moment
of couple 54
of force 7
of inertia 54
momentum
angular 54
conservation of angular 55
conservation of linear 39,
40
linear 38ff
of a photon 419
monkey and hunter 22
moon, motion of 223
motion
Newton's laws of 6, 7, 12

in circle 50ff, 222, 223
in straight line 20, 39
of projectile 21
simple harmonic 77ff
motors
d.c. (back e.m.f.) 285, 286
d.c. (efficiency) 286, 287
d.c. (structure) 291
Mount Palomar
telescope 324
moving coil
ammeter 108, 131
galvanometer 130, 263
loudspeaker 262
voltmeter 131
multiple slit
interference 331, 332
mutual
inductance 282
induction 284, 285, 287,
289, 290

NAND-gate 179
near point of eye 373
negative feedback 184, 185
Nernst's calorimeter 474
Neumann's equation 272
neutron 396
Newton's
law of cooling 469, 473, 476
law of gravitation 205, 222
laws of motion
first 7, 12
second 37, 38
third 6, 7, 12
nodes 98, 327
in pipes 100
in strings 98
non-ohmic conductors 114
NOR-gate 179
NOT-gate 179
npn transistor 127
n-type semiconductor 123,
126
nuclear
binding energy 398
cross-section 405
fission 399
fusion 401ff
reactions 396
reactor 400, 406
stability 397
structure 396
transmutations 389, 390
nucleon 381ff
nucleus 381ff

Ohm's law (electricity) 114
oil films 328
operational amplifier 191ff
optical activity 309
optical instruments 373ff
OR-gate 179
organ pipes 100, 104

oscillation 77
  damped 87
  electric 282, 305, 306
overtones
of pipes 100, 104
of strings 99

pair production 417
parallel-plate capacitor 243, 244
parabolic mirror 359
parallelogram
  of forces 4
  of velocities 18
particle-wave duality 416, 417
Paschen series 421
peak value 165, 166
pendulum (simple) 84
periscope 358
permeability
  relative 254
  vacuum 254
permittivity 244
  relative 245
  vacuum 244
phase difference 80
photon 415, 416
photon momentum 419
photo-cells 414, 418
photoelasticity 309
photoelectricity 413ff
piano tuning 303
pinhole camera 354
pipe organ 100
pitch (of sound) 101
Planck constant 415
plane mirror 345, 346
planetary motion 222ff
plastic behaviour 68
platinum resistance
  thermometer 451
pn junction 126
Poisson distribution 386
polarimeter 309
polarisation
  (electromagnetic waves) 306
  Brewster's angle 309
  by microwaves 312
  by Polaroid 308
  by reflection 308
  Malus's law 308
  optical activity 309
  stress analysis 309
polarisation of dielectric 228, 245
polarising angle 309
potential difference (p.d.) 111
potential divider 132
potential (electric) 233, 234
potential energy
  elastic 26
  gravitational 26

potentiometer 132, 133
power
  electrical 113, 329
  mechanical 28
pressure
  atmospheric 439
  gas 440, 441
  law 438, 442
principle of reversibility 361
prism 362, 366
  minimum deviation by 363, 367
projectiles 21
proton 381, 382
pulsatance ω 80
pumped storage (power station) 145

quality of sound 101
quantisation of energy 415ff, 421, 425
quantum shuffling 459, 460
quantum theory of light 413ff
quenching agent (G.M. tube) 427, 428

radiation (thermal) 472, 473
radioactivity, biological effects 404
radio telescope 323, 324
radius of curvature
  of lens 369
  of mirror 356
rainbow 367, 368
range (radioactivity) 391
rarefaction 298
ratio of molar heats, γ 455, 456
  numerical values 456, 457, 462
ray of light 353
Rayleigh's criterion 320, 332, 341
RC circuits 159, 160, 168
reactance
  of capacitor 67
  of coil 170
real gases 442
  critical temperature 444
  diffusion 447, 448
  finite molecular size 443
  intermolecular forces 443
rectification (electrical) 118, 175, 176
rectilinear propagation of light 353, 354
red shift in spectra 220, 303
reflection 345, 346
  of light (laws of) 354
refraction
  at plane surface 346
  at spherical surface 363, 368
  through prism 362, 363
  wave theory of 346

refraction of sound 348
refractive index 361, 362
  of a gas 349, 350
  of glass 310, 347, 361, 362, 366, 367
  of a lens 369
reinforced concrete 72
relative
  permeability 254
  velocity 42
relativistic mass 385
reluctance, magnetic 259
resistance 114
resistance thermometer 451
resistivity 120, 121
resistors
  in parallel 129
  in series 129
resolution of forces 45
resolution of an eye 320
  of a grating 33
  of a radio telescope 341
  of a slit 320
resolving power (telescope) 320
resonance
  atomic 430, 433
  Barton's pendula 97
  car suspension 93
  electrical (LC) 170ff
  mechanical 95ff
  sound 100
  tube 100
root-mean-square (r.m.s)
  current 166
  speed 441
rotational dynamics 54, 55
rotational energy 54, 456, 457
Rutherford, scattering 409
Rutherford and Royds' experiment 384
Rydberg constant 442

satellites 222, 223
scales of temperature 450, 451
  disagreement between scales 452, 463, 464
  Kelvin scale 451
  international scale 453
Searle's bar 471
Seebeck effect 451, 452
self inductance 280, 281
self induction 161, 283, 287, 288
semiconductors
  intrinsic 123
  n-type 123, 126
  p-type 123, 126
  pure 123
series a.c. circuits 168, 169
shells (energy) 421, 425, 433
silicon diode 126
simple harmonic motion (s.h.m.) 77ff

effects of damping 86
simple pendulum 84
slide-wire bridge 133, 139, 141
Snell's law 309, 361, 362
soap bubble raft 70
soap film 349
solenoid 253, 254
solid carbon dioxide 381, 382
sound waves 298
  supposition of 328, 329
specific charge 385
specific heat capacity 468, 473, 474, 475, 477
  constant pressure 455, 462
  constant volume 455, 462
  of a solid 474
specific heats, difference of 455
specific latent heat 469, 473, 476
spectra
  absorption 333, 426, 427
  emission 333, 420
  hydrogen 421
  line 333
  red shifted 220
  X-ray 422, 426
spectrometer 334
spectrum, hydrogen 421
speed of light (measurement) 309, 314, 315
speed of sound
  in air (measurement of) 299
  in a gas 298
  in a solid 297
spherical abberation
  of a lens 365
  of mirror 355
spooning charge 154
spring constant 64
square pulses
  into RC circuits 159, 160
  into RL circuits 161f
stability (nuclear) 397
S.T.P. 445
standard atmospheric pressure 439, 445
standard deviation 386
standard temperature 439, 445
standing waves 296, 297
statics 1
stationary (standing waves)
  pipe 100, 101
  string 98
stiffness 67
strain gauge 143
  hardening 67
  tensile 65
strength (breaking stress) 67
stress (tensile) 65

stretched wire energy 26, 66
strings
  harmonic in 98
  resonance in 98
struts and ties 10
superposition of waves 326ff
suppressor circuit 292
synchroton 268

telescope
  astronomical 374
  eye-ring 375
  magnification 374
  Newtonian reflecting 377
  refracting 373, 374, 376
  terrestial 376
temperature 449
temperature scale
  Celsius 450
  international practical 465
  kelvin 451
  coefficient of
    resistance 452, 462
  ideal gas scale 450, 451
  gradient 470, 471
tensile
  strain 65
  stress 65
terminal velocity 24, 31
tesla (T), the 256
thermal
  capacity 468, 473
  conduction (model of) 472
  conductivity 469ff, 473, 472
  detectors 472, 473
  equilibrium 449
  insulation 478, 479
  radiation 472
thermodynamic scale 451
thermocouple 134, 451, 452,
  472
thermodynamics 449
  Zeroth law of 453

first law of 454
  second law of 457, 461
thermoelectricity 452
thermometers 463
  ideal gas 451
  lower fixed point 450
  mercury-in-glass 450, 463
  platinum resistance 451
  thermocouple 451, 452
  upper fixed point 450
thermometric
  properties 450, 463
thermonuclear reaction 402
thermopile 472
thermos flask 476
thin oil films 328
timbre 101
toroidal coil, 287
torque 54
total internal reflection 358,
  362
transformers 282ff, 289
transistor 177ff
  amplifier 178, 184, 189
  amplifier with
    feedback 185
  npn 127
translational energy 456, 457
transport equation
  (electrical) 110
trans-uranic elements 394
transverse waves 295, 296
triangle
  of forces 4
  of velocities 18
tuning (piano) 303
tuning fork 303
turbines and dynamos 134
TV interference
  suppression 292
two-slit interference 329, 330

unified atomic mass unit 398

uniform acceleration 20
universal gas constant 439
unstable equilibrium 12
uranium
  fission 399, 406
  series 394

Van de Graaff generator 238,
  239
variation of G 206
vectors, addition of 4
velocities, addition of 18
velocity
  angular 52
  relative 12
  terminal 24, 31
  time graph 20
velocity of gamma rays 314
velocity of light
  Fizeau 314
  Foucault 309
  optical fibre 310
velocity selector 385, 393
velocity of sound
  in air 298, 299, 445
  in a gas 298
  in hydrogen 445
  in oxygen 445
vibrating reed switch 243
vibrational energy 86
vibrations
  forced 95
  in pipes 100
  in strings 296
  longitudinal 296
  resonant 296
  transverse 296
virtual
  image 364
  object 370
voltage amplification 191ff
voltmeter, CRO 112
vortex shedding 103

Voyager 1 at Jupiter 217, 218,
  225

wave
  intensity 329
  nature (particle) 428ff
  speeds 295
  theory (of light) 328, 332
wavefront 345, 346
wavelength 98, 295
  of sound, 100
wave-particle duality 416,
  417
weightlessness 53, 224
Wheatstone bridge 133, 141
Wien's law 479
work 25
  done by couple 54
  done by gas 454
  function 415
  hardening 67
  in stretching wire 26, 27,
    66, 67

X-rays
  diffraction 62
  spectra 422, 426

Yield point 67, 68
Young's
  experiment 65, 66
  modulus 68
Young's slits
  by light 329, 330, 416
  by microwaves 336
  by sound 336

zero potential (atom) 68, 76
  electrical 235
  gravitational 213
Zeroth law of
  thermodynamics 453